Wildlife Protection and Habitat Management: Practice and Perspectives

Wildlife Protection and Habitat Management: Practice and Perspectives

Editors

Juan F. Beltrán
Pedro Abellán
John A. Litvaitis

MDPI • Basel • Beijing • Wuhan • Barcelona • Belgrade • Manchester • Tokyo • Cluj • Tianjin

Editors

Juan F. Beltrán
University of Seville
Sevilla
Spain

Pedro Abellán
University of Seville
Sevilla
Spain

John A. Litvaitis
University of New
Hampshire
Durham, NH
USA

Editorial Office
MDPI
St. Alban-Anlage 66
4052 Basel, Switzerland

This is a reprint of articles from the Special Issue published online in the open access journal *Land* (ISSN 2073-445X) (available at: https://www.mdpi.com/journal/land/special_issues/wildlife_protection).

For citation purposes, cite each article independently as indicated on the article page online and as indicated below:

LastName, A.A.; LastName, B.B.; LastName, C.C. Article Title. *Journal Name* **Year**, *Volume Number*, Page Range.

ISBN 978-3-0365-4871-5 (Hbk)
ISBN 978-3-0365-4872-2 (PDF)

Cover image courtesy of Iñaki Relanzón

Contents

About the Editors

Juan F. Beltrán

Dr. Juan F. Beltrán is a Professor at the University of Seville (Spain) since 2000, where he teaches courses for pre- and postgraduates at the College of Biology as well as the College of Geography and History. He has published some 100 scientific papers, some of them in high-impact journals such as *Nature, Current Biology, Global Change Biology,* and *Methods in Ecology and Evolution.* His current research interests are dealing with evolutionary biology (amphibian and reptiles) and the conservation biology of vertebrates.

Pedro Abellán

Dr. Pedro Abellán is a Professor at the University of Seville (Spain). His research interests are in the interface of biological conservation, biogeography, and evolutionary biology. This multidisciplinary research aims to understand how biodiversity is generated and maintained, how biodiversity responds to global changes, both ancient and recent, and how biodiversity can be conserved given current and future challenges. He has authored around 100 scientific publications, including more than 70 scientific articles in peer-reviewed journals and several books, book chapters, and educational works. He has also mentored a number of undergraduate and graduate students.

John A. Litvaitis

Dr. John A. Litvaitis is an Emeritus Professor of Wildlife Ecology at the University of New Hampshire (USA). He has mentored hundreds of undergraduate students and more than 30 graduate students. His research largely focused on vertebrate responses to human-caused landscape changes, and he has published in a wide range of journals, including *Conservation Biology, Biological Conservation, Oecologia, The Journal of Wildlife Management, and Environmental Management.* Currently, he writes about wildlife-related subjects for nonscientists and advocates for greater public involvement in developing wildlife management policy.

land MDPI

Editorial

Seeking Sustainable Solutions in a Time of Change

Juan F. Beltrán [1,*], John A. Litvaitis [2] and Pedro Abellán [1]

[1] Departamento de Zoología, Universidad de Sevilla, 41012 Seville, Spain; pabellan@us.es
[2] Department of Natural Resources and the Environment, University of New Hampshire, Durham, NH 03824, USA; jal1092@unh.edu
* Correspondence: beltran@us.es

Citation: Beltrán, J.F.; Litvaitis, J.A.; Abellán, P. Seeking Sustainable Solutions in a Time of Change. *Land* **2022**, *11*, 851. https://doi.org/10.3390/land11060851

Received: 27 May 2022
Accepted: 2 June 2022
Published: 6 June 2022

Global change is impacting our lives in many ways. This is not only limited to climate change but also worldwide changes in land use, pollution, the over-exploitation of natural resources, biotic invasions, changes in plant–animal interactions, species extinctions, and emergent diseases [1]. These changes can have synergistic effects and are changing our environment and our lives in profound ways.

A dominant paradigm of wildlife and habitat conservation in the 1970s was built on the protection of enough land to assure the survival of at-risk taxa (i.e., the SLOSS debate [2]). In the 1980s, some of the SLOSS concepts were challenged (e.g., National Parks for faunal conservation [3]). We have learned some difficult lessons (e.g., declines and extinctions of amphibians [4]) that warn us that the impacts on species and habitats are not restricted to a single area or continent. Solutions need to be more comprehensive in the light of what we now know about species, habitats, and global dynamics. The effects of global change are operating at much larger scales. Our approaches need to go beyond protected areas. The role of the public and stakeholders has become recognized and implicated in conservation actions [5].

This Special Issue summarizes examples of evolving actions to conservation. Sixteen contributions showing how the needs of specific species and habitats are being addressed. In some way, all of them deal with the delicate relations between humans and wildlife [6] and movements toward sustainable human-dominated landscapes [7–10]. Major topics in this Special Issue include: (1) the impact of roads on fauna mortality (rates, detection, and modelling [11–13]; (2) the protection and conservation values of rural areas and species responses to habitat changes such as farming [8,9,14]; (3) the effects of changes in populations and habitats [7,15,16]; and (4) modelling species abundance and distribution [13,17]. Other notable subjects are reviewed, such as the role of private lands in species conservation [18] and the management of habitat in global bear conservation [19]. An interesting paper deals with understanding the environmental awareness of visitors to protected areas in Northern England [20]. From a taxonomic point of view, vertebrates (amphibians, reptiles, and mammals such as rodents, lagomorphs, bats, carnivores, ungulates, and humans) are the focal species. Lastly, we included the ins and outs of the startup and development of a LIFE project [21].

Our work gathering these studies in a Special Issue of *Land* is our humble contribution to addressing the escalating threats to our environment and the associated challenges. Our aim is that this collection will be of help to all who are aiming to work and study habitat and species conservation with their feet firmly planted in sound theory and practice. This is a growing, interdisciplinary field, and all hands and minds are welcome. In our opinion, this determined vision is needed to make a difference in times of change.

Conflicts of Interest: The authors declare no conflict of interest.

References

1. Matesanz, S.; Gianoli, E.; Valladares, F. Global change and the evolution of phenotypic plasticity in plants. *Ann. N. Y. Acad. Sci.* **2020**, *1206*, 35–55. [CrossRef]
2. Soulé, M.E.; Simberloff, D. What do genetics and ecology tell us about the design of nature reserves? *Biol. Cons.* **1986**, *35*, 19–40. [CrossRef]
3. Lunney, D. A history of a contested ideal: National Parks for fauna conservation. *Aust. Zool.* **2017**, *39*, 371–396. [CrossRef]
4. Wake, D.B.; Wredenburg, V.T. Are we in the midst of the sixth mass extinction? A view from the world of amphibians. *Proc. Natl. Acad. Sci. USA* **2008**, *105*, 11466–11473. [CrossRef] [PubMed]
5. Macdonald, D.W.; Willis, K.J. *Key Topics in Conservation Biology 2*; John Wiley and Sons, Ltd.: West Sussex, UK, 2013.
6. Feng, L.; Cai, Q.; Bai, Y.; Liao, W. China's Wildlife Management Policy Framework: Preferences, Coordination and Optimization. *Land* **2021**, *10*, 909. [CrossRef]
7. Hong, H.J.; Kim, C.K.; Lee, H.W.; Lee, W.K. Conservation, Restoration, and Sustainable Use of Biodiversity Based on Habitat Quality Monitoring: A Case Study on Jeju Island, South Korea (1989–2019). *Land* **2021**, *10*, 774. [CrossRef]
8. Albero, L.; Martínez-Solano, I.; Arias, A.; Lizana, M.; Bécares, E. Amphibian Metacommunity Responses to Agricultural Intensification in a Mediterranean Landscape. *Land* **2021**, *10*, 924. [CrossRef]
9. García-Solís, F.; Rau, J.R.; Niklitschek, E.J. Occurrence and Abundance of an Apex Predator and a Sympatric Mesopredator in Rural Areas of the Coastal Range of Southern Chile. *Land* **2022**, *11*, 40. [CrossRef]
10. Li, H.; Petric, R.; Alazzawi, Z.; Kauzlarich, J.; Mahmoud, R.H.; McFadden, R.; Perslow, N.; Rodriguez, A.; Soufi, H.; Morales, K.; et al. Four Years Continuous Monitoring Reveals Different Effects of Urban Constructed Wetlands on Bats. *Land* **2021**, *10*, 1087. [CrossRef]
11. Kučas, A.; Balčiauskas, L. Impact of Road Fencing on Ungulate-Vehicle Collisions and Hotspot Patterns. *Land* **2021**, *10*, 338. [CrossRef]
12. Kučas, A.; Balčiauskas, L. Roadkill-Data-Based Identification and Ranking of Mammal Habitats. *Land* **2021**, *10*, 477. [CrossRef]
13. Hallisey, N.; Buchanan, S.W.; Gerber, B.D.; Corcoran, L.S.; Karraker, N.E. Addressing Imperfect Detection in Road Mortality Hotspot Models: A Case Study with Amphibians and Reptiles. *Land* **2022**, *11*, 739. [CrossRef]
14. Schmitz, M.F.; Arnaiz-Schmitz, C.; Sarmiento-Mateos, P. High Nature Value Farming Systems and Protected Areas: Conservation Opportunities or Land Abandonment? A Study Case in the Madrid Region (Spain). *Land* **2021**, *10*, 721. [CrossRef]
15. Beltrán, J.F.; Rau, J.R.; Soriguer, R.C.; Kufner, M.B.; Delibes, M.; Carro, F. Effects of Population Declines on Habitat Segregation and Activity Patterns of Rabbits and Hares in Doñana National Park, Spain. *Land* **2022**, *11*, 461. [CrossRef]
16. Rammou, D.L.; Astaras, C.; Migli, D.; Boutsis, G.; Galaki, A.; Kominos, T.; Youlatos, D. European Ground Squirrels at the Edge: Current Distribution Status and Anticipated Impact of Climate on Europe's Southernmost Population. *Land* **2022**, *11*, 301. [CrossRef]
17. Li, X.; Li, N.; Li, B.; Sun, Y.; Gao, E. Abundance R: A Novel Method for Estimating Wildlife Abundance Based on Distance Sampling and Species Distribution Models. *Land* **2022**, *11*, 660. [CrossRef]
18. Litvaitis, J.A.; Larkin, J.L.; McNel, D.J.; Keirstead, D.; Constanzo, B. Addressing the Early-Successional Habitat Needs of At-Risk Species on Private Owned Lands in the Eastern United States. *Land* **2021**, *10*, 1116. [CrossRef]
19. Garshelis, D.L. Understanding Species-Habitat Associations: A case Study with the World's Bears. *Land* **2022**, *11*, 180. [CrossRef]
20. Gosal, A.S.; McMahon, J.A.; Bowgen, K.M.; Hoppe, C.H.; Ziv, G. Identifying and Mapping Groups of Protected Area Visitors by Environmental Awareness. *Land* **2021**, *10*, 560. [CrossRef]
21. Guinart, D.; Solórzano, S.; Amat, F.; Grau, J.; Fernández-Guiberteau, D.; Montori, A. Habitat Management of the Endemic and Critical Endangered Montsey Brook Newt (*Calotriton arnoldi*). *Land* **2022**, *11*, 449. [CrossRef]

land

MDPI

Article

Impact of Road Fencing on Ungulate–Vehicle Collisions and Hotspot Patterns

Andrius Kučas † and Linas Balčiauskas *

Nature Research Centre, Akademijos str. 2, LT-08412 Vilnius, Lithuania; andrius.kucas@gamtc.lt
* Correspondence: linas.balciauskas@gamtc.lt; Tel.: +370-68534141
† Present address: Territorial Development Unit, Growth and Innovation Directorate, Joint Research Centre, European Commission, Via Fermi 2749, I-21027 Ispra, Italy.

Abstract: The number of road traffic accidents decreased in Lithuania from 2002 to 2017, while the ungulate–vehicle collision (UVC) number increased and accounted for approximately 69% of all wildlife–vehicle collisions (WVC) in the country. Understanding the relationship between UVCs, traffic intensity, and implemented mitigation measures is important for the assessment of UVC mitigation measure efficiency. We assessed the effect of annual average daily traffic (AADT) and wildlife fencing on UVCs using regression analysis of changes in annual UVCs and UVC hotspots on different categories of roads. At the highest rates, annual UVC numbers and UVC hotspots increased on lower category (national and regional) roads, forming a denser network. Lower rates of UVC increase occurred on higher category (main) roads, forming sparser road networks and characterized by the highest AADT. Before 2011, both UVC occurrence and fenced road sections were most common on higher-category roads. However, as of 2011, the majority of UVCs occurred on lower-category roads where AADT and fencing had no impact on UVCs. We conclude that wildlife fencing on roads characterized by higher speed and traffic intensity may decrease UVC numbers and at the same time shifting UVC occurrence towards roads characterized by lower speed and traffic intensity. Wildlife fencing re-allocates wildlife movement pathways toward roads with insufficient or no mitigation measures.

Keywords: road safety; roadkill; clustering; growth rate; traffic intensity; mitigation measures

Citation: Kučas, A.; Balčiauskas, L. Impact of Road Fencing on Ungulate–Vehicle Collisions and Hotspot Patterns. *Land* **2021**, *10*, 338. https://doi.org/10.3390/land10040338

Academic Editor: Juan F. Beltrán

Received: 1 March 2021
Accepted: 23 March 2021
Published: 25 March 2021

1. Introduction

Systems that record regular road traffic accidents, including those with animals, are continually evolving and becoming highly integrated [1]. In Lithuania, computerized wildlife–vehicle collision (WVC) reporting started in 2002, which integrates data stored at the Lithuanian Police Traffic Supervision Service with road data maintained by the Lithuanian Road Administration under the Ministry of Transport and Communication [2]. In the period 2002–2017, over 73,211 road traffic accident records were registered. The number of records decreased from 6090 in 2002 to 3192 in 2017 [3], constituting a 4.2% compound annual reduction (Table A1). In the period 2002–2017, total ungulate–vehicle collision (UVC) numbers in Lithuania have been constantly growing, constituting a 16.4% compound annual increase (Table A1).

WVCs present a serious problem and an increasing threat to traffic safety, socio-economics, animal welfare, and wildlife management and conservation in many countries [4–7]. The number of WVCs is steadily increasing in many countries [1,8–12] and in Lithuania [2,13].

Collisions with large mammals are a global and persuasive problem [14]. The rate of collision numbers has increased significantly over time, suggesting the growing importance of traffic in ungulate management. UVC is of particular importance from the perspective of drivers because of the large body size of the animals, resulting in a strong impact and consequences [15]. In contrast, small animals rarely cause traffic crashes, and the only evidence of WVCs in such cases is the carcasses of the animals along roads, which can be a proxy for the extent of such events [1,8].

Knowledge of sites where UVCs occur more frequently is important for the effective application of mitigation measures [6,16–21] and identifying significant locations where animal pathways intersect with roads. Collision risks may be associated with linear landscape features that funnel animals alongside or across the road and with artificial infrastructure [4,22].

A variety of UVC mitigation measures may be implemented to modify the behavior of drivers and/or animals in order to reduce the number of UVCs. These include measures to physically block animal movement on roads. Wildlife fencing along roads in conjunction with the construction of wildlife passages has been widely accepted as an effective way to minimize collisions with animals [23]. However, there is limited information on the effectiveness of mitigation [24,25].

We analyzed UVCs in Lithuania on different categories of roads in the period 2002–2017, aiming to (i) map UVCs for each year, (ii) identify yearly UVC hotspots (short significant road segments where UVCs occurs more frequently than expected), and (iii) analyze UVCs and UVC hotspot relationships with annual average daily traffic (AADT) and the length of fenced road sections accounting for yearly changes.

We tested the following two hypotheses:

1. The occurrence of UVCs (and consequently UVC hotspots) directly depends on transport intensity (that is, UVC numbers will be bigger on the main roads, which are characterized by higher levels of speed and traffic).
2. Wildlife fences are sufficient measures for UVC prevention (that is, no UVCs or UVC hotspots will be recorded within the fenced road sections).

2. Materials and Methods

We collected and mapped UVC data in our study area. Using this data we identified UVC hotspots that allowed us to identify UVC spatial locations on different categories of roads. Compound annual growth rates used to identify long term change patterns of UVCs, UVC hotspots, AADT, and fence length on different categories of roads. Finally we assessed the effect of AADT and wildlife fencing on UVCs using regression analysis of changes in annual UVCs and UVC hotspots on different categories of roads.

2.1. Study Area

Our study area covered the entire territory of Lithuania (Figure 1). The country is located in northern Europe and borders the Baltic Sea. The flat area (with the highest point of ~294 m above sea level) of the country extends to 65,286 square kilometers. Lithuania's climate is transitional between maritime and continental regions. The average air temperatures are –4.9 °C in January and 17.2 °C in July. Annual rainfall average is from 570 mm to 902 mm, depending on the location [26].

The country is located in a mixed-forest zone; 33% of the surface is occupied by arable land and permanent crops, 27% by semi-natural vegetation, 33% by forested land, 3% by artificial areas, and 4% by water bodies and other land [27]. The country is inhabited by 68 species of mammals, including eight species of ungulates [28].

In 2017 there were 21,244 km of state-owned roads of national significance (excluding roads in cities). While sources differ in terms of their exact nomenclature, the basic hierarchy comprises freeways, arterials, collectors, and local roads [29]. In Lithuania, main roads or highways can be considered as freeways or motorways (total length 1865 km with AADT 3000–20,000 cars per day), national roads as arterials and collectors/distributors (5006 km, 500–3000 cars per day), and regional roads as local roads (14,600 km, up to 500 cars per day) [30]. The AADT values for different road categories are provided in Table A1.

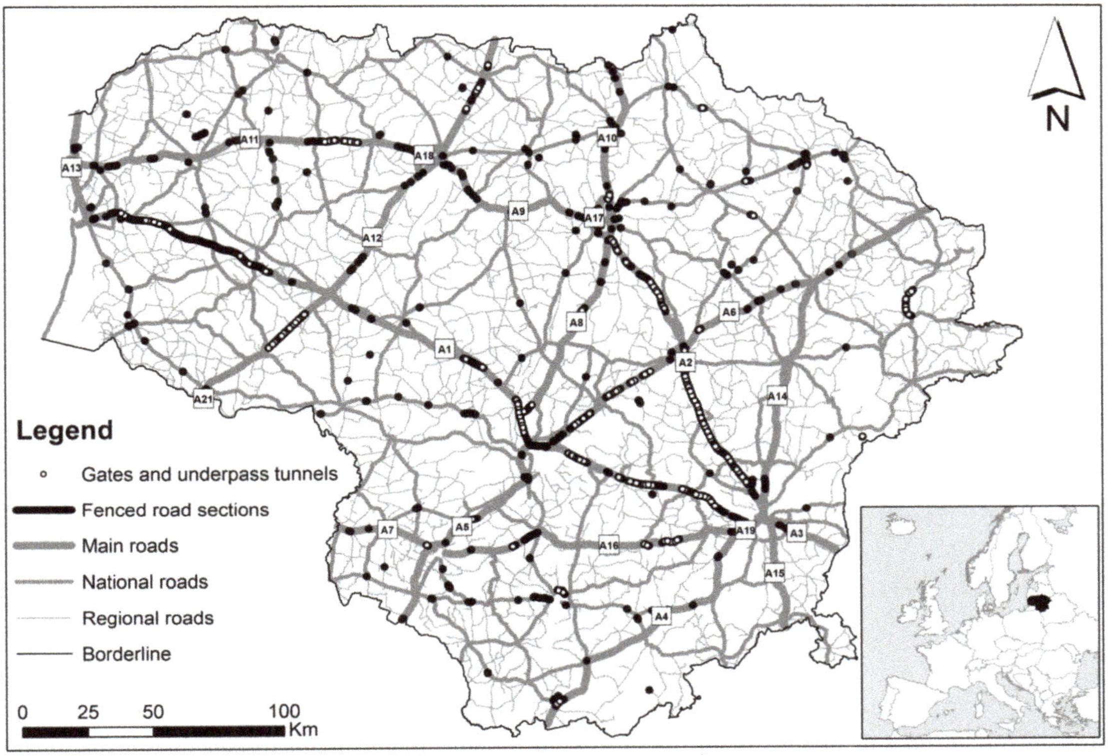

Figure 1. Study area, roads by categories, and fenced road sections in conjunction with additional mitigation measures such as underground passes, gates, etc., in 2002–2017. Labels show the unique identification numbers of main roads/highways.

The width of traffic lane ranges from 2.0 m to 3.75 m. The width of road lanes (traffic lanes including shoulders and safety rails) depends on five different road categories and ranges from 18 m (1–2 lanes) to 39 m (4–6 lanes separated by grass line).

Wildlife fences are the most common UVC mitigation measure in Lithuania. In addition to wildlife fences, other UVC mitigation measures (underground passages, tunnels, gates and jump outs) have been implemented [3].

In 2017 only 3.78% of all roads were fenced in Lithuania. There were 1088 (total length 803.5 km) segments of wildlife fences, 680 of which (743.8 km) were implemented on main roads, 256 (48.6 km) on national regional roads, and 152 (11.1 km) on regional roads (Figure 1).

2.2. Ungulate–Vehicle Collision Data

UVCs occur in approximately 47% of the entire Lithuanian road network (not including cities). According to data from the Lithuanian Police Traffic Supervision Service and Nature Research Centre, a total of 21,847 WVCs (15,006 UVCs) were recorded over the period of 2002–2017 in Lithuania (Table A1). These numbers may not account for all accidents, as the reporting of accidents with all WVCs to the authorities is not mandatory in Lithuania. However UVCs are reported in most cases, since reporting is mandatory in cases where animals and/or people involved in the accident are killed and/or injured, or vehicles and/or road infrastructure are damaged.

We selected 13,762 UVC reports that had coordinates and involved six species of large mammals (Figure 2, Tables A1 and A2): 1340 moose (*Alces alces*), 248 red deer (*Cervus elaphus*), 10,741 roe deer (*Capreolus capreolus*), 1416 wild boar (*Sus scrofa*), 11 fallow deer

(*Dama dama*), and 6 European bison (*Bison bonasus*). These large animals caused the most serious accidents and in the vast majority of cases were registered by the police [31]. So far there have been no UVCs reported with European mouflon (*Ovis aries*) or sika deer (*Cervus nippon*).

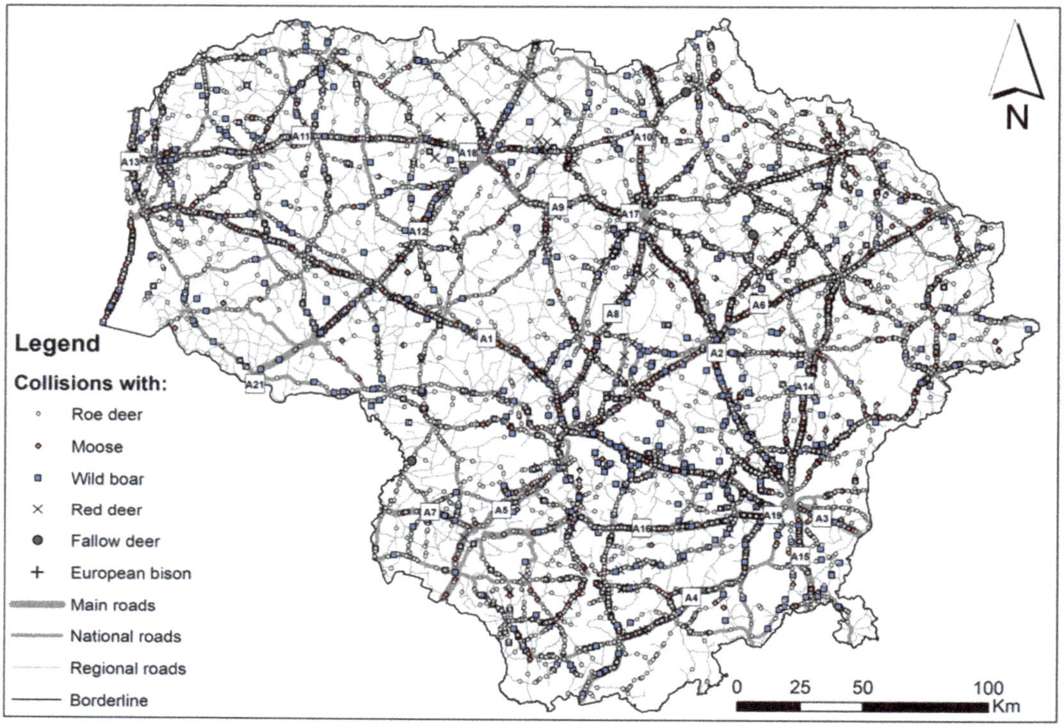

Figure 2. Locations of the ungulate–vehicle collisions (UVCs) in Lithuania, 2002–2017.

2.3. Identification of Hotspots

The literature reports many different spatial techniques to identify relatively short road segments where road accidents or hazards occur more frequently than expected [4,16,32–38]. We performed UVC data clustering using the KDE+ tool [16,39], which analyses UVCs, represented as point features, located along the roads that are represented as line features. The KDE + tool identifies short significant road sections (so-called "clusters") where accidents occur more frequently than expected. The tool also assigns strength values (measured from 0 to 1) to the clusters, showing the highest probability of a crash from the driver's point of view [16]. We created UVC clusters (hereinafter "UVC hotspots") using the recommended parameters [16] by following road network properties (KDE+ bandwidth, 150 m; Monte Carlo simulations, 800; and minimal cluster strength, 0.2).

2.4. Compound Annual Growth Rates

We analyzed the UVC, UVC hotspot, AADT, and fencing compound annual growth rate (CAGR) on the different categories of roads. CAGR is defined as [40]

$$\text{CAGR}\,(t_0, t_n) = \left(\frac{V(t_n)}{V(t_0)}\right)^{\frac{1}{t_n - t_0}} - 1, \quad (1)$$

where $V(t_0)$ is the initial value of AADT, UVC, and UVC hotspot numbers and fence lengths, $V(t_n)$ is the end value of the same parameters, and $t_n - t_0$ is the number of years. We used CAGR to smooth variable returns so that they could more easily be used for evaluation of long-term UVC, UVC hotspot, AADT, and fence length changes that occur in different road categories.

2.5. Multiple Linear Regression Analysis

We assessed the effect of wildlife fencing on UVC in Lithuania from 2002 to 2017 by performing multiple linear regression (MLR) analysis [41]. The UVC and UVC hotspots were the dependent variables. We checked the UVCs and UVC hotspots against AADT and fence lengths on main, national, and regional roads.

We checked how UVCs and UVC hotspots on one category of roads depend on UVCs and UVC hotspots on the other two categories of roads. Then we checked how UVCs on one category of roads depend on AADT on all categories of roads. Last we checked how UVCs and UVC hotspots on one category of roads depend on fence lengths on all categories of roads. We ran models separately for every category of roads.

In all MLR, regression coefficients b were treated as indicators of strength of the effect of each individual independent variable to the dependent variable.

Finally, we used a unified modelling language (UML) collaboration diagram [42] to describe the results of the multiple regression analyses.

3. Results

We identified UVC, UVC hotspot, AADT, and fencing change patterns on different categories of roads. To test our hypotheses we used regression analyses. We identified relationships between UVCs and UVC hotspots on different categories of roads. In addition, we identified UVC and UVC hotspot relationships with fencing and AADT on different categories of roads.

3.1. Roadkill Hotspots

We identified UVC hotspots every year from 2002 to 2017. In total, we found 691 unique UVC hotspot locations (Figure 3) for all categories of roads (261, 373, and 57 hotspots on main, national, and regional roads, respectively), with the hotspots having an average length of 133 m (Table A3) and a length range of 100–471 m. The range of strength (KDE+) of UVC hotspots on main, national, and regional roads was 0.27–0.50, 0.39–0.50, and 0.33–0.48, respectively (Table A3). The UVC hotspots resulting from the analysis are also accessible online as a web map service [43].

3.2. Patterns of UVC, AADT, and Fencing Changes

On average, 938 UVCs were recorded each year for the period 2002–2017 (Table A1). UVCs including roe deer, red deer, moose, fallow deer, bison, and wild boar accounted for approximately 69% of WVCs. Roe deer alone accounted for approximately 54% of all UVCs.

The regional road network was the largest and densest, while the national road network was denser than the main road network but sparser then the regional road network (Figure 1). The highest AADT was found on the main roads and the lowest on the regional roads (Table A3). The largest share of UVCs and UVC hotspots occurred on national roads (Figure 4). Decreasing AADT and increasing UVC hotspots on regional roads suggest that AADT does not impact UVCs on regional roads characterized by lower speed and traffic intensity.

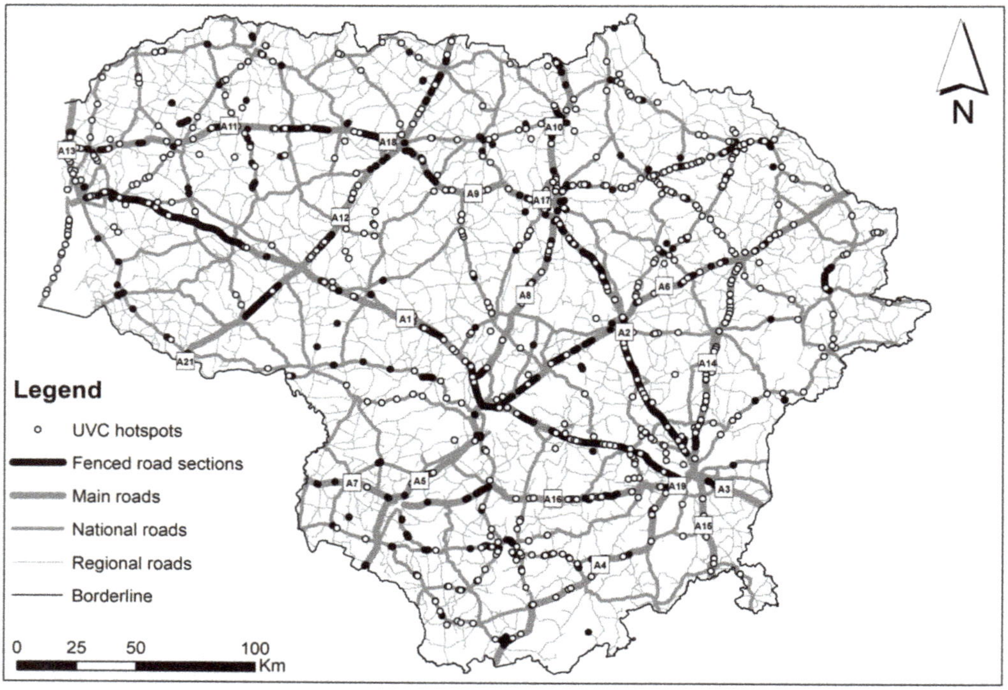

Figure 3. UVC hotspots and fenced road sections in the 2002–2017 period.

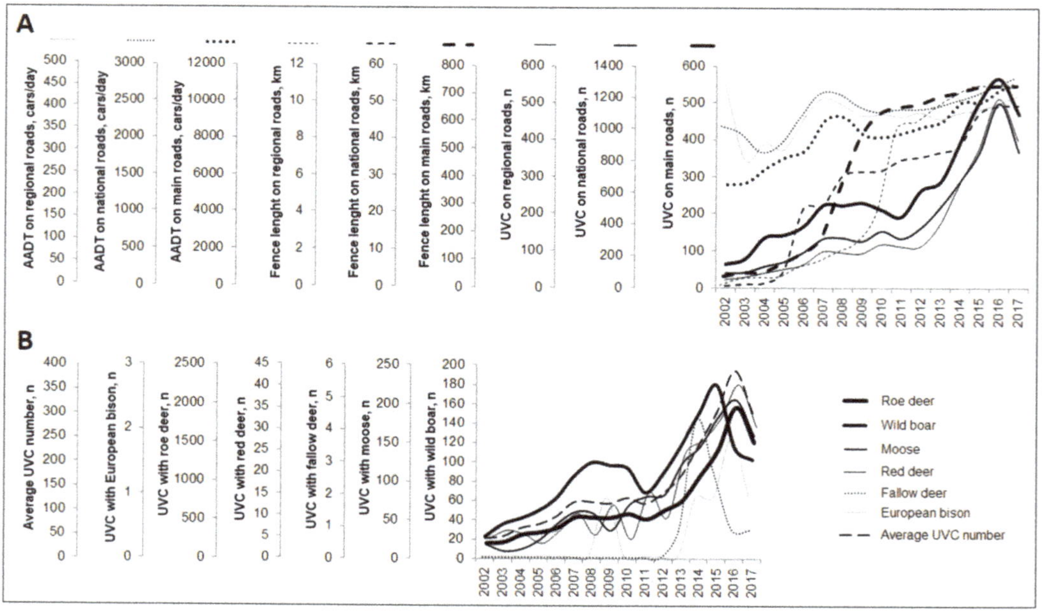

Figure 4. Change patterns of annual average daily traffic (AADT), including all types of vehicles, total UVCs (**A**, Table A1), fencing (**A**, Table A4), and species-specific UVCs (**B**, Table A2) in 2002–2017.

CAGR analysis revealed that AADT increased on main and national roads, while it decreased on regional roads (Figure 5, Table A1). UVC hotspot number, length, and average strength increased the most on regional roads (Table A3). The UVC hotspot average strength decreased on the main and regional roads (Table A3). The length of the new fences decreased the most on the main roads and increased on regional roads only (Table A4). The total length of fences increased the most on national roads (Figure 5). However, the share of new fences was decreased on the national and main roads (Figure 5, Table A1).

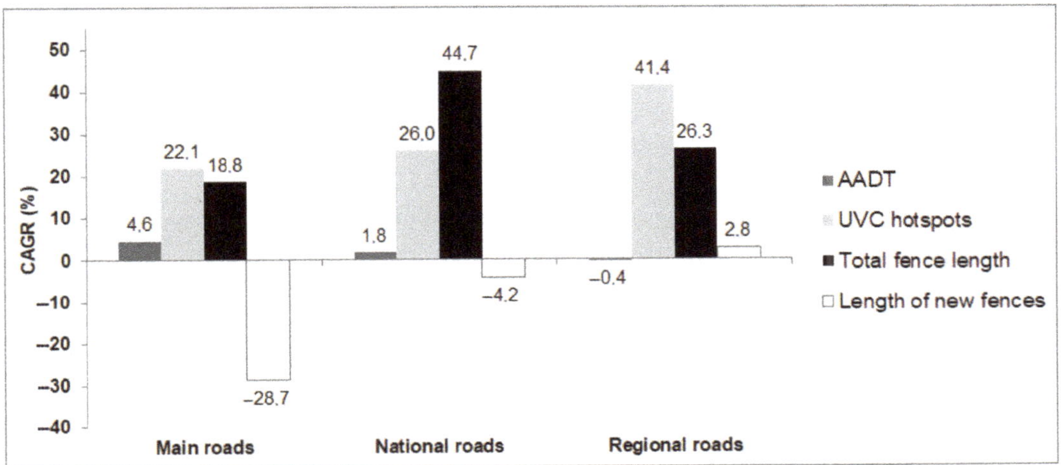

Figure 5. Compound annual growth rate (CAGR) of AADT (Table A1), UVC hotspot (Table A3), and fence length (Table A4) distributions within the different types of road networks in 2002–2017.

CAGR analysis suggested that wildlife fencing on roads characterized by higher speed and higher traffic intensity (sparser network) may shift collision occurrence towards roads (denser road network) characterized by lower speed and lower traffic intensity. To confirm or reject this assumption, we performed an additional MLR analysis.

3.3. Factors Influencing Roadkills

The patterns of relationship of UVCs and UVC hotspots on different road categories are shown in Table 1. We checked how changes of UVC and UVC hotspots on regional roads were influenced by dynamics of UVC and UVC hotspots on national and main roads, repeating the model for all road categories.

Table 1. Hypotheses that there are no relationships between UVC and UVC hotspot patterns on different categories of roads declined with * $p < 0.10$, ** $p < 0.05$, *** $p < 0.01$, **** $p < 0.001$. NS: not significant. Response variables are UVCs and UVC hotspots on regional, national, and main roads; b_{UVC} show the degree of change in the response variable for every 1-unit of change in the predictor variable.

Target		Intercept ± SE	b_{UVC} ± SE		F	R^2
UVC	Regional	−34.17 ± 13.08 **	National	Main	477.29 ****	0.99
			0.54 ± 0.09 ****	−0.15 ± 0.20 *		
	National	18.62 ± 24.69 NS	Regional	Main	864.69 ****	0.99
			1.33 ± 0.23 ****	0.77 ± 0.23 ***		
	Main	40.42 ± 19.48 ****	Regional	National	246.63 ****	0.97
			−0.23 ± 0.38 NS	0.61 ± 0.18 ***		

Table 1. *Cont.*

Target		Intercept ± SE	b_{UVC} ± SE		F	R^2
			b_{UVC} **hotspots** ± SE			
UVC hotspots	Regional	−1.47 ± 0.89 ^{NS}	National: 0.07 ± 0.08 ^{NS}	Main: −0.21 ± 0.14 ^{NS}	34.05 ****	0.84
	National	−0.94 ± 3.13 ^{NS}	Regional: 6.69 ± 0.89 ^{NS}	Main: 1.34 ± 0.30 ****	79.06 ****	0.92
	Main	3.20 ± 1.59 **	Regional: 0.74 ± 0.47 ^{NS}	National: 0.45 ± 0.10 ****	90.20 ****	0.93

The relationship patterns of UVC and UVC hotspots with AADT and fencing length on different categories of roads are shown in Table 2. We checked if UVC number on the regional, national, and main roads was related to AADT of all these road types. MLR with UVC hotspot number regressed to AADT were all not significant and therefore are not presented in Table 2. Then we checked if UVC numbers and UVC hotspot numbers were related to the length of fencing on all road categories.

Table 2. Hypotheses that there are no relationship between UVC and UVC hotspots with AADT and fencing length patterns on different categories of roads declined with * $p < 0.10$, ** $p < 0.05$, *** $p < 0.01$, **** $p < 0.001$. NS: not significant. Response variables are UVCs and UVC hotspots on regional, national, and main roads; b_{AADT} and $b_{length\ of\ fences}$ show the degree of change in the response variable for every 1-unit of change in the predictor variable.

Target		Intercept ± SE	Regional	National	Main	F	R^2
				b_{AADT} ± SE			
UVC	Regional	−405.17 ± 230.81 *	0.03 ± 0.80 ^{NS}	−0.08 ± 0.03 ^{NS}	0.09 ± 0.03 **	10.49 **	0.72
	National	−828.65 ± 432.42 *	0.13 ± 1.50 ^{NS}	−0.16 ± 0.44 ^{NS}	0.19 ± 0.06 ***	14.13 ****	0.78
	Main	−320.62 ± 172.45 ^{NS}	0.03 ± 0.60 ^{NS}	−0.12 ± 0.18 ^{NS}	0.10 ± 0.02 ***	22.30 ****	0.84
				$b_{lenght\ of\ fences}$ ± SE			
UVC	Regional	12.25 ± 38.84 ^{NS}	29.09 ± 10.83 **	8.05 ± 3.35 **	−0.49 ± 0.24 *	14.80 **	0.79
	National	86.55 ± 72.99 ^{NS}	53.82 ± 20.35 **	17.35 ± 6.29 **	−0.90 ± 0.45 *	19.39 ****	0.83
	Main	83.02 ± 32.34 **	23.07 ± 9.01 **	9.15 ± 2.79 ***	−0.43 ± 0.20 **	23.47 ****	0.85
UVC hotspots	Regional	−1.03 ± 1.61 ^{NS}	1.45 ± 0.45 ***	0.25 ± 0.14 *	−0.02 ± 0.01 **	11.65 ****	0.74
	National	1.64 ± 8.58 ^{NS}	4.03 ± 2.39 ^{NS}	1.74 ± 0.74 **	−0.10 ± 0.05 *	7.96 **	0.67
	Main	1.46 ± 4.76 ^{NS}	3.30 ± 1.33 **	0.97 ± 0.41 **	−0.06 ± 0.03 **	22.30 **	0.73

In order to simplify the interpretation of the MLR results (Tables 1 and 2), we used the UML collaboration diagram (Figure 6), which showed only significant relationships. The vertical swim lanes in Figure 6 represent the different road categories. The dependent and independent variables were represented as rectangles. The first sign on the right (+ or –) adjacent to the line indicates changes within the source variables, while the second sign on the left (+ or –) indicates changes in the target variables (effect). Arrows show significant source–target relationships (Tables 1 and 2). For instance, an increase (+) in fence length on main roads was significantly related to the decrease (–) of UVC and UVC hotspots on main roads (Table 2). The relationship lines are absent in cases where no significant relationship was found between variables (for instance, between AADT and UVC on national and regional roads).

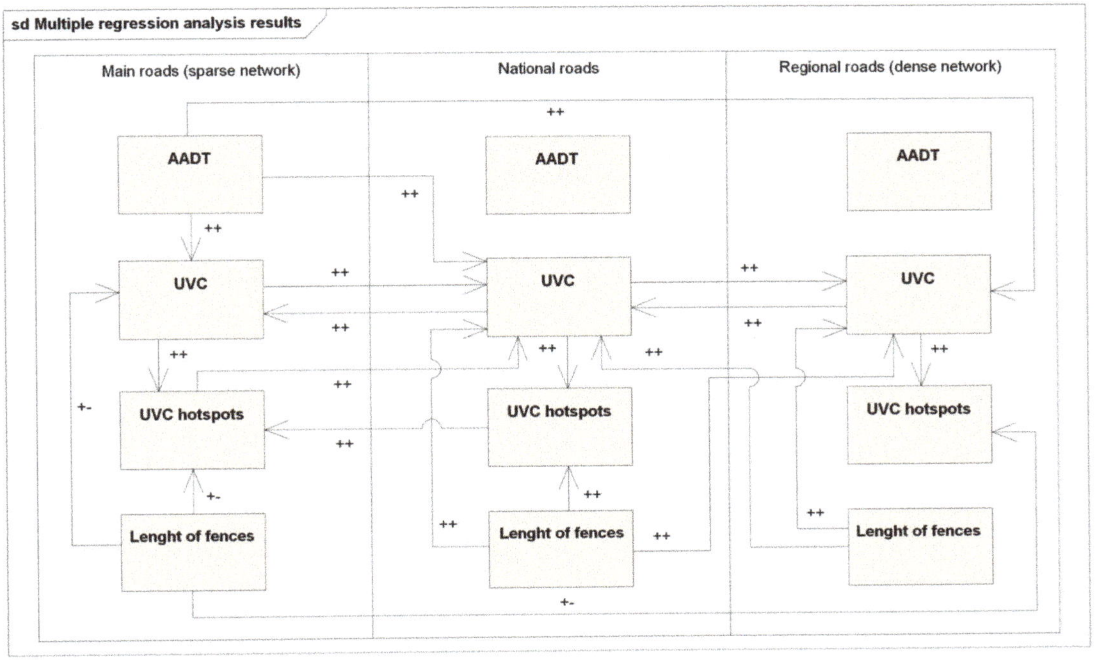

Figure 6. Unified modelling language (UML) collaboration diagram that shows MLR analysis results.

MLR analysis results show that while AADT increased on main roads, the UVC numbers increased on all lower category roads (Table 2, Figure 6). These relationships show that AADT had no effect on the UVC on national and regional roads (no relationship lines in Figure 6). However, an increase in AADT on higher-category roads may shift collision occurrence towards roads characterized by lower speed and lower traffic intensity.

As UVC increased on main roads, UVC became more frequent on both national and then regional roads. However, the increase in UVC on main roads did not directly impact UVC growth on regional roads (Table 1, Figure 6). As UVC increased on regional roads, UVC became more frequent on both national and main roads. However, the increase in UVC on regional roads did not directly impact UVC growth on the main roads (Table 1, Figure 6). These relationships show that changes in UVC on higher category roads have a direct relationship with changes in UVC on lower-category roads and vice versa.

An increase in UVCs increased UVC hotspots on the main roads. The same UVC and UVC hotspot relationships were also observed on national and regional roads. An increase in UVC hotspots on main roads increased UVCs on national roads. However, an increase in UVC hotspots on national roads increased UVC hotspots on main roads (Table 1, Figure 6). These relationships showed that changes in UVCs have a direct relationship with changes in UVC hotspots on all types of roads. However, an increase in UVC hotspots on lower-category roads might increase UVC hotspots on higher-category roads.

An increase in the length of wildlife fencing on the main roads diminished the number of UVCs and UVC hotspots (Table 2, Figure 6). This relationship showed that fences are effective mitigation measures for main roads.

The length of wildlife fencing on national and regional roads increased the UVC on national and regional roads (Table 2, Figure 6). These relationships showed that fencing had no effect on the UVCs on lower-category roads.

As the length of the fences increased, so did the UVC hotspots on national roads. As the length of the fences increased, so did the UVCs on regional roads. These relationships showed that fencing was not as effective as expected on lower-category national and regional roads (Table 2, Figure 6).

4. Discussion

First we discuss relationship between UVCs, AADT, and fencing. We partly confirm our hypotheses and explain changes in roadkill and fencing patterns, and evaluate wildlife fences on different categories of roads. We also provide recommendations on how to improve UVC mitigation measures on different categories of roads.

4.1. Relationship between UVC, AADT, and Fencing

MLR results confirm both tested hypotheses for the main roads. We acknowledge that the pattern of UVC numbers on the highest-category (main) roads is explained by changes in AADT. Moreover, the growth of AADT on main roads might increase the UVCs on national and regional roads (Table 2, Figure 6). Correlations between the number of UVCs and AADT on lower category (national and regional) roads were insignificant, and this is in agreement with [12,44].

We also found that the growing length of wildlife fencing on the main roads effectively diminished the number of UVCs and UVC hotspots (see Table 2 and Figure 6, links between UVCs, UVC hotspots, and fence length on main roads). However, main road fencing redirected wildlife towards lower category roads and dispelled UVC hotspots on regional roads (see Table 2 and Figure 6, link between fences on main roads and UVC hotspots on regional roads).

4.2. Changes in Roadkill and Fencing Patterns

A pattern is regular in the UVC, UVC hotspot, and fencing locations on different categories of roads over time. During 2002–2017 in Lithuania the number of UVCs increased in all categories of roads. At the beginning of the period, both UVCs and fenced road sections were most common on main roads characterized by higher speed and traffic intensity. The same pattern was observed in Spain [45]. However, as of 2011, the numbers of both UVC and fenced road sections started to grow on roads characterized by lower speed and traffic intensity (Figure 4A). Similar to other countries [46,47] this UVC increase can be an effect of increasing wildlife populations in Lithuania [13].

Changes in UVC patterns may be related to blocked wildlife pathways due to frequent fencing on the main roads. While the number of UVCs on fenced road sections has been reduced, it is growing on non-fenced road sections and adjacent roads (Figure 6, Table 2). It might require time for wildlife to rebuild new pathways (e.g., Figure 4A, fencing before 2011, UVC 2008–2012 period). Scattering of the new UVCs is the reason why UVCs do not form hotspots (Figure 6). After new wildlife paths are established, UVC hotspots start to occur in new locations on lower-category roads.

However, we assume that the formation of hotspots shows stability in ungulate pathways. We found that annually only 16% of all UVCs occurred within hotspots, which suggests that a large part of wildlife pathways in the country are scattered and not permanent. We partly confirmed the results of [7,48–50] that UVCs are not spatially random, since 84% of UVCs (2002–2017) were not located in the hotspots.

From the above we conclude that in the short-term, wildlife fencing can decrease UVC numbers on main roads, but as a result of altered wildlife pathways, UVC locations shift towards the denser lower category road network. At the same time, limited movements may reduce the importance of adjacent habitats for wildlife and may amplify the importance of more distant habitat patches [51].

4.3. Evaluation of Wildlife Fences

WVCs are a cause of serious concern for road planners and biologists in terms of traffic safety, species conservation, and animal welfare [4]. Collisions numbers can be significant to species conservation, wildlife management, and traffic safety, thus creating ethical, social, economic, and even political tensions. Putting in place mitigation measures is challenging, due to the lack of knowledge on the local spatial–temporal patterns of wildlife dynamics, including population, behavior, pathways, and habitat suitability [15,52,53].

Wildlife fences in conjunction with underpasses, gates, and jump outs are the most common WVC mitigation measures in Lithuania [54]. So far, no overpasses or advanced dynamic wildlife warning systems have been deployed, and the effectiveness of mitigation measures is still discussed in the country. The reason is that the location of fences is fixed, while the behavior of different species constantly changes [20,55]. A better understanding of the spatial distribution of WVC [4,20] requires consideration of many factors. They include, but are not limited to, understanding of wildlife movements and behaviors, localization of wildlife corridors [51,56–58], knowledge of population density [59], population dynamics and habitat properties. In line with other authors [15,52,53,60] we also confirm that placing mitigation measures is challenging, because of the lack of knowledge on the local spatiotemporal patterns of wildlife dynamics, including population, behavior, pathways and habitat suitability [13,20].

Exponential growth of the length of installed wildlife fences occurred in 2008–2011, and since 2011 fencing intensity considerably decreased (Figure 4A). As the outcome, an increase in UVCs was observed after 2012, when ungulates adapted to existing fencing. The longest sections of the wildlife fences on the main roads of Lithuania were installed in 2004–2010, on national roads in 2005–2008, and on regional roads in 2009–2012 (Figure 4A, Table A4). Thus, safety measures targeting roads with lower traffic intensity and speed were introduced at the end of the analyzed period. UVC and UVC hotspot occurrences constantly increased during 2004–2016. In 2017, the longest fenced road sections were on main roads, while regional roads had minimal fencing.

The highest rate of UVC on national roads (50.7%, Table A1) conforms to the fact that only 6.1% of all wildlife fences were installed there (Table A4). We concluded that the increase in the AADT and length of wildlife fences on the main roads shifted UVCs, first towards national roads and later towards regional road networks (Figure 6). The highest number of UVCs (especially with roe deer) was on national roads. In contrast to another study [61,62], fewer accidents were caused by ungulates on the main roads where the traffic volume was greater and speed was higher, as 93% of wildlife fences were installed along them (Table A4). Fencing is effective and may reduce roadkill rates [4,24], but on highest-category roads only (Table 2, Figure 6). Thus, the efficiency of fences decreases in the lower categories of the road network (Table 2, Figure 6 national and regional roads).

We found only a few new fences in the locations where UVC hotspots occurred in the previous year. This may be due to the fact that wildlife fencing in Lithuania is not based on WVC data [2] and is organized according to the strictly defined road infrastructure reconstruction programs. Such programs address the road safety standards for main roads/highways, rather than adequately responding to constantly changing UVC situations on lower-category roads.

Short wildlife fences may not sufficiently reduce the risk of accidents [63–66]; however, they are economically more affordable. Long fences are less economically efficient, but may perform better [18,63,66] on roads with the highest traffic intensity [67]. Building longer fences because of traffic safety reasons may be unduly costly [63], especially on dense road networks. From a wildlife perspective, longer fences cause landscape and habitat fragmentation and isolation of populations [66,68]. Therefore, they require additional measures to enable safe road crossings for wildlife.

Fencing not only prevents ungulates from crossing roads, but also directs them to the passage infrastructure. This might force animals to avoid roads with higher traffic and speed intensity. Mitigation measures may redirect animal pathways towards more attractive and distant habitat patches; however, they inevitably contribute to increasing UVC numbers on the lower category roads characterized by lower speed and traffic intensity (Table 2, Figure 6).

As a standard, all highways characterized with the highest traffic and speed intensity have to be fenced due to traffic safety reasons. Consequently, UVC rates grew on the lower category roads where no UVC mitigation measures were deployed (Table 2, Figure 6). Continuing building fences on unprotected main road sections without proper planning [69] can shift the problems to unfenced national and regional road sections. In addition, it may disconnect important habitats and may become the reason for serious ecological problems such as the extinction of local populations [70] and discontinuity of important ecological networks [51].

In line with [23,24,65] we confirm that wildlife fences are an effective long-term UVC mitigation measure on highways. However, this measure can only be effective if fences are planned in a timely manner [68,69,71], carefully inserted in the landscape [72], properly maintained, and in conjunction with other permanent UVC mitigation measures such as underpasses, overpasses, and driver warning systems. If not, habitat isolation may be amplified, and costs of construction and maintenance may be too high without any positive effect on the drivers and wildlife, especially on lower-category roads [65].

Modifying the natural behavior of ungulates is almost impossible [10]. Consequently mitigation should focus on the modification of human behavior and changing drivers' attitudes [73], introducing novel car safety systems [74], and improving road engineering [10,75].

There were no significant relationships between fences and UVC hotspots on regional roads; moreover, increases of fencing length resulted in increases of UVC hotspot numbers on national roads (Table 2, Figure 6). Therefore, the effectiveness of wildlife fences on national and regional roads is limited. We recommend less restrictive types of mitigation measures (e.g., advanced dynamic wildlife warning systems not preventing wildlife crossings) that should be applied for short significant road sections (hotspots) on lower (national and regional) category roads.

5. Conclusions

Analysis of the relationships between UVCs, UVC hotspots, fencing, and AADT in Lithuania showed the following:

1. Wildlife fences are an effective mitigation measure for the main roads characterized by the highest traffic intensity. Fencing is not effective on lower-category roads where traffic intensity has a less significant impact on UVCs.
2. Increased amounts of wildlife fencing may reduce the number of UVCs on the main roads and shift UVCs toward national and regional roads, characterized by lower speed and traffic intensity (denser road network).
3. We recommend that efforts to reduce wildlife collision occurrence on lower-category roads should focus on driver attitudes and road conditions, rather than animal movement and behavior.

Author Contributions: Conceptualization, A.K. and L.B.; methodology, A.K.; software, A.K.; validation, A.K.; formal analysis, A.K.; investigation, A.K. and L.B.; resources, L.B.; data curation, L.B.; writing—original draft preparation, A.K.; writing—review and editing, L.B.; visualization, A.K.; supervision, L.B.; project administration, L.B.; funding acquisition, L.B. Both authors have read and agreed to the published version of the manuscript.

Funding: This research received no external funding.

Data Availability Statement: Spatial data for this article are published as a web map journal https://www.arcgis.com/apps/MapJournal/index.html?appid=d81195212a4b4bcc9c5aab34a0037609 (accessed on 1 March 2021) and can also be found online at http://dx.doi.org/10.17632/8gdpz5845x.4 (accessed on 1 March 2021).

Acknowledgments: The authors would like to sincerely thank to K. Tóth, for comments and anonymous reviewers for their suggestions and contributions to this manuscript.

Conflicts of Interest: The authors declare no conflict of interest.

Appendix A

Table A1. The total number of road accidents, traffic intensity (AADT), and ungulate–vehicle collisions (UVCs) on different categories of roads.

Year	Total Road Accidents	AADT			Total UVC	UVC [1]		
		Main	National	Regional		Main	National	Regional
2002	6090	5610	2178	451	197	63	108	27
2003	5963	5729	2082	282	221	77	111	33
2004	6372	6519	1833	319	328	135	149	44
2005	6771	7107	1932	306	376	142	178	54
2006	6658	7488	2288	375	470	166	239	66
2007	6448	9100	2624	422	700	223	325	102
2008	4795	9240	2630	406	689	222	329	99
2009	3827	8293	2457	382	684	229	309	97
2010	3530	8196	2372	386	793	210	372	122
2011	3266	8415	2410	380	734	192	326	116
2012	3392	8744	2410	389	881	264	394	116
2013	3418	9036	2446	375	1120	288	525	175
2014	3255	10,086	2527	389	1548	395	689	285
2015	3033	10,083	2610	395	1897	505	874	381
2016	3201	10,802	2729	409	2457	567	1180	516
2017	3192	11,062	2845	428	1911	468	872	403
CAGR	−4.2	4.6	1.8	−0.4	16.4	14.3	14.9	19.7

[1] wild boar, moose, fallow deer, red deer, roe deer, European bison (decomposed in Table A2).

Table A2. Distribution of the annual ungulate–vehicle collisions (UVCs) by animal species.

Year	Wild Boar	Moose	Fallow Deer	Red Deer	Roe Deer	European Bison
2002	23	19	0	5	150	0
2003	36	11	0	7	166	0
2004	42	14	0	6	263	0
2005	50	26	0	4	291	0
2006	63	47	0	7	352	0
2007	86	62	0	11	490	0
2008	100	60	0	6	485	0
2009	97	38	0	13	483	1
2010	94	71	0	5	532	0
2011	69	82	0	16	464	0
2012	89	85	0	10	583	0
2013	118	126	1	25	712	0
2014	151	147	5	28	1031	1
2015	180	185	3	33	1367	1
2016	115	207	1	41	1912	2
2017	103	160	1	31	1460	1
CAGR	10.5	15.3	NS	12.9	16.4	NS

Table A3. The number, length, and strength of ungulate–vehicle collision (UVC) hotspots on different categories of roads.

Year	Number			Total Length (km)			Average Strength (KDE+)		
	Main	National	Regional	Main	National	Regional	Main	National	Regional
2002	2	2	0	279.00	267.00	0.00	0.49	0.47	0.00
2003	6	3	0	748.95	383.00	0.00	0.37	0.44	0.00
2004	2	4	0	244.67	492.94	0.00	0.27	0.40	0.00
2005	6	2	0	762.85	283.00	0.00	0.37	0.50	0.00
2006	8	10	0	972.35	1278.03	0.00	0.40	0.40	0.00
2007	10	16	0	1239.00	2105.55	0.00	0.40	0.42	0.00
2008	11	18	0	1559.51	2238.31	0.00	0.44	0.39	0.00
2009	8	14	0	1157.00	1840.59	0.00	0.50	0.43	0.00
2010	10	17	0	1347.00	2173.23	0.00	0.47	0.42	0.00
2011	6	10	3	734.78	1239.49	395.96	0.26	0.40	0.40
2012	8	11	2	1058.00	1437.90	200.34	0.46	0.40	0.33
2013	16	24	4	2046.77	2994.77	513.87	0.42	0.40	0.43
2014	33	35	7	4818.19	4533.86	916.35	0.42	0.42	0.33
2015	40	44	9	5715.89	6033.05	1236.00	0.45	0.45	0.48
2016	55	99	14	7543.99	13,230.43	1717.12	0.41	0.40	0.36
2017	40	64	18	5313.26	8573.58	2473.20	0.43	0.42	0.41
CAGR	22.1	26.0	41.4	21.7	26.0	96.3	−0.9	−0.7	9.9

Table A4. The number and length of new fences built on different categories of roads.

Year	Number			Length of New Fences (km)		
	Main	National	Regional	Main	National	Regional
2002	75	1	5	56.13	0.19	0.34
2003	14	10	3	12.67	0.42	0.33
2004	7	6	5	0.34	0.29	0.07
2005	32	9	0	32.23	3.17	0.00
2006	23	18	2	36.74	16.83	0.51
2007	73	20	9	56.62	0.73	0.31
2008	100	35	4	184.42	8.15	0.52
2009	99	28	15	178.02	1.23	0.53
2010	58	8	11	85.81	0.30	1.59
2011	48	27	24	22.43	2.87	4.52
2012	22	9	20	11.01	0.80	0.39
2013	41	33	11	28.26	1.27	0.80
2014	32	23	17	14.35	1.78	0.68
2015	47	21	11	20.50	8.61	0.49
2016	3	8	15	3.90	2.01	0.50
2017	6	0	0	0.35	0.00	0.00
CAGR	−15.5	−14.2	−23.0	−28.7	−4.2	2.8

References

1. Bíl, M.; Kubeček, J.; Sedoník, J.; Andrášik, R. Srazenazver.cz: A system for evidence of animal-vehicle collisions along transportation networks. *Biol. Conserv.* **2017**, *213*, 167–174. [CrossRef]
2. Balčiauskas, L. Distribution of species-specific wildlife–vehicle accidents on Lithuanian roads, 2002–2007. *Est. J. Ecol.* **2009**, *58*, 157. [CrossRef]
3. Lithuanian Road Administration under the Ministry of Transport and Communications. Distribution of Road Accidents by Type of Accident 2000–2017 in Lithuania. Available online: http://gissrv.eismoinfo.lt/server/rest/services/ARCGIS_ONLINE/OpenData/FeatureServer/9 (accessed on 16 March 2020).
4. Seiler, A. Predicting locations of moose-vehicle collisions in Sweden. *J. Appl. Ecol.* **2005**, *42*, 371–382. [CrossRef]
5. Seiler, A.; Helldin, J.O. Mortality in wildlife due to transportation. In *The Ecology of Transportation: Managing Mobility for the Environment*; Springer: Berlin/Heidelberg, Germany, 2006; pp. 165–189.
6. Sáenz-de-Santa-María, A.; Tellería, J.L. Wildlife-vehicle collisions in Spain. *Eur. J. Wildl. Res.* **2015**, *61*, 399–406. [CrossRef]
7. Favilli, F.; Bíl, M.; Sedoník, J.; Andrášik, R.; Kasal, P.; Agreiter, A.; Streifeneder, T. Application of KDE+ software to identify collective risk hotspots of ungulate-vehicle collisions in South Tyrol, Northern Italy. *Eur. J. Wildl. Res.* **2018**, *64*, 59. [CrossRef]
8. Sullivan, J.M. Trends and characteristics of animal-vehicle collisions in the United States. *J. Saf. Res.* **2011**, *42*, 9–16. [CrossRef] [PubMed]
9. Schwartz, A.L.W.; Shilling, F.M.; Perkins, S.E. The value of monitoring wildlife roadkill. *Eur. J. Wildl. Res.* **2020**, *66*, 18. [CrossRef]
10. Putzu, N.; Bonetto, D.; Civallero, V.; Fenoglio, S.; Meneguz, P.G.; Preacco, N.; Tizzani, P. Temporal patterns of ungulate-vehicle collisions in a subalpine Italian region. *Ital. J. Zool.* **2014**, *81*, 463–470. [CrossRef]
11. Gilbert, S.L.; Sivy, K.J.; Pozzanghera, C.B.; DuBour, A.; Overduijn, K.; Smith, M.M.; Zhou, J.; Little, J.M.; Prugh, L.R. Socioeconomic Benefits of Large Carnivore Recolonization Through Reduced Wildlife-Vehicle Collisions. *Conserv. Lett.* **2017**, *10*, 431–439. [CrossRef]
12. Madsen, A.B.; Strandgaard, H.; Prang, A. Factors causing traffic killings of roe deer Capreolus capreolus in Denmark. *Wildl. Biol.* **2002**, *8*, 55–61. [CrossRef]
13. Wierzchowski, J.; Kučas, A.; Balčiauskas, L. Application of Least-Cost Movement Modeling in Planning Wildlife Mitigation Measures along Transport Corridors: Case Study of Forests and Moose in Lithuania. *Forests* **2019**, *10*, 831. [CrossRef]
14. Seiler, A. Trends and spatial patterns in ungulate-vehicle collisions in Sweden. *Wildl. Biol.* **2004**, *10*, 301–313. [CrossRef]
15. Morelle, K.; Lehaire, F.; Lejeune, P. Spatio-temporal patterns of wildlife-vehicle collisions in a region with a high-density road network. *Nat. Conserv.* **2013**, *5*, 53–73. [CrossRef]
16. Bíl, M.; Andrášik, R.; Svoboda, T.; Sedoník, J. The KDE+ software: A tool for effective identification and ranking of animal-vehicle collision hotspots along networks. *Landsc. Ecol.* **2016**, *31*, 231–237. [CrossRef]
17. Bíl, M.; Andrášik, R.; Nezval, V.; Bílová, M. Identifying locations along railway networks with the highest tree fall hazard. *Appl. Geogr.* **2017**, *87*, 45–53. [CrossRef]
18. Ford, A.T.; Clevenger, A.P.; Huijser, M.P.; Dibb, A. Planning and prioritization strategies for phased highway mitigation using wildlife-vehicle collision data. *Wildl. Biol.* **2011**, *17*, 253–265. [CrossRef]
19. Keken, Z.; Sedoník, J.; Kušta, T.; Andrášik, R.; Bíl, M. Roadside vegetation influences clustering of ungulate vehicle collisions. *Transp. Res. Part D Transp. Env.* **2019**, *73*, 381–390. [CrossRef]
20. Diaz-Varela, E.R.; Vazquez-Gonzalez, I.; Marey-Pérez, M.F.; Álvarez-López, C.J. Assessing methods of mitigating wildlife–vehicle collisions by accident characterization and spatial analysis. *Transp. Res. Part D Transp. Env.* **2011**, *16*, 281–287. [CrossRef]

21. Bíl, M.; Kušta, T.; Andrášik, R.; Cícha, V.; Brodská, H.; Ježek, M.; Keken, Z. No clear effect of odour repellents on roe deer behaviour in the vicinity of roads. *Wildl. Biol.* **2020**. [CrossRef]
22. Clevenger, A.P.; Chruszcz, B.; Gunson, K.E. Spatial patterns and factors influencing small vertebrate fauna road-kill aggregations. *Biol. Conserv.* **2003**, *109*, 15–26. [CrossRef]
23. Ascensão, F.; Clevenger, A.; Santos-Reis, M.; Urbano, P.; Jackson, N. Wildlife–vehicle collision mitigation: Is partial fencing the answer? An agent-based model approach. *Ecol. Model* **2013**, *257*, 36–43. [CrossRef]
24. Clevenger, A.P.; Chruszcz, B.; Gunson, K.E. Highway mitigation fencing reduces wildlife-vehicle collisions. *Wildl. Soc. Bull.* **2001**, *29*, 646–653. [CrossRef]
25. Rytwinski, T.; Soanes, K.; Jaeger, J.A.G.; Fahrig, L.; Findlay, C.S.; Houlahan, J.; van der Ree, R.; van der Grift, E.A. How Effective Is Road Mitigation at Reducing Road-Kill? A Meta-Analysis. *PLoS ONE* **2016**, *11*, e0166941. [CrossRef]
26. Holmberg, M. *Uncertainty of Critical Loads in the Baltic Countries: Report from a Project Funded by the Nordic Council of Ministers*; TemaNord: Environment; Nordic Council of Ministers: Copenhagen, Denmark, 2000; ISBN 9789289304634.
27. European Environmental Agency. Lithuania Land Cover Country Fact Sheet. 2012. Available online: https://www.eea.europa.eu/ds_resolveuid/1ca731f33d0c48a0a4ed60abe67fa6bf (accessed on 16 March 2020).
28. Balčiauskas, L.; Trakimas, G.; Juškaitis, R.; Ulevičius, A.; Balčiauskienė, L. *Lietuvos Žinduolių, Varliagyvių ir Roplių Atlasas. Antras Papildytas Leidimas [Atlas of Lithuanian Mammals, Amphibians and Reptiles]*, 2nd ed.; Akstis: Vilnius, Lithuania, 1999; ISBN 9986-759-07-2.
29. Xie, F.; Levinson, D. Measuring the Structure of Road Networks. *Geogr. Anal.* **2007**, *39*, 336–356. [CrossRef]
30. Lithuanian Road Administration under the Ministry of Transport and Communications. Web Map of Annual Average Daily Traffic Data in Lithuania. Available online: http://gissrv.eismoinfo.lt/server/rest/services/LAKD/VMPEI/MapServer?f=jsapi (accessed on 16 March 2020).
31. Balčiauskas, L.; Stratford, J.; Balčiauskienė, L.; Kučas, A. Importance of professional roadkill data in assessing diversity of mammal roadkills. *Transp. Res. Part D Transp. Env.* **2020**, *87*, 102493. [CrossRef]
32. Clevenger, A.P.; Wierzchowski, J.; Chruszcz, B.; Gunson, K. GIS-Generated, Expert-Based Models for Identifying Wildlife Habitat Linkages and Planning Mitigation Passages. *Conserv. Biol.* **2002**, *16*, 503–514. [CrossRef]
33. Gunson, K.E.; Mountrakis, G.; Quackenbush, L.J. Spatial wildlife-vehicle collision models: A review of current work and its application to transportation mitigation projects. *J. Environ. Manag.* **2011**, *92*, 1074–1082. [CrossRef] [PubMed]
34. Santos, R.A.L.; Mota-Ferreira, M.; Aguiar, L.M.S.; Ascensão, F. Predicting wildlife road-crossing probability from roadkill data using occupancy-detection models. *Sci. Total Environ.* **2018**, *642*, 629–637. [CrossRef] [PubMed]
35. Borda-de-Água, L.; Ascensão, F.; Sapage, M.; Barrientos, R.; Pereira, H.M. On the identification of mortality hotspots in linear infrastructures. *Basic Appl. Ecol.* **2018**. [CrossRef]
36. Santos, S.M.; Marques, J.T.; Lourenço, A.; Medinas, D.; Barbosa, A.M.; Beja, P.; Mira, A. Sampling effects on the identification of roadkill hotspots: Implications for survey design. *J. Environ. Manag.* **2015**, *162*, 87–95. [CrossRef] [PubMed]
37. Okabe, A.; Okunuki, K.; Shiode, S. SANET: A Toolbox for Spatial Analysis on a Network. *Geogr. Anal.* **2006**, *38*, 57–66. [CrossRef]
38. Okabe, A.; Satoh, T.; Sugihara, K. A kernel density estimation method for networks, its computational method and a GIS-based tool. *Int. J. Geogr. Inf. Sci.* **2009**, *23*, 7–32. [CrossRef]
39. Bíl, M.; Andrášik, R.; Janoška, Z. Identification of hazardous road locations of traffic accidents by means of kernel density estimation and cluster significance evaluation. *Accid. Anal. Prev.* **2013**, *55*, 265–273. [CrossRef]
40. Anson, M.J.P.; Fabozzi, F.J.; Jones, F.J. *The Handbook of Traditional and Alternative Investment Vehicles: Investment Characteristics and Strategies*; John Wiley & Sons, Ltd.: Hoboken, NJ, USA, 2011; ISBN 978-0-470-60973-6.
41. Jobson, J.D. Multiple Linear Regression. In *Applied Multivariate Data Analysis*; Springer: New York, NY, USA, 1991; pp. 219–398.
42. ISO/IEC JTC 1/SC 7 ISO/IEC 19501:2005. *Information Technology—Open Distributed Processing—Unified Modeling Language (UML) Version 1.4.2*; International Standardization Organization: Geneva, Switzerland, 2005.
43. Kučas, A. Web Map of Ungulate-Vehicle Collision Hotspots in Lithuania 2002–2017. Available online: https://www.arcgis.com/apps/MapJournal/index.html?appid=d81195212a4b4bcc9c5aab34a0037609 (accessed on 16 March 2020).
44. Kang, W.; Minor, E.S.; Woo, D.; Lee, D.; Park, C.-R. Forest mammal roadkills as related to habitat connectivity in protected areas. *Biodivers. Conserv.* **2016**, *25*, 2673–2686. [CrossRef]
45. Canal, D.; Camacho, C.; Martín, B.; de Lucas, M.; Ferrer, M. Magnitude, composition and spatiotemporal patterns of vertebrate roadkill at regional scales: A study in southern Spain. *Anim. Biodivers. Conserv.* **2018**, *41*, 281–300. [CrossRef]
46. Seiler, A.; Helldin, J.O.; Seiler, C.H. Road mortality in Swedish mammals: Results of a drivers' questionnaire. *Wildl. Biol.* **2004**, *10*, 183–191. [CrossRef]
47. Jasińska, K.D.; Żmihorski, M.; Krauze-Gryz, D.; Kotowska, D.; Werka, J.; Piotrowska, D.; Pärt, T. Linking habitat composition, local population densities and traffic characteristics to spatial patterns of ungulate-train collisions. *J. Appl. Ecol.* **2019**, *56*, 2630–2640. [CrossRef]
48. Girardet, X.; Conruyt-Rogeon, G.; Foltête, J.-C. Does regional landscape connectivity influence the location of roe deer roadkill hotspots? *Eur. J. Wildl. Res.* **2015**, *61*, 731–742. [CrossRef]
49. Bartonička, T.; Andrášik, R.; Dul'a, M.; Sedoník, J.; Bíl, M. Identification of local factors causing clustering of animal-vehicle collisions. *J. Wildl. Manag.* **2018**, *82*, 940–947. [CrossRef]

50. Bíl, M.; Andrášik, R.; Sedoník, J. A detailed spatiotemporal analysis of traffic crash hotspots. *Appl. Geogr.* **2019**, *107*, 82–90. [CrossRef]
51. Kučas, A. Graph-based multi-attribute decision making: Impact of road fencing on ecological network. *Balt. J. Road Bridg. Eng.* **2015**, *10*, 105–111. [CrossRef]
52. Zimmermann Teixeira, F.; Kindel, A.; Hartz, S.M.; Mitchell, S.; Fahrig, L. When road-kill hotspots do not indicate the best sites for road-kill mitigation. *J. Appl. Ecol.* **2017**, *54*, 1544–1551. [CrossRef]
53. Gunson, K.E.; Clevenger, A.P.; Ford, A.T.; Bissonette, J.A.; Hardy, A. A Comparison of Data Sets Varying in Spatial Accuracy Used to Predict the Occurrence of Wildlife-Vehicle Collisions. *Env. Manag.* **2009**, *44*, 268–277. [CrossRef] [PubMed]
54. Lithuanian Road Administration under the Ministry of Transport and Communications. Web Map of Animal-Vehicle Collision Distributions in Lithuania. Available online: http://gissrv.eismoinfo.lt/server/rest/services/LAKIS/su_gyvunais/MapServer?f=jsapi (accessed on 16 March 2020).
55. Clevenger, A.P.; Waltho, N. Performance indices to identify attributes of highway crossing structures facilitating movement of large mammals. *Biol. Conserv.* **2005**, *121*, 453–464. [CrossRef]
56. Andrews, A. Fragmentation of Habitat by Roads and Utility Corridors: A Review. *Aust. Zool.* **1990**, *26*, 130–141. [CrossRef]
57. McRae, B.H.; Dickson, B.G.; Keitt, T.H.; Shah, V.B. Using circuit theory to model connectivity in ecology, evolution, and conservation. *Ecology* **2008**, *89*, 2712–2724. [CrossRef] [PubMed]
58. McRae, B.H. Isolation by resistance. *Evolution* **2006**, *60*, 1551–1561. [CrossRef] [PubMed]
59. Tatewaki, T.; Koike, F. Synoptic scale mammal density index map based on roadkill records. *Ecol. Indic.* **2018**, *85*, 468–478. [CrossRef]
60. Abra, F.D.; da Costa Canena, A.; Garbino, G.S.T.; Medici, E.P. Use of unfenced highway underpasses by lowland tapirs and other medium and large mammals in central-western Brazil. *Perspect. Ecol. Conserv.* **2020**, *18*, 247–256. [CrossRef]
61. Zuberogoitia, I.; del Real, J.; Torres, J.J.; Rodríguez, L.; Alonso, M.; Zabala, J. Ungulate Vehicle Collisions in a Peri-Urban Environment: Consequences of Transportation Infrastructures Planned Assuming the Absence of Ungulates. *PLoS ONE* **2014**, *9*, e107713. [CrossRef]
62. Valero, E.; Picos, J.; Lagos, L.; Álvarez, X. Road and traffic factors correlated to wildlife–vehicle collisions in Galicia (Spain). *Wildl. Res.* **2015**, *42*, 25. [CrossRef]
63. Huijser, M.P.; Fairbank, E.R.; Camel-Means, W.; Graham, J.; Watson, V.; Basting, P.; Becker, D. Effectiveness of short sections of wildlife fencing and crossing structures along highways in reducing wildlife–vehicle collisions and providing safe crossing opportunities for large mammals. *Biol. Conserv.* **2016**, *197*, 61–68. [CrossRef]
64. Davenport, J.; Davenport, J.L. (Eds.) *The Ecology of Transportation: Managing Mobility for the Environment*; Environmental Pollution; Springer: Dordrecht, The Netherlands, 2006; Volume 10, ISBN 978-1-4020-4503-5.
65. Spanowicz, A.G.; Teixeira, F.Z.; Jaeger, J.A.G. An adaptive plan for prioritizing road sections for fencing to reduce animal mortality. *Conserv. Biol.* **2020**, cobi.13502. [CrossRef] [PubMed]
66. Lima Santos, R.A.; Ascensão, F.; Ribeiro, M.L.; Bager, A.; Santos-Reis, M.; Aguiar, L.M.S. Assessing the consistency of hotspot and hot-moment patterns of wildlife road mortality over time. *Perspect. Ecol. Conserv.* **2017**, *15*, 56–60. [CrossRef]
67. Jaeger, J.A.G.; Fahring, L. Effects of Road Fencing on Population Persistence. *Conserv. Biol.* **2004**, *18*, 1651–1657. [CrossRef]
68. Lesbarrères, D.; Fahrig, L. Measures to reduce population fragmentation by roads: What has worked and how do we know? *Trends Ecol. Evol.* **2012**, *27*, 374–380. [CrossRef] [PubMed]
69. Kroll, G. An Environmental History of Roadkill: Road Ecology and the Making of the Permeable Highway. *Env. Hist. Durh. N. C.* **2015**, *20*, 4–28. [CrossRef]
70. Shepard, D.B.; Kuhns, A.R.; Dreslik, M.J.; Phillips, C.A. Roads as barriers to animal movement in fragmented landscapes. *Anim. Conserv.* **2008**, *11*, 288–296. [CrossRef]
71. Hepenstrick, D.; Thiel, D.; Holderegger, R.; Gugerli, F. Genetic discontinuities in roe deer (Capreolus capreolus) coincide with fenced transportation infrastructure. *Basic Appl. Ecol.* **2012**, *13*, 631–638. [CrossRef]
72. Malo, J.E.; Suarez, F.; Diez, A. Can we mitigate animal-vehicle accidents using predictive models? *J. Appl. Ecol.* **2004**, *41*, 701–710. [CrossRef]
73. Joyce, T.L.; Mahoney, S.P. Spatial and Temporal Distributions of Moose-Vehicle Collisions in Newfoundland. *Wildl. Soc. Bull.* **2001**, *29*, 281–291. [CrossRef]
74. Anderson, S.J.; Karumanchi, S.B.; Iagnemma, K. Constraint-based planning and control for safe, semi-autonomous operation of vehicles. In Proceedings of the 2012 IEEE Intelligent Vehicles Symposium, Alcala de Henares, Madrid, Spain, 3–7 June 2012; pp. 383–388.
75. Ford, A.T.; Clevenger, A.P.; Bennett, A. Comparison of Methods of Monitoring Wildlife Crossing-Structures on Highways. *J. Wildl. Manag.* **2009**, *73*, 1213–1222. [CrossRef]

land

MDPI

Article

Roadkill-Data-Based Identification and Ranking of Mammal Habitats

Andrius Kučas [1,2] and Linas Balčiauskas [1,*]

1 Nature Research Centre, Akademijos Street 2, LT-08412 Vilnius, Lithuania; andrius.kucas@gamtc.lt
2 Joint Research Centre, European Commission, Via Fermi 2749, I-21027 Ispra, Italy
* Correspondence: linas.balciauskas@gamtc.lt; Tel.:+370-685-34141

Abstract: Wildlife–vehicle collisions, as well as environmental factors that affect collisions and mitigation measures, are usually modelled and analysed in the vicinity of or within roads, while habitat attractiveness to wildlife along with risk to drivers remain mostly underestimated. The main goal of this study was the identification, characterisation, and ranking of mammalian habitats in Lithuania in relation to 2002–2017 roadkill data. We identified habitat patches as areas (varying from 1 to 1488 square kilometres) isolated by neighbouring roads characterised by at least one wildlife–vehicle collision hotspot. We ranked all identified habitats on the basis of land cover, the presence of an ecological corridor, a mammalian pathway, and roadkill hotspot data. A ranking scenario describing both habitat attractiveness to wildlife and the risk to drivers was defined and applied. Ranks for each habitat were calculated using multiple criteria spatial decision support techniques. Multiple regression analyses were used to identify the relationship between habitat ranks, species richness, and land cover classes. Strong relationships were identified and are discussed between the habitat patch ranks in five (out of 28) land cover classes and in eight (out of 28) species (97% of all mammal road kills). We conclude that, along with conventional roadkill hotspot identification, roadkill-based habitat identification and characterisation as well as species richness analysis should be used in road safety infrastructure planning.

Keywords: roadkill analysis; movement patterns; habitat characterisation; multiple criteria; multi-objective ranking; mitigation measures

Citation: Kučas, A.; Balčiauskas, L. Roadkill-Data-Based Identification and Ranking of Mammal Habitats. *Land* **2021**, *10*, 477. https://doi.org/10.3390/land10050477

Academic Editors: Juan F. Beltrán, Pedro Abellán and John Litvaitis

Received: 16 April 2021
Accepted: 30 April 2021
Published: 2 May 2021

1. Introduction

Wildlife–vehicle collisions (WVCs) pose a threat to human life and biological diversity and result in damage to property [1–6]. Over the last two decades in Lithuania, while the overall number of road traffic accidents has continuously decreased, road accidents involving wildlife have increased [7].

To mitigate mammal–vehicle collisions (MVCs), fencing, underpasses, gates, and jump-out ramps are used as the most common mitigation measures in the country [8]. Additional road safety infrastructure elements such as repellents, reflectors, noise, and natural predators can also be used; these focus on a single and/or multiple wildlife species. They repel, attract, or redirect wildlife with different ecological and financial efficiencies [9–18]. The selection of tangible multi-scale [19], multi-objective, and multi-functional WVC mitigation measures is the focus of a considerable research challenge [20].

The identification of roadkill hotspots (road sections where collisions occur more frequently than expected) is the first step of the highway safety management process. However, erroneous hotspot identification [21] as well as gaps in roadkill data [1] may result in inefficient use of resources for safety improvements [22]. There are many generalised linear models [23] that can be used to identify hotspots, such as ecological niche modelling [24], kernel density estimation [25–27], distance-based approaches [28], and methods based on modelling the number of collisions in a road section assuming a Poisson distribution [21,29–32]. These methods use roadkill data to detect collision hotspots as well as their risk to drivers. In order to assess habitat attractiveness to wildlife and the associated habitat risk to drivers, it is important to understand where mammals cross roads more frequently.

Multiple habitat suitability [33–35] is determined by spatial [36,37] and temporal [7,38–41] factors that might help us to identify and characterise wildlife habitats, animal movement patterns, and corridors [42–46]. Habitat suitability, together with spatial and temporal factors, helps us to obtain knowledge on how and when mammals traverse the landscape.

Field research usually brings disparate results [47] of differential scale and quality [1]. Consequently, the results are frequently not fit for deriving habitat patch characteristics and assessing habitat attractiveness to wildlife. This would require standardised habitat data that are usually lacking.

Habitats can be characterised using behavioural and spatiotemporal events, landscape connectivity and fragmentation, species richness, animal abundance, and other field research data. Large scale, long-term, and accurate data that can characterise habitats usually require methodologically robust and expensive research. Employing the available roadkill data from police reports would decrease (not replace) the amount of field research required in cases when there is insufficient habitat data available. Multiple, long-term, and standardised habitat characteristics (criteria) can help us to identify MVC mitigation measures focused on single or multiple species.

Decision-makers often deal with problems that involve multiple criteria [48–51]. Identification of the primary sources of MVCs, namely the habitats that are highly attractive to wildlife and simultaneously of high risk to drivers, is also a multiple criteria analysis problem. Therefore, we selected Simple Additive Weighting (SAW) and Technique for Order of Preference by Similarity to Ideal Solution (TOPSIS) [49,52] multi-criteria spatial decision support techniques for the ranking of habitats. The habitat ranking outcomes can be considered reliable if both methods generate similar ranking results [8].

Wildlife–vehicle collisions, as well as environmental factors that affect collisions and mitigation measures, are usually modelled and analysed at the level of the roads themselves [7,8,53–56], while the wider issues of adjacent habitat attractiveness to wildlife and its risk to drivers remain underestimated. MVCs with wild species accounted for about 91% of all WVCs in Lithuania in 2002–2017 [57]. There is a need for a framework that helps us to unify the disparate results emerging from different data sources and field studies on MVC occurrence, which allows for roadkill-based identification, characterisation, and multi-objective ranking of mammalian habitats by their attractiveness to wildlife and their risk to drivers. Here, we understand "risk to the driver" as a derivative of the cluster strength in KDE+ [26,27].

In this study, habitat identification, characterisation, and ranking are based on the definition of habitats as "areas isolated by neighbouring roads that have at least one hotspot (a road section where MVCs occur more frequently than expected)" and the assumptions that (1) highly attractive habitats for wildlife increase the risk of MVCs on adjacent roads; (2) habitats that are surrounded by roads with an abundance of MVC hotspots are of high attractiveness to wildlife movement; and (3) road kills in the hotspots can help us to identify species richness within adjacent habitats. However, the accuracy of such estimations depends on the completeness of MVC data [1].

The overall purpose of this study, therefore, is to:

- Identify habitat patches that are surrounded by roads with kernel density estimation (KDE+)-based [27] MVC hotspots;
- Characterise habitat patches using the properties of adjacent habitats, hypothetical corridors and wildlife pathways, hotspots, and land cover data;
- Define ranking scenarios (criteria utility functions and criteria weights) to detect habitat patches that are highly attractive to wildlife [37] and pose a risk to drivers [27];
- Rank habitat patches using two different multiple criteria spatial decision support techniques: SAW and TOPSIS; and

- Find relationships between habitat ranks, species richness, and land cover classes for use in the planning of multispecies MVC mitigation measures using multiple linear regressions.

2. Materials and Methods

2.1. Study Area

Our study area covers the entire territory of Lithuania (Figure 1), which can be characterised as mostly a plain. It represents a surface area of 65,286 square kilometres. In 2012, 33% of the surface was covered by arable land and permanent crops, 27% by semi-natural vegetation, 33% by forested land, 3% by developed (artificial) areas, and 4% by water bodies and other land. The land cover change (a 0.48% change rate per year) in the country is slowing, mainly due to a rapid decrease in the intensity of forest conversion [58].

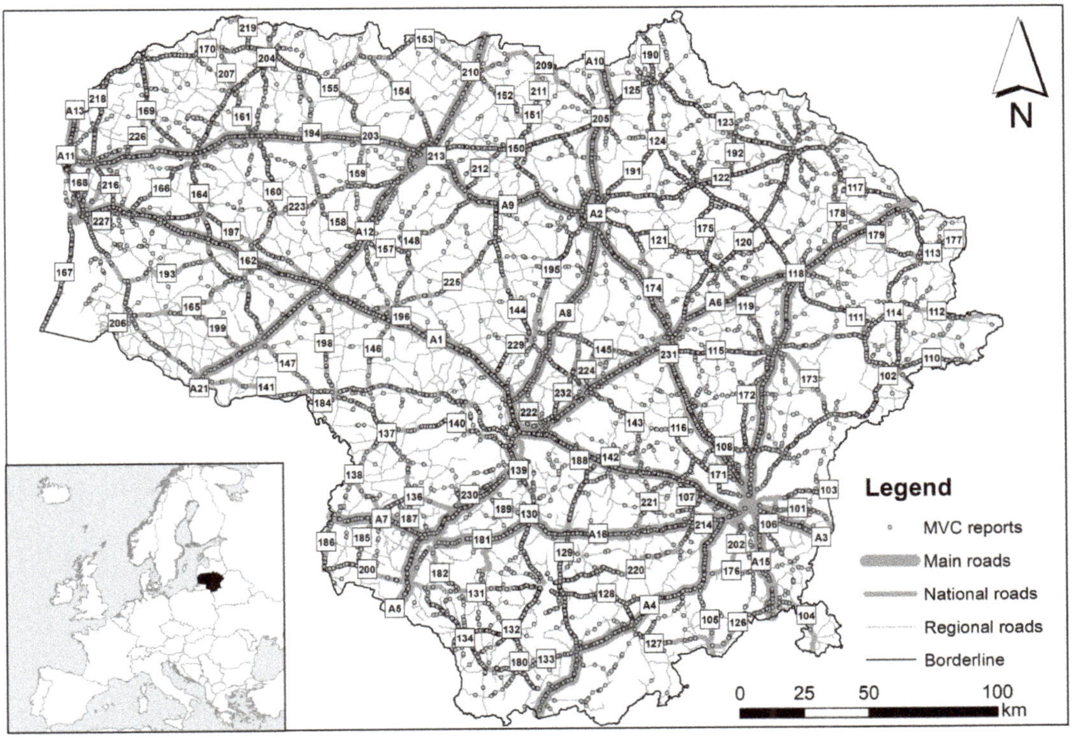

Figure 1. The study area, roads (main roads/highways, national and regional), and locations of MVCs in 2002–2017.

In 2017, there were 21,244 km of State-owned roads of national significance (excluding roads in cities): 1751 km of main roads; 4925 km of national roads; and 14,568 km of regional roads [59]. In this study, we analysed 1784 roads (21 main/highway, 13 national, and 1631 regional) as shown in Figure 1.

In the period 2002–2017, the annual average daily traffic (AADT) increased from 5600 to 11,000 vehicles a day on main roads, from 2200 to 2900 vehicles a day on national roads, and remained at up to 500 vehicles a day on regional roads [60].

2.2. Mammal–Vehicle Collision Data

According to the data from the Lithuanian Police Traffic Supervision Service and the Nature Research Centre, a total of 24,083 WVCs were recorded over the period 2002–2017 in

Lithuania [57]. These numbers may, however, have a bias regarding taxonomic groups and not account for all accidents as reporting to the authorities is not mandatory in Lithuania. The Traffic Supervision Service registers only road kills from those accidents that were reported by drivers; therefore, their data are biased to larger species. Small mammals are represented exclusively in the data from the Nature Research Centre, which registered all road kills.

Out of all WVCs, we selected 21,911 WVC reports that involved mammals. A total of 19,622 reports included accurate information relating to 32 wild mammal species (Table 1). Of these reports, we mapped the 18,218 reports that included precise information on location (Figure 1, Table 1).

Table 1. Numbers of MVCs with wild mammals in Lithuania with precise location and species information. Species included in the national Red data list are marked with an asterisk.

Species	Mapped (MVC)
Roe deer (*Capreolus capreolus*)	10,741
Wild boar (*Sus scrofa*)	1416
Moose (*Alces alces*)	1340
Raccoon dog (*Nyctereutes procyonoides*)	1331
Eastern European hedgehog (*Erinaceus concolor*)	993
Red fox (*Vulpes vulpes*)	829
European hare (*Lepus europaeus*)	456
Marten (*Martes* sp.)	405
Red deer (*Cervus elaphus*)	248
European polecat (*Mustela putorius*)	160
Badger (*Meles meles*)	89
Pine marten (*Martes martes*)	40
Beaver (*Castor fiber*)	25
Red squirrel (*Sciurus vulgaris*)	30
American mink (*Neovison vison*)	26
European mole (*Talpa europaea*)	19
Stone marten (*Martes foina*)	14
Eurasian otter (*Lutra lutra*) *	13
Fallow deer (*Dama dama*)	11
Norway rat (*Rattus norvegicus*)	6
European bison (*Bison bonasus*) *	6
Grey wolf (*Canis lupus*)	3
Bank vole (*Myodes glareolus*)	3
Lynx (*Lynx lynx*) *	1
Stoat (*Mustela erminea*) *	2
Least weasel (*Mustela nivalis*)	2
Common shrew (*Sorex araneus*)	2
Yellow-necked mouse (*Apodemus flavicollis*)	2
Muskrat (*Ondatra zibethicus*)	2
Water shrew (*Neomys fodiens*)	1
Mountain hare (*Lepus timidus*) *	1
Black rat (*Rattus rattus*)	1

2.3. Clustering of Collision Data

Using a clustering method, habitats were identified according to the location of hotspots. The literature contains many different spatial techniques for identifying short, significant road segments where collisions occur more frequently than usual [21,27,32,61–66]. We utilized the KDE+, which analyses MVCs that are represented as point features and are located along the roads represented as line features (Figure 2a). The KDE+ algorithm finds locations (clusters) with statistically significant concentrations of collisions and assigns strength values (measured from 0 to 1) showing the risk severity to drivers [27,36] (Figure 2b). We performed MVC clustering analysis and created MVC clusters using the KDE+ parameters derived from the road network properties (KDE+ bandwidth—150 metres, Monte Carlo simulations—800, and minimal cluster strength—0.2).

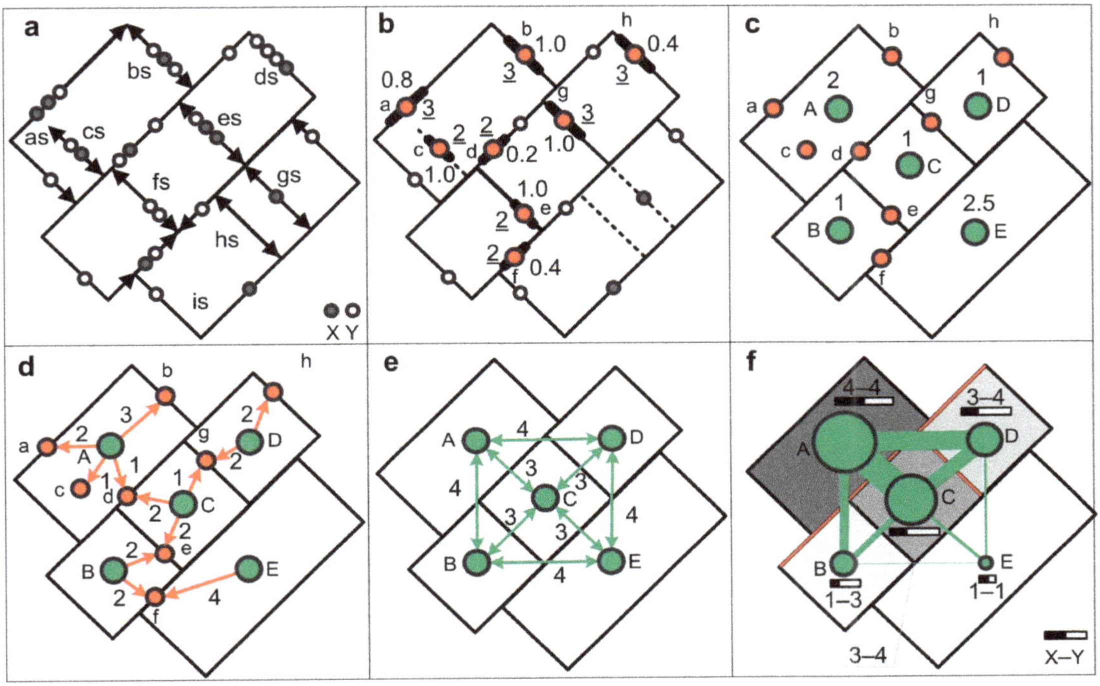

Figure 2. Roadkill-data-based identification, characterisation, and ranking of mammalian habitats: (**a**) MVC reports (small dots) with different species (dots marked as X and Y) placed within a road network (double-arrowed and labelled lines); (**b**) KDE+ clusters (short thick lines) labelled with underlined integer numbers show the length and non-integer numbers the strength of a cluster, small grey and white dots represent MVCs that did not form a cluster, and dashed lines represent the roads without clusters that did not form habitat patches; (**c**) Numbers show areas of habitat patches; (**d**) Numbers represent the length of hypothetical wildlife pathways (single-arrowed lines); (**e**) Numbers represent the length of hypothetical mammal corridors (double-arrowed lines); (**f**) Larger dots (habitat patches), darker colours of habitat patches, and thicker lines (corridors) represent a higher risk to drivers and higher attractiveness to wildlife, red lines (roads) highlight the highly ranked adjacent habitat patches, white- and black-coloured bars illustrate the share of species richness (for species X and Y), and labels show the number of mammals involved in MVCs (within the clusters); (**b**–**d**) Red dots represent the centre points of clusters; (**c**–**f**) Habitat patches (large green dots and polygons) labelled as ABCDE are represented by centre points.

2.4. *Definition of Mammalian Habitats and Movement Patterns*

Our conceptual model for the identification of mammalian habitats is shown in Figure 2a–c, characterization in Figure 2d,e, and ranking of habitat patches in Figure 2e. MVC reports with different species were mapped on the road network. Road sections where MVCs occurred more frequently were identified using the KDE+ clustering method [27].

We assumed that roadkill clusters are important indicators not only of risk to drivers [27], but also indicate locations where important mammalian pathways and roads intersect. We identified habitat patches as areas that are bounded (surrounded) by neighbouring road sections characterised by at least one cluster. We merged habitats having no clusters with neighbouring habitats iteratively until a merged habitat patch had a road with at least one neighbouring cluster. In our study, urban areas and urban clusters were excluded and not used for the identification of habitats. Identified habitats were used for their characterisation and, later, for ranking.

Hypothetical wildlife pathways were created by connecting the Clementini [67] centroids of habitat patches and cluster centroids using spider lines illustrating the shortest (Euclidean) distances. Hypothetical mammal corridors were created using the triangulated irregular network (TIN) between the Clementini [67] habitat patch centroids as peaks [37,42].

2.5. Characterisation of Mammalian Habitats

The habitat patches (Figure 2c) were characterised using topological relationships between habitat patches, hypothetical pathways, and corridors. Each cluster centroid illustrates a "gateway" that mammals use to traverse from one habitat patch to another.

Following this conceptual framework (Figure 2), we identified and collected the necessary network-based criteria (Table 2) for each habitat patch. Later, the habitat patches were ranked according to their attractiveness to wildlife and risk severity to drivers.

Table 2. Criteria used for ranking the habitat patches described in Figure 2.

Criteria Name *	Variable	Habitat Patches					Objective Function	Weight (Index) **
		A	B	C	D	E		
Total number of collisions within adjacent clusters [i] (b)	count	10.0	4.0	7.0	7.0	2.0	Max	0.102
Average strength of adjacent clusters [ii] (b)	index	0.8	0.7	0.7	0.7	0.4	Max	0.098
Total length of adjacent clusters [ii] (b)	km	10.0	4.0	7.0	6.0	2.0	Max	0.103
Number of species within adjacent clusters [iii] (b)	count	2.0	2.0	2.0	2.0	2.0	Max	0.097
Habitat patch area [i] (c)	ha	2.0	1.0	1.0	1.0	2.5	Max	0.102
Number of adjacent clusters/pathways [ii] (d)	count	4.0	2.0	3.0	2.0	1.0	Max	0.102
Total length of adjacent pathways [ii] (d)	km	7.0	4.0	5.0	4.0	4.0	Min	0.098
Number of adjacent corridors [i] (e)	count	3.0	3.0	4.0	3.0	3.0	Max	0.100
Total length of adjacent corridors [i] (e)	km	11.0	11.0	12.0	11.0	11.0	Min	0.097
Total area of adjacent habitat patches [i] (e)	ha	3.0	5.5	6.5	5.5	3.0	Max	0.101
SAW values (f) ***	index	0.86	0.72	0.82	0.78	0.64		
TOPSIS values (f) ***	index	0.69	0.37	0.58	0.48	0.32		

* Figure 2, part identifier provided in the brackets. ** A higher weight value shows higher criterion importance. *** The higher the resulting SAW and TOPSIS values are, the larger are the green dots in Figure 2f. [i] Habitat patches with a larger area connected to other larger habitat patches by very short and abundant ecological corridors show habitat patches that are less fragmented (high component connectivity). They are considered as attractive to wildlife. [ii] Habitat patches with a higher number of shorter mammalian pathways and longer and stronger KDE+ clusters are characterised by higher numbers of collisions. They are considered as being a more severe risk to drivers. [iii] The number of species is an important indicator for both ([i],[ii]) modelling assumptions, since a higher number of species within a certain habitat patch (species richness) simultaneously indicates a higher attractiveness to wildlife and a higher risk to drivers.

2.6. Objective Functions and Criteria Importance

The objective criterion importance (weights) for all criteria was calculated based on criteria utility (minimisation/maximisation) functions using SortViz for the ESRI inc. ArcGIS desktop software add-in [37,68].

Using the same ArcGIS desktop software add-in, we ranked habitat patches based on criteria derived from the individual (Figure 2) and spatial connectivity properties (Table 2) of the habitat patch. In order to find habitat patches that were simultaneously the most attractive to wildlife and of most severe risk to drivers, modelling assumptions (see Table 2's footnote) and objective (utility) functions (Table 2) were set.

2.7. Ranking of Habitats and Ecological Corridors

Criterion importance values, defined as weights (Table 2), were then used as an input for ranking the habitat patches using the SAW and TOPSIS [49,52] methods. Both ranking approaches use the same input habitat data (Table 2). The final SAW and TOPSIS values ranged from 0 (worst) to 1 (best) and altogether built the so-called 'composite indicator' of habitat attractiveness to wildlife (mammals) and risk to drivers. A higher rank value means higher attractiveness to wildlife and a higher risk to drivers. The SAW and TOPSIS values for each habitat patch were separately calculated and compared with each other (Table 2).

Average SAW and average TOPSIS rank values, derived from the habitat patches connecting the two ends of the corridor, were allocated to each ecological corridor to determine the relative importance of the corridor (Figure 2f, Table 3).

Table 3. Average SAW and TOPSIS rank values (Figure 2f).

Corridor Identification (Figure 2e Cases)	Average SAW Value *	Average TOSPIS Value *
A–B	0.78	0.53
A–C	0.84	0.63
B–C	0.77	0.47
A–D	0.82	0.59
C–D	0.80	0.53
D–E	0.71	0.40
E–C	0.73	0.45
B–E	0.68	0.35

* The higher the average SAW and TOPSIS values are, the thicker are the lines in Figure 2f.

Different average rank values assigned to ecological corridors show different degrees of importance to mammals and drivers. Higher average SAW and TOPSIS rank values (Figure 2f, Table 3) illustrate higher and more intense mammalian locomotion patterns [69] and risk to drivers.

2.8. Identification of Key Habitat Characteristics

We assessed the relationship (a correlation matrix using Pearson's correlation index) between habitat patch ranks, number of species, and CLC land cover [70,71] classes [72]. Interpretation of r: 0—no association; 0 to 0.25 (0 to −0.25)—weak association; 0.25 to 0.50 (−0.25 to −0.50)—moderate association; 0.50–0.75 (−0.50 to −0.75)—strong association; 0.75 to 1.00 (−0.75 to −1.00)—very strong association; and 1 (−1) perfect association [73].

We analysed land cover classes and species that had a strong ($r > 0.50$) relationship to habitat ranks (SAW and TOPSIS values). Habitat ranks were used as intercept and land cover classes and species as independent regressors.

3. Results

3.1. Habitats and Habitat Characteristics

We identified 281 state-owned roads with at least one KDE+ cluster (Figure 3): 18 main roads/highways; 107 national roads; and 156 regional roads (85.7%, 81.1%, and 9.6% of all roads in their respective category). The rest of the roads (thin grey lines in Figure 4) were not taken into account.

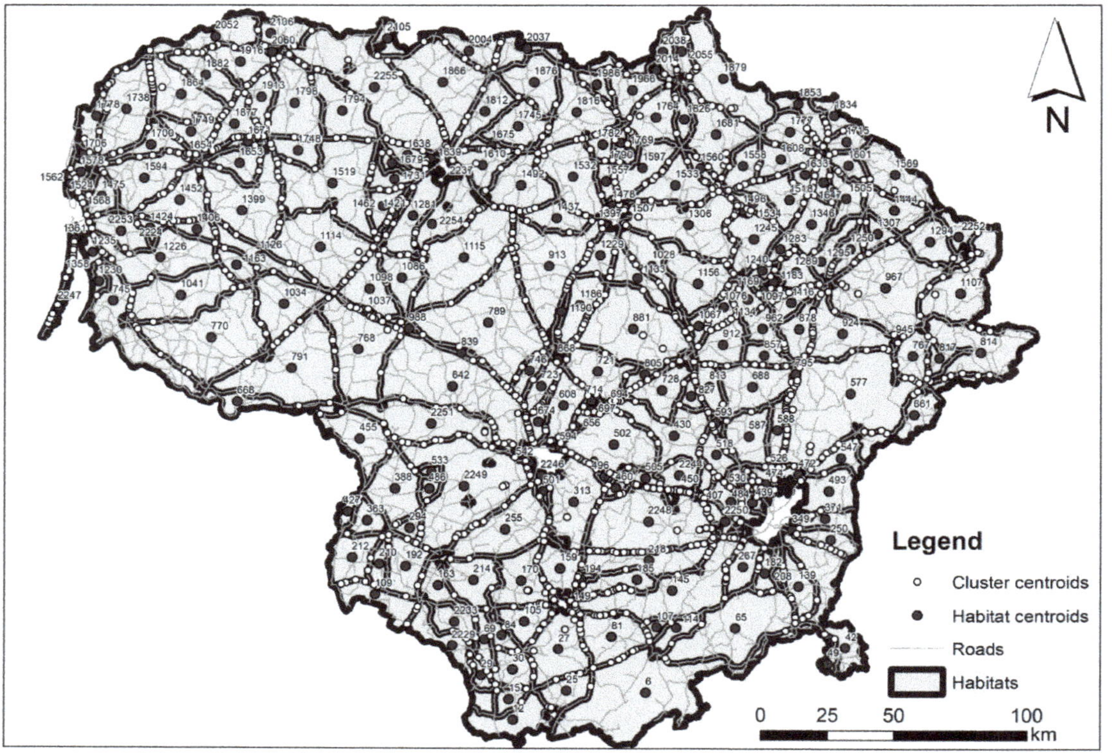

Figure 3. KDE+ cluster centroids, habitats, and their Clementini centroids (corresponding to Figure 2a,c). Labels show unique identifiers (id) used for the identification of the habitats (Tables 1 and 2).

Using the KDE+ method, we found 1642 mammalian clusters (Figure 3), of which 22 (1.3%) were located in urban areas and therefore were excluded from further analyses. A total of 28 out of the 32 road-killed mammal species were identified within the clusters. Four small-sized mammals (*M. glareolus, S. araneus, R. rattus, N. fodiens*) were only registered as road kills outside the clusters. However, small numbers of these species registered in the road kills (Table 1) had no impact on the location and number of identified clusters.

We identified 3171 hypothetical mammalian pathways (thin grey lines in Figure 4), 672 corridors (dashed lines in Figure 4), and 243 habitat patches (Figure 4). The hypothetical mammalian pathways (Figures 2d and 4) and corridors (Figures 2e and 4) were used for the characterisation and collection of criteria (Tables 1 and 2) for habitat patches (Figure 4).

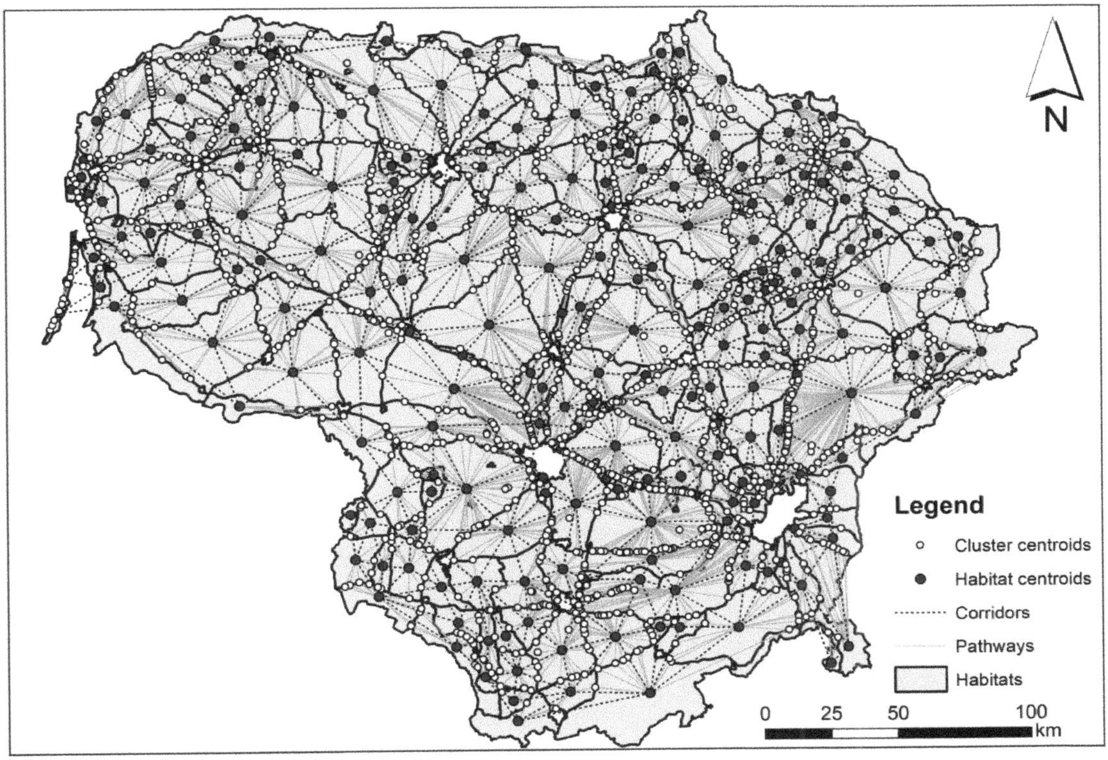

Figure 4. Topologically connected hypothetical mammalian pathways (spider lines) and ecological corridors (the triangulated network) used for the characterisation of newly identified habitat patches.

3.2. Criteria Weights, Habitat Ranks, Ecological Corridors, and Movement Patterns

In order to rank the identified habitat patches (Table 1), the criteria weights (Table 4) were calculated using objective functions (Table 2). The most important criterion for assessment was the shortest length of adjacent pathways, while the least important was the number of adjacent corridors.

Table 4. Criteria (Table 1) and criteria weights used for ranking (following the same objective functions as in Table 2) the habitat patches (Figure 3) in Lithuania.

Criteria Name *	Weight (Index)
Total number of collisions within adjacent clusters [i] (b)	0.107
Average strength of adjacent clusters [ii] (b)	0.085
Total length of adjacent clusters [ii] (b)	0.104
Number of species within adjacent clusters [iii] (b)	0.099
Habitat patch area [i] (c)	0.105
Number of adjacent clusters/pathways [ii] (d)	0.103
Total length of adjacent pathways [ii] (d)	0.109
Number of adjacent corridors [i] (e)	0.088
Total length of adjacent corridors [i] (e)	0.103
Total area of adjacent habitat patches [i] (e)	0.098

* Table 2, footnote identifier provided in the superscript. Figure 2, part identifier provided in the brackets.

Following objective functions, the SAW (Figure 5) and TOPSIS (Figure 6) ranks (Table 2) were assigned to each habitat patch. The average rank values were calculated for the corridors as well. The labels (Figures 5 and 6) identify the main roads.

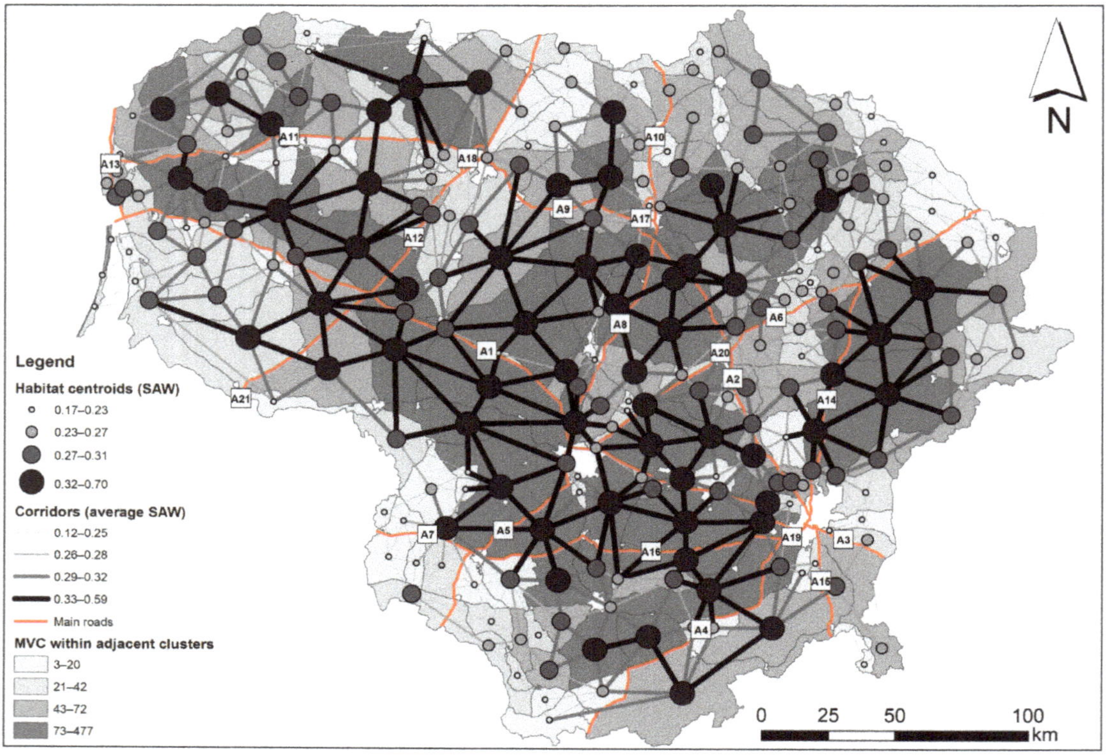

Figure 5. Habitat centroids, corridor links, and MVC quartiles derived using SAW values (the mapping approach is shown in Figure 2f).

The highest SAW and TOPSIS rank values assigned to the habitat patches were 0.7 and 0.6, respectively. Furthermore, the SAW and TOPSIS ranks of habitats had a very strong correlation (r = 0.86), which means that the ranking results are similar and can be trusted.

The habitat patches contained from 3 to 477 MVCs and from 1 to 20 road-killed mammal species (Table 2). The corridor links (Figures 5 and 6) indicate the most probable movement patterns. The highly ranked corridors that intersect main roads highlight the highest potential risk to drivers and wildlife. Consequently, the MCV clusters that are on the roads with such intersections are of the highest importance for MVC mitigation actions.

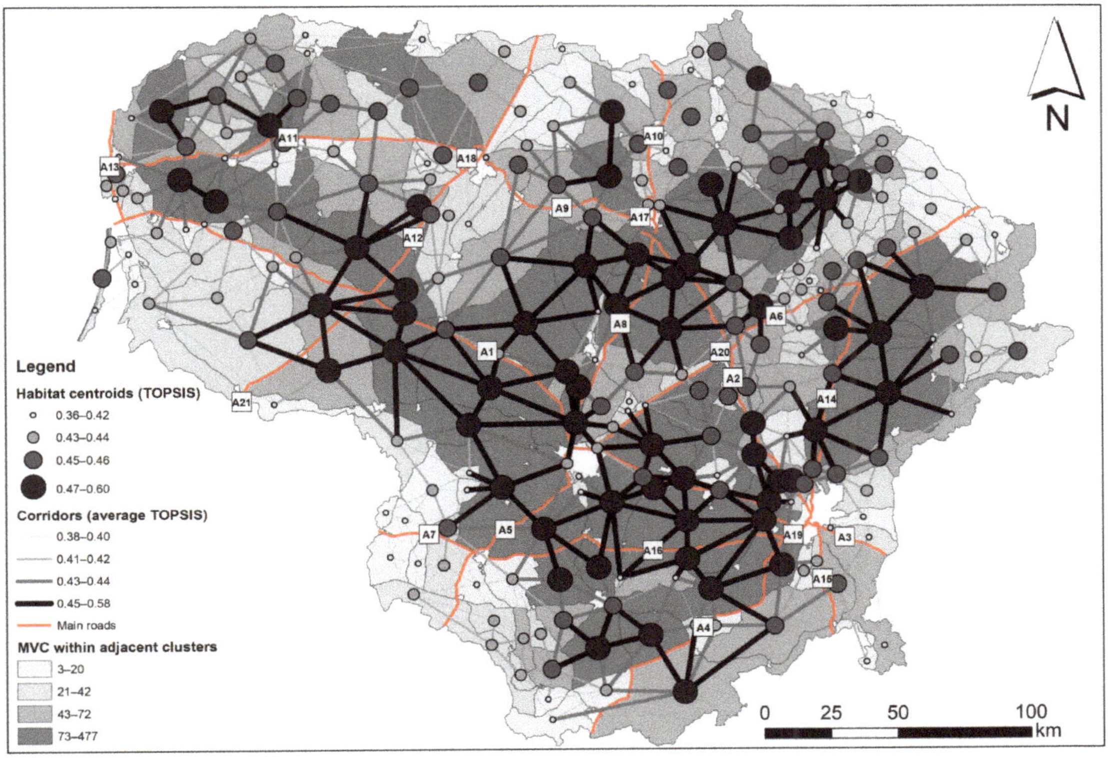

Figure 6. Habitat centroids, corridor links, and MVC quartiles derived using TOPSIS values (the mapping approach is shown in Figure 2f).

3.3. Relationship between Habitat Ranks, Species Richness, and Land Cover Classes

Inside the clusters, MVCs with *C. capreolus*, *S. scrofa*, *V. vulpes*, *L. europaeus*, *E. concolor*, *N. procyonoides*, *A. alces*, *M. putorius*, and *Martes* sp. had strong relationships ($r > \sim 0.5$) with habitat patch ranks, showing the high severity risk to drivers and wildlife (Figure 7). All other species had a weak or no relationship with habitat patch ranks. Five of these species, *B. bonasus*, *L. lynx*, *M. erminea*, *L. lutra*, and *L. timidus*, are rare in nature (Table 1), while others are small in size and their road kills were most probably under-registered.

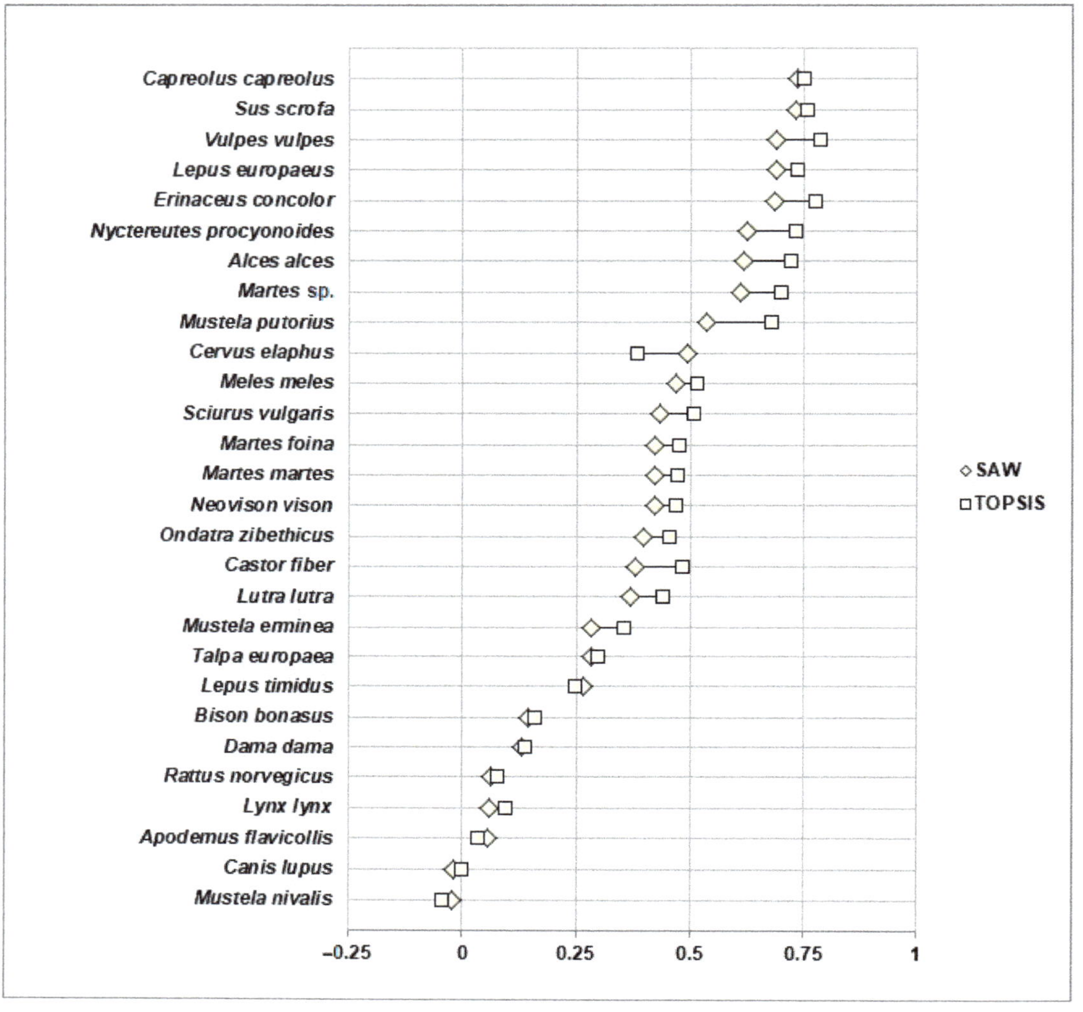

Figure 7. Relationships between different habitat patch ranks and species involved in MVC clusters.

Land cover classes such as road and rail networks, transitional woodland–shrub areas, mixed forest, broad-leaved forest, pastures, complex cultivation patterns, and discontinuous urban fabrics showed strong relationships ($r > 0.5$) with habitat patch ranks (Figure 8). All other land cover classes had a weak or no relationship with habitat patch ranks, indicating that these land cover classes do not pose a severe risk to drivers and wildlife.

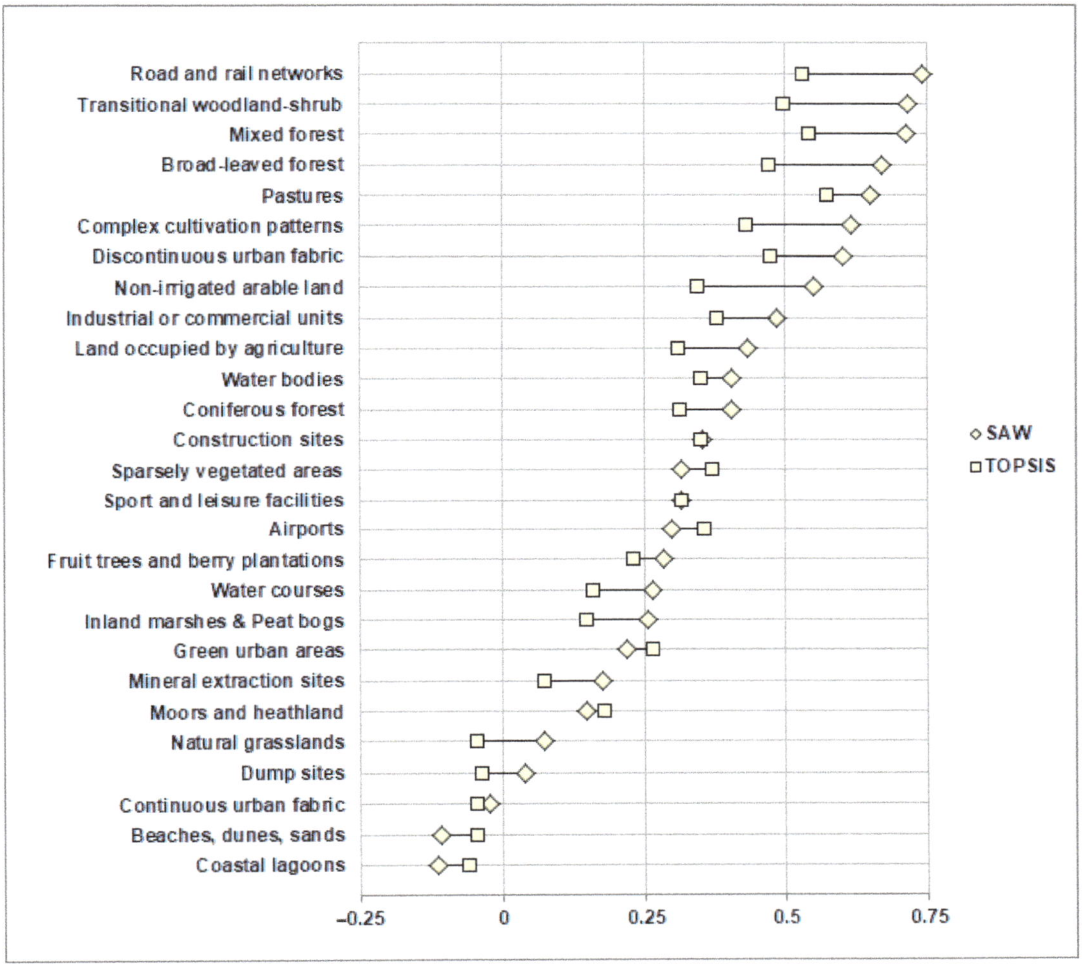

Figure 8. Relationships between different habitat patch ranks and land cover classes.

The results of multiple linear regression analyses (Table 5) indicate that broad-leaved forests and transitional woodland–shrub areas bordered by road and rail networks are characterised by the highest risk to drivers and wildlife. In the vicinity of such habitats, MVCs mostly occur with *C. capreolus* and *S. scrofa*. MVCs with other species such as *A. alces*, *V. vulpes*, *Martes* sp., *M. putorius*, *L. europaeus*, and *E. concolor* are also likely.

Table 5. Relationships between habitat ranks, land cover classes, and species involved in MVC clusters.

Independent ± SE\Dependent	SAW	TOPSIS
Intercept	0.208791 ± 0.000 ****	0.4098186 ± 0.000 ****
$b_{\text{discontinuous urban fabric}}$	−0.000003 ± 0.393 NS	−0.0000002 ± 0.896 NS
$b_{\text{road and rail networks and associated land}}$	0.000019 ± 0.000 ****	0.0000043 ± 0.025 **
b_{pastures}	−0.000002 ± 0.074 *	−0.0000004 ± 0.251 NS
$b_{\text{complex cultivation patterns}}$	−0.000001 ± 0.481 NS	−0.0000004 ± 0.502 NS
$b_{\text{broad-leaved forest}}$	0.000002 ± 0.005 ***	0.0000000 ± 0.871 NS
$b_{\text{mixed forest}}$	−0.000003 ± 0.053 *	0.0000006 ± 0.343 NS
$b_{\text{transitional woodland–shrub}}$	0.000015 ± 0.001 ***	−0.0000051 ± 0.003 **
$b_{\textit{Mustela putorius}}$	0.003593 ± 0.189 NS	0.0023776 ± 0.016 **
$b_{\textit{Martes}\text{ sp.}}$	0.000106 ± 0.939 NS	0.0011592 ± 0.021 **
$b_{\textit{Lepus europaeus}}$	0.003250 ± 0.041 **	0.0000625 ± 0.912 NS
$b_{\textit{Vulpes vulpes}}$	0.001217 ± 0.269 NS	0.0010652 ± 0.007 ***
$b_{\textit{Erinaceus concolor}}$	0.001905 ± 0.016 **	0.0003457 ± 0.222 NS
$b_{\textit{Nyctereutes procyonoides}}$	−0.000047 ± 0.928 NS	0.0003072 ± 0.106 NS
$b_{\textit{Alces alces}}$	0.000058 ± 0.898 NS	0.0005224 ± 0.001 ****
$b_{\textit{Sus scrofa}}$	0.002013 ± 0.001 ***	0.0005760 ± 0.006 ***
$b_{\textit{Capreolus capreolus}}$	0.000803 ± 0.000 ****	0.0004290 ± 0.000 ****
$F_{(16,226)}$	98.02606 ± 0.000 ****	132.64396 ± 0.000 ****
R^2	0.874	0.904

*—$p < 0.10$, **—$p < 0.05$, ***—$p < 0.01$, ****—$p < 0.001$. NS—not significant.

Using SAW (Figure 9) and TOPSIS (Figure 10) values, we created heat maps that show the potential risk severity to drivers and wildlife (urban areas were used as a reference). For better visual representation, the maps were created using the inverse distance weighed (IDW) interpolation method. The IDW method is used to interpolate spatial data and is based on the concept of distance weighting [74,75].

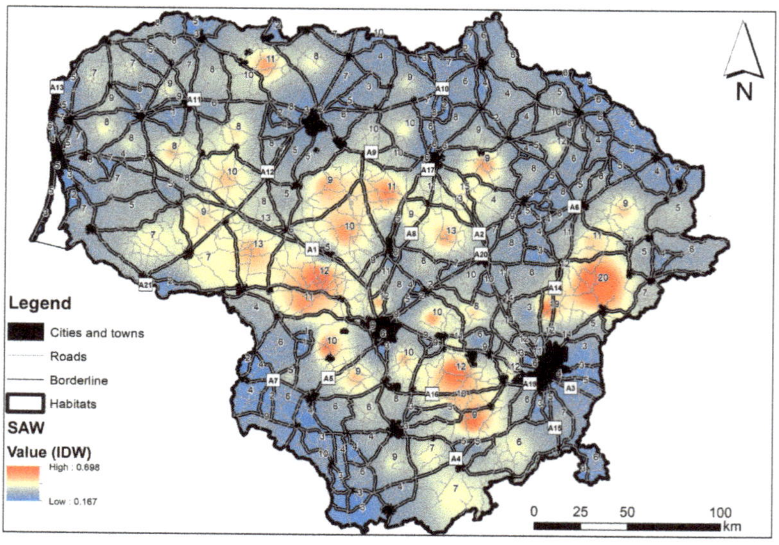

Figure 9. Overlay of habitat patch boundaries, roads (all categories), urban areas, the habitat rank (SAW) heat map, and the number of species. Labels within the square show the identification numbers of main roads, while other labels show the total number of species involved in the MVC clusters located in the vicinity of habitat patches.

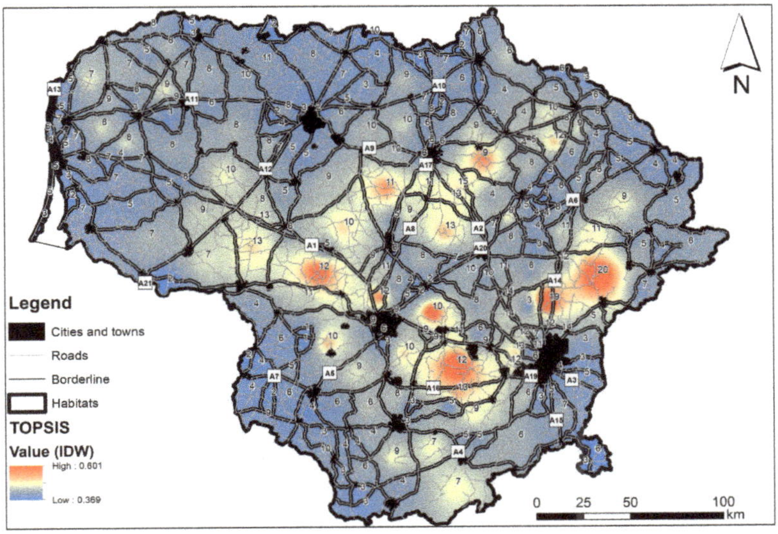

Figure 10. Overlay of habitat patch boundaries, roads (all categories), urban areas, the habitat rank (TOPSIS) heat map, and the number of species. Labels within the square show the identification numbers of main roads, while other labels show the total number of species involved in the MVC clusters located in the vicinity of habitat patches.

The SAW-based habitat patch heat map (Figure 9) shows more severe risk habitat patches than the TOPSIS-based heat map (Figure 10) due to the differences in the ranking methods. However, both heat maps identified the same highly severe locations for drivers and wildlife.

Following the results from both ranking methods (Figures 9 and 10), we identified that the habitat patches with the unique identification numbers 577 and 2248 (Figure 3, Tables 1 and 2) posed the most severe risk to drivers and wildlife. For instance, around the top-ranked habitat patch (id: 577), which is bordered by the A14 main road and national roads 114, 111, 102, and 108, we found MVC clusters including 20 different mammal species (Figure 11). Most of the MVC clusters were found on A14. Clusters on the roads at the edge and within the habitat patch were also present. Due to the low traffic intensity there, we did not find any cluster on the national road 173, which is within the habitat patch.

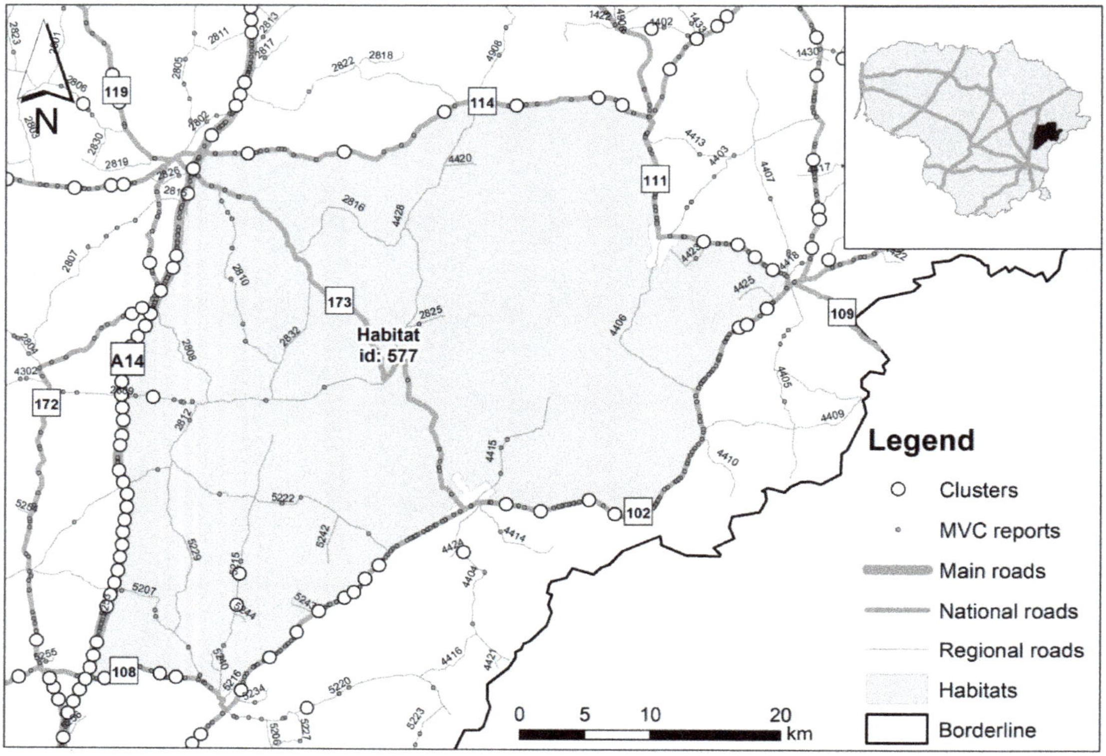

Figure 11. The habitat patch with id: 577 in the eastern part of Lithuania.

4. Discussion

4.1. Habitat Risk Severity to Drivers

In order to plan MVC mitigation measures, spatial habitat characteristics together with MVCs and MVC cluster data are needed [76]. Habitat characteristics and factors that allow us to predict MVCs are important, but are usually the missing component. This can be explained by the disparate character of field research data [47]. Thus, the framework we propose may help to identify and characterise the missing components in a unified form.

Our results on habitat risk severity to drivers and wildlife at the local level are based on a long-term mammal roadkill dataset [77,78]. The main A14 road, delimiting the top-ranked habitat patch (id: 577, see Figure 11), is one of the most frequently checked for roadkill [1]. Because of ongoing long-term reconstruction of the A14 road (until 2030), short-term redirection of traffic onto national road 173 might be foreseen, thereby increasing the traffic intensity on that road, thus also increasing the likelihood of more MVCs than before and a higher risk to drivers.

4.2. Habitat Attractiveness to Mammals

The rates of annual land cover change in Lithuania are decreasing, dropping from 0.48% in 1990 to 0.18% in 2012 [58]. This indicates that the habitat composition has remained stable over time. A growing rate of forest land (woodland) and a rapid decline in active farming [58] has improved habitat attractiveness to different wildlife species, especially for forest dwellers. The increasing MVC numbers in all categories of roads and the increase in annual average daily traffic [79] have coincided with an enlargement of wildlife populations in the country. Species richness (the number of species in Table 2) has a strong relationship (r = 0.72) with the number of MVCs (the number of MVCs within adjacent clusters in Table 1).

We assumed that larger values of species richness indicate higher habitat diversity [80], suitability [81], and attractiveness to wildlife. We found 20 different species within MVC clusters that are adjacent to habitat patch id: 577, which means that road 137 (Figure 11) is more dangerous than the roads adjacent to habitats with a smaller number of species. However, species richness does not take into account the abundances of the species or their relative abundance distributions. The proposed framework allows for the accurate identification of species richness in relation to MVCs that are in the vicinity of the particular habitat. This information is especially useful when wildlife observation data (ground-truth) are not available at all. However, the accuracy of the result is very much dependent on the quality of the available police registered reports and professional field research data [1].

Habitats were defined and characterised across all territory of Lithuania; therefore, the validation of our model is possible: (i) using data from a similar territory, such as a neighbouring country; or (ii) using data from Lithuania from a different time period, e.g., 2018–2021 (our model covered 2002–2017). At the moment, however, such a dataset is not available.

Species richness may be validated by intensive roadkill counts or using wildlife cameras to check for animal movement across roads.

4.3. Multi-Objective Mitigation Measures

The only effective mitigation of road kills in a multi-species animal community is a complex of wildlife fencing (with a sufficient number of wildlife underpasses and overpasses according to the length of the fence) and active driver warning systems on roads without wildlife fences. We did not manage to find tangible research on other effective multi-species and multi-objective mitigation measures for large and small mammal species.

Mitigating MVCs on road 173 (Figure 11) may be challenging, as the MVC-targeting measures are likely to focus de minimis on ungulates, namely *C. capreolus, S. scrofa,* and *A. alces* (Tables 2 and 5), rather than on the other 17 large and small body size mammal species recorded (Table 2). Numbers of carnivore road kills also grow in areas with a higher abundance of small mammal species [55]. MVC clusters found in different locations can help us to select species-specific mitigation measures. However, due to the high cost of the abovementioned complex of measures and the low traffic intensity on roads other than A14 and 102 (Figure 11), implementation of such measures is not possible in the near future. Therefore, our method currently may serve as part of the toolbox to identify the most dangerous roads and the most important habitat patches.

The observation of near misses (road 173 in Figure 11) might provide further input for the task. Field studies should incorporate long-term data collection, before the mitigation measure is applied [18]. Last, but not least, clearing vegetation along roads can also help to lower the MVC risk [54,82]. The mitigation measures for managing the risks to drivers and wildlife may be more challenging when many species are present. This may result in higher road reconstruction costs. The lack of data on the effectiveness of road mitigation measures [18,20] is a further obstacle to decision-making. The most common MVC mitigation measure in Lithuania is fencing. Short wildlife fences may not sufficiently reduce the risk of MVCs, but they are economically more affordable. Long fences are less efficient economically, but may perform better [9–11,17] on the roads with the highest traffic intensity. Therefore, we conclude that, even when involving all habitat data, the selection of multi-objective MVC mitigation measures in a dynamic environment still remains a considerable research challenge.

5. Conclusions

This study developed models that allow for the identification, characterisation, and ranking of habitats based on mammal roadkill data. The main conclusions are:

1. Habitats were characterised by connectivity, land cover, roadkill, roadkill cluster, and mammal species and ranked using multiple criteria for the identification of habitat risk severity to drivers and attractiveness to wildlife;
2. Despite the potential limitations of the scope of the roadkill data, our habitat ranking suggests that this procedure can provide information on habitats, habitat locations, species richness, habitat risk severity to drivers, and attractiveness to wildlife;
3. Strong relationships were identified and discussed between the habitat patch ranks, five (out of 28) land cover classes, and eight (out of 28) species (97% of all mammal road kills);
4. This methodology facilitates decision-making on the habitats that must be prioritized to preserve wildlife in the vicinity of roads that are prone to MVCs. It is also suitable for the planning of multi-objective mitigation measures to improve road security in a dynamic environment.

Author Contributions: Conceptualization, A.K. and L.B.; methodology, A.K.; software, A.K.; validation, A.K.; formal analysis, A.K.; investigation, A.K. and L.B.; resources, L.B.; data curation, L.B.; writing—original draft preparation, A.K.; writing—review and editing, L.B.; visualization, A.K.; supervision, L.B.; project administration, L.B. All authors have read and agreed to the published version of the manuscript.

Funding: This research received no external funding.

Institutional Review Board Statement: Not applicable.

Data Availability Statement: Supplementary spatial data for this article can be found online at http://dx.doi.org/10.17632/4c58n345h5.1 (last accessed on 1 May 2021).

Acknowledgments: The authors thank Jos Stratford for linguistic revision of the manuscript as well as the anonymous reviewers for their suggestions and contributions.

Conflicts of Interest: The authors declare no conflict of interest.

Appendix A

Table 1. The habitat characterisation data (criteria) used to rank the habitat patches.

Unique Identification Number of Habitat (Figure 3)	Total Area of (ha):		Number of Adjacent:		Total Length of Adjacent (km):			Number of Collisions (MVC) within Adjacent Clusters	Average Strength of Adjacent Clusters (KDE+)
	Habitat Patch	Adjacent Habitat Patches	Clusters	Corridors	Corridors	Pathways to Clusters	Clusters		
6	139,650.90	200,454.05	10	6	202.50	214.45	3.60	45	0.4081
12	32,970.44	177,828.03	3	4	102.88	37.15	0.59	6	0.3919
15	9886.41	82,018.06	5	4	55.59	30.78	0.84	11	0.3530
25	20,544.02	293,639.62	7	6	140.00	53.80	1.64	18	0.3777
27	49,493.57	105,316.53	25	5	90.43	301.53	7.06	107	0.4034
29	7746.70	110,394.31	9	5	76.99	49.82	2.67	37	0.4663
30	20,756.90	133,555.22	12	7	114.14	95.74	3.84	56	0.3973
42	56,222.02	45,595.47	19	3	81.38	578.92	3.64	44	0.3999
49	8263.08	78,347.16	3	2	41.45	33.45	0.54	6	0.3491
65	86,540.34	287,351.74	12	6	168.75	267.29	3.67	45	0.4512
69	13,269.35	123,512.66	11	6	85.77	99.30	3.26	43	0.4903
81	40,881.40	237,462.40	26	6	133.16	297.33	6.90	113	0.3696
84	10,394.58	88,278.91	5	4	55.15	29.58	2.31	34	0.3914
105	22,220.59	153,376.31	17	6	106.67	142.58	4.79	64	0.4851
107	6097.13	295,799.24	8	5	86.22	38.80	1.81	26	0.3532
109	26,971.83	141,639.80	11	6	135.03	125.33	2.40	33	0.4293
114	13,420.71	313,371.47	6	4	72.12	40.04	1.26	20	0.4183
139	22,125.14	183,833.68	16	6	137.47	143.62	4.36	63	0.4556

Table 1. Cont.

Unique Identification Number of Habitat (Figure 3)	Total Area of (ha):		Number of Adjacent:		Total Length of Adjacent (km):			Number of Collisions (MVC) within Adjacent Clusters	Average Strength of Adjacent Clusters (KDE+)
	Habitat Patch	Adjacent Habitat Patches	Clusters	Corridors	Corridors	Pathways to Clusters	Clusters		
145	81,083.09	199,083.28	54	7	147.61	1330.48	14.44	230	0.4003
149	913.63	202,816.14	8	7	130.44	15.42	2.82	48	0.4873
159	26,302.43	156,552.24	19	5	91.64	177.86	4.92	74	0.3987
163	19,150.25	116,087.17	3	5	91.54	22.00	0.49	8	0.3044
170	39,785.55	139,555.23	26	5	91.86	288.08	6.47	77	0.4501
182	10,805.79	135,726.79	8	4	52.26	49.21	2.00	20	0.4316
185	20,763.14	172,284.99	17	6	113.47	240.50	3.85	49	0.3928
192	29,897.38	65,613.76	10	4	54.50	103.37	1.90	26	0.3510
194	3369.45	237,614.49	4	6	130.76	9.11	1.46	23	0.5280
208	6795.19	69,431.17	5	5	66.65	17.48	1.75	27	0.5321
210	11,376.97	107,910.32	7	5	67.54	53.89	1.56	16	0.4786
212	28,632.79	59,050.92	8	4	60.40	78.05	1.80	20	0.4757
214	32,032.08	190,092.71	6	8	175.94	61.15	1.31	17	0.4390
218	39,940.29	212,569.11	35	5	96.67	532.34	10.46	177	0.4332
250	15,207.25	131,842.62	11	6	127.65	76.62	2.11	27	0.4054
255	58,086.51	285,182.80	25	8	202.54	375.49	5.85	105	0.4294
267	20,266.12	202,654.73	18	6	117.49	141.77	5.35	76	0.4140
294	8114.69	332,292.67	12	9	199.44	65.72	2.81	32	0.4470
313	54,397.11	197,333.56	26	9	194.20	390.66	7.77	149	0.3597
349	1026.87	37,640.64	2	3	42.03	4.42	0.42	5	0.4798
363	14,293.62	96,397.55	4	5	72.93	28.90	0.77	9	0.4270
371	15,638.20	54,135.88	7	4	51.56	37.97	1.54	16	0.4737
388	41,864.59	79,652.04	8	6	103.05	97.75	1.59	17	0.4521
407	16,006.40	41,733.83	24	5	62.46	182.59	7.09	127	0.4176
427	6408.49	84,791.01	2	3	45.25	10.00	0.42	4	0.4930
430	53,447.47	140,843.94	29	8	153.38	541.35	7.49	101	0.4734
439	5374.78	67,230.93	6	6	71.85	22.70	1.49	25	0.3304
450	406.03	161,433.60	9	6	95.49	21.48	2.35	54	0.4300
455	40,050.24	279,808.91	8	5	152.33	97.57	2.34	23	0.4668
460	4956.26	159,646.28	9	4	52.22	34.46	3.78	72	0.3962
472	7866.55	87,828.86	11	6	90.31	70.93	2.88	68	0.4175
474	8684.84	15,464.01	12	4	32.74	66.13	2.91	46	0.3780
484	8547.01	69,712.63	14	5	54.50	75.37	3.92	95	0.3897
486	6332.88	173,689.79	2	4	49.39	9.39	0.38	4	0.4377
493	30,035.21	55,487.94	8	4	55.55	70.16	2.06	27	0.4608
496	2851.82	126,623.42	8	5	69.66	27.55	2.50	54	0.3609
501	4449.84	114,707.35	5	4	50.59	14.77	1.14	14	0.4892
502	59,532.25	72,106.47	34	7	120.92	448.42	12.26	233	0.4272
505	7099.46	232,115.55	14	7	107.84	67.28	4.60	95	0.4451
518	15,350.47	130,369.46	13	7	111.52	98.47	4.56	72	0.3834
526	551.67	65,675.38	7	5	55.70	9.24	1.93	42	0.4503
530	1671.02	54,468.27	11	5	44.23	37.72	3.16	78	0.4152
533	4452.12	271,067.83	3	5	83.56	15.29	0.59	6	0.4561
542	842.71	255,407.87	4	6	127.53	8.53	1.12	23	0.4925
547	16,775.94	239,527.33	10	5	114.00	71.23	4.50	66	0.4707
577	148,761.11	139,830.71	64	7	180.59	1623.02	19.96	477	0.4834
587	31,309.97	91,225.40	8	7	108.87	79.98	1.73	19	0.4083
588	16,143.01	250,902.96	30	7	143.51	265.72	8.98	294	0.4930
593	5028.73	167,525.08	11	6	89.96	78.10	3.64	71	0.3318
594	638.35	132,279.96	4	5	85.21	7.39	1.84	32	0.3771
608	27,848.17	73,135.06	17	7	89.06	181.17	3.89	50	0.4399
642	120,160.25	163,452.54	42	6	159.99	1052.79	14.11	291	0.4690
656	6828.41	97,816.57	12	5	57.81	87.08	2.89	36	0.4691
661	36,721.46	258,703.50	14	5	138.30	254.92	3.38	44	0.4192
668	22,672.16	259,011.55	5	3	111.59	150.84	1.22	12	0.4795
674	8670.38	246,371.53	22	8	163.20	125.72	7.85	178	0.5224
688	43,933.50	109,036.29	16	7	123.69	203.36	3.64	44	0.3900
694	113.52	191,806.33	2	7	108.51	1.71	0.62	8	0.4044
697	1127.42	97,322.29	4	5	45.04	8.33	0.92	9	0.4802
714	2999.93	64,663.82	5	4	32.25	15.65	0.96	11	0.3500
721	35,574.69	160,770.09	18	7	124.19	187.91	4.90	66	0.4099
723	14,918.98	51,106.71	16	4	44.41	132.84	4.19	92	0.4803

Table 1. *Cont.*

Unique Identification Number of Habitat (Figure 3)	Total Area of (ha): Habitat Patch	Total Area of (ha): Adjacent Habitat Patches	Number of Adjacent: Clusters	Number of Adjacent: Corridors	Total Length of Adjacent (km): Corridors	Total Length of Adjacent (km): Pathways to Clusters	Total Length of Adjacent (km): Clusters	Number of Collisions (MVC) within Adjacent Clusters	Average Strength of Adjacent Clusters (KDE+)
728	29,942.96	94,841.87	19	6	105.29	241.89	4.83	82	0.4239
745	59,675.07	253,380.91	11	6	180.15	176.09	2.19	24	0.4566
746	11,572.92	276,983.34	15	7	144.76	109.63	4.22	68	0.3886
767	21,779.78	197,364.85	15	4	68.84	129.11	3.79	54	0.4355
768	73,317.38	373,589.09	20	7	206.77	283.63	5.23	103	0.3799
770	102,723.38	298,359.55	16	6	178.11	326.31	3.46	41	0.4190
789	123,084.90	189,787.84	33	6	155.53	726.81	9.52	130	0.4333
791	96,613.09	284,751.11	18	5	144.74	297.36	4.47	55	0.4172
795	1704.22	289,397.87	6	6	115.97	22.12	1.51	49	0.4971
805	9181.59	163,837.55	11	5	90.35	75.59	2.35	37	0.4059
813	15,330.96	122,295.05	11	6	96.95	82.02	3.14	66	0.3218
814	60,899.32	106,904.68	13	3	74.10	177.67	3.04	31	0.4411
817	10,487.36	180,491.33	8	5	93.20	55.64	1.79	26	0.4502
827	8152.71	103,750.13	7	4	49.23	39.81	2.09	40	0.3644
839	3269.34	257,878.86	5	4	77.02	29.51	1.21	19	0.3707
857	12,897.86	111,902.63	8	5	69.55	53.26	1.82	24	0.3679
868	3015.25	121,799.88	4	6	98.02	15.75	0.86	17	0.3900
878	22,351.00	90,552.99	16	5	74.46	154.28	4.22	103	0.4157
881	89,590.76	150,862.11	29	6	135.03	500.15	7.93	141	0.4503
912	26,621.53	110,432.37	12	6	95.02	108.57	3.26	49	0.3266
913	81,476.51	281,052.45	36	6	139.18	658.06	11.69	159	0.4182
924	45,311.39	304,598.21	20	7	158.74	265.97	5.37	120	0.4233
945	1394.92	384,991.16	2	6	124.12	3.54	0.48	5	0.4912
962	17,292.37	103,458.26	8	7	98.88	78.87	2.26	31	0.4659
967	98,955.66	175,763.59	24	7	164.13	413.74	6.13	94	0.4509
988	3060.79	436,613.07	5	7	170.72	21.79	1.49	23	0.4690
1028	28,199.78	161,276.22	20	6	115.64	290.72	6.78	140	0.3881
1034	45,987.96	427,134.53	16	8	222.17	193.22	4.25	58	0.3733
1037	4155.04	181,101.39	16	5	90.50	99.08	4.74	89	0.3982
1041	59,317.61	243,963.67	12	5	112.18	226.55	2.34	25	0.3946
1067	8615.61	229,667.24	7	7	144.12	35.62	2.99	51	0.3589
1076	12,362.06	109,332.11	19	5	65.93	135.30	6.33	107	0.3906
1086	26,082.13	205,227.06	13	7	148.29	114.91	3.02	35	0.4076
1097	13,347.14	126,268.91	13	7	95.26	99.40	3.76	59	0.4014
1098	32,653.14	159,838.18	19	5	109.33	197.80	6.16	78	0.4310
1107	59,695.86	212,908.34	13	6	153.40	189.37	3.06	44	0.4832
1114	77,091.89	289,196.36	22	7	190.06	350.43	6.37	83	0.4176
1115	86,544.04	356,520.16	26	8	244.87	499.88	6.05	72	0.4151
1116	787.12	38,519.26	3	4	26.58	5.12	0.55	8	0.3514
1126	26,486.94	209,006.42	15	4	71.82	230.05	3.16	44	0.3920
1133	22,831.99	210,541.11	14	5	89.80	141.13	5.17	112	0.3376
1134	10,165.22	88,243.02	10	5	61.36	52.30	2.80	38	0.4820
1156	46,637.37	282,819.38	16	9	195.23	206.97	3.83	57	0.3521
1163	14,093.66	316,790.63	6	6	127.59	38.48	1.23	17	0.2998
1169	5713.44	50,558.89	7	6	52.52	33.05	1.86	25	0.5095
1183	2729.25	37,035.79	2	5	38.27	4.85	0.39	4	0.4581
1186	28,020.86	254,445.78	19	7	142.02	176.61	4.62	94	0.4864
1190	3864.25	247,170.44	7	5	97.64	75.05	1.97	25	0.4169
1226	57,029.73	166,243.04	10	7	127.21	199.43	2.83	30	0.4426
1229	18,092.33	182,645.42	13	6	110.63	87.64	5.95	92	0.4375
1230	6752.07	156,443.64	2	6	97.55	12.95	0.37	4	0.4233
1235	8685.78	33,343.23	7	4	46.77	36.13	2.12	20	0.4726
1240	6237.78	105,562.83	6	6	74.03	26.95	2.04	34	0.4520
1245	32,749.54	171,518.38	16	7	121.81	174.41	4.65	83	0.3271
1250	15,274.28	176,484.33	16	5	80.88	131.18	4.66	61	0.4613
1281	33,658.14	88,145.48	14	5	77.94	163.87	3.00	34	0.4529
1283	7568.01	131,088.86	5	7	95.31	26.06	1.30	15	0.3860
1289	16,729.43	42,515.68	12	5	50.80	91.01	3.51	52	0.3805
1294	24,681.00	211,432.77	4	5	96.24	35.94	1.19	17	0.4695
1295	18,084.15	223,680.12	10	7	127.52	65.36	2.81	32	0.4252
1306	72,066.64	151,755.00	37	7	154.21	669.36	11.31	217	0.3624
1307	11,322.00	185,508.44	7	5	81.89	46.77	1.77	26	0.4137

Table 1. *Cont.*

Unique Identification Number of Habitat (Figure 3)	Total Area of (ha):		Number of Adjacent:		Total Length of Adjacent (km):			Number of Collisions (MVC) within Adjacent Clusters	Average Strength of Adjacent Clusters (KDE+)
	Habitat Patch	Adjacent Habitat Patches	Clusters	Corridors	Corridors	Pathways to Clusters	Clusters		
1346	26,494.20	100,497.63	8	6	93.29	84.91	2.00	34	0.3864
1358	3699.64	20,747.85	14	4	52.69	142.23	3.94	84	0.3620
1361	424.04	34,852.89	6	3	40.20	27.61	1.99	46	0.4280
1397	670.21	67,738.18	3	3	54.17	4.27	0.73	13	0.3525
1399	71,832.91	289,533.30	28	8	199.92	522.36	6.20	82	0.3709
1406	10,441.83	236,626.73	10	6	106.40	64.25	3.04	35	0.4599
1421	6521.17	199,715.34	17	6	111.34	95.68	4.06	55	0.3914
1424	7075.41	127,018.38	4	4	49.50	24.32	0.80	16	0.4225
1437	21,446.06	278,307.57	11	7	153.66	100.68	3.10	53	0.3250
1444	24,969.18	113,672.42	4	5	87.69	41.17	1.48	19	0.4627
1452	33,839.26	150,121.36	21	6	105.84	208.66	6.27	84	0.4571
1462	15,010.46	189,536.41	19	4	65.83	134.36	4.49	63	0.3714
1475	14,808.02	82,476.01	7	6	76.88	55.63	1.97	34	0.5179
1478	1949.18	81,757.01	6	4	53.73	18.19	1.79	32	0.3442
1492	40,178.40	217,814.35	17	6	131.45	187.58	3.75	63	0.3973
1496	1408.69	140,511.54	9	5	77.56	22.37	2.23	42	0.4113
1505	21,628.32	139,095.96	11	7	116.70	91.87	2.92	30	0.4774
1507	977.68	175,298.34	5	6	104.25	7.71	1.67	38	0.4843
1518	28,064.51	114,924.57	22	7	107.51	242.41	7.62	146	0.3569
1519	90,703.78	241,999.71	22	7	174.59	389.71	4.65	63	0.3948
1524	279.17	22,245.26	3	3	18.66	5.40	0.72	12	0.4726
1532	49,396.90	164,132.83	21	7	137.65	298.00	5.93	94	0.3801
1533	33,873.78	110,225.07	25	4	62.59	284.54	6.22	104	0.3429
1534	9535.38	101,188.48	13	5	66.96	84.17	4.68	97	0.4054
1557	9936.36	98,767.85	9	6	77.03	55.41	2.47	40	0.4222
1558	18,251.82	109,634.35	13	6	100.03	106.30	4.69	91	0.4045
1560	7908.16	214,037.46	14	7	128.26	80.78	3.19	36	0.4075
1562	3397.56	13,000.93	5	4	30.57	18.12	1.39	27	0.5100
1568	4039.68	24,080.89	6	4	23.87	21.96	1.77	26	0.5915
1569	39,551.03	86,889.66	4	5	108.00	38.76	1.43	18	0.4485
1578	5596.14	25,331.20	9	4	31.73	40.50	2.86	51	0.5361
1594	47,007.60	103,575.45	20	8	150.55	261.14	6.32	103	0.4576
1597	29,272.60	100,781.16	19	7	118.82	203.48	4.24	66	0.3757
1601	11,606.92	91,238.03	10	5	67.79	71.71	3.18	55	0.4186
1608	20,713.92	85,718.96	16	5	72.09	150.38	6.10	108	0.3762
1610	22,556.62	173,460.63	18	5	103.06	149.97	3.94	62	0.4065
1633	7985.15	104,340.98	13	6	83.77	67.19	3.44	71	0.3290
1638	12,344.12	212,857.39	10	5	85.90	69.94	2.35	36	0.3649
1639	1687.87	173,026.58	3	5	102.57	9.39	0.76	10	0.2541
1647	9878.36	95,779.10	20	5	60.78	112.47	5.33	84	0.4204
1653	19,429.28	152,497.77	5	6	110.43	44.90	1.07	11	0.5182
1654	896.17	161,786.45	6	6	100.72	9.37	1.29	21	0.3603
1671	6958.72	128,202.48	14	5	77.31	96.41	3.37	58	0.3581
1675	7633.48	121,198.20	10	5	68.96	62.58	2.07	34	0.3651
1679	5533.32	275,379.01	2	5	93.62	9.75	0.39	4	0.4524
1681	48,186.10	122,791.45	20	5	92.73	256.60	5.16	68	0.4215
1700	20,889.61	189,438.39	17	7	131.67	194.62	4.63	68	0.3667
1706	3085.94	111,060.99	3	6	101.86	7.74	0.81	14	0.4478
1715	12,195.17	94,176.12	7	5	84.10	52.02	2.35	45	0.3747
1731	15,219.57	165,458.85	10	7	96.29	76.21	2.24	27	0.4569
1738	55,927.96	116,427.39	28	4	92.87	387.47	8.44	99	0.3379
1745	30,237.24	209,558.57	7	6	117.77	70.76	1.84	21	0.4572
1748	17,446.66	279,990.97	6	6	131.88	45.34	1.30	19	0.3747
1749	19,096.63	87,371.87	14	4	56.83	108.10	2.92	37	0.3756
1764	30,762.74	78,861.47	17	6	94.85	175.03	4.13	57	0.3799
1769	6344.60	85,765.72	7	5	54.05	40.60	1.50	16	0.4538
1777	21,882.10	201,834.33	12	8	160.72	97.02	4.22	55	0.4025
1778	19,362.06	112,017.20	16	4	104.46	212.13	4.60	57	0.2977
1782	10,826.89	120,203.69	10	5	64.05	59.33	2.99	46	0.4271
1790	4967.14	132,758.39	4	6	92.82	17.89	0.80	12	0.3525
1794	39,864.90	284,582.45	11	6	146.02	156.82	2.41	32	0.4017
1798	51,201.38	85,322.04	22	5	93.40	317.22	5.45	68	0.4150
1812	27,755.98	164,965.06	6	6	113.93	64.07	1.33	17	0.3973

Table 1. *Cont.*

Unique Identification Number of Habitat (Figure 3)	Total Area of (ha):		Number of Adjacent:		Total Length of Adjacent (km):			Number of Collisions (MVC) within Adjacent Clusters	Average Strength of Adjacent Clusters (KDE+)
	Habitat Patch	Adjacent Habitat Patches	Clusters	Corridors	Corridors	Pathways to Clusters	Clusters		
1816	49,558.70	140,839.59	19	6	116.25	248.76	5.68	72	0.4426
1826	10,977.84	194,907.87	4	7	137.28	35.05	0.77	11	0.3532
1834	13,150.92	51,656.98	7	3	42.72	35.24	1.56	18	0.4388
1853	17,579.71	98,804.55	14	3	54.98	195.40	3.76	39	0.4100
1864	44,062.02	175,180.04	23	6	118.97	306.20	6.15	70	0.3894
1866	73,170.00	189,601.82	18	7	177.14	319.39	3.99	47	0.4651
1876	35,035.12	161,066.27	5	6	122.93	46.46	1.19	12	0.4917
1877	21,524.06	134,671.99	33	7	120.46	297.69	7.16	105	0.3517
1879	63,771.53	111,222.35	15	5	123.75	197.72	5.18	69	0.4853
1882	25,628.08	142,367.89	11	5	84.92	114.24	2.41	29	0.4335
1913	18,601.09	133,829.94	20	6	100.98	190.26	5.72	73	0.4415
1916	26,067.02	97,119.28	20	5	70.92	195.93	5.52	72	0.3984
1966	24,889.16	48,516.32	13	4	59.28	142.07	3.81	50	0.4004
1986	10,376.30	157,970.95	4	7	153.83	24.50	0.96	9	0.4402
2004	18,419.25	172,357.90	7	5	119.25	65.55	1.63	25	0.5463
2014	2410.14	105,994.53	1	6	93.61	3.55	0.35	3	0.4506
2037	24,718.80	63,830.67	10	3	64.44	216.47	2.25	36	0.4385
2038	16,391.90	25,383.04	12	3	42.24	127.44	3.04	40	0.3997
2052	32,113.69	189,372.89	17	6	172.21	198.68	4.09	53	0.3576
2055	12,596.60	93,551.41	11	4	64.24	81.57	3.93	47	0.5472
2060	2450.68	282,984.78	4	7	177.34	14.43	0.66	8	0.3172
2105	11,678.01	229,612.31	3	5	164.37	28.76	0.90	9	0.5434
2106	18,325.75	72,309.40	8	4	87.07	68.23	1.63	19	0.4210
2224	513.57	155,393.82	3	5	61.12	5.00	0.61	10	0.4559
2229	33,511.21	67,059.07	13	4	72.74	158.39	3.01	42	0.3636
2233	19,071.18	124,934.73	9	5	87.46	67.43	2.16	28	0.4049
2237	469.96	169,750.82	2	5	106.25	3.05	0.39	6	0.3960
2244	15,623.84	76,303.43	7	4	49.94	50.12	1.38	20	0.4367
2246	1381.02	59,689.66	3	3	28.55	10.64	0.67	10	0.3596
2247	4885.96	70,126.78	8	3	51.26	52.32	1.95	38	0.3126
2248	95,297.89	122,224.14	58	7	158.06	980.16	17.59	334	0.4143
2249	119,258.39	141,390.65	42	6	128.49	861.51	10.65	160	0.4427
2250	12,055.54	258,374.60	16	7	153.61	119.02	4.74	96	0.4157
2251	63,561.74	366,751.45	31	7	209.27	592.77	8.38	92	0.4794
2252	16,490.07	148,897.07	5	4	92.80	38.48	1.51	21	0.4779
2253	22,467.48	141,782.65	9	7	118.89	88.74	2.16	32	0.5217
2254	38,634.75	170,998.75	15	6	130.44	223.51	3.45	41	0.4515
2255	117,246.63	145,041.02	35	6	159.50	897.39	8.29	100	0.4424

Table 2. List of all identified habitats, their ranks, and wildlife species (species richness) identified within habitats.

Unique Identification Number of Habitat (Figure 3)	Ranks		Species	
	SAW	TOPSIS	Count	Latin Names
6	0.3508	0.4690	7	*M. meles, V. vulpes, E. concolor, N. procyonoides, A. alces, S. scrofa,* C. capreolus
12	0.2064	0.4166	3	*L. europaeus, S. scrofa, C. capreolus*
15	0.1856	0.4128	3	*A. alces, S. scrofa, C. capreolus*
25	0.2542	0.4239	4	*L. europaeus, V. vulpes, S. scrofa, C. capreolus*
27	0.3329	0.4683	9	*C. fiber, M. meles, M. putorius, C. elaphus, L. europaeus, V. vulpes, A. alces, S. scrofa, C. capreolus*
29	0.2264	0.4241	3	*A. alces, S. scrofa, C. capreolus*
30	0.2705	0.4363	6	*C. fiber, L. europaeus, E. concolor, A. alces, S. scrofa, C. capreolus*

Table 2. *Cont.*

Unique Identification Number of Habitat (Figure 3)	Ranks		Species	
	SAW	TOPSIS	Count	Latin Names
42	0.2498	0.3686	6	*L. europaeus, V. vulpes, E. concolor, A. alces, S. scrofa, C. capreolus*
49	0.1689	0.4089	3	*A. alces, S. scrofa, C. capreolus*
65	0.3394	0.4414	6	*M. meles, L. europaeus, E. concolor, A. alces, S. scrofa, C. capreolus*
69	0.2547	0.4256	4	*C. elaphus, L. europaeus, S. scrofa, C. capreolus*
81	0.3468	0.4687	7	*M. meles, C. elaphus, V. vulpes, N. procyonoides, A. alces, S. scrofa, C. capreolus*
84	0.2112	0.4240	4	*E. concolor, A. alces, S. scrofa, C. capreolus*
105	0.2843	0.4396	4	*M. putorius, A. alces, S. scrofa, C. capreolus*
107	0.2503	0.4290	5	*V. vulpes, E. concolor, A. alces, S. scrofa, C. capreolus*
109	0.2691	0.4238	9	*R. norvegicus, M. putorius, Martes* sp., *L. europaeus, V. vulpes, E. concolor, N. procyonoides, A. alces, C. capreolus*
114	0.2558	0.4285	5	*V. vulpes, E. concolor, N. procyonoides, A. alces, C. capreolus*
139	0.2935	0.4391	7	*T. europaea, L. europaeus, V. vulpes, E. concolor, A. alces, S. scrofa, C. capreolus*
145	0.4994	0.4687	9	*C. elaphus, Martes* sp., *L. europaeus, V. vulpes, E. concolor, N. procyonoides, A. alces, S. scrofa, C. capreolus*
149	0.2646	0.4308	3	*L. europaeus, S. scrofa, C. capreolus*
159	0.2943	0.4477	8	*M. putorius, C. elaphus, Martes* sp., *L. europaeus, V. vulpes, A. alces, S. scrofa, C. capreolus*
163	0.1906	0.4136	4	*Martes* sp., *L. europaeus, V. vulpes, A. alces,*
170	0.3171	0.4493	6	*C. elaphus, L. europaeus, V. vulpes, A. alces, S. scrofa, C. capreolus*
182	0.2224	0.4208	3	*A. alces, S. scrofa, C. capreolus*
185	0.2693	0.4163	5	*C. elaphus, L. europaeus, A. alces, S. scrofa, C. capreolus*
192	0.2240	0.4220	6	*Martes* sp., *L. europaeus, V. vulpes, E. concolor, S. scrofa, C. capreolus*
194	0.2594	0.4205	3	*L. europaeus, S. scrofa, C. capreolus*
208	0.2332	0.4219	5	*T. europaea, E. concolor, A. alces, S. scrofa, C. capreolus*
210	0.2147	0.4146	2	*L. europaeus, C. capreolus*
212	0.2217	0.4195	4	*Martes* sp., *E. concolor, N. procyonoides, C. capreolus*
214	0.2705	0.4209	6	*L. europaeus, E. concolor, N. procyonoides, A. alces, S. scrofa, C. capreolus*
218	0.4073	0.4991	10	*M. putorius, C. elaphus, Martes* sp., *L. europaeus, V. vulpes, E. concolor, N. procyonoides, A. alces, S. scrofa, C. capreolus*
250	0.2342	0.4192	5	*L. europaeus, E. concolor, A. alces, S. scrofa, C. capreolus*
255	0.3937	0.4584	9	*C. elaphus, Martes* sp., *L. europaeus, V. vulpes, E. concolor, N. procyonoides, A. alces, S. scrofa, C. capreolus*
267	0.2992	0.4501	6	*Martes* sp., *E. concolor, N. procyonoides, A. alces, S. scrofa, C. capreolus*
294	0.3109	0.4310	5	*M. meles, L. europaeus, E. concolor, S. scrofa, C. capreolus*

Table 2. *Cont.*

Unique Identification Number of Habitat (Figure 3)	Ranks			Species
	SAW	TOPSIS	Count	Latin Names
313	0.3980	0.4812	10	*N. vison, M. putorius, Martes* sp., *L. europaeus, V. vulpes, E. concolor, N. procyonoides, A. alces, S. scrofa, C. capreolus*
349	0.2215	0.4103	4	*E. concolor, A. alces, S. scrofa, C. capreolus*
363	0.2070	0.4140	4	*L. europaeus, V. vulpes, E. concolor, C. capreolus*
371	0.2103	0.4179	3	*A. alces, S. scrofa, C. capreolus*
388	0.2463	0.4208	6	*M. meles, L. europaeus, V. vulpes, N. procyonoides, S. scrofa, C. capreolus*
407	0.3297	0.4810	13	*C. fiber, N. vison, S. vulgaris, M. martes, C. elaphus, Martes* sp., *L. europaeus, V. vulpes, E. concolor, N. procyonoides, A. alces, S. scrofa, C. capreolus*
427	0.2010	0.4105	2	*L. europaeus, C. capreolus*
430	0.3764	0.4324	8	*M. putorius, C. elaphus, V. vulpes, E. concolor, N. procyonoides, A. alces, S. scrofa, C. capreolus*
439	0.2127	0.4203	6	*L. europaeus, V. vulpes, N. procyonoides, A. alces, S. scrofa, C. capreolus*
450	0.2691	0.4374	9	*M. putorius, Martes* sp., *L. europaeus, V. vulpes, E. concolor, N. procyonoides, A. alces, S. scrofa, C. capreolus*
455	0.2716	0.4273	4	*D. dama, A. alces, S. scrofa, C. capreolus*
460	0.2725	0.4493	9	*M. putorius, Martes* sp., *L. europaeus, V. vulpes, E. concolor, N. procyonoides, A. alces, S. scrofa, C. capreolus*
472	0.2851	0.4444	14	*L. lutra, N. vison, S. vulgaris, M. martes, M. meles, M. putorius, C. elaphus, L. europaeus, V. vulpes, E. concolor, N. procyonoides, A. alces, S. scrofa, C. capreolus*
474	0.2440	0.4312	7	*C. elaphus, L. europaeus, V. vulpes, E. concolor, A. alces, S. scrofa, C. capreolus*
484	0.2880	0.4575	12	*C. fiber, M. martes, M. putorius, C. elaphus, Martes* sp., *L. europaeus, V. vulpes, E. concolor, N. procyonoides, A. alces, S. scrofa, C. capreolus*
486	0.2152	0.4131	1	*C. capreolus*
493	0.2208	0.4228	3	*A. alces, S. scrofa, C. capreolus*
496	0.2481	0.4365	9	*M. putorius, Martes* sp., *L. europaeus, V. vulpes, E. concolor, N. procyonoides, A. alces, S. scrofa, C. capreolus*
501	0.2209	0.4173	3	*A. alces, S. scrofa, C. capreolus*
502	0.4247	0.5521	10	*N. vison, M. putorius, Martes* sp., *L. europaeus, V. vulpes, E. concolor, N. procyonoides, A. alces, S. scrofa, C. capreolus*
505	0.3324	0.4649	11	*S. vulgaris, M. meles, M. putorius, Martes* sp., *L. europaeus, V. vulpes, E. concolor, N. procyonoides, A. alces, S. scrofa, C. capreolus*
518	0.3129	0.4537	14	*C. fiber, S. vulgaris, M. martes, M. meles, M. putorius, C. elaphus, Martes* sp., *L. europaeus, V. vulpes, E. concolor, N. procyonoides, A. alces, S. scrofa, C. capreolus*
526	0.2735	0.4355	12	*L. lutra, N. vison, S. vulgaris, M. martes, M. meles, M. putorius, V. vulpes, E. concolor, N. procyonoides, A. alces, S. scrofa, C. capreolus*

Table 2. *Cont.*

Unique Identification Number of Habitat (Figure 3)	Ranks		Species	
	SAW	TOPSIS	Count	Latin Names
530	0.2765	0.4482	11	*C. fiber, M. martes, M. putorius, C. elaphus, Martes* sp., *V. vulpes, E. concolor, N. procyonoides, A. alces, S. scrofa, C. capreolus*
533	0.2278	0.4157	1	*C. capreolus*
542	0.2862	0.4264	9	*M. meles, Martes* sp., *L. europaeus, V. vulpes, E. concolor, N. procyonoides, A. alces, S. scrofa, C. capreolus*
547	0.2807	0.4453	5	*L. europaeus, N. procyonoides, A. alces, S. scrofa, C. capreolus*
577	0.6981	0.5850	20	*O. zibethicus, M. erminea, L. lutra, M. foina, T. europaea, C. fiber, N. vison, S. vulgaris, M. martes, M. meles, M. putorius, C. elaphus, Martes* sp., *L. europaeus, V. vulpes, E. concolor, N. procyonoides, A. alces, S. scrofa, C. capreolus*
587	0.2307	0.4174	3	*A. alces, S. scrofa, C. capreolus*
588	0.4762	0.5779	19	*O. zibethicus, M. erminea, L. lutra, M. foina, T. europaea, C. fiber, N. vison, S. vulgaris, M. martes, M. meles, M. putorius, Martes* sp., *L. europaeus, V. vulpes, E. concolor, N. procyonoides, A. alces, S. scrofa, C. capreolus*
593	0.2931	0.4495	14	*C. fiber, S. vulgaris, M. martes, M. meles, M. putorius, C. elaphus, Martes* sp., *L. europaeus, V. vulpes, E. concolor, N. procyonoides, A. alces, S. scrofa, C. capreolus*
594	0.2487	0.4269	9	*M. putorius, Martes* sp., *L. europaeus, V. vulpes, E. concolor, N. procyonoides, A. alces, S. scrofa, C. capreolus*
608	0.2888	0.4316	8	*Martes* sp., *L. europaeus, V. vulpes, E. concolor, N. procyonoides, A. alces, S. scrofa, C. capreolus*
642	0.5255	0.5489	12	*L. timidus, M. meles, M. putorius, C. elaphus, Martes* sp., *L. europaeus, V. vulpes, E. concolor, N. procyonoides, A. alces, S. scrofa, C. capreolus*
656	0.2457	0.4247	5	*Martes* sp., *L. europaeus, A. alces, S. scrofa, C. capreolus*
661	0.2924	0.4197	7	*S. vulgaris, M. martes, L. europaeus, V. vulpes, N. procyonoides, S. scrofa, C. capreolus*
668	0.2178	0.4031	2	*S. scrofa, C. capreolus*
674	0.4036	0.5114	12	*L. lutra, M. meles, M. putorius, C. elaphus, Martes* sp., *L. europaeus, V. vulpes, E. concolor, N. procyonoides, A. alces, S. scrofa, C. capreolus*
688	0.2807	0.4270	6	*C. elaphus, V. vulpes, E. concolor, A. alces, S. scrofa, C. capreolus*
694	0.3145	0.4123	2	*E. concolor, C. capreolus*
697	0.2287	0.4143	2	*S. scrofa, C. capreolus*
714	0.2128	0.4155	4	*V. vulpes, A. alces, S. scrofa, C. capreolus*
721	0.3139	0.4454	8	*C. elaphus, Martes* sp., *V. vulpes, E. concolor, N. procyonoides, A. alces, S. scrofa, C. capreolus*

Table 2. *Cont.*

Unique Identification Number of Habitat (Figure 3)	Ranks			Species
	SAW	TOPSIS	Count	Latin Names
723	0.2975	0.4526	11	*L. lutra, M. putorius, C. elaphus, Martes* sp., *L. europaeus, V. vulpes, E. concolor, N. procyonoides, A. alces, S. scrofa, C. capreolus*
728	0.3046	0.4410	10	*N. vison, M. putorius, C. elaphus, Martes* sp., *V. vulpes, E. concolor, N. procyonoides, A. alces, S. scrofa, C. capreolus*
745	0.3042	0.4272	7	*Martes* sp., *L. europaeus, V. vulpes, E. concolor, N. procyonoides, S. scrofa, C. capreolus*
746	0.3155	0.4479	9	*M. putorius, Martes* sp., *L. europaeus, V. vulpes, E. concolor, N. procyonoides, A. alces, S. scrofa, C. capreolus*
767	0.2715	0.4380	5	*L. europaeus, N. procyonoides, A. alces, S. scrofa, C. capreolus*
768	0.4155	0.4828	13	*T. europaea, M. martes, M. meles, M. putorius, C. elaphus, Martes* sp., *L. europaeus, V. vulpes, E. concolor, N. procyonoides, A. alces, S. scrofa, C. capreolus*
770	0.3573	0.4448	7	*M. putorius, Martes* sp., *V. vulpes, E. concolor, N. procyonoides, S. scrofa, C. capreolus*
789	0.4444	0.4831	10	*L. timidus, C. elaphus, Martes* sp., *L. europaeus, V. vulpes, E. concolor, N. procyonoides, A. alces, S. scrofa, C. capreolus*
791	0.3541	0.4581	7	*M. putorius, Martes* sp., *E. concolor, N. procyonoides, A. alces, S. scrofa, C. capreolus*
795	0.3092	0.4406	12	*M. foina, N. vison, M. martes, M. meles, C. elaphus, L. europaeus, V. vulpes, E. concolor, N. procyonoides, A. alces, S. scrofa, C. capreolus*
805	0.2471	0.4270	7	*C. elaphus, Martes* sp., *E. concolor, N. procyonoides, A. alces, S. scrofa, C. capreolus*
813	0.2686	0.4422	11	*N. vison, M. meles, M. putorius, Martes* sp., *L. europaeus, V. vulpes, E. concolor, N. procyonoides, A. alces, S. scrofa, C. capreolus*
814	0.2601	0.4320	6	*L. europaeus, V. vulpes, N. procyonoides, A. alces, S. scrofa, C. capreolus*
817	0.2333	0.4208	4	*N. procyonoides, A. alces, S. scrofa, C. capreolus*
827	0.2449	0.4323	10	*M. meles, M. putorius, Martes* sp., *L. europaeus, V. vulpes, E. concolor, N. procyonoides, A. alces, S. scrofa, C. capreolus*
839	0.2272	0.4219	5	*V. vulpes, E. concolor, N. procyonoides, A. alces, C. capreolus*
857	0.2097	0.4190	3	*C. elaphus, A. alces, C. capreolus*
868	0.2200	0.4159	6	*Martes* sp., *V. vulpes, E. concolor, N. procyonoides, A. alces, C. capreolus*
878	0.2971	0.4560	11	*L. lutra, M. putorius, C. elaphus, Martes* sp., *L. europaeus, V. vulpes, E. concolor, N. procyonoides, A. alces, S. scrofa, C. capreolus*
881	0.4191	0.4936	13	*R. norvegicus, T. europaea, M. martes, M. putorius, C. elaphus, Martes* sp., *L. europaeus, V. vulpes, E. concolor, N. procyonoides, A. alces, S. scrofa, C. capreolus*

Table 2. Cont.

Unique Identification Number of Habitat (Figure 3)	Ranks		Species	
	SAW	TOPSIS	Count	Latin Names
912	0.2580	0.4346	8	*C. elaphus*, *Martes* sp., *V. vulpes*, *E. concolor*, *N. procyonoides*, *A. alces*, *S. scrofa*, *C. capreolus*
913	0.4624	0.5094	11	*B. bonasus*, *L. lutra*, *M. martes*, *C. elaphus*, *Martes* sp., *L. europaeus*, *V. vulpes*, *N. procyonoides*, *A. alces*, *S. scrofa*, *C. capreolus*
924	0.3840	0.4740	11	*L. lutra*, *M. putorius*, *C. elaphus*, *Martes* sp., *L. europaeus*, *V. vulpes*, *E. concolor*, *N. procyonoides*, *A. alces*, *S. scrofa*, *C. capreolus*
945	0.2964	0.4194	1	*C. capreolus*
962	0.2440	0.4198	4	*N. procyonoides*, *A. alces*, *S. scrofa*, *C. capreolus*
967	0.3909	0.4678	9	*M. meles*, *Martes* sp., *L. europaeus*, *V. vulpes*, *E. concolor*, *N. procyonoides*, *A. alces*, *S. scrofa*, *C. capreolus*
988	0.3064	0.4315	6	*M. meles*, *Martes* sp., *V. vulpes*, *N. procyonoides*, *S. scrofa*, *C. capreolus*
1028	0.3610	0.4766	15	*M. foina*, *T. europaea*, *S. vulgaris*, *M. martes*, *M. meles*, *M. putorius*, *C. elaphus*, *Martes* sp., *L. europaeus*, *V. vulpes*, *E. concolor*, *N. procyonoides*, *A. alces*, *S. scrofa*, *C. capreolus*
1034	0.3753	0.4533	9	*M. putorius*, *C. elaphus*, *Martes* sp., *V. vulpes*, *E. concolor*, *N. procyonoides*, *A. alces*, *S. scrofa*, *C. capreolus*
1037	0.3075	0.4597	13	*T. europaea*, *M. martes*, *M. meles*, *M. putorius*, *C. elaphus*, *Martes* sp., *L. europaeus*, *V. vulpes*, *E. concolor*, *N. procyonoides*, *A. alces*, *S. scrofa*, *C. capreolus*
1041	0.2820	0.4222	5	*Martes* sp., *L. europaeus*, *N. procyonoides*, *S. scrofa*, *C. capreolus*
1067	0.2692	0.4353	7	*M. erminea*, *C. elaphus*, *Martes* sp., *N. procyonoides*, *A. alces*, *S. scrofa*, *C. capreolus*
1076	0.3009	0.4665	9	*M. erminea*, *Martes* sp., *L. europaeus*, *V. vulpes*, *E. concolor*, *N. procyonoides*, *A. alces*, *S. scrofa*, *C. capreolus*
1086	0.2753	0.4272	5	*V. vulpes*, *N. procyonoides*, *A. alces*, *S. scrofa*, *C. capreolus*
1097	0.2793	0.4382	8	*C. lupus*, *M. meles*, *L. europaeus*, *V. vulpes*, *N. procyonoides*, *A. alces*, *S. scrofa*, *C. capreolus*
1098	0.3081	0.4531	8	*C. elaphus*, *Martes* sp., *V. vulpes*, *E. concolor*, *N. procyonoides*, *A. alces*, *S. scrofa*, *C. capreolus*
1107	0.3030	0.4332	5	*L. europaeus*, *N. procyonoides*, *A. alces*, *S. scrofa*, *C. capreolus*
1114	0.3951	0.4655	10	*M. meles*, *C. elaphus*, *Martes* sp., *L. europaeus*, *V. vulpes*, *E. concolor*, *N. procyonoides*, *A. alces*, *S. scrofa*, *C. capreolus*
1115	0.4213	0.4460	9	*C. elaphus*, *Martes* sp., *L. europaeus*, *V. vulpes*, *E. concolor*, *N. procyonoides*, *A. alces*, *S. scrofa*, *C. capreolus*
1116	0.2367	0.4130	4	*L. europaeus*, *A. alces*, *S. scrofa*, *C. capreolus*
1126	0.2787	0.4232	8	*M. nivalis*, *Martes* sp., *L. europaeus*, *V. vulpes*, *N. procyonoides*, *A. alces*, *S. scrofa*, *C. capreolus*

Table 2. *Cont.*

Unique Identification Number of Habitat (Figure 3)	Ranks			Species
	SAW	TOPSIS	Count	Latin Names
1133	0.3224	0.4713	13	*R. norvegicus, S. vulgaris, M. meles, M. putorius, C. elaphus, Martes* sp., *L. europaeus, V. vulpes, E. concolor, N. procyonoides, A. alces, S. scrofa, C. capreolus*
1134	0.2440	0.4287	5	*L. europaeus, N. procyonoides, A. alces, S. scrofa, C. capreolus*
1156	0.3240	0.4357	4	*L. europaeus, V. vulpes, S. scrofa, C. capreolus*
1163	0.2474	0.4246	5	*M. nivalis, Martes* sp., *V. vulpes, N. procyonoides, C. capreolus*
1169	0.2358	0.4206	4	*L. europaeus, N. procyonoides, S. scrofa, C. capreolus*
1183	0.2246	0.4091	1	*C. capreolus*
1186	0.3501	0.4594	9	*M. putorius, C. elaphus, Martes* sp., *L. europaeus, V. vulpes, E. concolor, N. procyonoides, A. alces, C. capreolus*
1190	0.2411	0.4193	5	*C. elaphus, L. europaeus, V. vulpes, S. scrofa, C. capreolus*
1226	0.2979	0.4253	7	*L. europaeus, V. vulpes, E. concolor, N. procyonoides, A. alces, S. scrofa, C. capreolus*
1229	0.3218	0.4666	11	*B. bonasus, M. martes, M. putorius, C. elaphus, Martes* sp., *L. europaeus, V. vulpes, N. procyonoides, A. alces, S. scrofa, C. capreolus*
1230	0.2094	0.4098	2	*S. scrofa, C. capreolus*
1235	0.2085	0.4182	3	*V. vulpes, A. alces, C. capreolus*
1240	0.2323	0.4236	4	*L. europaeus, A. alces, S. scrofa, C. capreolus*
1245	0.3021	0.4492	8	*L. lynx, D. dama, L. europaeus, V. vulpes, E. concolor, A. alces, S. scrofa, C. capreolus*
1250	0.2941	0.4440	8	*M. meles, C. elaphus, Martes* sp., *E. concolor, N. procyonoides, A. alces, S. scrofa, C. capreolus*
1281	0.2539	0.4223	5	*C. elaphus, V. vulpes, A. alces, S. scrofa, C. capreolus*
1283	0.2289	0.4175	5	*L. europaeus, E. concolor, A. alces, S. scrofa, C. capreolus*
1289	0.2537	0.4356	8	*C. lupus, M. meles, L. europaeus, V. vulpes, N. procyonoides, A. alces, S. scrofa, C. capreolus*
1294	0.2426	0.4230	4	*Martes* sp., *E. concolor, A. alces, C. capreolus*
1295	0.2731	0.4301	5	*C. elaphus, N. procyonoides, A. alces, S. scrofa, C. capreolus*
1306	0.4299	0.5150	9	*D. dama, M. foina, C. fiber, L. europaeus, E. concolor, N. procyonoides, A. alces, S. scrofa, C. capreolus*
1307	0.2365	0.4236	5	*L. europaeus, V. vulpes, A. alces, S. scrofa, C. capreolus*
1346	0.2389	0.4237	6	*M. meles, Martes* sp., *L. europaeus, N. procyonoides, A. alces, C. capreolus*
1358	0.2228	0.4302	5	*C. fiber, V. vulpes, A. alces, S. scrofa, C. capreolus*
1361	0.2086	0.4238	5	*C. fiber, V. vulpes, A. alces, S. scrofa, C. capreolus*
1397	0.2159	0.4147	6	*L. lutra, V. vulpes, N. procyonoides, A. alces, S. scrofa, C. capreolus*
1399	0.3925	0.4367	8	*M. meles, Martes* sp., *L. europaeus, V. vulpes, N. procyonoides, A. alces, S. scrofa, C. capreolus*

Table 2. *Cont.*

Unique Identification Number of Habitat (Figure 3)	Ranks			Species
	SAW	TOPSIS	Count	Latin Names
1406	0.2807	0.4334	7	*Martes* sp., *L. europaeus, E. concolor, N. procyonoides, A. alces, S. scrofa, C. capreolus*
1421	0.2791	0.4401	7	*T. europaea, S. vulgaris, L. europaeus, V. vulpes, A. alces, S. scrofa, C. capreolus*
1424	0.2287	0.4200	7	*C. fiber, L. europaeus, V. vulpes, E. concolor, A. alces, S. scrofa, C. capreolus*
1437	0.3024	0.4411	10	*A. flavicollis, M. putorius, C. elaphus, Martes* sp., *V. vulpes, E. concolor, N. procyonoides, A. alces, S. scrofa, C. capreolus*
1444	0.2281	0.4208	5	*Martes* sp., *E. concolor, A. alces, S. scrofa, C. capreolus*
1452	0.3259	0.4577	8	*Martes* sp., *L. europaeus, V. vulpes, E. concolor, N. procyonoides, A. alces, S. scrofa, C. capreolus*
1462	0.2839	0.4474	8	*T. europaea, S. vulgaris, C. elaphus, L. europaeus, V. vulpes, A. alces, S. scrofa, C. capreolus*
1475	0.2542	0.4254	7	*L. europaeus, V. vulpes, E. concolor, N. procyonoides, A. alces, S. scrofa, C. capreolus*
1478	0.2101	0.4234	6	*Martes* sp., *V. vulpes, N. procyonoides, A. alces, S. scrofa, C. capreolus*
1492	0.3191	0.4457	10	*M. putorius, C. elaphus, Martes* sp., *L. europaeus, V. vulpes, E. concolor, N. procyonoides, A. alces, S. scrofa, C. capreolus*
1496	0.2290	0.4279	4	*V. vulpes, A. alces, S. scrofa, C. capreolus*
1505	0.2615	0.4239	4	*V. vulpes, A. alces, S. scrofa, C. capreolus*
1507	0.2658	0.4268	6	*L. europaeus, E. concolor, N. procyonoides, A. alces, S. scrofa, C. capreolus*
1518	0.3513	0.4863	12	*L. lynx, T. europaea, C. fiber, M. meles, M. putorius, C. elaphus, Martes* sp., *L. europaeus, E. concolor, A. alces, S. scrofa, C. capreolus*
1519	0.3684	0.4461	8	*C. elaphus, L. europaeus, V. vulpes, E. concolor, N. procyonoides, A. alces, S. scrofa, C. capreolus*
1524	0.2710	0.4144	4	*L. europaeus, E. concolor, A. alces, C. capreolus*
1532	0.3446	0.4553	10	*A. flavicollis, T. europaea, C. elaphus, Martes* sp., *V. vulpes, E. concolor, N. procyonoides, A. alces, S. scrofa, C. capreolus*
1533	0.3089	0.4592	9	*C. fiber, M. meles, C. elaphus, L. europaeus, E. concolor, N. procyonoides, A. alces, S. scrofa, C. capreolus*
1534	0.2593	0.4502	5	*D. dama, V. vulpes, A. alces, S. scrofa, C. capreolus*
1557	0.2429	0.4274	6	*C. elaphus, Martes* sp., *N. procyonoides, A. alces, S. scrofa, C. capreolus*
1558	0.2608	0.4455	4	*M. putorius, A. alces, S. scrofa, C. capreolus*
1560	0.2551	0.4260	2	*N. procyonoides, C. capreolus*
1562	0.2419	0.4213	6	*T. europaea, V. vulpes, E. concolor, A. alces, S. scrofa, C. capreolus*
1568	0.2708	0.4229	5	*L. europaeus, E. concolor, A. alces, S. scrofa, C. capreolus*
1569	0.2161	0.4211	3	*E. concolor, A. alces, C. capreolus*
1578	0.2664	0.4338	7	*T. europaea, L. europaeus, V. vulpes, E. concolor, A. alces, S. scrofa, C. capreolus*

Table 2. *Cont.*

Unique Identification Number of Habitat (Figure 3)	Ranks		Species	
	SAW	TOPSIS	Count	Latin Names
1594	0.3417	0.4572	8	*Martes* sp., *L. europaeus*, *V. vulpes*, *E. concolor*, *N. procyonoides*, *A. alces*, *S. scrofa*, *C. capreolus*
1597	0.2899	0.4354	8	*M. meles*, *C. elaphus*, *L. europaeus*, *E. concolor*, *N. procyonoides*, *A. alces*, *S. scrofa*, *C. capreolus*
1601	0.2434	0.4327	6	*C. elaphus*, *L. europaeus*, *E. concolor*, *A. alces*, *S. scrofa*, *C. capreolus*
1608	0.2960	0.4638	10	*T. europaea*, *C. fiber*, *C. elaphus*, *Martes* sp., *L. europaeus*, *V. vulpes*, *E. concolor*, *A. alces*, *S. scrofa*, *C. capreolus*
1610	0.2751	0.4384	6	*C. elaphus*, *V. vulpes*, *E. concolor*, *A. alces*, *S. scrofa*, *C. capreolus*
1633	0.2551	0.4421	8	*M. meles*, *C. elaphus*, *L. europaeus*, *E. concolor*, *N. procyonoides*, *A. alces*, *S. scrofa*, *C. capreolus*
1638	0.2589	0.4318	8	*M. meles*, *C. elaphus*, *V. vulpes*, *E. concolor*, *N. procyonoides*, *A. alces*, *S. scrofa*, *C. capreolus*
1639	0.1898	0.4117	3	*M. meles*, *V. vulpes*, *C. capreolus*
1647	0.2944	0.4575	9	*M. meles*, *C. elaphus*, *L. europaeus*, *V. vulpes*, *E. concolor*, *N. procyonoides*, *A. alces*, *S. scrofa*, *C. capreolus*
1653	0.2237	0.4130	1	*C. capreolus*
1654	0.2223	0.4180	3	*E. concolor*, *S. scrofa*, *C. capreolus*
1671	0.2585	0.4368	9	*M. meles*, *Martes* sp., *L. europaeus*, *V. vulpes*, *E. concolor*, *N. procyonoides*, *A. alces*, *S. scrofa*, *C. capreolus*
1675	0.2241	0.4232	5	*C. elaphus*, *E. concolor*, *A. alces*, *S. scrofa*, *C. capreolus*
1679	0.2353	0.4155	2	*S. scrofa*, *C. capreolus*
1681	0.2866	0.4383	4	*V. vulpes*, *A. alces*, *S. scrofa*, *C. capreolus*
1700	0.3052	0.4384	9	*C. elaphus*, *Martes* sp., *L. europaeus*, *V. vulpes*, *E. concolor*, *N. procyonoides*, *A. alces*, *S. scrofa*, *C. capreolus*
1706	0.2300	0.4141	5	*L. europaeus*, *E. concolor*, *A. alces*, *S. scrofa*, *C. capreolus*
1715	0.2222	0.4263	6	*M. meles*, *C. elaphus*, *L. europaeus*, *A. alces*, *S. scrofa*, *C. capreolus*
1731	0.2580	0.4230	4	*C. elaphus*, *A. alces*, *S. scrofa*, *C. capreolus*
1738	0.3203	0.4617	7	*C. elaphus*, *L. europaeus*, *V. vulpes*, *N. procyonoides*, *A. alces*, *S. scrofa*, *C. capreolus*
1745	0.2521	0.4228	3	*N. procyonoides*, *S. scrofa*, *C. capreolus*
1748	0.2567	0.4245	6	*Martes* sp., *L. europaeus*, *V. vulpes*, *E. concolor*, *N. procyonoides*, *C. capreolus*
1749	0.2375	0.4277	6	*M. meles*, *E. concolor*, *N. procyonoides*, *A. alces*, *S. scrofa*, *C. capreolus*
1764	0.2654	0.4327	6	*M. meles*, *V. vulpes*, *N. procyonoides*, *A. alces*, *S. scrofa*, *C. capreolus*
1769	0.2105	0.4150	2	*S. scrofa*, *C. capreolus*
1777	0.2891	0.4370	5	*V. vulpes*, *N. procyonoides*, *A. alces*, *S. scrofa*, *C. capreolus*
1778	0.2272	0.4203	5	*C. elaphus*, *N. procyonoides*, *A. alces*, *S. scrofa*, *C. capreolus*
1782	0.2547	0.4340	7	*C. elaphus*, *L. europaeus*, *V. vulpes*, *E. concolor*, *N. procyonoides*, *S. scrofa*, *C. capreolus*

Table 2. *Cont.*

Unique Identification Number of Habitat (Figure 3)	Ranks		Species	
	SAW	TOPSIS	Count	Latin Names
1790	0.2266	0.4179	8	*M. meles, C. elaphus, L. europaeus, V. vulpes, E. concolor, N. procyonoides, S. scrofa, C. capreolus*
1794	0.3097	0.4325	10	*M. meles, C. elaphus, Martes* sp., *L. europaeus, V. vulpes, E. concolor, N. procyonoides, A. alces, S. scrofa, C. capreolus*
1798	0.3036	0.4373	8	*C. elaphus, Martes* sp., *V. vulpes, E. concolor, N. procyonoides, A. alces, S. scrofa, C. capreolus*
1812	0.2472	0.4218	7	*C. elaphus, L. europaeus, V. vulpes, E. concolor, A. alces, S. scrofa, C. capreolus*
1816	0.3270	0.4500	9	*T. europaea, M. meles, C. elaphus, V. vulpes, E. concolor, N. procyonoides, A. alces, S. scrofa, C. capreolus*
1826	0.2227	0.4127	4	*V. vulpes, A. alces, S. scrofa, C. capreolus*
1834	0.2018	0.4180	3	*L. europaeus, A. alces, C. capreolus*
1853	0.2393	0.4175	6	*M. meles, L. europaeus, N. procyonoides, A. alces, S. scrofa, C. capreolus*
1864	0.3212	0.4423	7	*M. meles, C. elaphus, V. vulpes, N. procyonoides, A. alces, S. scrofa, C. capreolus*
1866	0.3408	0.4310	8	*C. elaphus, L. europaeus, V. vulpes, E. concolor, N. procyonoides, A. alces, S. scrofa, C. capreolus*
1876	0.2466	0.4210	4	*C. elaphus, N. procyonoides, S. scrofa, C. capreolus*
1877	0.3395	0.4639	9	*M. meles, Martes* sp., *L. europaeus, V. vulpes, E. concolor, N. procyonoides, A. alces, S. scrofa, C. capreolus*
1879	0.3014	0.4502	6	*D. dama, L. europaeus, N. procyonoides, A. alces, S. scrofa, C. capreolus*
1882	0.2470	0.4225	5	*C. elaphus, Martes* sp., *N. procyonoides, S. scrofa, C. capreolus*
1913	0.2983	0.4438	7	*L. europaeus, V. vulpes, E. concolor, N. procyonoides, A. alces, S. scrofa, C. capreolus*
1916	0.2911	0.4465	8	*C. elaphus, L. europaeus, V. vulpes, E. concolor, N. procyonoides, A. alces, S. scrofa, C. capreolus*
1966	0.2455	0.4302	7	*C. elaphus, L. europaeus, V. vulpes, E. concolor, N. procyonoides, S. scrofa, C. capreolus*
1986	0.2229	0.4107	3	*M. meles, L. europaeus, C. capreolus*
2004	0.2580	0.4220	7	*N. vison, M. putorius, C. elaphus, E. concolor, N. procyonoides, S. scrofa, C. capreolus*
2014	0.2359	0.4075	2	*S. scrofa, C. capreolus*
2037	0.2402	0.4105	10	*N. vison, M. putorius, C. elaphus, Martes* sp., *V. vulpes, E. concolor, N. procyonoides, A. alces, S. scrofa, C. capreolus*
2038	0.2298	0.4229	7	*C. elaphus, L. europaeus, V. vulpes, N. procyonoides, A. alces, S. scrofa, C. capreolus*
2052	0.2875	0.4299	8	*C. elaphus, Martes* sp., *L. europaeus, E. concolor, N. procyonoides, A. alces, S. scrofa, C. capreolus*
2055	0.2583	0.4334	6	*D. dama, L. europaeus, N. procyonoides, A. alces, S. scrofa, C. capreolus*
2060	0.2295	0.4129	3	*V. vulpes, N. procyonoides, C. capreolus*

Table 2. *Cont.*

Unique Identification Number of Habitat (Figure 3)	Ranks			Species
	SAW	TOPSIS	Count	Latin Names
2105	0.2264	0.4107	2	*S. scrofa, C. capreolus*
2106	0.2040	0.4154	5	*C. elaphus, L. europaeus, N. procyonoides, A. alces, C. capreolus*
2224	0.2483	0.4160	4	*L. europaeus, E. concolor, A. alces, C. capreolus*
2229	0.2532	0.4296	10	*R. norvegicus, M. putorius, C. elaphus, Martes* sp., *V. vulpes, E. concolor, N. procyonoides, A. alces, S. scrofa, C. capreolus*
2233	0.2202	0.4210	3	*C. elaphus, S. scrofa, C. capreolus*
2237	0.2398	0.4091	2	*A. alces, C. capreolus*
2244	0.2249	0.4206	6	*L. europaeus, V. vulpes, E. concolor, N. procyonoides, S. scrofa, C. capreolus*
2246	0.2204	0.4152	6	*Martes* sp., *L. europaeus, V. vulpes, E. concolor, N. procyonoides, C. capreolus*
2247	0.1809	0.4188	3	*A. alces, S. scrofa, C. capreolus*
2248	0.5543	0.6014	12	*S. vulgaris, M. meles, M. putorius, C. elaphus, Martes* sp., *L. europaeus, V. vulpes, E. concolor, N. procyonoides, A. alces, S. scrofa, C. capreolus*
2249	0.4619	0.4924	10	*M. meles, C. elaphus, Martes* sp., *L. europaeus, V. vulpes, E. concolor, N. procyonoides, A. alces, S. scrofa, C. capreolus*
2250	0.3401	0.4618	12	*C. fiber, N. vison, S. vulgaris, M. martes, Martes* sp., *L. europaeus, V. vulpes, E. concolor, N. procyonoides, A. alces, S. scrofa, C. capreolus*
2251	0.4426	0.4503	11	*D. dama, M. meles, C. elaphus, Martes* sp., *L. europaeus, V. vulpes, E. concolor, N. procyonoides, A. alces, S. scrofa, C. capreolus*
2252	0.2188	0.4191	4	*L. europaeus, A. alces, S. scrofa, C. capreolus*
2253	0.2819	0.4262	8	*C. fiber, M. putorius, V. vulpes, E. concolor, N. procyonoides, A. alces, S. scrofa, C. capreolus*
2254	0.2808	0.4197	5	*C. elaphus, V. vulpes, E. concolor, A. alces, C. capreolus*
2255	0.4262	0.4315	11	*T. europaea, M. meles, C. elaphus, Martes* sp., *L. europaeus, V. vulpes, E. concolor, N. procyonoides, A. alces, S. scrofa, C. capreolus*

References

1. Balčiauskas, L.; Stratford, J.; Balčiauskienė, L.; Kučas, A. Importance of professional roadkill data in assessing diversity of mammal roadkills. *Transp. Res. Part D Transp. Environ.* **2020**, *87*, 102493. [CrossRef]
2. Cherry, C.C.; Dietz, S.; Sauber-Schatz, E.; Russell, S.; Proctor, J.; Buttke, D. Characteristics of animal-related motor vehicle crashes in select National Park Service units—United States, 1990–2013. *Traffic Inj. Prev.* **2019**, *20*, 58–63. [CrossRef] [PubMed]
3. Abra, F.D.; Granziera, B.M.; Huijser, M.P.; Ferraz, K.M.P.M.d.B.; Haddad, C.M.; Paolino, R.M. Pay or prevent? Human safety, costs to society and legal perspectives on animal-vehicle collisions in São Paulo state, Brazil. *PLoS ONE* **2019**, *14*, e0215152. [CrossRef]
4. Coffin, A.W. From roadkill to road ecology: A review of the ecological effects of roads. *J. Transp. Geogr.* **2007**, *15*, 396–406. [CrossRef]
5. González-Gallina, A.; Benítez-Badillo, G.; Rojas-Soto, O.R.; Hidalgo-Mihart, M.G. The small, the forgotten and the dead: Highway impact on vertebrates and its implications for mitigation strategies. *Biodivers. Conserv.* **2013**, *22*, 325–342. [CrossRef]
6. Gilbert, S.L.; Sivy, K.J.; Pozzanghera, C.B.; DuBour, A.; Overduijn, K.; Smith, M.M.; Zhou, J.; Little, J.M.; Prugh, L.R. Socioeconomic Benefits of Large Carnivore Recolonization Through Reduced Wildlife-Vehicle Collisions. *Conserv. Lett.* **2017**, *10*, 431–439. [CrossRef]
7. Kučas, A.; Balčiauskas, L. Temporal patterns of ungulate-vehicle collisions in Lithuania. *J. Environ. Manag.* **2020**, *273*, 111172. [CrossRef]
8. Kučas, A.; Balčiauskas, L. Impact of Road Fencing on Ungulate-Vehicle Collisions and Hotspot Patterns. *Land* **2021**, *10*, 338. [CrossRef]

9. Ford, A.T.; Clevenger, A.P.; Huijser, M.P.; Dibb, A. Planning and prioritization strategies for phased highway mitigation using wildlife-vehicle collision data. *Wildl. Biol.* **2011**, *17*, 253–265. [CrossRef]
10. Huijser, M.P.; Fairbank, E.R.; Camel-Means, W.; Graham, J.; Watson, V.; Basting, P.; Becker, D. Effectiveness of short sections of wildlife fencing and crossing structures along highways in reducing wildlife–vehicle collisions and providing safe crossing opportunities for large mammals. *Biol. Conserv.* **2016**, *197*, 61–68. [CrossRef]
11. Davenport, J.; Davenport, J.L. (Eds.) *The Ecology of Transportation: Managing Mobility for the Environment*; Environmental Pollution; Springer: Dordrecht, The Netherlands, 2006; Volume 10, ISBN 978-1-4020-4503-5.
12. Ford, A.T.; Clevenger, A.P.; Bennett, A. Comparison of Methods of Monitoring Wildlife Crossing-Structures on Highways. *J. Wildl. Manag.* **2009**, *73*, 1213–1222. [CrossRef]
13. Clevenger, A.P.; Waltho, N. Factors Influencing the Effectiveness of Wildlife Underpasses in Banff National Park, Alberta, Canada. *Conserv. Biol.* **2000**, *14*, 47–56. [CrossRef]
14. Bager, A.; Fontoura, V. Evaluation of the effectiveness of a wildlife roadkill mitigation system in wetland habitat. *Ecol. Eng.* **2013**, *53*, 31–38. [CrossRef]
15. Ascensão, F.; Clevenger, A.; Santos-Reis, M.; Urbano, P.; Jackson, N. Wildlife–vehicle collision mitigation: Is partial fencing the answer? An agent-based model approach. *Ecol. Model.* **2013**, *257*, 36–43. [CrossRef]
16. Little, S.J.; Harcourt, R.G.; Clevenger, A.P. Do wildlife passages act as prey-traps? *Biol. Conserv.* **2002**, *107*, 135–145. [CrossRef]
17. Plante, J.; Jaeger, J.A.G.; Desrochers, A. How do landscape context and fences influence roadkill locations of small and medium-sized mammals? *J. Environ. Manag.* **2019**, *235*, 511–520. [CrossRef] [PubMed]
18. Rytwinski, T.; Soanes, K.; Jaeger, J.A.G.; Fahrig, L.; Findlay, C.S.; Houlahan, J.; van der Ree, R.; van der Grift, E.A. How Effective Is Road Mitigation at Reducing Road-Kill? A Meta-Analysis. *PLoS ONE* **2016**, *11*, e0166941. [CrossRef]
19. Spanowicz, A.G.; Teixeira, F.Z.; Jaeger, J.A.G. An adaptive plan for prioritizing road sections for fencing to reduce animal mortality. *Conserv. Biol.* **2020**, *34*, 1210–1220. [CrossRef] [PubMed]
20. Rytwinski, T.; van der Ree, R.; Cunnington, G.M.; Fahrig, L.; Findlay, C.S.; Houlahan, J.; Jaeger, J.A.G.; Soanes, K.; van der Grift, E.A. Experimental study designs to improve the evaluation of road mitigation measures for wildlife. *J. Environ. Manag.* **2015**, *154*, 48–64. [CrossRef]
21. Borda-de-Água, L.; Ascensão, F.; Sapage, M.; Barrientos, R.; Pereira, H.M. On the identification of mortality hotspots in linear infrastructures. *Basic Appl. Ecol.* **2019**, *34*, 25–35. [CrossRef]
22. Montella, A. A comparative analysis of hotspot identification methods. *Accid. Anal. Prev.* **2010**, *42*, 571–581. [CrossRef] [PubMed]
23. Gomes, L.; Grilo, C.; Silva, C.; Mira, A. Identification methods and deterministic factors of owl roadkill hotspot locations in Mediterranean landscapes. *Ecol. Res.* **2009**, *24*, 355–370. [CrossRef]
24. Hirzel, A.H.; Hausser, J.; Chessel, D.; Perrin, N. Ecological-Niche Factor Analysis: How to Compute Habitat-Suitability Maps without Absence Data? *Ecology* **2002**, *83*, 2027–2036. [CrossRef]
25. Gitman, I.; Levine, M.D. An Algorithm for Detecting Unimodal Fuzzy Sets and Its Application as a Clustering Technique. *IEEE Trans. Comput.* **1970**, *100*, 583–593. [CrossRef]
26. Bíl, M.; Andrášik, R.; Janoška, Z. Identification of hazardous road locations of traffic accidents by means of kernel density estimation and cluster significance evaluation. *Accid. Anal. Prev.* **2013**, *55*, 265–273. [CrossRef] [PubMed]
27. Bíl, M.; Andrášik, R.; Svoboda, T.; Sedoník, J. The KDE+ software: A tool for effective identification and ranking of animal-vehicle collision hotspots along networks. *Landsc. Ecol.* **2016**, *31*, 231–237. [CrossRef]
28. Kolowski, J.M.; Nielsen, C.K. Using Penrose distance to identify potential risk of wildlife–vehicle collisions. *Biol. Conserv.* **2008**, *141*, 1119–1128. [CrossRef]
29. Malo, J.E.; Suarez, F.; Diez, A. Can we mitigate animal-vehicle accidents using predictive models? *J. Appl. Ecol.* **2004**, *41*, 701–710. [CrossRef]
30. Grilo, C.; Ascensão, F.; Santos-Reis, M.; Bissonette, J.A. Do well-connected landscapes promote road-related mortality? *Eur. J. Wildl. Res.* **2011**, *57*, 707–716. [CrossRef]
31. Planillo, A.; Kramer-Schadt, S.; Malo, J.E. Transport Infrastructure Shapes Foraging Habitat in a Raptor Community. *PLoS ONE* **2015**, *10*, e0118604. [CrossRef]
32. Santos, S.M.; Marques, J.T.; Lourenço, A.; Medinas, D.; Barbosa, A.M.; Beja, P.; Mira, A. Sampling effects on the identification of roadkill hotspots: Implications for survey design. *J. Environ. Manag.* **2015**, *162*, 87–95. [CrossRef] [PubMed]
33. Dussault, C.; Courtois, R.; Ouellet, J.-P. A habitat suitability index model to assess moose habitat selection at multiple spatial scales. *Can. J. For. Res.* **2006**, *36*, 1097–1107. [CrossRef]
34. Andrén, H.; Andren, H. Effects of Habitat Fragmentation on Birds and Mammals in Landscapes with Different Proportions of Suitable Habitat: A Review. *Oikos* **1994**, *71*, 355–366. [CrossRef]
35. Balčiauskas, L.; Wierzchowski, J.; Kučas, A.; Balčiauskienė, L. Habitat Suitability Based Models for Ungulate Roadkill Prognosis. *Animals* **2020**, *10*, 1345. [CrossRef] [PubMed]
36. Bíl, M.; Andrášik, R.; Sedoník, J. A detailed spatiotemporal analysis of traffic crash hotspots. *Appl. Geogr.* **2019**, *107*, 82–90. [CrossRef]
37. Kučas, A. Graph-based multi-attribute decision making: Impact of road fencing on ecological network. *Balt. J. Road Bridg. Eng.* **2015**, *10*, 105–111. [CrossRef]

38. Putzu, N.; Bonetto, D.; Civallero, V.; Fenoglio, S.; Meneguz, P.G.; Preacco, N.; Tizzani, P. Temporal patterns of ungulate-vehicle collisions in a subalpine Italian region. *Ital. J. Zool.* **2014**, *81*, 463–470. [CrossRef]
39. Zhang, W.; Shu, G.; Li, Y.; Xiong, S.; Liang, C.; Li, C. Daytime driving decreases amphibian roadkill. *PeerJ* **2018**, *6*, e5385. [CrossRef]
40. Hastings, H.; Barr, J.; Bateman, P.W. Spatial and temporal patterns of reptile roadkill in the north-west Australian tropics. *Pac. Conserv. Biol.* **2019**, *25*, 370–376. [CrossRef]
41. Kammerle, J.-L.; Brieger, F.; Kroschel, M.; Hagen, R.; Storch, I.; Suchant, R. Temporal patterns in road crossing behaviour in roe deer (Capreolus capreolus) at sites with wildlife warning reflectors. *PLoS ONE* **2017**, *12*, e0184761. [CrossRef]
42. Sepp, K.; Kaasik, A. *Development of National Ecological Networks in the Baltic Countries in the Framework of the Pan-European Ecological Network*; IUCN Office for Central Europe: Warsaw, Poland, 2002; ISBN 2-8317-0679-3.
43. McRae, B.H.; Hall, S.A.; Beier, P.; Theobald, D.M. Where to Restore Ecological Connectivity? Detecting Barriers and Quantifying Restoration Benefits. *PLoS ONE* **2012**, *7*, e52604. [CrossRef] [PubMed]
44. McRae, B.H. Isolation by resistance. *Evolution* **2006**, *60*, 1551–1561. [CrossRef] [PubMed]
45. McRae, B.H.; Dickson, B.G.; Keitt, T.H.; Shah, V.B. Using circuit theory to model connectivity in ecology, evolution, and conservation. *Ecology* **2008**, *89*, 2712–2724. [CrossRef]
46. Wierzchowski, J.; Kučas, A.; Balčiauskas, L. Application of Least-Cost Movement Modeling in Planning Wildlife Mitigation Measures along Transport Corridors: Case Study of Forests and Moose in Lithuania. *Forests* **2019**, *10*, 831. [CrossRef]
47. Clevenger, A.P.; Barrueto, M.; Gunson, K.E.; Caryl, F.M.; Ford, A.T. Context-dependent effects on spatial variation in deer-vehicle collisions. *Ecosphere* **2015**, *6*, 7. [CrossRef]
48. Jankowski, P. Integrating geographical information systems and multiple criteria decision-making methods. *Int. J. Geogr. Inf. Syst.* **1995**, *9*, 251–273. [CrossRef]
49. Hwang, C.-L.; Yoon, K. *Multiple Attribute Decision Making*; Lecture Notes in Economics and Mathematical Systems; Springer: Berlin/Heidelberg, Germany, 1981; Volume 186, ISBN 978-3-540-10558-9.
50. Mardani, A.; Jusoh, A.; Zavadskas, E.K. Fuzzy multiple criteria decision-making techniques and applications—Two decades review from 1994 to 2014. *Expert Syst. Appl.* **2015**, *42*, 4126–4148. [CrossRef]
51. Malczewski, J. GIS-based multicriteria decision analysis: A survey of the literature. *Int. J. Geogr. Inf. Sci.* **2006**, *20*, 703–726. [CrossRef]
52. Wira Trise Putra, D.; Agustian Punggara, A. Comparison Analysis of Simple Additive Weighting (SAW) and Weigthed Product (WP) In Decision Support Systems. *MATEC Web Conf.* **2018**, *215*, 01003. [CrossRef]
53. Keken, Z.; Sedoník, J.; Kušta, T.; Andrášik, R.; Bíl, M. Roadside vegetation influences clustering of ungulate vehicle collisions. *Transp. Res. Part D Transp. Environ.* **2019**, *73*, 381–390. [CrossRef]
54. Ascensão, F.; Clevenger, A.P.; Grilo, C.; Filipe, J.; Santos-Reis, M. Highway verges as habitat providers for small mammals in agrosilvopastoral environments. *Biodivers. Conserv.* **2012**, *21*, 3681–3697. [CrossRef]
55. Silva, C.; Simões, M.P.; Mira, A.; Santos, S.M. Factors influencing predator roadkills: The availability of prey in road verges. *J. Environ. Manag.* **2019**, *247*, 644–650. [CrossRef] [PubMed]
56. Bíl, M.; Andrášik, R.; Duľa, M.; Sedoník, J. On reliable identification of factors influencing wildlife-vehicle collisions along roads. *J. Environ. Manag.* **2019**, *237*, 297–304. [CrossRef] [PubMed]
57. Lithuanian Road Administration under the Ministry of Transport and Communications. Web Map of Animal-Vehicle Collision Distributions in Lithuania. Available online: http://gissrv.eismoinfo.lt/server/rest/services/LAKIS/su_gyvunais/MapServer?f=jsapi (accessed on 16 March 2020).
58. European Environmental Agency. Lithuania Land Cover Country Fact Sheet 2012. Available online: https://www.eea.europa.eu/ds_resolveuid/1ca731f33d0c48a0a4ed60abe67fa6bf (accessed on 16 March 2020).
59. Jakimavičius, M. Analysis and Assessment of Lithuanian Road Accidents by AHP Method. *Balt. J. Road Bridg. Eng.* **2018**, *13*, 238–260. [CrossRef]
60. Lithuanian Road Administration under the Ministry of Transport and Communications. Web Map of Annual Average Daily Traffic Data in Lithuania. Available online: http://gissrv.eismoinfo.lt/server/rest/services/LAKD/VMPEI/MapServer?f=jsapi (accessed on 16 March 2020).
61. Clevenger, A.P.; Wierzchowski, J.; Chruszcz, B.; Gunson, K. GIS-Generated, Expert-Based Models for Identifying Wildlife Habitat Linkages and Planning Mitigation Passages. *Conserv. Biol.* **2002**, *16*, 503–514. [CrossRef]
62. Gunson, K.E.; Mountrakis, G.; Quackenbush, L.J. Spatial wildlife-vehicle collision models: A review of current work and its application to transportation mitigation projects. *J. Environ. Manag.* **2011**, *92*, 1074–1082. [CrossRef]
63. Santos, R.A.L.; Mota-Ferreira, M.; Aguiar, L.M.S.; Ascensão, F. Predicting wildlife road-crossing probability from roadkill data using occupancy-detection models. *Sci. Total Environ.* **2018**, *642*, 629–637. [CrossRef]
64. Seiler, A. Predicting locations of moose-vehicle collisions in Sweden. *J. Appl. Ecol.* **2005**, *42*, 371–382. [CrossRef]
65. Okabe, A.; Okunuki, K.; Shiode, S. SANET: A Toolbox for Spatial Analysis on a Network. *Geogr. Anal.* **2006**, *38*, 57–66. [CrossRef]
66. Okabe, A.; Satoh, T.; Sugihara, K. A kernel density estimation method for networks, its computational method and a GIS-based tool. *Int. J. Geogr. Inf. Sci.* **2009**, *23*, 7–32. [CrossRef]
67. Clementini, E.; Di Felice, P.; van Oosterom, P. A small set of formal topological relationships suitable for end-user interaction. In *International Symposium on Spatial Databases*; Springer: Berlin/Heidelberg, Germany, 1993; pp. 277–295.

68. Kučas, A. Location prioritization by means of multicriteria spatial decision-support systems: A case study of forest fragmentation-based ranking of forest administrative areas. *J. Environ. Eng. Landsc. Manag.* **2010**, *18*, 312–320. [CrossRef]
69. Kušta, T.; Keken, Z.; Ježek, M.; Holá, M.; Šmíd, P. The effect of traffic intensity and animal activity on probability of ungulate-vehicle collisions in the Czech Republic. *Saf. Sci.* **2017**, *91*, 105–113. [CrossRef]
70. Feranec, J.; Jaffrain, G.; Soukup, T.; Hazeu, G. Determining changes and flows in European landscapes 1990–2000 using CORINE land cover data. *Appl. Geogr.* **2010**, *30*, 19–35. [CrossRef]
71. European Environmental Agency. CORINE Land Cover—Copernicus Land Monitoring Service. Available online: https://land.copernicus.eu/pan-european/corine-land-cover (accessed on 16 March 2020).
72. European Environmental Agency. CORINE Land Cover Nomenclature Guidelines. Available online: https://land.copernicus.eu/user-corner/technical-library/corine-land-cover-nomenclature-guidelines/html (accessed on 12 November 2020).
73. Akoglu, H. User's guide to correlation coefficients. *Turk. J. Emerg. Med.* **2018**, *18*, 91–93. [CrossRef] [PubMed]
74. Maleika, W. Inverse distance weighting method optimization in the process of digital terrain model creation based on data collected from a multibeam echosounder. *Appl. Geomat.* **2020**, *12*, 397–407. [CrossRef]
75. Chen, F.-W.; Liu, C.-W. Estimation of the spatial rainfall distribution using inverse distance weighting (IDW) in the middle of Taiwan. *Paddy Water Environ.* **2012**, *10*, 209–222. [CrossRef]
76. Luque, S.; Saura, S.; Fortin, M.-J. Landscape connectivity analysis for conservation: Insights from combining new methods with ecological and genetic data. *Landsc. Ecol.* **2012**, *27*, 153–157. [CrossRef]
77. Kučas, A. Web Map of Ungulate-Vehicle Collision Hotspots in Lithuania 2002–2017. Available online: https://www.arcgis.com/apps/MapJournal/index.html?appid=d81195212a4b4bcc9c5aab34a0037609 (accessed on 16 March 2020).
78. Kučas, A.; Balčiauskas, L. Mammal Roadkill Clusters, Corridors, Pathways and Habitats. [WWW Document]. Mendeley Data, V1. Available online: https://doi.org/10.17632/4c58n345h5.1 (accessed on 1 May 2021).
79. Lithuanian Road Administration under the Ministry of Transport and Communications. Annual Average Daily Traffic Volumes in Lithuania. Available online: https://lakd.lrv.lt/en/sector-activities/traffic-volumes (accessed on 16 March 2020).
80. Hortal, J.; Triantis, K.A.; Meiri, S.; Thébault, E.; Sfenthourakis, S. Island Species Richness Increases with Habitat Diversity. *Am. Nat.* **2009**, *174*, E205–E217. [CrossRef]
81. Hernández-Quiroz, N.S.; Badano, E.I.; Barragán-Torres, F.; Flores, J.; Pinedo-Álvarez, C. Habitat suitability models to make conservation decisions based on areas of high species richness and endemism. *Biodivers. Conserv.* **2018**, *27*, 3185–3200. [CrossRef]
82. Ramp, D.; Wilson, V.K.; Croft, D.B. Assessing the impacts of roads in peri-urban reserves: Road-based fatalities and road usage by wildlife in the Royal National Park, New South Wales, Australia. *Biol. Conserv.* **2006**, *129*, 348–359. [CrossRef]

land

MDPI

Article

Identifying and Mapping Groups of Protected Area Visitors by Environmental Awareness

Arjan S. Gosal [1,*], Janine A. McMahon [1], Katharine M. Bowgen [2], Catherine H. Hoppe [3] and Guy Ziv [1]

1 School of Geography, Faculty of Environment, University of Leeds, Leeds LS2 9JT, UK; janine@ecologyresources.co.uk (J.A.M.); G.Ziv@leeds.ac.uk (G.Z.)
2 British Trust for Ornithology, The Nunnery, Thetford IP24 2PU, UK; katharine.bowgen@bto.org
3 Independent Researcher, 31515 Wunstorf, Germany; hoppec82@gmail.com
* Correspondence: a.gosal@gmx.co.uk

Abstract: Protected areas worldwide receive billions of visitors annually. The positive impact of nature on health and wellbeing, in addition to providing opportunities for cultural activities such as recreation and aesthetic appreciation, is well documented. Management to reduce negative impacts to biodiversity and conservation aims whilst providing amenities and access to visitors is important. Understanding environmental awareness of visitors and their on-site spatial patterns can assist in making effective management decisions within often constrained resources. However, there is a lack of strategies for site-specific identification and predictive mapping of visitors by environmental awareness. Here, we demonstrate a method to map on-site visitation by latent groups of visitors based on their environmental awareness of on-site issues. On-site surveys and participatory mapping were used to collect data on environmental awareness on bird nesting and spatial visitation patterns in an upland moor in northern England. Latent class analysis (LCA), a structural equation model, was used to discover underlying groups of environmental awareness, with random forest (RF) modelling, a machine learning technique, using a range of on-site predictors (bioclimatic, land cover, elevation, viewshed, and proximity to paths and freshwater) to predict and map visitation across the site by each group. Visitors were segmented into 'aware' and 'ambiguous' groups and their potential spatial visitation patterns mapped. Our results demonstrate the ability to uncover groups of users by environmental awareness and map their potential visitation across a site using a variety of on-site predictors. Spatial understanding of the movement patterns of differently environmentally aware groups of visitors can assist in efficient targeting of conservation education endeavours (i.e., signage, positioning of staff, monitoring programmes, etc.), therefore maximising their efficacy. Furthermore, we anticipate this method will be of importance to environmental managers and educators when deploying limited resources.

Keywords: environmental awareness; machine learning; random forest model; structural equation modelling; latent class analysis; visitor mapping

Citation: Gosal, A.S.; McMahon, J.A.; Bowgen, K.M.; Hoppe, C.H.; Ziv, G. Identifying and Mapping Groups of Protected Area Visitors by Environmental Awareness. *Land* **2021**, *10*, 560. https://doi.org/10.3390/land10060560

Academic Editors: Juan F. Beltrán, Pedro Abellán and John Litvaitis

Received: 24 April 2021
Accepted: 22 May 2021
Published: 27 May 2021

1. Introduction

Much of the environmental deterioration evidenced around the world, including the alarming rate of biodiversity loss, can be attributed to humans [1,2]. Indeed, Schultz [2] argues that only through changing human behaviour can conservation have a chance of success and that identifying those behaviours that need to change, rather than broad education and awareness-raising campaigns, should be the priority. With visitation to protected areas expected to increase [3], already stretched resources for improving visitors' awareness of key site objectives, such as protecting bird species, will only need to go further.

The impact of human disturbance on birds can manifest in multiple ways, through a combination of changing distributions, behaviour, demography or population size [4]. For example, during sensitive periods of the year, such as territory establishment, even low levels of human disturbance can alter bird behaviour [5]. Bird disturbance caused by humans has been studied in various habitats, including forests [6,7], coastal areas [8], and upland habitats such as moorland [9,10]. The risk of negative impacts to birds is considered serious enough to be written into international law prohibiting deliberate disturbance (EU Birds Directive 2009).

Separating nature area visitors into groups has been explored for several scenarios in literature, including segmentation by motivation for visiting a site, to best respond to visitor needs [11,12]. Several studies have used a market segmentation approach to group visitors [13,14], with Halpenny [15] investigating place attachments as a predictor of pro-environmental behaviours. Conservation social sciences can be very useful in guiding conservation actions and outcomes that are effective and robust [16]. For example, Booth et al. [17] found substantial variation in protected status awareness in site visitors, with only conservation organisation membership being of predictive significance for being more informed.

Context-specific maps can improve and/or enable replication of strategic environmental assessment mapping to aid conservation decision-making [18], and knowledge of visitor preferences offers insights to inform destination development and promotion of ecotourism within a site [19]. In particular, mapping to support environmental decision-making provides an approach to understand where policy interventions could be most effective [20] and is a valid mechanism to translate scientific findings into tangible products that can be used by local on-site practitioners. Conservation education is the mechanism through which awareness and concern for the environment are raised and can take several forms, from free-choice learning to structured initiatives [16,21,22]. Understanding environmental awareness on-site can help in personal delivery of on-site minimum impact education strategies, as messages to park visitors have been found to minimise off-trail behaviour, whilst physical signs were found to be ineffective in preventing this behaviour [23].

Identifying visitation patterns of nature site users lacking in awareness of on-site issues is important. As lower awareness may result in unwanted behaviour, such as disturbance to birds or other wildlife, predicting areas these users may visit on-site is vital for both management and mitigation strategies (e.g., limiting access to sensitive areas or targeted education campaigns). In this study, we aim to provide a replicable method that can be used to both define and spatially map different types of protected area visitors, based on their environmental awareness of on-site issues. We use latent class analysis (LCA), a type of structural equation model, to group visitors based on environmental awareness and random forest (RF) modelling, a machine learning method, with participatory mapping data to produce predictive maps of visitation. This is demonstrated for a protected English upland moor that has importance for protected bird populations.

2. Materials and Methods

2.1. Study Site and Context

Ilkley Moor, an upland moor in northern England, UK (53°54′11.1″ N, 1°49′36.9″ W), Figure 1, is dominated by heather moorland over acidic soils from the Millstone Grit underlying it [22]. It is 402 m above sea level, which exposes it to harsh winds in the winter. Its nature conservation importance is recognised through international and national designations, including Site of Special Scientific Interest, Special Area of Protection, and Special Area of Conservation. In addition to being of high ecological importance, the site provides cultural ecosystem services such as recreation and sense of place, as well as providing regulating and supporting services including carbon storage in peat, water and air purification, and floodwater retention.

Figure 1. The study site in multiple dimensions: (**a**) aerial view using Sentinel 2 natural colour (bands 432) mosaic imagery [23], (**b**) location of site in relation to the British Isles, (**c**) example photographs taken on-site, with surrounding landscape views, and (**d**) Red List Species of Conservation Concern [24,25] recorded on the site between 2015 and 2019 during the British Trust for Ornithology (BTO) British Breeding Bird Surveys [26]. The BTO/JNCC/RSPB Breeding Bird Survey is a partnership jointly funded by the BTO, RSPB, and JNCC, with fieldwork conducted by volunteers. All bird images are published on www.wikimedia.org and licensed under Creative Commons; see Table S1 for full attributions. © contains Copernicus data (2019) and JNCC/NE/NRW/SNH/NIEA data, © copyright and database right 2019.

In this study, we specifically consider environmental and ecological awareness regarding nesting birds on Ilkley Moor, which experiences large numbers of visitors. During the bird breeding season, the implications of disturbance for outcomes of nesting attempts range across multiple impacts, including nest failure [24], impaired nestling growth [7], reduction in the areas suitable for breeding [25], and immunosuppression of fledglings [26], which all put pressure on individual birds and future recruitment into the local population [27]. The strength of these impacts can vary widely with species and type of disturbance; they may only have mild effects on fitness or cause total breeding season failure [28,29]. Hence, it is important for disturbance at this time to be kept to a minimum.

The BTO's Breeding Bird Survey [30] has recorded 70 bird species on Ilkley Moor in the last 10 years, declining to 63 in the last 5 years. The seven lost species include the chiffchaff (*Phylloscopus collybita*), coal tit (*Periparus ater*), collared dove (*Streptopelia decaocto*), and hobby (*Falco subbuteo*); two Red listed species, the grasshopper warbler (*Locustella naevia*) and tree sparrow (*Passer montanus*); and one Amber listed species, the house martin (*Delichon urbicum*) [31]. In the last five years, 16 Amber listed species have been recorded (see Table S2 for full lists of bird species that can be found on-site).

2.2. Independent Variables; Survey and Participatory Mapping

A total of 124 surveys were conducted in the summer of 2019 with visitors to the study site. The survey contained questions covering demographics and questions related to participants' experiences on-site. Here we focus on five environmental awareness questions (Table 1) structured in the five-point Likert format [32,33] and a participatory mapping exercise, where the participants were asked to mark on a map the areas of the site they had visited on that specific visit. Through permission of the site owner (Bradford Council), 11 access points were used for surveying, where 10 out of the 11 had car parks. Stratified sampling was used to identify a minimum of one access point per grid square by draping a 4 × 4 grid over an Ordnance Survey map of the site. Surveys took place between 09:00 and 17:00 BST, with people interviewed by a surveyor transcribing the responses to the questionnaire.

Table 1. Environmental awareness Likert questions used in the surveys and the abbreviated form referred to in this study.

Environmental Awareness Question	Abbreviated Form
Are you aware that Ilkley Moor is a South Pennine Moors Site of Special Scientific Interest (SSSI), a Special Protection Area (SPA), and a Special Area of Conservation (SAC)?	Site designations
Are you aware that the nesting season for birds is between March and July every year (5 months)?	Nesting season
Are you aware of the ecological importance of the nesting season in March–July every year (5 months) for birds such as the golden plover, red grouse, lapwing, and short eared owls?	Eco. bird nesting
Are you aware of the issues that off-lead dogs can cause to the nesting birds?	Issues of off-leading
Were you previously aware of the request to keep dogs on a lead during the period of March–July (5 months)?	Requested on-leading

The responses to the environmental questions (Table 1) were dichotomously recoded (with Likert ratings 1, 2, and 3 being "unaware" and 4 and 5 being "aware"). As a rating of 3 could be perceived as a neutral response, it was assumed this indicated a lack of awareness of the issue, rather than awareness. During the participatory mapping exercise, participants marked points, lines, or polygons with a pen on a map to indicate the areas they visited. Maps were digitised in ESRI ArcMap v10.6. Points and lines were converted to polygons using a 250 m buffer before all polygons for all participants were converted to a binary raster (25 m per pixel) to indicate visitation per participant. These were summed later by the allocated LCA group to produce visitation maps.

2.3. Secondary Variables; Independent Secondary Data

To upscale from the participatory survey data to a map covering the entire region, we used an RF algorithm with a set of climatic and physical factors as predictors. Bioclimatic variables were sourced from WorldClim version 2.1 [34], in the form of 19 variables derived from monthly rainfall and temperature (1970–2000) at the finest resolution available of 30 s as GeoTiff files. Variables were extracted as values from the centroid point per 25 m pixel across the site.

Land cover (such as forest) was used as the basis for the 25 m pixel size for this study and sourced from CEH LCM2019 [35] based on 21 land cover types at a 25 m by 25 m resolution raster map. Dummy binary variables were created for each pixel for the classes present over the site (deciduous woodland, coniferous woodland, improved grassland, acid grassland, heather, heather grassland, bog, freshwater, and suburban). Elevation was calculated from Ordnance Survey 50 m digital terrain model [36]. Proximity to paths and freshwater was calculated per pixel in meters in QGIS v3.16.0 [37] using OpenStreetMap data retrieved from Geofabrik.de [38].

Viewshed analysis (delineation of the area visible from any given location on a map) was calculated in QGIS v3.16.0 [37] using the "Advanced Viewshed Analysis" v1.4 plugin [39]. The site visibility index was calculated as a cumulative binary viewshed utilising only visibility points from within the site and the Ordnance Survey 50 m digital terrain model [36]. Visibility points were created at the centroid location of each 25 m pixel, with the default values of observer height set at 1.6 m and a 5 km radius used. This allowed consideration of whether on-site visibility of an area contributed to visitation by differently environmentally aware users.

2.4. Statistic Analysis

All statistical analysis was conducted in R (R Core Team, 2020). See Figure 2 for a flowchart of the study methodology.

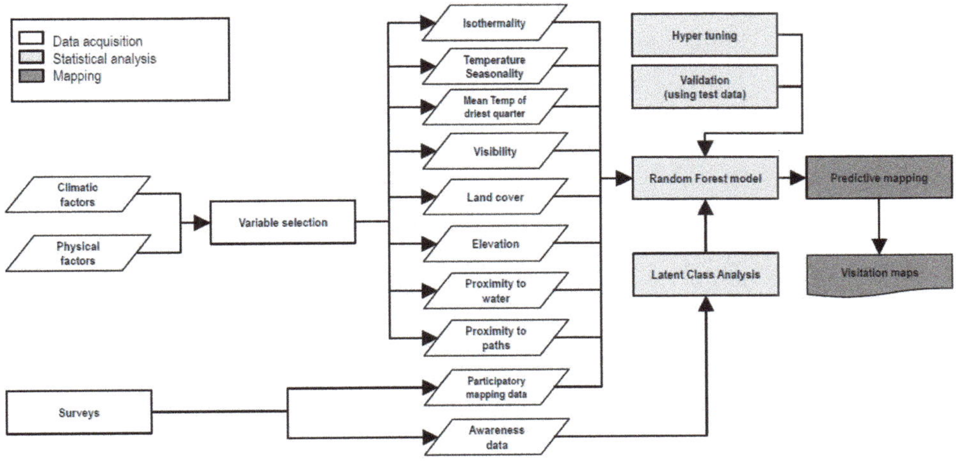

Figure 2. Flowchart of study methodology, showing the progression from data acquisition and selection through to the latent class analysis (LCA) and random forest (RF) model and finishing in mapping. White boxes refer to data acquisition and selection, light grey indicates statistical analysis, and dark grey highlights the mapping stages.

2.4.1. Variable Selection

While RF models suffer less from collinearity than other traditional statistical methods, multiple predictors that are weakly correlated to the response, and strongly with each other, can cause unstable results, with it being unlikely that the averaging that occurs across all trees in an RF being able to overcome this [40,41]. We dealt with collinearity through several steps. Firstly, the *Caret* R package [42] was used to identify variables that were correlated at 0.75 or higher with each other, and the variable with the largest mean absolute correlation amongst all variables was removed. Secondly, variance inflation factors (VIFs) were calculated, and the variable with the highest VIF was removed until the cohort all had values under 10, resulting in fifteen predictors. Thirdly, the *Boruta* R package [43] was used to check for "variable importance" (the utilisation by the RF model to use a given variable to make accurate predictions; higher usage to make predictions implies a greater importance for the model). *Boruta* does this through iterative removal of variables found to be statistically less relevant than random probes using RF [43].

2.4.2. Latent Class Analysis

Latent class analysis (LCA) has been used in the study of environmental issues, including environmentally sustainable food choices [44], environmental concern of the UK population [45], and recreational demand based on attitudes towards water resources [46]. Here we use LCA to segment visitors into different environmental awareness groups. The *poLCA* R package [47] was used for the LCA analysis. Multiple classes were explored, with the lowest BIC dictating the class size of two (see Figure S1). LCA posterior probabilities were used to segment the surveyed individuals into the two classes ('aware' and 'ambiguous'). Classes were named on the resultant probability of binary response to any given environmental question.

2.4.3. Random Forest Modelling

Since its first introduction by Breiman in 2001 random forest (RF) modelling, a machine learning technique, has been used extensively in the literature, e.g., in using semantic information to classify urban buildings [48], forecasting power consumption using hybrid models including RF [49], and modelling travel mode choice behaviour [50]. Braun, Cottrell, and Dierkes [51] used RF to investigate the effect of outdoor education programmes in school children across multiple countries. The predictive accuracy of RF lends the method to producing potentiality or susceptibility maps. Naghibi, Pourghasemi, and Dixon [52] used RF with boosted, classification, and regression trees to produce groundwater potential maps. Elsewhere, RF has been used for landslide susceptibility mapping [53–55], soil parent material and carbon mapping [56,57], and vegetation mapping and land cover classification [58–60]. In this study, we use RF to produce potential maps of visitation for different environmental awareness groups.

The ranger R package [61] was used to run separate RF models for 'all visitors' and both LCA classes, using the sum of the binary visitation maps per pixel as the response variable. Only pixels visited by at least one survey respondent were included. Each model was hyper-tuned to find the model with the best predictive accuracy running a grid search for the number of variables sampled from at each split (1–15), node size (3, 5, 7, and 9), and sample size (0.550, 0.632, 0.700, and 0.800), resulting in 240 variants for each model. All models were run with 500 trees and showed stabilisation of out-of-bag (OOB) error within this number (see Table S3). The models with the lowest OOB error were chosen.

2.4.4. Validation and Mapping

RF models are robust in providing out-of-bag estimation of error, though as an additional layer of validation, data for the RF models were randomly split into testing (20%) and training datasets (80%). Predicted values from the RF models were mapped at 25 m pixel resolution using ESRI ArcGIS Pro 2.7.0. Due to heavy-tail distribution of the data, an m-out-of-n bootstrap was used to discern statistical difference between the maps using the R package distillery [62] with Pearson correlation tests for a random set of 1000 subsampled points, bootstrapped over 1000 iterations with resampling.

3. Results

The LCA analysis resulted in two classes, as this presented the lowest AIC/BIC (Figure S1); these were named 'aware' and 'ambiguous' following the probability of the first group being likely to be aware across all questions, with the latter having more ambiguity in probability of awareness (Figure 3). Posterior probabilities showed a split of 63.5% "aware" and 35.5% "ambiguous" amongst the surveyed visitors.

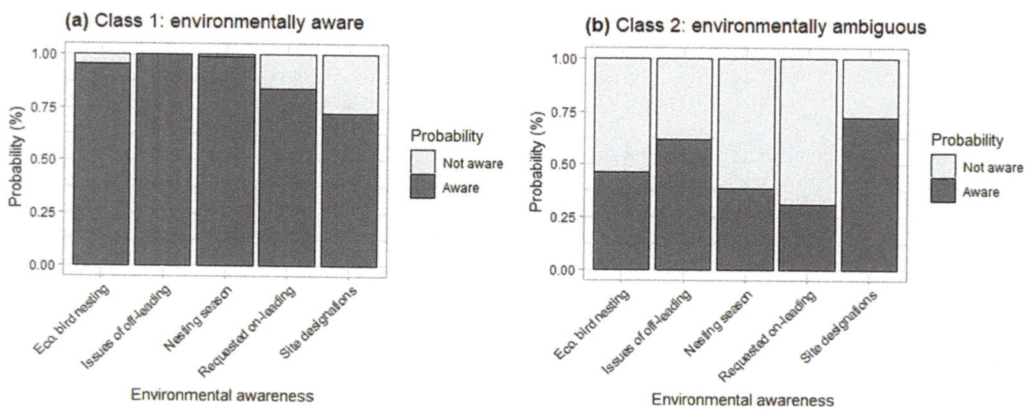

Figure 3. Two latent classes of environmental awareness were identified amongst the visitors: (**a**) environmentally aware and (**b**) environmentally ambiguous.

Demographic and behavioural responses in full can be found in Table 2. The RF model was performed for all visitors, and the 'aware' and 'ambiguous' groups were identified from the LCA. All three models had high accuracy (all visitors and 'ambiguous' groups had an OOB R^2 of 0.97, with 'aware' having an OOB R^2 of 0.96). OOB prediction error

(MSE) was lowest for the 'ambiguous' group (0.36), followed by 'aware' (1.20) and the 'all visitors' group (2.56), with the error showing stabilisation within 500 RF trees (see Figure S2 and Table S3). Root mean square error (RMSE) was calculated on the 20% holdback validation dataset for all models: 'all visitors' (1.61), 'aware' (1.10), and 'ambiguous' (0.60).

Variable importance was computed for all RF models; variables with higher importance levels contribute to explaining the outcome the most (Figure 4). Mean temperature in the driest quarter was the most important for all models. 'All visitors' and 'aware' visitors shared similar levels and rank of variable importance, with temperature seasonality being second most important, whereas this was moderately important for the other group. For the 'ambiguous' group the second most important variable was isothermality. Isothermality quantifies the extent to which day-to-night temperatures oscillate relative to summer-to-winter annual oscillations [63]. Proximity to water, elevation, and viewshed were also moderately important for all groups. Heather grassland and proximity to paths were important to a lesser extent, and the remaining variables (deciduous woodland, coniferous woodland, suburban, improved grassland, freshwater, bog, and arid grassland) showed very low importance.

Table 2. Demographic and behavioural responses within environmentally aware and ambiguous visitors in percentages. Percentages within each variable group for each environmental awareness group sum to 100.

Variable	Environmentally Aware (%)	Environmentally Ambiguous (%)
Age		
18–30	5.00	22.70
31–40	7.50	6.80
41–50	17.50	15.90
51–60	27.50	20.50
61–70	18.80	25.00
70+	23.70	9.10
Gender		
Female	47.50	29.55
Male	52.50	70.45
Travel distance		
Less than 1 mile	31.25	25.00
1 to 5 miles	28.75	31.82
More than 5 miles	40.00	43.18
Frequency of visit		
Daily	26.25	9.09
2–3 times a week	20.00	20.45
Once a week	10.00	15.91
Twice a month	2.50	2.27
Once a month	7.50	13.64
1–2 times a year	5.00	9.09
2–3 times a year	8.75	4.55
Once every 2–3 years	1.25	2.27
Sporadically	16.25	9.09
First visit	2.50	13.64

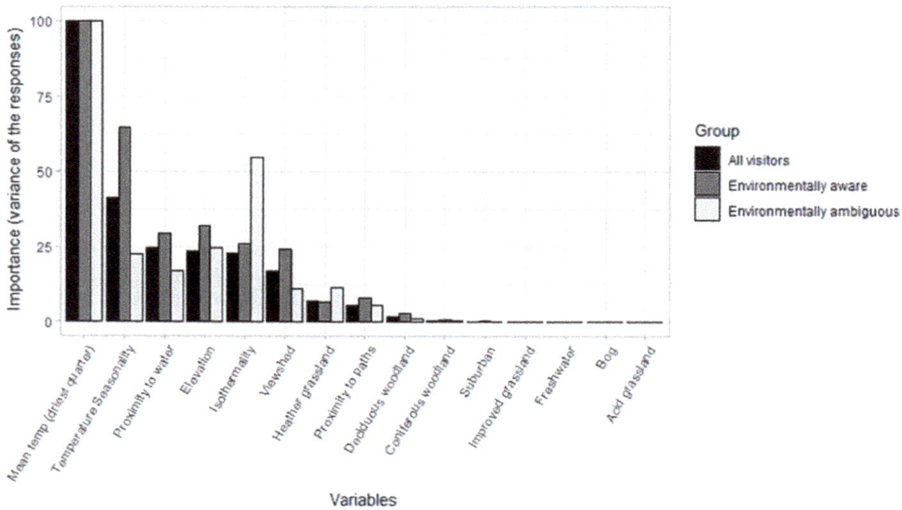

Figure 4. Random forest variable importance (values standardised between 0 and 100 within each group for comparability, ordered by descending 'all visitors' values). 'All visitors' group is shown in black, 'environmentally aware' is shown in dark grey, and 'environmentally ambiguous' is shown in light grey.

Mapped values showed distinct spatial patterns across all groups (Figure 5). Potential visitation was predicted, at least for low levels, across the entirety of the site for the 'all visitors' and 'aware' groups, with activity concentrated in the north to north-east areas of the sites, where popular access points are situated, with north–south and east–west areas of higher visitation clearly visible. In the 'ambiguous' group, there are distinct areas where no visitation was predicted, with activity once again highest in the north-east of the site, though only clear north–south channels could be identified, rather than the east–west channels also visible in the 'aware' group map.

Bootstrapped Pearson correlation was significant for all iterations for all maps at the $p < 0.001$ threshold; 'aware' vs. 'ambiguous' (correlation mean: 0.893, 95% CI: 0.892, 0.894), 'all visitors' vs. 'aware' (correlation mean: 0.984, 95% CI: 0.984, 0.984), and 'all visitors' vs. 'ambiguous' (correlation mean: 0.955, 95% CI: 0.954, 0.955).

4. Discussion

Visitors to the protected area studied here can be segmented into 'aware' and 'ambiguous' in their on-site environmental awareness from survey data using LCA. Information provided by each group on their use of the site can be upscaled using RF models and mapped as shown in Figure 5. Over two-thirds of the visitors surveyed could be classed as 'aware', similarly to the study of Beh and Bruyere [11], where visitors were segmented by their motivation for site visitation, and most were found to have a high awareness of the environment. Our study spatially maps visitation patterns and provides a means to target 'ambiguous' groups.

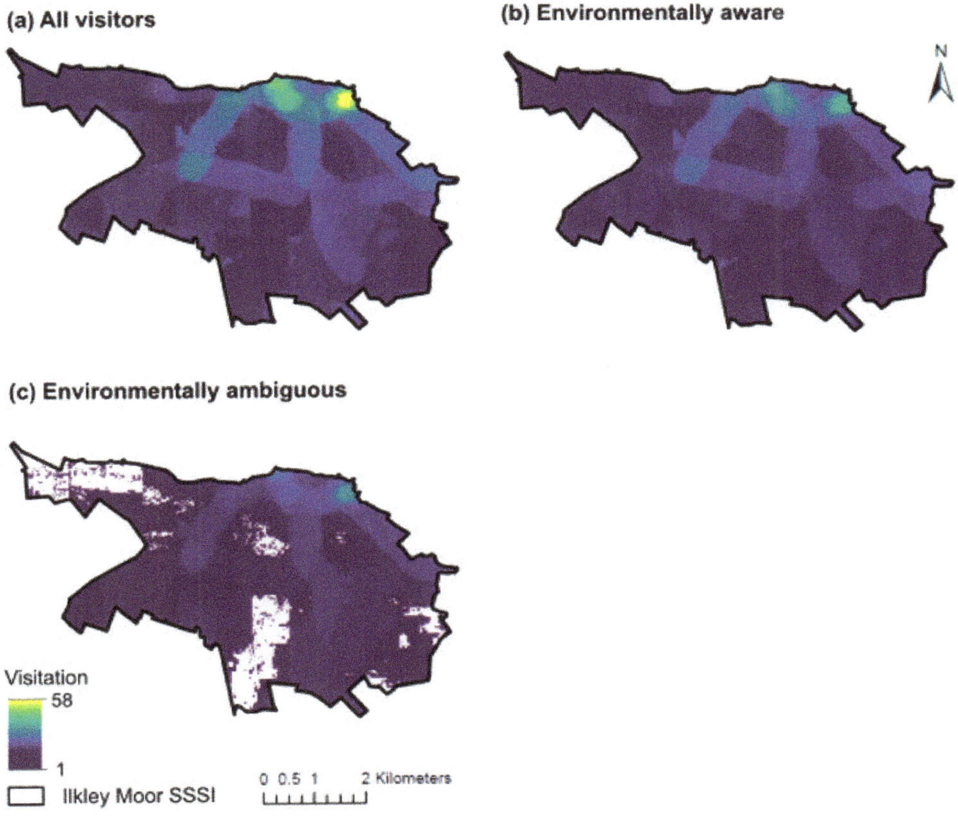

Figure 5. Maps of potential visitation by different groups to Ilkley Moor: (**a**) all visitors, (**b**) environmentally aware, and (**c**) environmentally ambiguous. Zero values predicted by the random forest models are set as white background values for clarity, allowing identification of non-visitation areas for the environmentally ambiguous group. The scale begins at 1 as purple and moves through green to the maximum value of 58 as yellow. The scale is calibrated the same across all three maps, making them comparable.

The LCA showed that awareness levels were higher in the 'aware' group, apart from awareness of site designations being similar for both groups. Surprisingly, awareness of site designations had the lowest probability of all the questions in the 'aware' group and the highest in the 'ambiguous' group (Figure 3). This may infer that being generally 'aware' may not necessarily translate to being educated on all environmental aspects of a site; for example, Booth et al. [17] showed a variation of 8–43% in understanding what an SSSI was amongst site visitors.

Variable importance in the random forest models showed that land cover predictors were of lower importance than other factors. The 'aware' group had a high importance related to mean temperature and seasonality, which could be attributed to people potentially visiting the site when the weather is favourable. The 'ambiguous' group had a relatively low importance for all factors apart from mean temperature and isothermality. Outdoor sensory experiences attract all types of visitors [64], which helps to explain the importance of these factors; hence, when the weather has been constant, visitors could potentially feel more confident in outside pursuits.

A larger proportion of the members of the 'aware' group were over 50 years old. Older generations have grown up in the increased presence of nature, and it has been suggested that they have been sensitised by those experiences they had when they were younger [65], whereas technological advancements may have distanced younger generations from nature [66]. More 'aware' visitors identified as visiting daily. This is supported by Halpenny [15] and Maguire et al. [67], who suggest that a sense of place promotes pro-environmental behaviour and stewardship. We found that within the 'ambiguous' group most visited at a frequency of 2–3/week, but altogether visited more at the lower frequencies. Ballantyne et al. [64] found that infrequent visitors were more likely to be motivated by learning, and thus the more infrequent visitors in the ambiguous group could be more receptive to conservation education delivery. In the aware group, 31.25% travelled less than from one mile away, with 25% in the ambiguous group, yet the highest proportion for both groups was more than five miles away (40% compared to 43.18% within each group respectively). When nature is nearby, it has been suggested that visitation increases, which fosters place dependence [68]. Nevertheless, close proximity can be linked with convenience, being part of an individual's residential environment, which they regularly use [69]. Lower visitation from those nearer the site could have been attributed to the timing of the survey periods in this study, with locals visiting earlier or later in the day. Alcock et al. [70] found that frequency of visitation was lower in individuals living in green-space-abundant areas compared to those living in areas that lack green spaces, who were potentially compensating for the lack of nature, which may also explain the higher number of visitators from further away in this study.

Visitors from the 'ambiguous' group were found to venture less from the northern access points on-site, and then in a linear north–south pattern, as opposed to 'aware' visitors who used these channels and also adopted east–west patterns of movement. This may be due to 'aware' visitors having a larger proportion of daily visitors (indicating familiarity with the site) and included many dog walkers, who tend to spend more time on-site and therefore take longer and more exploratory routes [71]. Bootstrapped Pearson correlation revealed relatively high similarity between the 'aware' and 'ambiguous' potential visitation maps when comparing the entire site (correlation mean: 0.893), which highlights the overlap in visitation across the sites between both groups. Though as can be seen from Figure 5, specific areas, and differences in where visitors from both groups are more likely to visit, can readily be discerned.

By understanding which areas are most visited by different groups, site management can be implemented in association with the results of areas of conservation priority within the site, for example alongside bird survey maps to reduce potentially negative impacts on breeding birds. Kim and Weiler [14] recommend differing on-site communication strategies for visitors with low and high environmental awareness (in relation to fossil collecting). This can be taken further through differentiating strategies for different age cohorts. Personal delivery has been found to be more successful in educating and changing behaviours [23], and hence different events and inclusive educational activities can be targeted for different ages. Habitat management could be focussed on areas with fewer visitors and/or areas with better-informed visitors to attract sensitive species into areas where they are less likely to experience the negative impacts of disturbance. Signage has been shown to work in multiple conservation scenarios [72], though the style of such signs needs to be carefully considered for greatest impact in terms of "behaviour change" outcomes [8,73]. An understanding of where visitors are most concentrated, especially those that may require more conservation education, will allow more targeted education efforts. This could be in the form of increased signage or targeted personal delivery as demonstrated by Kidd et al. [23]. Weaver and Lawton [74] suggest that adhering to a completely biocentric approach that sees visitors as an inherent threat to protected areas can lead to suboptimal sustainability outcomes, whereas seeing visitors as an opportunity rather than a threat for the park can allow visitor mobilisation towards park enhancement, such as pro-environmental activities and citizen science (e.g., bird surveys). Therefore, conservation education combined with these activities could provide a multitude of benefits.

To further assess the robustness of this study, the behaviour of the 'ambiguous' visitors would need validation through replication of this study in other protected areas. Other sites may have site-specific characteristics, e.g., coastal sites, woodland, or different biomes, that will need to be assessed for relevant predictor variables, though the same approach elucidated in this study could be used. The variables chosen in this study were limited to bioclimatic and physical, and they could be expanded to include a wider range, both within and outside these categories. Utilisation for machine learning technologies requires specialised skills; thus, the techniques demonstrated in this study are reliant on access to these resources. However, the resources saved in other areas, e.g., wider-scale conservation education, may provide greater savings in the long term. This study used stated responses as part of the survey and participatory GIS exercise. This could be improved using revealed behaviour methodologies, such as voluntary GPS tracking using independent sensors and data loggers or mobile telephones, as seen in Wolf et al. [75]. Alternatively, the accuracy of the participatory mapping could be improved, for example, by studies using online participatory exercise which can capture additional information such as aesthetic appeal of visited areas, as demonstrated by Gosal et al. [76].

5. Conclusions

The methodology elucidated in this study can be readily applied to other areas for which suitable spatial data are available, allowing the development of highly site-specific maps of visitors with differing environmental awareness. Predictors used in this study, e.g., the digital terrain model, calculated viewshed analysis, and bioclimatic variables, are easily found for many regions of the world, albeit at varying resolutions. There is a paramount need for conservation resources to be channelled into the most effective management strategies, and this study demonstrates a method to spatially define those areas that attract visitors with lower environmental and ecological awareness so that on-site resources can be efficiently targeted to where they are needed the most.

Supplementary Materials: The following are available online at https://www.mdpi.com/article/10.3390/land10060560/s1, Table S1: Full attribution for bird images in Figure 1 under Creative Commons licenses; Table S2: Birds recorded on Ilkley Moor with Birds of Conservation Concern (BoCC) designations; Table S3: Parameters giving the best out-of-bag (OOB) root mean square error (RMSE) for all RF models using hyper-tuning; Figure S1: Latent class analysis (LCA) AIC and BIC for multiple group (class) sizes; Figure S2: Random forest tree out-of-bag (OOB) stabilisation for random forest models.

Author Contributions: Conceptualization, A.S.G. and G.Z.; methodology, A.S.G., J.A.M., and G.Z.; writing—original draft preparation, review, and editing, A.S.G., J.A.M., K.M.B., C.H.H., and G.Z. All authors have read and agreed to the published version of the manuscript.

Funding: A.S.G. and G.Z. were supported by the Natural Environment Research Council [grant number NE/T002115/1] under the ADVANCES (ADVancing Analysis of Natural Capital in LandscapE DecisionS) project. J.A.M. was supported by Q-Step programme (University of Leeds), which is funded by the Nuffield Foundation, the Economic and Social Research Council (ESRC), and the Higher Education Funding Council for England (HEFCE).

Institutional Review Board Statement: The study was approved by the Social Sciences, Environment and Leeds University Business School Faculty Research Ethics Committee of the University of Leeds on 20 May 2019.

Informed Consent Statement: Informed consent was obtained from all subjects involved in the study.

Acknowledgments: The authors are grateful for the support of City of Bradford Metropolitan District Council, England, for allowing surveying at Ilkley Moor, Bradford. The authors thank the photographers of the bird species licensed under Creative Commons in Figure 1d (see Table S1) and Kulvinder Kaur for photographs of Ilkley Moor used in Figure 1c. The authors are grateful to the visitors to the study site that participated in the surveys.

Conflicts of Interest: The authors declare no conflict of interest.

References

1. Mascia, M.B.; Brosius, J.P.; Dobson, T.A.; Forbes, B.C.; Horowitz, L.; McKean, M.A.; Turner, N.J. Conservation and the social sciences. *Conserv. Biol.* **2003**, *17*, 649–650. [CrossRef]
2. Schultz, P.W. Conservation Means Behavior. *Conserv. Biol.* **2011**, *25*, 1080–1083. [CrossRef]
3. Fisichelli, N.A.; Schuurman, G.W.; Monahan, W.B.; Ziesler, P.S. Protected Area Tourism in a Changing Climate: Will Visitation at US National Parks Warm Up or Overheat? *PLoS ONE* **2015**, *10*, e0128226. [CrossRef]
4. Gill, J.A. Approaches to measuring the effects of human disturbance on birds. *Ibis* **2007**, *149*, 9–14. [CrossRef]
5. Bötsch, Y.; Tablado, Z.; Jenni, L. Experimental evidence of human recreational disturbance effects on bird-territory establishment. *Proc. R. Soc. B Biol. Sci.* **2017**, *284*. [CrossRef] [PubMed]
6. Alwis, N.S.; Perera, P.; Dayawansa, N.P. Response of tropical avifauna to visitor recreational disturbances: A case study from the Sinharaja World Heritage Forest, Sri Lanka. *Avian Res.* **2016**, *7*, 15. [CrossRef]
7. Remacha, C.; Delgado, J.A.; Bulaic, M.; Pérez-Tris, J. Human Disturbance during Early Life Impairs Nestling Growth in Birds Inhabiting a Nature Recreation Area. *PLoS ONE* **2016**, *11*, e0166748. [CrossRef]
8. Allbrook, D.L.; Quinn, J.L. The effectiveness of regulatory signs in controlling human behaviour and Northern gannet (Morus bassanus) disturbance during breeding: An experimental test. *J. Nat. Conserv.* **2020**, *58*, 125915. [CrossRef] [PubMed]
9. Finney, S.K.; Pearce-Higgins, J.W.; Yalden, D.W. The effect of recreational disturbance on an upland breeding bird, the golden plover Pluvialis apricaria. *Biol. Conserv.* **2005**, *121*, 53–63. [CrossRef]
10. Calladine, J.; Critchley, C.N.R.; Baker, D.; Towers, J.; Thiel, A. Conservation management of moorland: A case study of the effectiveness of a combined suite of management prescriptions which aim to enhance breeding bird populations. *Bird Study* **2014**, *61*, 56–72. [CrossRef]
11. Beh, A.; Bruyere, B.L. Segmentation by visitor motivation in three Kenyan national reserves. *Tour. Manag.* **2007**, *28*, 1464–1471. [CrossRef]
12. Fung, C.K.W.; Jim, C.Y. Segmentation by motivation of Hong Kong Global Geopark visitors in relation to sustainable nature-based tourism. *Int. J. Sustain. Dev. World Ecol.* **2015**, *22*, 76–88. [CrossRef]
13. Formica, S.; Uysal, M. Segmentation of travelers based on environmental attitudes. *J. Hosp. Leis. Mark.* **2001**, *9*, 35–49. [CrossRef]
14. Kim, A.K.; Weiler, B. Visitors' attitudes towards responsible fossil collecting behaviour: An environmental attitude-based segmentation approach. *Tour. Manag.* **2013**, *36*, 602–612. [CrossRef]
15. Halpenny, E.A. Pro-environmental behaviours and park visitors: The effect of place attachment. *J. Environ. Psychol.* **2010**, *30*, 409–421. [CrossRef]
16. Bennett, N.J.; Roth, R.; Klain, S.C.; Chan, K.; Christie, P.; Clark, D.A.; Cullman, G.; Curran, D.; Durbin, T.J.; Epstein, G.; et al. Conservation social science: Understanding and integrating human dimensions to improve conservation. *Biol. Conserv.* **2017**, *205*, 93–108. [CrossRef]
17. Booth, J.E.; Gaston, K.J.; Armsworth, P.R. Public understanding of protected area designation. *Biol. Conserv.* **2009**, *142*, 3196–3200. [CrossRef]
18. González Del Campo, A. Mapping environmental sensitivity: A systematic online approach to support environmental assessment and planning. *Environ. Impact Assess. Rev.* **2017**, *66*, 86–98. [CrossRef]
19. Dhami, I.; Deng, J.; Burns, R.C.; Pierskalla, C. Identifying and mapping forest-based ecotourism areas in West Virginia—Incorporating visitors' preferences. *Tour. Manag.* **2014**, *42*, 165–176. [CrossRef]
20. Rieb, J.T.; Chaplin-Kramer, R.; Daily, G.C.; Armsworth, P.R.; Böhning-Gaese, K.; Bonn, A.; Cumming, G.S.; Eigenbrod, F.; Grimm, V.; Jackson, B.M.; et al. When, Where, and How Nature Matters for Ecosystem Services: Challenges for the Next Generation of Ecosystem Service Models. *Bioscience* **2017**, *67*, 820–833. [CrossRef]
21. Hammitt, W.E.; Cole, D.N.; Monz, C.A. *Free-Choice Environmental Learning: Framing the Discussion*; Taylor and Francis Ltd.: New York, NY, USA, 2005; Volume 11.
22. Salata, T.L.; Ostergren, D.M. Evaluating Forestry Camps with National Standards in Environmental Education: A Case Study of the Junior Forester Academy, Northern Arizona University. *Appl. Environ. Educ. Commun.* **2010**, *9*, 50–57. [CrossRef]
23. Kidd, A.M.; Monz, C.; D'Antonio, A.; Manning, R.E.; Reigner, N.; Goonan, K.A.; Jacobi, C. The effect of minimum impact education on visitor spatial behavior in parks and protected areas: An experimental investigation using GPS-based tracking. *J. Environ. Manag.* **2015**, *162*, 53–62. [CrossRef] [PubMed]
24. Langston, R.H.W.; Liley, D.; Murison, G.; Woodfield, E.; Clarke, R.T. What effects do walkers and dogs have on the distribution and productivity of breeding European Nightjar Caprimulgus europaeus? *Ibis* **2007**, *149*, 27–36. [CrossRef]
25. Liley, D.; Sutherland, W.J. Predicting the population consequences of human disturbance for Ringed Plovers Charadrius hiaticula: A game theory approach. *Ibis* **2007**, *149*, 82–94. [CrossRef]
26. Bötsch, Y.; Tablado, Z.; Almasi, B.; Jenni, L. Human recreation decreases antibody titre in bird nestlings: An overlooked transgenerational effect of disturbance. *J. Exp. Biol.* **2020**, *223*. [CrossRef] [PubMed]
27. Steven, R.; Pickering, C.; Guy Castley, J. A review of the impacts of nature based recreation on birds. *J. Environ. Manag.* **2011**, *92*, 2287–2294. [CrossRef]
28. Mayer-Gross, H.; Crick, H.Q.P.; Greenwood, J.J.D. Bird Study The effect of observers visiting the nests of passerines: An experimental study. *Bird Study* **2010**, *44*, 53–65. [CrossRef]
29. Burger, J.; Leonard, J. Conflict resolution in coastal waters: The case of personal watercraft. *Mar. Policy* **2000**, *24*, 61–67. [CrossRef]

30. Harris, S.J.; Massimino, D.; Balmer, D.E.; Eaton, M.A.; Noble, D.G.; Pearce-Higgins, J.W.; Woodcock, P.; Gillings, S. *The Breeding Bird Survey 2019*; Thetford: Ann Arbor, MI, USA, 2020.
31. Eaton, M.; Aebischer, N.; Brown, A.; Hearn, R.; Lock, L.; Musgrove, A.; Noble, D.; Stroud, D.; Gregory, R.; Powell, R. Birds of Conservation Concern 4: The Population Status of Birds in the UK, Channel Islands and Isle of Man. *Br. Birds* **2015**, *108*, 708–746.
32. Likert, R. A technique for the measurement of attitudes. *Arch. Psychol.* **1932**, *140*, 44–53.
33. Joshi, A.; Kale, S.; Chandel, S.; Pal, D. Likert Scale: Explored and Explained. *Br. J. Appl. Sci. Technol.* **2015**, *7*, 396–403. [CrossRef]
34. Fick, S.E.; Hijmans, R.J. WorldClim 2: New 1-km spatial resolution climate surfaces for global land areas. *Int. J. Climatol.* **2017**, *37*, 4302–4315. [CrossRef]
35. Centre for Ecology and Hydrology the UKCEH Land Cover Maps for 2017, 2018 and 2019. Available online: https://www.ceh.ac.uk/services/lcm2019-lcm2018-and-lcm2017 (accessed on 15 December 2020).
36. Ordnance Survey OS Terrain®50. Available online: https://osdatahub.os.uk/downloads/open/Terrain50 (accessed on 13 November 2020).
37. QGIS. Development Team QGIS Geographic Information System. Available online: https://qgis.org/ (accessed on 4 January 2021).
38. Geofabrik OpenStreetMap Data Extracts. Available online: https://download.geofabrik.de/ (accessed on 15 December 2020).
39. Cuckovic, Z. Advanced viewshed analysis: A Quantum GIS plug-in for the analysis of visual landscapes. *J. Open Source Softw.* **2016**, *1*, 32. [CrossRef]
40. Berk, R.A. An introduction to ensemble methods for data analysis. *Sociol. Methods Res.* **2006**, *34*, 263–295. [CrossRef]
41. Segal, M.R. Machine Learning Benchmarks and Random Forest Regression. *UCSF: Center for Bioinformatics and Molecular Biostatistics*. 2004. Available online: https://escholarship.org/uc/item/35x3v9t4 (accessed on 4 January 2021).
42. Kuhn, M. Building predictive models in R using the caret package. *J. Stat. Softw.* **2008**, *28*, 1–26. [CrossRef]
43. Kursa, M.B.; Rudnicki, W.R. Feature selection with the boruta package. *J. Stat. Softw.* **2010**, *36*, 1–13. [CrossRef]
44. Peschel, A.O.; Grebitus, C.; Steiner, B.; Veeman, M. How does consumer knowledge affect environmentally sustainable choices? Evidence from a cross-country latent class analysis of food labels. *Appetite* **2016**, *106*, 78–91. [CrossRef]
45. Rhead, R.; Elliot, M.; Upham, P. Using latent class analysis to produce a typology of environmental concern in the UK. *Soc. Sci. Res.* **2018**, *74*, 210–222. [CrossRef]
46. Ehrlich, O.; Bi, X.; Borisova, T.; Larkin, S. A latent class analysis of public attitudes toward water resources with implications for recreational demand. *Ecosyst. Serv.* **2017**, *28*, 124–132. [CrossRef]
47. Linzer, D.A.; Lewis, J.B. poLCA: An R package for polytomous variable latent class analysis. *J. Stat. Softw.* **2011**, *42*, 1–29. [CrossRef]
48. Du, S.; Zhang, F.; Zhang, X. Semantic classification of urban buildings combining VHR image and GIS data: An improved random forest approach. *ISPRS J. Photogramm. Remote Sens.* **2015**, *105*, 107–119. [CrossRef]
49. Moon, J.; Kim, Y.; Son, M.; Hwang, E. Hybrid Short-Term Load Forecasting Scheme Using Random Forest and Multilayer Perceptron. *Energies* **2018**, *11*, 3283. [CrossRef]
50. Cheng, L.; Chen, X.; De Vos, J.; Lai, X.; Witlox, F. Applying a random forest method approach to model travel mode choice behavior. *Travel Behav. Soc.* **2019**, *14*, 1–10. [CrossRef]
51. Braun, T.; Cottrell, R.; Dierkes, P. Fostering changes in attitude, knowledge and behavior: Demographic variation in environmental education effects. *Environ. Educ. Res.* **2018**, *24*, 899–920. [CrossRef]
52. Naghibi, S.A.; Pourghasemi, H.R.; Dixon, B. GIS-based groundwater potential mapping using boosted regression tree, classification and regression tree, and random forest machine learning models in Iran. *Environ. Monit. Assess.* **2016**, *188*, 1–27. [CrossRef]
53. Dou, J.; Yunus, A.P.; Tien Bui, D.; Merghadi, A.; Sahana, M.; Zhu, Z.; Chen, C.W.; Khosravi, K.; Yang, Y.; Pham, B.T. Assessment of advanced random forest and decision tree algorithms for modeling rainfall-induced landslide susceptibility in the Izu-Oshima Volcanic Island, Japan. *Sci. Total Environ.* **2019**, *662*, 332–346. [CrossRef]
54. Kim, J.-C.; Lee, S.; Jung, H.-S.; Lee, S. Landslide susceptibility mapping using random forest and boosted tree models in Pyeong-Chang, Korea. *Geocarto Int.* **2018**, *33*, 1000–1015. [CrossRef]
55. Park, S.; Kim, J. Landslide Susceptibility Mapping Based on Random Forest and Boosted Regression Tree Models, and a Comparison of Their Performance. *Appl. Sci.* **2019**, *9*, 942. [CrossRef]
56. Heung, B.; Bulmer, C.E.; Schmidt, M.G. Predictive soil parent material mapping at a regional-scale: A Random Forest approach. *Geoderma* **2014**, *214–215*, 141–154. [CrossRef]
57. Mascaro, J.; Asner, G.P.; Knapp, D.E.; Kennedy-Bowdoin, T.; Martin, R.E.; Anderson, C.; Higgins, M.; Chadwick, K.D. A Tale of Two "Forests": Random Forest Machine Learning Aids Tropical Forest Carbon Mapping. *PLoS ONE* **2014**, *9*, e85993. [CrossRef]
58. Feng, Q.; Liu, J.; Gong, J. UAV Remote sensing for urban vegetation mapping using random forest and texture analysis. *Remote Sens.* **2015**, *7*, 1074–1094. [CrossRef]
59. Hayes, M.M.; Miller, S.N.; Murphy, M.A. High-resolution landcover classification using Random Forest. *Remote Sens. Lett.* **2014**, *5*, 112–121. [CrossRef]
60. Rodriguez-Galiano, V.F.; Ghimire, B.; Rogan, J.; Chica-Olmo, M.; Rigol-Sanchez, J.P. An assessment of the effectiveness of a random forest classifier for land-cover classification. *ISPRS J. Photogramm. Remote Sens.* **2012**, *67*, 93–104. [CrossRef]

61. Wright, M.N.; Ziegler, A. ranger: A Fast Implementation of Random Forests for High Dimensional Data in C++ and R. *JSS J. Stat. Softw.* **2017**, *77*. [CrossRef]
62. Gilleland, E. Bootstrap methods for statistical inference. Part i: Comparative forecast verification for continuous variables. *J. Atmos. Ocean. Technol.* **2020**, *37*, 2117–2134. [CrossRef]
63. O'Donnell, M.S.; Ignizio, D.A. Bioclimatic Predictors for Supporting Ecological Applications in the Conterminous United States. *US Geol. Surv. Data Ser.* **2012**, *691*, 4–9.
64. Ballantyne, R.; Packer, J.; Hughes, K. Environmental awareness, interests and motives of botanic gardens visitors: Implications for interpretive practice. *Tour. Manag.* **2008**, *29*, 439–444. [CrossRef]
65. Finger, M. From Knowledge to Action? Exploring the Relationships Between Environmental Experiences, Learning, and Behavior. *J. Soc. Issues* **1994**, *50*, 141–160. [CrossRef]
66. Schultz, P.W. Inclusion with nature: The psychology of human-nature relations. In *Psychology of Sustainable Development*; Springer: Cham, Switzerland, 2002; pp. 61–78.
67. Maguire, G.; Rimmer, J.; Weston, M. Stakeholder Perceptions of Threatened Species and Their Management on Urban Beaches. *Animals* **2013**, *3*, 1002–1020. [CrossRef]
68. Halpenny, E.A. *Environmental Behaviour, Place Attachment and Park Visitation: A Case Study of Visitors to Point Pelee National Park*; University of Waterloo: Waterloo, ON, Canada, 2006.
69. Sterl, P.; Brandenburg, C.; Arnberger, A. Visitors' awareness and assessment of recreational disturbance of wildlife in the Donau-Auen National Park. *J. Nat. Conserv.* **2008**, *16*, 135–145. [CrossRef]
70. Alcock, I.; White, M.P.; Pahl, S.; Duarte-Davidson, R.; Fleming, L.E. Associations between pro-environmental behaviour and neighbourhood nature, nature visit frequency and nature appreciation: Evidence from a nationally representative survey in England. *Environ. Int.* **2020**, *136*, 105441. [CrossRef]
71. Reeves, M.J.; Rafferty, A.P.; Miller, C.E.; Lyon-Callo, S.K. The Impact of Dog Walking on Leisure-Time Physical Activity: Results From a Population-Based Survey of Michigan Adults. *J. Phys. Act. Health* **2011**, *8*, 436–444. [CrossRef]
72. Williams, D.R.; Child, M.F.; Dicks, L.V.; Ockendon, N.; Pople, R.G.; Showler, D.A.; Walsh, J.C.; zu Ermgassen, E.K.H.J.; Sutherland, W.J. Bird Conservation. In *What Works in Conservation 2020*; Sutherland, W.J., Dicks, L.V., Petrovan, S.O., Smith, R.K., Eds.; Open Book Publishers: Cambridge, UK, 2020; pp. 137–281.
73. Ballantyne, R.; Hughes, K. Using front-end and formative evaluation to design and test persuasive bird feeding warning signs. *Tour. Manag.* **2006**, *27*, 235–246. [CrossRef]
74. Weaver, D.B.; Lawton, L.J. A new visitation paradigm for protected areas. *Tour. Manag.* **2017**, *60*, 140–146. [CrossRef]
75. Wolf, I.D.; Hagenloh, G.; Croft, D.B. Visitor monitoring along roads and hiking trails: How to determine usage levels in tourist sites. *Tour. Manag.* **2012**, *33*, 16–28. [CrossRef]
76. Gosal, A.S.; Newton, A.C.; Gillingham, P.K. Comparison of methods for a landscape-scale assessment of the cultural ecosystem services associated with different habitats. *Int. J. Biodivers. Sci. Ecosyst. Serv. Manag.* **2018**, *14*, 91–104. [CrossRef]

land

MDPI

Article

High Nature Value Farming Systems and Protected Areas: Conservation Opportunities or Land Abandonment? A Study Case in the Madrid Region (Spain)

María F. Schmitz [1,*], Cecilia Arnaiz-Schmitz [2] and Patricio Sarmiento-Mateos [3]

[1] Department of Biodiversity, Ecology and Evolution, Universidad Complutense de Madrid, 28040 Madrid, Spain
[2] Department of Civil Engineering, Transport, Territory and Urbanism, Higher Technical School of Engineering of Roads, Channels and Ports, Universidad Politécnica de Madrid, 28040 Madrid, Spain; cecilia.arnaiz@upm.es
[3] Regional Ministry of Education and Youth of the Government of Madrid, 28014 Madrid, Spain; patricio.sarmiento@educa.madrid.org
* Correspondence: ma296@ucm.es

Abstract: European rural landscapes contain high nature value farmlands that, in addition to being the main economic activity in many rural areas, host habitats and species of great conservation value. The maintenance of these farming systems largely depends on traditional ecological knowledge and the rural lifestyles of the local populations. However, they have not been sufficiently appreciated and protected, and as a result, they are currently threatened. In this study, which was performed in the Madrid region (central Spain), we analyse the social-ecological changes of the rural landscape after the establishment of a protected natural area network. The obtained results highlight a significant loss of these high nature value farming systems and a marked increase in the rewilding processes characterised by scrub–forest transition and the development of forest systems. These processes are linked to the disruption of the transmission of traditional ecological knowledge, which may imply negative consequences for both the high biocultural diversity that these systems host and the cultural identity and the socioeconomics of the rural populations that live there. A useful methodological tool is provided for social–ecological land planning and the design of effective management strategies for the conservation of rural cultural landscapes.

Keywords: social–ecological systems; cultural rural landscape; protected areas; rewilding; rural socioeconomics; forest expansion; rural to urban land conversion; biocultural heritage; biodiversity; naturalness

Citation: Schmitz, M.F.; Arnaiz-Schmitz, C.; Sarmiento-Mateos, P. High Nature Value Farming Systems and Protected Areas: Conservation Opportunities or Land Abandonment? A Study Case in the Madrid Region (Spain). *Land* **2021**, *10*, 721. https://doi.org/10.3390/land10070721

Academic Editors: Juan F. Beltrán, Pedro Abellán and John Litvaitis

Received: 16 June 2021
Accepted: 6 July 2021
Published: 8 July 2021

Publisher's Note: MDPI stays neutral with regard to jurisdictional claims in published maps and institutional affiliations.

1. Introduction

The theory of social–ecological systems arose from the recognition of the close interaction between nature and society [1]. Among the different types of systems, rural cultural landscapes are characterized by the conservation and protection of ecological processes, natural resources, and cultural biodiversity [2–4]. Secular interactions between humans and nature have given rise to a wide variety of sustainable cultural landscapes that have primarily been maintained over time with traditional ecological knowledge (TEK). This cultural process is based on the transmission of the deep empirical knowledge of the sustainable use of natural resources and, therefore, of the conservation of ecological processes and biodiversity [5].

In recent decades, significant efforts have been made to preserve TEK and cultural landscapes in Europe [6]. The European countryside is characterized by a rich array of rural cultural landscapes that have been shaped over millennia by traditional land uses [7,8]. The dynamic interrelationship between social and ecological systems has given rise to a broad range of cultural agricultural landscapes that, in addition to their primary functions of producing food and fibre, are rich in natural and/or semi-natural vegetation and often harbour habitats and species of great conservation value [9,10]. These "high natural value farming systems" (HNVF), with their strong heritage significance and outstanding aesthetic and recreational qualities, favour the conservation of biodiversity and provide society with a great variety of essential ecosystem services that have improved the living standards of local populations and have resulted in valuable cultural landscapes [7,11,12]. The HNVF concept emphasizes the role of certain types of farmlands in the conservation of biodiversity in rural areas [13]. At present, European agricultural biodiversity is considered to be as valuable as wild biodiversity [14].

Despite the indicated characteristics, and probably also due to their everyday use, rural cultural landscapes and their associated HNVF have not been sufficiently valued and protected. This has involved the rapid and radical transformations of traditional land-use systems across Europe in recent decades, mainly as a consequence of the continuous process of polarization derived from the main land use change trajectories: extensification and rural abandonment, on the one hand, and intensification and urban expansion, on the other [15–19]. The rate and extent of technological, economic, and cultural changes threaten cultural landscapes and the rural societies associated with them [7,16,20].

In this context, protected areas (PAs) could play a key role in the protection and maintenance of European HNVF because they are social–ecological systems whose sustainability and management are strongly influenced by people [21]. PAs are central to conservation strategies, but the effectiveness of some measures included in their regulatory schemes can be questioned [22]. Therefore, several studies highlight that PAs often restrict rural activities and neglect local populations with respect to their TEK, their historical and cultural context, and their important contribution to the maintenance of these cultural landscapes. These restrictions promote the abandonment of agricultural land and traditional management practices, causing the loss of biocultural diversity [20,23–26]. Thus, nature conservation strategies supported by PAs have often promoted the abandonment of farmlands, pastures, and cultural forests that host high biodiversity and that are being transformed into mosaics of scrub and mixed forests and forest systems. These processes of rewilding (returning ecosystems to a higher level of naturalness, seeking wildlife comeback without human intervention), cause the disappearance of HNVF, fostering spatial homogeneity and the degradation of the cultural landscape [7,25,27]. This restrictive approach to conservation favours "inside-out" processes, generating the development of different opportunities inside and outside the limits of the parks [20,26,28–31]. So far, no land planning and management schemes have been achieved that can provide effective designs and responses to safeguard the values of cultural landscapes and traditional land-use systems, which are still at risk today. Therefore, the future of HNVF is very uncertain [17,32,33].

The overall objective of this paper is contextualized within the conceptual framework of social–ecological systems and, specifically, within the study of the effectiveness of PA guidelines and management in relation to the protection and maintenance of HNVF. The baseline for this study is the research carried out by Sarmiento-Mateos et al. (2019) [22], which, from a scientific perspective, focused on the evaluation of normative documents and the guidelines for the planning and management of two Spanish PAs under different protection. The findings from that research highlight various weaknesses and inconsistencies in the zoning design and regulation schemes of the studied PAs, which mainly promote uses and activities more related to the nature of the areas than to the culture, causing negative consequences for the cultural landscape that, contradictorily, these legal instruments claim to protect (Figure 1). On this basis, the specific objectives of our research are are: (i) to find out through empirical evidence based on the social–ecological evolution of the study area whether rural land planning and nature conservation strategies by means of the establishment of PAs consider traditional agricultural systems as valuable components of rural cultural landscapes; (ii) to detect inside-out processes related to the dynamics of HNVF in territories with established PAs (i.e., restrictions on agricultural uses and practices within the boundaries of PAs and opportunities outside them, or vice versa);

and (iii) to identify the social and economic impacts of PAs on rural populations living inside of and outside of their boundaries.

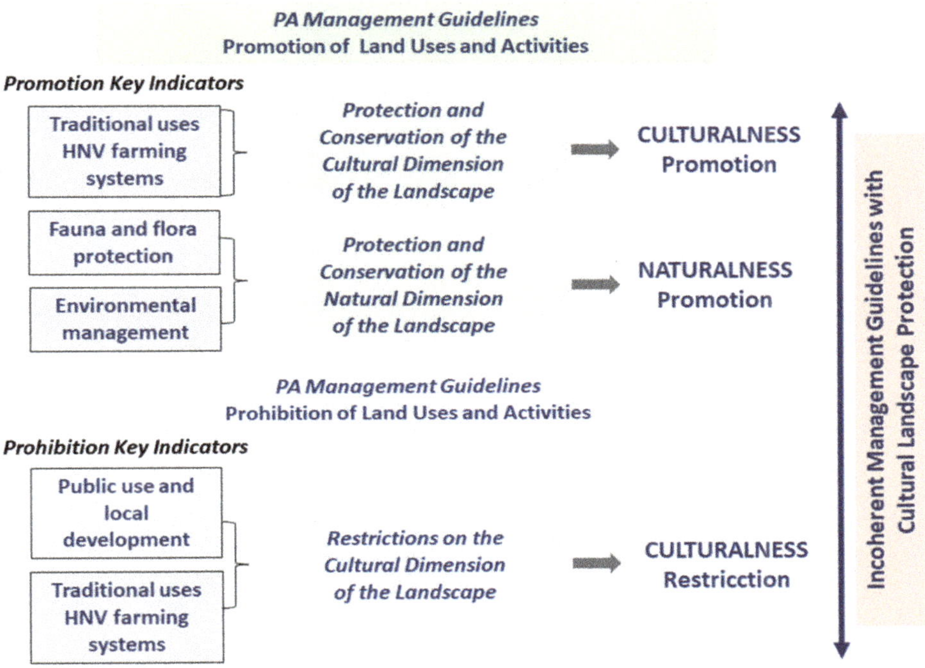

Figure 1. General outline of the main indicators of the PA management guidelines, identified from the analysis of the corresponding normative documents. Modified from Sarmiento-Mateos et al. 2019 [22].

2. Materials and Methods

2.1. Study Area

The study area is located in the north-northwest area of the Madrid region (central Spain) and covers 38 municipalities. As a result of the recognition of the natural and cultural values of this territory, a few decades ago, a PA network began to establish and expand PAs in this area, which consist of the "Cuenca Alta del Manzanares Regional Park" (52,800 ha; declared in 1985), the "Cumbre, Circo and Lagunas de Peñalara Natural Park" (15,030 ha; declared in 1990) and, later, the "Sierra de Guadarrama National Park" (33,960 ha, of which 21,714 ha are in the Madrid region; declared in 2013). After the declaration of the National Park, the limits of this space partially overlapped with not only thousands of hectares of the Regional Park but also overlapped practically the entire area belonging to the Natural Park, which became integrated into the former. As a consequence of the establishment of the national park, the natural park was derogated as a protection category of the territory (Figure 2).

Because of its location, is the study area is of a continental Mediterranean climate, characterised by hot, dry summers and cold, wet winters (according with the Köppen classification). In the study area, the average annual temperature ranges from 4 °C in the summits of the mountains to 13.5 °C in the foothills. Average rainfall ranges from 450 mm to 1615 mm per year. The substrate is formed by granitoid and gneiss rocks with lithic and dystric leptosols. The main environmental characteristic is the marked altitudinal variation, which is between 600 and 2383 m a.s.l. and favours the development of different vegetation belts. The natural vegetation corresponds to Mediterranean forests, with different species of trees such as Quercus ilex, Q. pyrenaica, Q. faginea, and Fraxinus angustifolia as well as scrubs such as Cistus ladanifer, Cytisus scoparius, Lavandula stoechas, and Genista cinerea, among others.

(a)

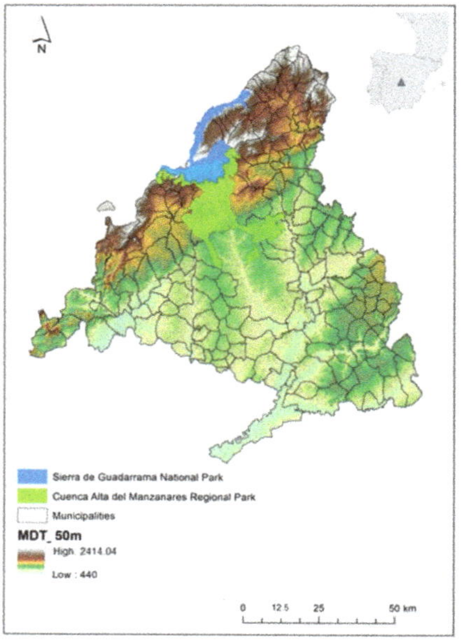

(b)

Figure 2. (**a**) Location of the study area in the Madrid region (central Spain). The two main categories of landscape protection at present (Regional Park and National Park) are shown, as well as the overlapping area between both parks; (**b**) HNVF characteristics of the study area (mainly silvopastoral systems).

The area is an ancestral mountainous cultural landscape that is mainly composed of traditional agrosilvopastoral land uses with relevant HNVF, mainly pastures and "dehesas" at low and medium altitudes immersed in a bocage type rural matrix with drove roads, hedgerow networks, woodland, and enclosures of stone walls as unique and characteristic elements of the landscape [20,26]. Dehesas (open savannah-like woodlands used as pastures) are human-made systems that combine exploitation with habitat conservation and support high habitat heterogeneity and great natural and social values while providing relevant ecosystem services and biodiversity conservation [34,35]. For these reasons, dehesa systems are considered a model for sustainable agriculture, and their conservation depends on the maintenance and effective management of traditional land uses [25,36]. Highlands have traditionally been used as summer pastures for native livestock breeds, such as the Avileña-Black Iberian cattle, which moves seasonally from the pasture systems of the valleys to the upland grasslands (this altitudinal movement of livestock is called trasterminance) through a wide network of drove roads of great cultural value [25,37,38]. The result of this complex social–ecological relationship is a multifunctional rural landscape that provides numerous provisioning and regulating ecosystem services and high biocultural diversity [20,39].

For centuries, these HNVF have been the main economic activity for local populations and have constituted a major factor in the shaping of the current landscape [22]. Recognition of the value of this multifunctional landscape and its accelerated dynamics of change, similar to many other European rural landscapes, has led to the design and application of nature conservation strategies through the establishment of different categories of PAs in the last 40 years. Regional and Natural Parks combine the protection and use of sustainable landscapes. Thus, among the objectives contained in their regulatory frameworks is the promotion of the maintenance, recovery, and implantation of the traditional productive activities of an agricultural, livestock, and forestry area as a means to ensure the survival of natural and cultural values [40,41]. Despite this, the evolution of the territory after the establishment of the first PAs has favoured the raising of the protection regime to that of a National Park, the highest level of protection allowed by the Spanish legal system. Its main objective is to ensure the conservation of the natural values of the protected territory. Spanish National Parks are places where "non-intervention" prevails and where the principle is to allow the free evolution of natural processes [42].

2.2. Selection of Land Analysis Units

This paper focuses on the study of the social-ecological evolution experienced by the studied rural cultural landscape, the relevant HNVF's of which were the main reason for the establishment of a PA network several decades ago. The spatial-temporal analysis units were both the municipalities included within the limits of the PA network and the ones closest to them in the influenced area surrounding the territory of the parks. We analysed a total of 38 municipalities, 15 of them totally or partially included within the PA network (with more than 25% of the municipal area) and 23 municipalities outside the network (with a municipal area < 25% or not included within its boundaries) (Appendix A).

Municipalities are an interesting and effective local scale of analysis, and several authors encourage their use in landscape studies [43–45]. In Spain, municipalities are the smallest administrative units of land management and the most detailed scale of decision-making. Furthermore, socioeconomic information is recorded at this level [19,20,46].

2.3. Data Collection

We registered quantitative descriptors of the social-ecological variation of the studied territory. These descriptors can be considered representative of HNVF and the socioeconomics of the local populations. Thus, for each municipality, we selected (a) a set of 10 descriptors of the HNVF in the study area that were linked to farmland dominated by low intensity farming practices and whose nature value results come primarily from (i) a high proportion of semi-natural vegetation; (ii) a mosaic of semi-natural and cultivated land; and (iii) a high diversity of land use–land cover (LULC) combined with semi-natural elements [47,48]. These HNVF descriptors were obtained from the reclassification of 27 LULC from pre-existing land use maps (SIGA public database, 1990–2010; Table 1a); (b) 11 socioeconomic descriptors (Table 1b) related to local population structure (population density, population aging degree), population dynamics (emigration), economic living conditions (income per capita), labour market (unemployment in the agricultural sector, agricultural workers), contribution of gross domestic product (GDP) to the local economy (agricultural GDP, industrial GDP and GDP from service sector), and characteristics of the land planning (urban land area, agricultural land area). Socioeconomic data were obtained from regional public censuses [49].

Data of the selected social-ecological descriptors (LULC and socioeconomic variables) were recorded in two periods, 1990–2000 and 2000–2010, prior to the declaration and establishment of the National Park in the study area. With the collected social and economic data, we elaborated four quantitative matrices, describing the 38 municipalities in the two time periods by means of the 10 descriptors based on the most representative land uses of HVN farming systems and 11 descriptors of the socioeconomic structure of local populations, respectively. LULC were quantified as the occupied area in relation to the municipal area (Table 1a). Socioeconomic data were recorded at the municipal level and their units of measurement varied depending on the type of descriptor used (see Table 1b). Data recorded in the matrices were the average values of each social-ecological descriptor in each period.

Table 1. Social-ecological descriptors recorded in each municipality of the study area. A brief description and units of measurement are indicated. (a) Land use–land cover descriptors considered as representative of high nature value farmlands of the studied cultural landscape. The unit of measurement was percentage area in relation to municipal area; (b) socioeconomic variables related to local population structure.

(a)

Land Use Variables	Description
Pastures	Pastureland; scrub-pastureland mosaics
Dehesas	Open formations with a mixture of pastures and isolated trees. Pastures with broadleaf tree species; pastures with coniferous species; pastures with mosaics of scrub and tree species (broadleaf trees and/or coniferous)
Herbaceous crops	Irrigated herbaceous crops; orchards and forced crops; rainfed herbaceous crops; mosaics of crops and broadleaf tree species
Woody crops	Rainfed olive groves; rainfed vineyard; rainfed mixed crops of olive grove and vineyard; irrigated fruit tree crops; mosaics of crops of fruit trees with conifers
Semi-natural meadows	Semi-natural meadows; mosaics of semi-natural meadows with broadleaf tree species
Shrubland	Mediterranean shrubland dominated by high cover of scrubs of different types and multiple uses (grazing of domestic and wild herbivores, honeybee colonies ...)
Systems in scrub-forest transition	Shrubby and woody vegetation. Associations of scrub-coniferous, scrub-broadleaf tree species, scrub-coniferous and broadleaf tree species
Mixed forests	Multi-specific and heterogeneous forests
Broad-leaved forests	Mediterranean broad-leaved sclerophyllous and deciduous forests. Forests of Holm oak (*Quercus ilex*), Pyrenee oak (*Q. pyrenaica*), juniper (*Juniperus oxycedrus*), and different types of scrubs
Coniferous formations	Montane pine forests and plantations of Scots pine (*Pinus sylvestris*) and black pine (*P. nigra*) in hillside slopes and pinaster pine (*P. pinaster*) in the valley bottoms

(b)

Socioeconomic Variables	Description	Units
Population density	Number of inhabitants per km^2	Inhabitants/km^2
Population aging	Population of 65 years and over in relation to total population	Percentage
Emigration	Number of people that have changed their home outside the municipality in relation to the total population	Percentage
Income per capita	Disposable income per capita	Euros
Agricultural workers	Number of people working in the agricultural sector in relation to the economically active population	Percentage
Agricultural unemployment	Number of unemployed in the agricultural sector in relation to the total of registered unemployed	Percentage
Agricultural sector GDP	Gross domestic product from the agricultural sector	Percentage
Industrial sector GDP	Gross domestic product from the industrial sector	Percentage
Service sector GDP	Gross domestic product from the service sector	Percentage
Urban land area	Municipal area that has all urban services	Percentage
Agricultural land area	Municipal area dedicated to farmland and pastures	Percentage

2.4. Data Analyses

In order to detect and quantify the main social-ecological characteristics of the studied rural landscape and its variation over time, four principal component analyses (PCA) were performed: one for each matrix corresponding to the different land uses and socioeconomic conditions of the municipalities in the different considered time periods. The factor loading of the initial sets of descriptors in the main axes of the PCAs allowed us to identify the main characteristics (tendencies of variation) and indicators of the studied rural landscape as well as their changes over time, both inside of and outside of the limits of the PAs.

To determine the statistical significance of the magnitude of the changes to the LULC and the considered socioeconomic descriptors over time, Student's *t*-tests on two paired samples were used (two series of quantitative measures on the same units) at regional scale and inside of and outside of the PAs.

3. Results

3.1. High Nature Value Farmlands and Landscape Dynamic

PCAs carried out from the data matrices of the municipalities x HNVF descriptors indicate that in the studied periods, the two main ordination axes obtained express the same landscape variation tendencies. Since the explained variance in the first dimension of the PCA is the highest, only the variation expressed by PCA-axis 1 has been considered in both cases. Figure 3a,b, show the distribution of the municipalities along the first ordination axis in the two considered timeframes. The analyses highlight that the structure of the landscape inside of and outside of the boundaries of the PAs established in the study area and its variation over time are very similar.

Table 2. Factor loadings of the land use descriptors on PCAs-Axes 1 of the analysed two periods (variance absorptions are shown in brackets). Loadings of the variables identified as key indicators of the cultural landscape over time are indicated in italics (see Figure 3).

Land Use Descriptors	PCA-Axis 1 1990–2000 (33.33%) Factor Loadings	PCA-Axis 1 2000–2010 (33.42%) Factor Loadings
Pastures	−0.531	−0.268
Dehesas	−0.286	−0.248
Herbaceous crops	−0.243	0.002
Woody crops	−0.166	0.115
Semi-natural meadows	0.767	0.591
Shrubland	0.650	0.503
Systems in scrub-forest transition	−0.325	−0.665
Mixed forests	−0.420	−0.776
Broad-leaved forests	0.306	0.349
Coniferous formations	0.705	0.653

According to the loading of the analysed variables (Table 2; Figure 3a), we can observe that in the first period, the traditional silvopastoral systems that prevail in this territory are only represented by pastures (highest factor loading at the negative end of the PCA-axis 1: −0.531). Similarly, mixed forests (−0.42) and uses linked to the scrub–forest transition (−0.325) are also associated with the negative end of this axis, indicating the great importance that processes such as shrub encroachment and forest expansion have acquired. The positive end of PCA-axis 1 corresponding to the period of 1990–2000 presents land uses characteristic of forest systems with a high degree of naturalness as landscape indicators (in order of importance, according to their weights in the axis formation: 0.767, semi-natural meadows; 0.705, coniferous formations; 0.650, shrubland; 0.306, broad-leaved forests).

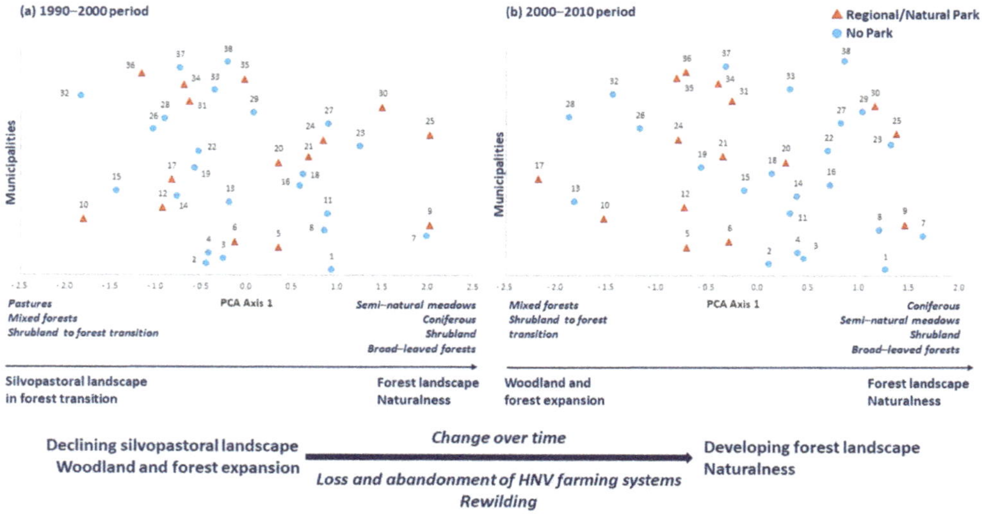

Figure 3. Land use dynamics. Coordinates of the municipalities of the study area along PCAs-axes 1: (**a**) period 1990–2000; (**b**) period 2000–2010. Land use indicators (variables with higher loadings in the PCAs) are shown at both ends of the axes (see Table 2). The codes of the municipalities are indicated in Appendix A.

The PCA calculated for the second period (2000–2010) also shows the interspersion between the municipalities of the study area, regardless of their location inside of or outside of the PA network (Figure 3b). The factor loadings of the indicator variables (Table 2) highlight the transformation of the traditional rural landscape structure towards an eminently forested and naturalized landscape. Thus, at the negative end of the axis, the variables with the greatest loading are mixed forests (−0.776) and systems in transition from scrub to forest (−0.665), while the positive end is characterized, in order of importance, by coniferous formations (0.653), semi-natural meadows (0.591), shrubland (0.503), and broad-leaved forests (0.349). The variation tendencies identified from the analyses indicate a rural landscape structure regression process, in which livestock systems have lost importance in favour of woodland and forest landscapes. This process implies a significant loss of HNVF. The degradation of traditional rural landscapes and their associated HNVF has occurred throughout the territory, both inside of and outside of PAs.

Figure 4 represents a significant decrease over time in the area occupied by the main silvopastoral land uses (pastures and dehesas) both on a regional scale (the entire study area; Figure 4a) and local scale (municipalities inside of and outside of the boundaries of the PA network; Figure 4b,c, respectively). At all scales, this abandonment trend and loss of high natural value pasture systems corresponds to the significant increase of the area occupied by forest systems with a lesser degree of human intervention (Figure 3).

3.2. *Socioeconomic Dimensions of the Rural Landscape*

Figure 5 shows the main socioeconomic characteristics of the territory and their variation over the study time. According to the loadings of the of the PCA variables performed on the data matrix corresponding to the 1990–2000 period (Table 3), the first variation tendency represented by PCA-axis 1 highlights a rural–urban socioeconomic gradient (Figure 5a). Its key indicators are the degree of population aging (factor loading: 0.929), the number of workers in the agricultural sector (0.892), and the agricultural GDP (0.875) at the positive end of the axis, and the income per capita of local people (−0.857), the urban land area of the municipalities (−0.753), and their population density (−0.625) at the negative end. The coordinates of the municipalities along PCA-axis 1 indicate similar characteristics between them, regardless of whether they belong to the PA network or not. However, it is noteworthy that the municipalities with the highest degree of rurality, according to their position on the axis, are not within the limits of the PA.

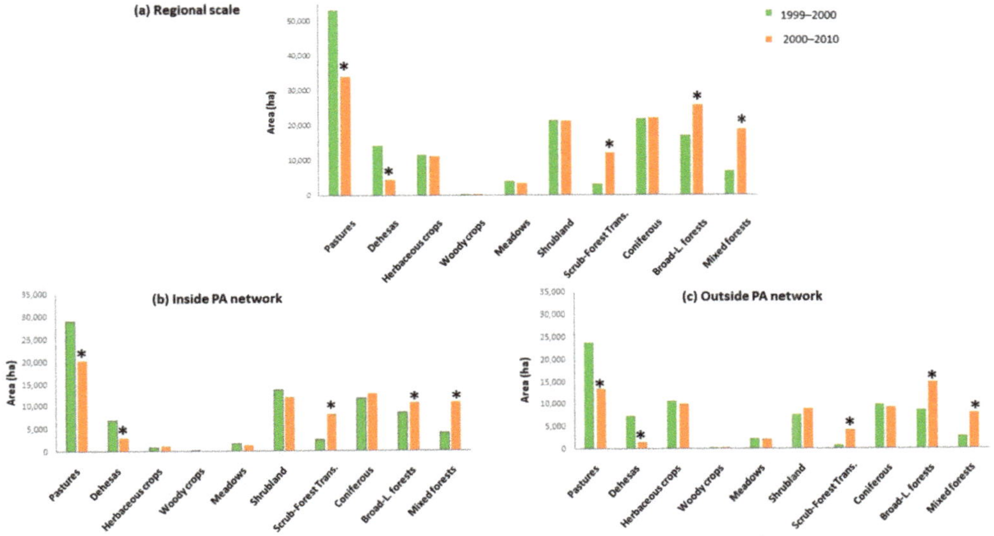

Figure 4. Temporal evolution of the land use descriptors of HNVF in the PA network and their surrounding area: (**a**) at a regional scale (study area); (**b**) inside of and (**c**) outside of the PA boundaries. Statistically significant changes ($p < 0.05$) analysed by means of Student's t test are indicated with an asterisk.

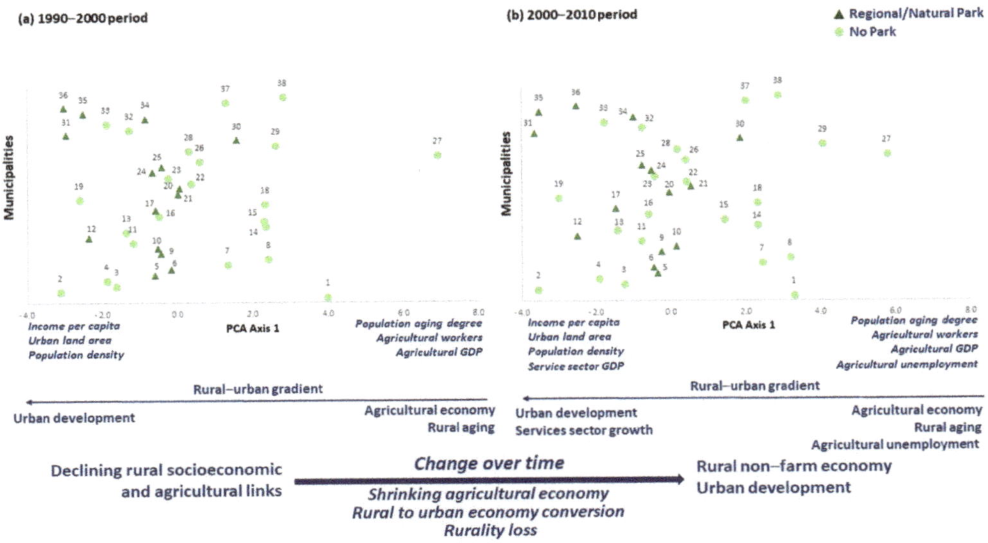

Figure 5. Socioeconomic dynamics. Coordinates of the municipalities of the study area along PCAs-axes 1: (**a**) period 1990–2000; (**b**) period 2000–2010. Socioeconomic indicators (variables with higher loadings in the PCAs) are shown at both ends of the axes (see Table 2). The codes of the municipalities are indicated in Appendix A.

Table 3. Factor loadings of the socioeconomic descriptors on PCAs-Axes 1 of the two analysed periods (variance absorptions are shown in brackets). Loadings of the variables identified as key indicators of the socioeconomic structure of local population over time are indicated in italics (see Figure 5).

Socioeconomic Descriptors	PCA-Axis 1 1990–2000 (40.60%) Factor Loadings	PCA-Axis 1 2000–2010 (43.61%) Factor Loadings
Population density	−0.625	−0.717
Population aging	0.929	0.843
Emigration	0.383	−0.198
Income per capita	−0.857	−0.885
Agricultural workers	0.892	0.818
Agricultural unemployment	0.042	0.687
Agricultural sector GDP	0.875	0.823
Industrial sector GDP	−0.226	0.211
Service sector GDP	−0.265	−0.507
Urban land area	−0.753	−0.782
Agricultural land area	−0.286	0.136

The change of this territory over time accentuates the identified tendency. Thus, both the main socioeconomic indicators detected by the PCA analysis on the 2000–2010 data matrix and the distribution of the municipalities along axis 1 are very similar to those identified in the previous period. The main difference between both time lapses is the development of the service sector in the 2000–2010 period (factor loading: −0.507), which is associated with urban expansion on the negative end of the axis, and the unemployment in the agricultural sector (0.687), which is linked to rural economy on its positive end. Those municipalities that belong to the PA network are preferably related to urban development conditions (Figure 5b). From these analyses, a marked trend of change can be observed from both inside of and outside of the boundaries of the PA network, with a declining rural socioeconomic link to agrarian systems with movement towards a new rural non-farm economy that is related to urban development and a growing economic service sector as well as the decrease of traditional rural activities (Figure 5).

The statistical significance of the changes in the key indicators of the socioeconomic structure obtained from PCA analyses are indicated in Figure 6. It shows the significant increase in urban land area and the decrease in agricultural GDP at both the regional scale and inside of and outside of the PA network (Figure 6a–c). Other statistically significant socioeconomic indicators are the number of workers dedicated to the agricultural sector throughout the territory, the influence area of the parks (Figure 6a,c respectively), and the increase in agricultural unemployment in the municipalities located inside the boundaries of the PA network (Figure 6b).

3.3. *National Park Establishment. Social-Ecological Conditions of a Changing Protected Landscape*

Figure 7 highlights the social-ecological characteristics of the territory that was proposed and later declared as a National Park and its surrounding area. The 2013 declaration that a portion of the studied territory be deemed a National Park involved 14 municipalities, which were considered those with more than 25% of the municipal area inside the boundaries of the park (Appendix A). At the time of the establishment of the National Park, eight of the municipalities belonged to the previous PA network (Regional or Natural Park) (Figure 7a, dark blue triangles within the light blue-shaded area), while the remaining six municipalities were unprotected (Figure 7a, dark blue circles within the light blue-shaded area). Based on the previously performed analyses (see Figure 3), we can verify how the territory prior to the establishment of the National Park was characterized by both the decline of silvopastoral systems and the expansion of woodland and forest as well as the promotion of naturalness. The selection of the area as a new National Park (Figure 7a, blue-shaded area) has prioritized the protection of forest landscapes over the conservation of silvopastoral landscapes of high natural and cultural value, and to a great extent, they have been left outside of the National Park (Figure 7a gray triangles and circles at the negative end of PCA Axis 1, outside the blue-shaded area).

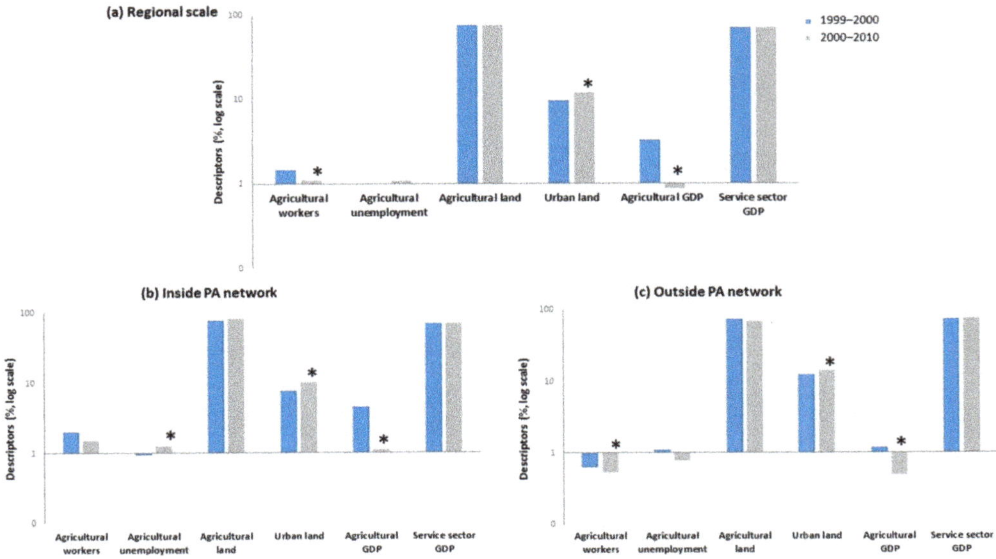

Figure 6. Temporal evolution of the socioeconomic descriptors of the local populations linked to HNVF (**a**) at a regional scale (study area); (**b**) inside of, and (**c**) outside of the PA boundaries. Statistically significant changes ($p < 0.05$) analysed by means of Student's *t* test are indicated with an asterisk.

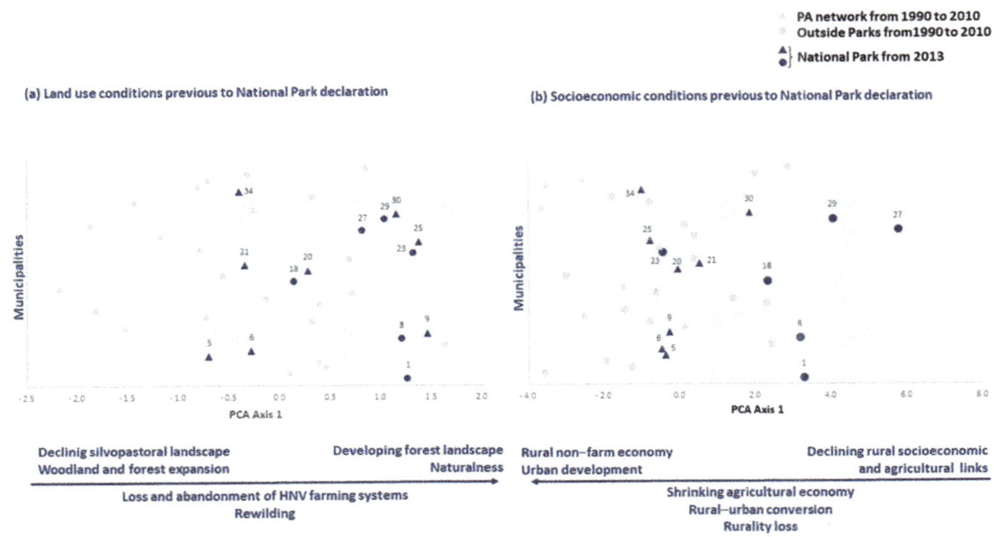

Figure 7. Social-ecological characteristics of the study territory prior to the establishment of the National Park (from the results represented in Figures 3 and 5). The blue-shaded area highlights the municipalities included inside the boundaries of the National Park. (**a**) Land use conditions; (**b**) socioeconomic conditions.

Regarding the socio-economic structure of the territory belonging to the National Park, protection has focused both on some municipalities with a high degree of rurality as well as on others characterized by a considerable level of urban development and the transition from an agriculture-based economy to a service-based one (see Figures 5 and 7b, blue shaded area).

From a social-ecological perspective, the obtained results reveal a certain inconsistency in the selection of the municipalities that would belong to the future National Park. In many cases there are no obvious differences (ecological and socioeconomic) between the selected municipalities and those that would remain outside of the park.

4. Discussion

The cultural character of European rural landscapes has given rise to a rich heritage built over centuries and is representative of both the historical interaction and co-evolution of the natural and social systems and the biocultural diversity of these landscapes [7]. These anthropogenic landscapes often have a high conservation value that depends on the maintenance of traditional agricultural systems and practices [33]. In this paper, the obtained results underline the trend of abandonment and degradation of the studied cultural landscape, which is immersed in an evident gradient of socioeconomic transition from traditional rural conditions to urban ones and is largely responsible for the pressure to which HVN farming systems are currently subjected. The establishment of a conspicuous PA network has not prevented this transformation process of the rural landscape, but rather, it seems to have accentuated it. Numerous studies emphasize that the lack of adequate and effective land planning and management aimed at the conservation of cultural landscapes and their associated HNVF together with institutional deficiencies in supporting local populations and their own TEK have favoured the abandonment of rural landscapes and traditional land uses and have placed the HNVF in a vulnerable position ([17,20,25], among others).

This study, carried out in a county of the Madrid region, with a wide PA network that has been established for years, was performed from a social-ecological approach using an easily replicable methodological development, which has allowed us to quantify the evolution of the territory and the degree of affectation of the HNVF over the last few decades since the declaration of the different PA categories. The design inside of and outside of the PAs, considering municipalities as units of analysis, has been effective in achieving the proposed objectives. In the area, we have detected a similar tendency of LULC change both inside of and outside of PAs (Figure 3). Throughout the studied period, there has been an important and statistically significant decrease in traditional silvopastoral uses (pastures and dehesas) as well as a notable rewilding process, with an evident increase in woodlands and systems in forest transition (Figures 3 and 4). This entails the change of the cultural landscape towards naturalness and the abandonment and loss of HNVF, regardless of their status or category of protection. Land abandonment and passive rewilding results in the degradation of the rural cultural landscape and the interruption of the TEK transmission, which is essential for its maintenance [50,51]. In Europe, the rewilding of cultural landscapes will be one of the most important landscape changes in the coming decades [52].

The changing landscape that has been described is driven by social and economic factors, but it also is dependent on current environmental policies and the lack of support for the rural population, both politically and economically, which hinders the profitability of traditional agricultural practices [17,53,54]. Consequently, forests are increasing in areas that have been abandoned by livestock. These afforestation and rewilding processes that are currently underway are probably induced and supported by the idea of human-caused environmental transformation and degradation and the need to recover a "natural state" [7,25]. The concepts of naturalness and wilderness have been widely used as a point of reference for the conservation, restoration, and management of ecosystems, especially in the nature conservation strategies promoted by the PA guidelines [22]. However, in cultural landscapes with a long history of interaction with humans, such as the one studied here, it seems incongruous to take naturalness as a reference to design conservation plans. The main arguments against rewilding include the loss of valued cultural landscapes, a decrease in landscape heterogeneity, negative impacts on biodiversity and ecosystem services, and an increase in human–wildlife conflicts [55–57]. In this regard, segregation between nature and culture, called "cultural severance", has been described as a serious problem in the conservation of natural and cultural heritage [58].

The socioeconomic conditions of local populations have shaped cultural landscapes and their associated HNVF over the previous centuries (Arnaiz-Schmitz et al., 2018b). Changes in the local socioeconomic system are the major driving forces of changes in land use, as landscape and socioeconomic components constitute a co-evolving system [59,60]. In this case study, a range of socioeconomic factors have allowed us to identify a marked process of urban–rural transition maintained over time, both inside of and outside of the boundaries of the PA network (Figure 5). The main indicators that were identified highlight the existence of an aging rural population that shows an evident decoupling with traditional agricultural socioeconomics. The loss of rurality is related to the increase in urban land area and the development of the service sector (Figures 5 and 6). The social-ecological change from rural to urban systems causes rural decoupling and its corresponding ecological, social, and economic consequences [15,19,25].

Protection efforts through the establishment of the PA network have not prevented the processes of degradation of the cultural landscape, the decline of HNVF, and the loss of rurality that prevail in the area, mainly due to both the abandonment of traditional land uses such as afforestation and urban expansion [20,61,62]. On the contrary, in 2013, three years after the last analysed period, a large part of the study area was declared a National Park, the highest Spanish protection category (Figure 1). The conditions of the territory at the time of the establishment of the National Park were typical of a cultural landscape immersed in a process of transformation, rewilding, and rural marginalization. After the declaration of the National Park, it is foreseeable that the HNVF characteristics of the area will accentuate their degradation process as a consequence of the application of restrictions to local development and the practice of traditional land uses included in the normative schemes of this management category in Spain to promote land uses and activities more related to naturalness than to the protection of the cultural aspects of the territory [22,42].

5. Conclusions

In this paper, we quantify the evolution and current situation of the HNVF belonging to a historically cultural landscape that houses an extensive network of Pas that have been established for decades and that continue to be developed.

The obtained results highlight a significant loss of HNVF, mainly represented by traditional livestock systems (pastures and dehesas), and a marked increase in rewilding processes characterised by scrub–forest transition and the development of forest systems. The observed decrease of HNVF is linked to the disruption of the transmission of TEK and the decline of traditional land uses and practices, which may imply negative consequences both for the high biocultural diversity that these systems host and the cultural identity and the socioeconomics of these rural populations. Thus, the identified socioeconomic indicators reveal the decoupling between the rural population and traditional agricultural socioeconomics. The loss of rurality is mainly related to the transition from an agriculture-based economy to a service-based economy and urban development.

This social–ecological dynamic has been identified both inside of and outside of the boundaries of the PAs, so the transformation of the rural cultural landscape and the abandonment and loss of their HNVF seems to be a generalized process independent of the status or protection category of the territory.

The used method is easily replicable and useful in social-ecological land planning and in the design and implementation of effective management plans for the conservation of rural cultural landscapes as well as in testing the effectiveness of PAs. The design inside of and outside of PAs has proven to be successful in achieving the proposed objectives. Thus, the degradation of the rural landscape and the vulnerability of the HNVF inside the limits of the established PAs reveal the ineffectiveness of their conservation plans, which do not favour the maintenance of traditional rural systems.

In this degraded cultural landscape, a National Park has recently been declared. In Spain, this land protection category is aimed at a type of conservation based on the restriction of human intervention in the environment. The establishment of the National Park has prioritized rewilding processes through land abandonment and the protection of forest landscapes over the conservation of traditional grassland systems of high natural value, which have hardly been considered.

Since rural cultural landscapes and their associated HNVF largely depend on the assesments and decisions of society, our results raise some relevant questions: (i) Are we designing and applying the appropriate management strategies to guarantee the sustainable future of cultural landscapes?; (ii) Are the current regulatory and normative frameworks for PAs really effective in conserving the cultural values and biodiversity of the landscape?

Author Contributions: Conceptualization, methodology, software, analysis, data curation, writing—original draft preparation, writing—review and editing, and supervision: M.F.S., C.A.-S. and P.S.-M.; funding acquisition: M.F.S. and C.A.-S. All authors have read and agreed to the published version of the manuscript.

Funding: This research was funded by the project LABPA-CM: Contemporary Criteria, Methods and Techniques for Landscape Knowledge and Conservation (H2019/HUM-5692), funded by the European Social Fund and the Madrid Regional Government.

Institutional Review Board Statement: Not applicable.

Informed Consent Statement: Not applicable.

Data Availability Statement: Not applicable.

Acknowledgments: The authors thank the Ministry of Science and Innovation and the State Research Agency as well as the postdoctoral support provided through the Juan de la Cierva training program (FECI2019-040562-I/AEI/10.13039/50110001103).

Conflicts of Interest: The authors declare no conflict of interest.

Appendix A

Table A1. Municipalities of the study area inside of and outside of the protected area network: "Cuenca Alta de Manzanares Regional Park", "Cumbre, Circo and Lagunas de Peñalara Natural Park", and "Sierra de Guadarrama National Park".

Municipalities	Municipality Code	Municipal Area within Regional Park Boundaries (%)	Municipal Area within Natural Park Boundaries (%)	Municipal Area within SG National Park Boundaries (%)	Municipal Area within Peripheral Protection Zone of National Park (%)
Alameda del Valle	1	0.0	0.0	25.5	72.4
Alcobendas *	2	10.2	-	0.0	0.0
Algete *	3	0.0	-	0.0	0.0
Alpedrete *	4	0.0	-	0.0	0.0
Becerril de la Sierra	5	57.6	-	20.9	16.2
El Boalo	6	75.3	-	16.8	26.0
Bustarviejo *	7	0.0	-	0.0	0.0
Canencia de la Sierra	8	0.0	-	0.0	82.6
Cercedilla	9	62.6	-	28.1	52.1
Colmenar Viejo	10	29.6	-	0.0	0.0
Collado Mediano *	11	0.0	-	0.0	0.0
Collado Villalba	12	38.1	-	0.0	0.0
Galapagar *	13	7.3	-	0.0	0.0
Garganta de los Montes *	14	0.0	-	0.0	0.0
Gargantilla del Lozoya *	15	0.0	-	0.0	0.0
Guadarrama *	16	0.0	-	0.0	1.3

Table A1. *Cont.*

Municipalities	Municipality Code	Municipal Area within Regional Park Boundaries (%)	Municipal Area within Natural Park Boundaries (%)	Municipal Area within SG National Park Boundaries (%)	Municipal Area within Peripheral Protection Zone of National Park (%)
Hoyo de Manzanares	17	100.0	-	0.0	0.0
Lozoya	18	0.0	-	22.3	74.1
Majadahonda *	19	0.0	-	0.0	0.0
Manzanares el Real	20	98.9	-	55.0	11.3
Miraflores de la Sierra	21	52.7	-	7.3	48.2
El Molar *	22	0.0	-	0.0	0.0
Los Molinos *	23	0.1	-	0.0	44.1
Moralzarzal	24	64.5	-	0.0	0.0
Navacerrada	25	64.1	-	32.9	36.6
Navalafuente *	26	0.0	-	0.0	0.0
Navarredonda y San Mamés	27	0.0	-	3.6	27.3
Pedrezuela *	28	0.0	-	0.0	0.0
Pinilla del Valle	29	0.0	-	27.6	71.4
Rascafría	30	0.0	100.0	52.8	43.7
Las Rozas de Madrid	31	37.0	-	0.0	0.0
San Agustín de Guadalix *	32	0.1	-	0.0	0.0
San Sebastián de los Reyes *	33	14.9	-	0.0	0.0
Soto del Real	34	43.9	-	3.6	27.3
Torrelodones	35	58.3	-	0.0	0.0
Tres Cantos	36	100.0	-	0.0	0.0
El Vellón *	37	0.0	-	0.0	0.0
Villanueva del Lozoya *	38	0.0	-	0.0	0.0

The municipality code used in analyses and graphs and the municipal area within the boundaries of the parks are indicated. An asterisk indicates municipalities with less than 25% of their area within the park network boundaries, or municipalities that are not included at all within the park boundaries.

References

1. Petrosillo, I.; Aretano, R.; Zurlini, G. Socioecological systems. *Encycl. Ecol.* **2015**, *4*, 419–425.
2. Antrop, M. Why landscapes of the past are important for the future. *Landsc. Urban Plan* **2005**, *70*, 21–34. [CrossRef]
3. Marull, J.; Tello, E.; Fullana, N.; Murray, I.; Jover, G.; Font, C.; Coll, F.; Domene, E.; Leoni, V.; Decolli, T. Long-term bio-cultural heritage: Exploring the intermediate disturbance hypothesis in agroecological landscapes (Mallorca, c. 1850–2012). *Biodivers. Conserv.* **2015**, *24*, 217–3251.
4. Vlami, V.; Kokkoris, I.P.; Zogaris, S.; Cartalis, C.; Kehayias, G.; Dimopoulos, P. Cultural landscapes and attributes of "culturalness" in protected areas: An exploratory assessment in Greece. *Sci. Total Environ.* **2017**, *595*, 229–243. [CrossRef] [PubMed]
5. Berkes, F.; Colding, J.; Folke, C. Rediscovery of Traditional Ecological Knowledge as adaptive management. *Ecol. Appl.* **2000**, *10*, 1251–1262. [CrossRef]
6. Hernández-Morcillo, M.; Hoberg, J.; Oteros-Rozas, E.; Plieninger, T.; Gómez-Baggethun, E.; Reyes-García, V. Traditional ecological knowledge in Europe: Status quo and insights for the environmental policy agenda. *Environ. Sci. Policy Sustain. Dev.* **2014**, *56*, 3–17. [CrossRef]
7. Agnoletti, M. Rural landscape, nature conservation and culture: Some notes on research trends and management approaches from a (southern) European perspective. *Landsc. Urban Plan* **2014**, *126*, 66–73. [CrossRef]
8. Agnoletti, M.; Tredici, M.; Santoro, A. Biocultural diversity and landscape patterns in three historical rural areas of Morocco, Cuba and Italy. *Biodivers. Conserv.* **2015**, *24*, 3387–3404. [CrossRef]
9. Fischer, J.; Hartel, T.; Kuemmerle, T. Conservation policy in traditional farming landscapes. *Conserv. Lett.* **2012**, *5*, 167–175. [CrossRef]
10. Sutcliffe, L.M.; Batáry, P.; Kormann, U.; Báldi, A.; Dicks, L.V.; Herzon, I.; Kleijn, D.; Tryjanowski, P.; Apostolova, I.; Arlettaz, R.; et al. Harnessing the biodiversity value of Central and Eastern European farmland. *Divers. Distrib.* **2015**, *21*, 722–730. [CrossRef]
11. Campedelli, T.; Calvi, G.; Rossi, P.; Trisorio, A.; Florenzano, G.T. The role of biodiversity data in High Nature Value Farmland areas identification process: A case study in Mediterranean agrosystems. *J. Nat. Conserv.* **2018**, *46*, 66–78. [CrossRef]
12. Plieninger, T.; Torralba, M.; Hartel, T.; Fagerholm, N. Perceived ecosystem services synergies, trade-offs, and bundles in European high nature value farming landscapes. *Landsc. Ecol.* **2019**, *34*, 1565–1581. [CrossRef]
13. Bignal, E.M.; Mccraken, D.I. The nature conservation value of European traditional farming systems. *Environ. Rev.* **2000**, *8*, 149–171. [CrossRef]
14. Phillips, A. The nature of cultural landscapes—A nature conservation perspective. *Landsc. Res.* **1998**, *23*, 21–38. [CrossRef]
15. Antrop, M. Rural-urban conflicts and opportunities. In *The New Dimensions of the European Landscape*; Jongman, R.H.G., Ed.; Springer: Dordrecht, The Netherlands, 2004; pp. 83–91.
16. Antrop, M. Sustainable landscapes: Contradiction, fiction or utopia? *Landsc. Urban Plan* **2006**, *75*, 187–197. [CrossRef]

17. Plieninger, T.; Höchtl, F.; Spek, T. Traditional land-use and nature conservation in European rural landscapes. *Environ. Sci. Policy* **2006**, *9*, 317–321. [CrossRef]
18. Antrop, M.; Van Eetvelde, V. Mechanisms in recent landscape transformation. *WIT Trans. Built Environ.* **2008**, *100*, 183–192.
19. Arnaiz-Schmitz, C.; Schmitz, M.F.; Herrero-Jáuregui, C.; Gutiérrez-Angonese, J.; Pineda, F.D.; Montes, C. Identifying socio-ecological networks in rural-urban gradients: Diagnosis of a changing cultural landscape. *Sci. Total Environ.* **2018**, *612*, 625–635. [CrossRef] [PubMed]
20. Arnaiz-Schmitz, C.; Herrero-Jáuregui, C.; Schmitz, M.F. Losing a heritage hedgerow landscape. Biocultural diversity conservation in a changing social-ecological Mediterranean system. *Sci. Total Environ.* **2018**, *637*, 374–384. [CrossRef] [PubMed]
21. Cumming, G.S.; Allen, C.R. Protected areas as social-ecological systems: Perspectives from resilience and social-ecological systems theory. *Ecol. Appl.* **2017**, *27*, 1709–1717. [CrossRef]
22. Sarmiento-Mateos, P.; Arnaiz-Schmitz, C.; Herrero-Jáuregui, C.; Pineda, F.D.; Schmitz, M.F. Designing Protected Areas for Social–Ecological Sustainability: Effectiveness of Management Guidelines for Preserving Cultural Landscapes. *Sustainability* **2019**, *11*, 2871. [CrossRef]
23. Verdú, J.R.; Crespo, M.B.; Galante, E. Conservation strategy of a nature reserve in Mediterranean ecosystems: The effects of protection from grazing on biodiversity. *Biodivers. Conserv.* **2000**, *9*, 1707–1721. [CrossRef]
24. Sims, K.R.E. Conservation and development: Evidence from Thai protected areas. *J. Environ. Econ. Manag.* **2010**, *60*, 94–114. [CrossRef]
25. Schmitz, M.F.; Matos, D.G.G.; De Aranzabal, I.; Ruiz-Labourdette, D.; Pineda, F.D. Effects of a protected area on land-use dynamics and socioeconomic development of local populations. *Biol. Conserv.* **2012**, *149*, 122–135. [CrossRef]
26. Schmitz, M.F.; Herrero-Jáuregui, C.; Arnaiz-Schmitz, C.; Sánchez, I.A.; Rescia, A.J.; Pineda, F.D. Evaluating the Role of a Protected Area on Hedgerow Conservation: The Case of a Spanish Cultural Landscape. *Land Degrad. Dev.* **2017**, *28*, 833–842. [CrossRef]
27. Pereira, H.M.; Navarro, L.M. *Rewilding European Landscapes*; Springer International Publishing: New York, NY, USA, 2015.
28. Wallsten, P. The "inside-out" process: A key approach for establishing Fulufjället National Park in Sweden. *Mt. Res. Dev.* **2003**, *23*, 227–229. [CrossRef]
29. Alo, C.A.; Pontius, R.G., Jr. Identifying systematic land-cover transitions using remote sensing and GIS: The fate of forests inside and outside protected areas of Southwestern Ghana. *Environ. Plan. B Plan. Des.* **2008**, *35*, 280–295. [CrossRef]
30. Western, D.; Russell, S.; Cuthill, I. The status of wildlife in protected areas compared to non-protected areas of Kenya. *PLoS ONE* **2009**, *4*, e6140. [CrossRef] [PubMed]
31. Leisher, C.; Touval, J.; Hess, S.M.; Boucher, T.M.; Reymondin, L. Land and forest degradation inside protected areas in Latin America. *Diversity* **2013**, *5*, 779–795. [CrossRef]
32. Plieninger, T.; Bieling, C. Resilience-based perspectives to guiding high-nature-value farmland through socioeconomic change. *Ecol. Soc.* **2013**, *18*, 4. [CrossRef]
33. Lomba, A.; Moreira, F.; Klimek, S.; Jongman, R.H.; Sullivan, C.; Moran, J.; Poux, X.; Honrado, J.P.; Pinto-Correia, T.; Plieninger, T.; et al. Back to the future: Rethinking socioecological systems underlying high nature value farmlands. *Front. Ecol. Environ.* **2020**, *18*, 36–42. [CrossRef]
34. Pineda, F.D.; Montalvo, J. Dehesa systems in the western mediterranean. Biological diversity in traditional land use systems. In *Conserving Biodiversity Outside Protected Areas. The Role of Traditional Agro-Ecosystems*; Halladay, P., Gilmour, D.A., Eds.; Forest Conservation Programme, IUCN: Gland, Switzerland, 1995; pp. 107–122.
35. Ferraz-de-Oliveira, M.I.; Azeda, C.; Pinto-Correia, T. Management of Montados and Dehesas for High Nature Value: An interdisciplinary pathway. *Agrofor. Syst* **2016**, *90*, 1–6. [CrossRef]
36. Martín-Vicente, A.; Fernández-Alés, R. Long term persistence of dehesas. Evidences from history. *Agrofor. Syst.* **2006**, *67*, 19–28. [CrossRef]
37. Bunce, R.H.G.; Perez-Soba, M.; Gomez-Sal, A.; Herzog, F.; Austad, I. (Eds.) *Transhumance and Biodiversity in European Mountains*; Iale-Alterra: Wageningen, The Netherlands, 2004.
38. Caballero, R.; Fernandez-Gonzalez, F.; Badia, R.P.; Molle, G.; Roggero, P.P.; Bagella, S.; Papanastasis, V.P.; Fotiadis, G.; Sidiropoulou, A.; Ispikoudis, I. Grazing systems and biodiversity in Mediterranean areas: Spain, Italy and Greece. *Pastos* **2011**, *39*, 9–154.
39. Herrero-Jáuregui, C.; Arnaiz-Schmitz, C.; Herrera, L.; Smart, S.M.; Montes, C.; Pineda, F.D.; Schmitz, M.F. Aligning landscape structure with ecosystem services along an urban–rural gradient. Trade-offs and transitions towards cultural services. *Landsc. Ecol.* **2019**, *34*, 1525–1545. [CrossRef]
40. Madrid Regional Government. *Steering Plan for Use and Management of Parque Regional de la Cuenca Alta del Manzanares. Direccion General de Educacion y Prevencion Ambiental*; Comunidad de Madrid: Madrid, Spain, 1997.
41. Madrid Regional Government. *Steering Plan for Use and Management of Parque Natural de la Cumbre, Circo y Lagunas de Peñalara. Consejería de Medio Ambiente*; Comunidad de Madrid: Madrid, Spain, 2003.
42. Government of Spain Law 30/2014, of December 3, on National Parks, 2014. Reference: BOE-A-2014-12588. Available online: https://www.boe.es/eli/es/l/2014/12/03/30/con (accessed on 8 July 2021).
43. Salvati, L.; Serra, P. Estimating rapidity of change in complex urban systems: A multidimensional, local-scale approach. *Geogr. Anal.* **2016**, *48*, 132–156. [CrossRef]

44. Sarra, A.; Mazzocchitti, M.; Rapposselli, A. Evaluating joint environmental and costperformance in municipal waste management systems through data envelopment analysis: Scale effects and policy implications. *Ecol. Indic.* **2017**, *73*, 756–771. [CrossRef]
45. Zúñiga-Upegui, P.; Arnaiz-Schmitz, C.; Herrero-Jáuregui, C.; Smart, S.M.; López-Santiago, C.A.; Schmitz, M.F. Exploring social-ecological systems in the transition from war to peace: A scenario-based approach to forecasting the post-conflict landscape in a Colombian region. *Sci. Total Environ.* **2019**, *695*, 133874. [CrossRef] [PubMed]
46. De Aranzabal, I.; Schmitz, M.F.; Aguilera, P.; Pineda, F.D. Modelling of landscape changes derived from the dynamics of socio-ecological systems: A case of study in a semiarid Mediterranean landscape. *Ecol. Indic* **2008**, *8*, 672–685. [CrossRef]
47. Cooper, T.; Arblaster, K.; Baldock, D.; Farmer, M.; Beaufoy, G.; Jones, G.; Poux, X.; McCracken, D.I.; Bignal, E.M.; Elbersen, B.; et al. *Final Report for the Study on HNV Indicators for Evaluation*; Institute for European Environmental Policy: London, UK, 2017.
48. Brunbjerg, A.K.; Bladt, J.; Brink, M.; Fredshavn, J.; Mikkelsen, P.; Moeslund, J.E.; Nygaard, B.; Skov, F.; Ejrnæs, R. Development and implementation of a high nature value (HNV) farming indicator for Denmark. *Ecol. Indic.* **2016**, *61*, 274–281. [CrossRef]
49. ALMUDENA (1990–2010). Municipal and zonal data base. Madrid Government. Available online: http://www.madrid.org/desvan/Inicio.icm?enlace=almudena (accessed on 8 July 2021).
50. Berkes, F.; Folke, C.; Gadgil, M. Traditional Ecological Knowledge, Biodiversity, Resilience and Sustainability. In *Biodiversity Conservation. Ecology, Economy & Environment*; Perrings, C.A., Mäler, K.G., Folke, C., Holling, C.S., Jansson, B.O., Eds.; Springer: Dordrecht, The Netherlands, 1994; Volume 4, pp. 269–287.
51. Carver, S. Rewilding through land abandonment. In *Rewilding*; Pettorelli, N., Duran, S.M., du Toit, J.T., Eds.; Cambridge University Press: Cambridge, UK, 2019; pp. 99–122.
52. Bauer, N.; Wallner, A.; Hunziker, M. The change of European landscapes: Human-nature relationships, public attitudes towards rewilding, and the implications for landscape management in Switzerland. *J. Environ. Manage.* **2009**, *90*, 2910–2920. [CrossRef]
53. Ruiz-Labourdette, D.; Schmitz, M.F.; Montes, C.; Pineda, F.D. Zoning a protected area: Proposal based on a multi-thematic approach and final decision. *Environ. Model. Assess.* **2010**, *15*, 531–547. [CrossRef]
54. Hodge, I.; Hauck, J.; Bonn, A. The alignment of agricultural and nature conservation policies in the European Union. *Conserv. Biol.* **2015**, *29*, 996–1005. [CrossRef] [PubMed]
55. Conti, G.; Fagarazzi, L. Forest expansion in mountain ecosystems:"environmentalist's dream" or societal nightmare. *Planum* **2005**, *11*, 1–20.
56. Rocchini, D.; Perry, G.L.W.; Salerno, M.; Maccherini, S.; Chiarucci, A. Landscape change and the dynamics of open formations in a natural reserve. *Landsc. Urban Plan.* **2006**, *77*, 167–177. [CrossRef]
57. Corlett, R.T. The role of rewilding in landscape design for conservation. *Curr. Landsc. Ecol. Rep.* **2016**, *1*, 127–133. [CrossRef]
58. Bridgewater, P.; Rotherham, I.D. A critical perspective on the concept of biocultural diversity and its emerging role in nature and heritage conservation. *People Nat.* **2019**, *1*, 291–304. [CrossRef]
59. Lacitignola, D.; Petrosillo, I.; Cataldi, M.; Zurlini, G. Modelling socio-ecological tourism-based systems for sustainability. *Ecol. Model.* **2007**, *206*, 191–204. [CrossRef]
60. Ropero, R.F.; Aguilera, P.A.; Fernández, A.; Rumí, R. Regression using hybrid Bayesian networks: Modelling landscape–socioeconomy relationships. *Environ. Modell. Softw.* **2014**, *57*, 127–137. [CrossRef]
61. Petanidou, T.; Kizos, T.; Soulakellis, N. Socioeconomic dimensions of changes in the agricultural landscape of the Mediterranean basin: A case study of the abandonment of cultivation terraces on Nisyros Island, Greece. *J. Environ. Manage.* **2008**, *41*, 250–266. [CrossRef]
62. Anderson, E.; Mammides, C. Changes in land-cover within high nature value farmlands inside and outside Natura 2000 sites in Europe: A preliminary assessment. *AMBIO* **2020**, *49*, 1958–1971. [CrossRef] [PubMed]

land

MDPI

Article

Conservation, Restoration, and Sustainable Use of Biodiversity Based on Habitat Quality Monitoring: A Case Study on Jeju Island, South Korea (1989–2019)

Hyun-Jung Hong [1], Choong-Ki Kim [1,*], Hyun-Woo Lee [1] and Woo-Kyun Lee [2]

[1] Korea Environment Institute, Sejong 30147, Korea; hjhong@kei.re.kr (H.-J.H.); hwlee@kei.re.kr (H.-W.L.)
[2] Department of Environmental Science and Ecological Engineering, Korea University, Seoul 02841, Korea; leewk@korea.ac.kr
* Correspondence: ckkim@kei.re.kr; Tel.: +82-44-415-7007

Abstract: Biodiversity loss is progressing despite biodiversity being essential for human survival, prosperity, and well-being. Conservation, restoration, and sustainable use of the habitat, given that its change is the most prominent factor causing the deterioration of biodiversity, represents a highly effective way of securing biodiversity. Therefore, we assessed and monitored habitat quality as a proxy for biodiversity with habitat quantity in Jeju Island, South Korea. We used an InVEST model with data on the habitat type, suitability, sensitivity, accessibility, and threat factors. Natural habitats throughout Jeju had rapidly decreased in area by 24.9% from 1989 to 2019, and this change contributed to the degradation of habitat quality by 15.8%. We provided significant evidence on the critical degradation of habitat for a long period of over 30 years and highlighted the urgent need for policies and behaviors that enhance biodiversity. We proposed appropriate strategies to prompt people to conserve better, restore effectively, and use biodiversity sustainably. We expect that our findings will provide scientific and evidence-based guidance for policy-making on biodiversity enhancement and will further support achievement of the Sustainable Development Goals and Aichi Biodiversity Targets, in addition to compliance with the New Deal for Nature and People.

Keywords: land-cover change; habitat quality; InVEST; ecosystem-based approach and assessment; nature-based solution; decision-making support; national park management

Citation: Hong, H.-J.; Kim, C.-K.; Lee, H.-W.; Lee, W.-K. Conservation, Restoration, and Sustainable Use of Biodiversity Based on Habitat Quality Monitoring: A Case Study on Jeju Island, South Korea (1989–2019). *Land* **2021**, *10*, 774. https://doi.org/10.3390/land10080774

Academic Editors: Juan F. Beltrán, Pedro Abellán and John Litvaitis

Received: 9 June 2021
Accepted: 20 July 2021
Published: 23 July 2021

1. Introduction

Biodiversity is defined as the diversity of organisms arising from land, aquatic, and other complex ecosystems and includes the diversity within species, diversity between species, and ecosystem diversity [1]. Healthy biodiversity is a source of various resources and services necessary to support human survival, prosperity, and well-being [2,3]. Biodiversity is, however, declining globally at a historically fast rate, as a consequence of continued human population and economic growth [4–6]. International societies have recognized the rapid rate of anthropogenic-induced ecological degradation and have encouraged ecosystem-based approaches, scientifically credible assessments of biodiversity, and nature-based solutions [6,7]. Accordingly, individual countries, including parties to the Convention on Biodiversity, aim to assess biodiversity comprehensively and systematically for supporting decision-making based on these assessment results and to prompt public and private actions to ensure that current and future generations are able to use biological resources [6].

Biodiversity is affected by habitat change, presence of invasive species, climate change, overfishing, deforestation, disease, nutrient loading, and pollution [8–10]. A habitat is defined as "the resources and conditions present in an area that produce occupancy—including survival and reproduction—by a given organism" [11], and habitat change is the most prominent factor causing the deterioration of biodiversity [10,12–14]. Conserving habitats quantitatively and qualitatively is a highly effective way of securing the sustenance of biodiversity [15–18]. Habitat quantity can be regarded as the area of a habitat, and habitat quality as the ability of a habitat to provide conditions that facilitate the persistence of a species, a population, or a community in terms of the resources available within a given ecosystem [11]. Habitat quality determines the composition, reproduction, distribution, behavior, and movement of the habitat's species, as well as its maintenance, as it affects the acquisition and usage of resources essential for survival [19,20]. Therefore, the enhancement of habitat quality is as necessary as increasing the habitat quantity [21].

However, habitat quality is rarely accounted for sufficiently, when devising biodiversity-focused strategies and action plans, due to the following reasons: limited investigation of habitat quality within specific spaces such as protected areas, insufficient data on factors determining habitat quality, difficulty in quantification by the lack of standard measurement units for habitat quality, complex assessment process of habitat quality, and verification shortage of assessment results. Nevertheless, new approaches [22–33] that evaluate biodiversity in terms of habitat quality have recently been established. Nelson et al. [22], Polasky et al. [23], Terrado et al. [24], Ntshane et al. [25], and Salata et al. [26] assessed and mapped habitat quality. Xu et al. [27] monitored the temporal and spatial evolution of habitat quality. Based on the assessment of habitat quality, Terrado et. al. [28] and Nematollahi et al. [29] identified vulnerable habitats, and Wang et al. [30] and Duarte et al. [31] prioritized habitats for protection. Zhu et al. [32] evaluated the impact of urbanization and landscape pattern on habitat quality. Gibson and Quinn [33] assessed habitat quality using land development scenarios and utilized them for landscape planning. Thus, the authors of the aforementioned studies (1) assessed habitat quality; (2) identified, in advance, significant or vulnerable zones of biodiversity, particularly in relation with development plans; and (3) assisted decision makers in formulating actions to be prioritized for the protection and restoration of biodiversity, as well as for mitigation and adaption to biodiversity loss.

South Korea is under severe developmental pressure, propelled by national, local, and infrastructure development projects designed to meet the needs of its large population, as well as by the prioritization of economic growth. This has resulted in physical changes to habitats and continuous deterioration to their quantity and quality [34]. The country has implemented various policies for the enhancement of biodiversity-containing habitats, such as expanding protected areas, securing habitats, and restoring ecological networks. However, the qualitative improvements in habitats are yet to be realized, principally because of the lack of accurate assessment and proper management of habitat quality. Therefore, we aimed to: (1) examine the land-cover changes having the largest relative negative impact on habitat [6], which occurred between 1989 and 2019 on Jeju Island, South Korea, where rich ecosystems and development pressures coexist; (2) assess the change in habitat quality over 30 years; and (3) propose management plans to improve biodiversity by connecting policy alternatives and application target zones based on these assessment results.

2. Materials and Methods

2.1. Study Area

Jeju is a volcanic island in South Korea and has a rich natural heritage. It is the only area in the world where various internationally designated protected areas are located within the same location: World Heritage, Biosphere Reserve, Global Geopark, and RAMSAR Wetland. Jeju is inhabited by approximately 4600 species and was recently voted as a Wonder of Nature [35]. Jeju is composed of (1) mountainous, (2) semi-mountainous, (3) low-lying, and (4) coastal zones. Halla Mountain is located in the mountainous zone (from 600 to 1950 m) and contains a national park. The national park is surrounded by the semi-mountainous zone (between 200 and 600 m) that connects the protected area to the urban district, a low-lying zone (below 200 m; excluding the coastal zone) that predominantly encompasses residential and business districts and a coastal zone that is composed of a tourism and business district (with a standard 500 m coastline buffer zone) (Figure 1).

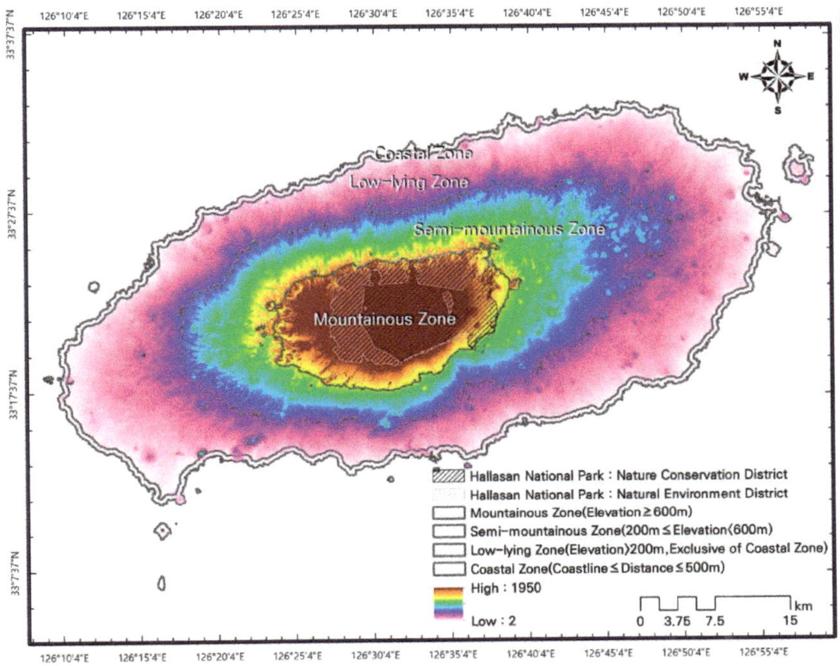

Figure 1. Study area.

The biodiversity of Jeju is subject to social and economic pressures from development that have already adversely affected habitats and damaged biodiversity. This has been exacerbated by the release of green belts that had once prevented urban sprawl and ensured a healthy living environment for citizens through the conservation of the biodiversity surrounding cities [36]. This pressure on biological resources is being intensified by large-scale projects and related real estate development in the semi-mountainous, low-lying, and coastal zones. These developments cause a major influx of tourists and immigrants. The resulting demand for biological resources threatens the sustainability of the biodiversity of Jeju.

2.2. InVEST Habitat Quality Model

The Natural Capital Project has developed a decision-making support software model, named InVEST (Integrated Valuation of Environmental Services and Tradeoffs; copyright by Natural Capital Project; version 3.9.0), which is based on the assessment of the services provided by ecosystems and biodiversity [37]. InVEST assesses habitat quality within a spatial context by combining habitat types with threat factors, to evaluate the spatial extent and degradation of a habitat. We used a corresponding model for habitat quality assessment governed by Equation (1), where habitat suitability (H_j) was converted to a numerical value (Q_{xj}) using a half-saturation function, thereby scoring the degree of the threat of habitat degradation (D_{xj}).

$$Q_{xj} = H_j(1 - \left(\frac{D_{xj}{}^z}{D_{xj}{}^z + k^Z}\right)) \quad (1)$$

where Q_{xj} represents the habitat quality in grid cell x with land use/land cover (LULC) type j; H_j represents the habitat suitability with LULC type j for survivability of a species, a population, or a community; D_{xj} represents the impact of threats on habitat in grid cell x with LULC type j; z represents the scaling constant (2.5); and k represents the half-saturation constant (0.5).

2.3. Construction of Input Data

Habitat quality is affected by habitat type, habitat suitability, factors threatening habitat quality, habitat sensitivity to threat factors, and habitat accessibility [28,33,34]. Habitat type was categorized as water, urban land, barren land, wetland, grassland, forest land, and agricultural land. Habitat suitability (0.00 (uninhabitable)–1.00 (ideally habitable)) for each habitat type was set based on literature review [24,38,39] and local conditions (Table 1). We set the habitat suitability to consider characteristics of Jeju based on a field survey and an interview with local experts, while ensuring, as much as possible, that the range of values used in existing research is not exceeded. The habitat suitability of forest land, habitable for wildlife due to the presence of various resources, and that of urban land, uninhabitable for wildlife due to the predominance of human activities, was set at 0.86 and 0.00, respectively. Agricultural land provides habitat space and food for wildlife. In South Korea, however, agricultural practices are typically intensive, and generally, have a detrimental effect on ecosystems and biodiversity. Accordingly, the suitability of agricultural land was set at 0.30, which was lower than that of grassland.

Habitats are altered by farming, urbanization, development, deforestation, natural disasters, and pollution [12]. We selected agriculture, urbanization, building of roads, forest fragmentation, vulnerability to natural disasters, and pollution treatment facilities as the elements threatening habitat quality and constructed these as spatial data. Agricultural land, urban land, and roads were identified from a land-cover map (issued by the Ministry of Environment, South Korea). Fragmented forests were constructed by extracting fragmented areas of less than 10 ha from the forest patches that were previously larger than 10 ha in area, based on a time-series analysis of land-cover maps (1989, 2019). In South Korea, forest patches 10 ha in area do not act as base habitats for the stable inhabitation of wildlife, but they serve as ecological stepping stones where wildlife rest while migrating or where they come to for food [40,41]. Areas that have been designated as districts requiring improvements to prevent natural disasters or have a history of flooding were selected as areas vulnerable to disasters. We selected sewage treatment plants, wastewater disposal plants, wastewater treatment plants, filtration plants, waste disposal facilities, and landfills as pollution treatment facilities. As in the case of habitat suitability, we established the properties (relative intensity, maximum impact distance, and type of decay based on distance) of threats to habitat sensitivity, to decisively analyze the impact on habitat sensitivity of these threats, based on literature review [24,38,39] and assessment of the local condition (Table 1).

Protected areas in South Korea are designated in keeping with laws specific to the area of interest. Wetland protection areas, specified islands, natural parks, natural monument designation areas, absolute preservation areas, ecosystem preservation areas, river zones, water source protection areas, green zones, and preserved mountainous areas are public interest areas for the conservation of biodiversity. Habitat accessibility for these areas can be categorized according to the level of acceptable activity and restricted access, as stipulated in the area-specific laws. We set the accessibility value of each protected area based on individually categorized levels of protection and usage intensity. We set the accessibility of strictly protected areas as 0.1 and that of non-protected areas as 1.0.

Table 1. Habitat suitability and habitat sensitivity to threat factors on Jeju, South Korea.

Habitat Type	Habitat Suitability [1]	Sensitivity [2] to Threat Factors by Habitat Type					
		Agriculture (0.57, 3.40, Linear) [3]	Urbanization (0.88, 5.90, Exponential) [3]	Roads (0.59, 2.40, Linear) [3]	Forest Fragmentation (0.43, 1.50, Exponential) [3]	Vulnerability to Natural Disasters (0.40, 1.20, Linear) [3]	Pollution Treatment Facilities (0.50, 5.00, Linear) [3]
Water	0.65 (0.65–1.00) [4]	0.65 (0.51–0.78) [4]	0.73 (0.35–0.91) [4]	0.55 (0.24–0.73) [4]	0.00	0.28	0.73
Urban Land	0.00 (0.00–0.30) [4]	0.00 (0.16) [4]	0.00 (0.00–0.01) [4]	0.00 (0.00–0.10) [4]	0.00	0.00	0.00
Barren Land	0.08 (0.00–0.50) [4]	0.15 (0.10–0.20) [4]	0.13 (0.10–0.30) [4]	0.05 (0.30–0.50) [4]	0.00	0.10	0.00
Wetland	0.70 (0.60–0.70) [4]	0.75 (0.30–0.70) [4]	0.70 (0.20–0.30) [4]	0.55 (0.10) [4]	0.00	0.28	0.65
Grass land	0.42 (0.30–0.80) [4]	0.46 (0.20–0.67) [4]	0.45 (0.20–0.80) [4]	0.33 (0.20–0.70) [4]	0.00	0.45	0.00
Forest Land	0.86 (0.50–1.00) [4]	0.66 (0.10–0.70) [4]	0.75 (0.20–0.90) [4]	0.52 (0.20–0.78) [4]	0.45	0.78	0.00
Agricultural Land	0.30 (0.10–1.00) [4]	0.31 (0.01–0.40) [4]	0.41(0.00–0.72) [4]	0.25 (0.20–0.59) [4]	0.00	0.60	0.00

[1] The ability of a habitat to support a viable population over an ecological time-scale [42]. [2] The combination of its ability to tolerate an external pressure without changing significantly its biotic and abiotic characteristics and the time needed to recover once the pressure in question has been alleviated [43]. [3] Relative intensity, maximum impact distance (km), and type of decay associated with threat factor. [4] Range of values (minimum–maximum) of habitat suitability in previous studies [24,38,39].

2.4. Habitat Quality Assessment

We assessed habitat quality throughout Jeju by running the model using our input data. As biodiversity of Jeju has different characteristics according to elevation, we categorized the physical geography of the island into mountainous, semi-mountainous, low-lying, and coastal zones according to altitude (Figure 1) and analyzed the habitat quality for each zone.

The Ministry of Environment, South Korea conducted a field investigation of topography, geology, natural landscapes, green zones, vegetation, and soil, as a part of efforts toward the conservation of biodiversity. This investigation involved a demarcation of the spots where wildlife, including mammals, birds, reptiles, amphibians, fish, insects, and benthic invertebrates, was found. We used the average habitat quality of locations where wildlife was found as a criterion for designating habitats with high quality or low quality. We categorized Jeju based on the average (reference) value of habitat quality in identified sites as follows: areas with habitat quality above 0.56 had favorable habitat quality and were categorized as "conservation zones", whereas areas with habitat quality below 0.56 required improvements to prevent a decline in biodiversity and were categorized as "management zones".

2.5. Habitat Quality Verification

An environmental conservation value assessment map (issued by Ministry of Environment, South Korea) rates land according to its environmental value. It is based on assessments of nationally integrated environmental information and the ranking of items (protected area, diversity, natural characteristics, richness, rarity, fragility, stability, and connectivity) in natural (ecosystem, natural scenery, etc.), living (air, water, soil, etc.), and human/social environments. It is primarily used for land management. First-grade zones on this map are marked as top priority conservation areas, where no development activities are permitted, and only ecological restoration projects can be implemented. Second-grade zones are marked as priority conservation areas, where only small-scale development activities can be implemented. Third-grade zones are marked as main conservation areas, where conditional development is permitted. Fourth-grade zones are marked as already developed or developing areas, where conservation is accomplished partially. Fifth-grade zones are permitted development areas.

An ecological and natural map (issued by Ministry of Environment, South Korea) rates the natural environment of a region, detailing mountains, rivers, inland wetlands, lakes, farmland, and cities according to the ecological, natural, and landscape values. It is generated through field investigation assessments and the ranking of items (vegetation, endangered wildlife, wetlands, and topography) in the natural environment and is used for planning and implementing national/local land use. Separately managed zones on this map are protected areas that have historical, cultural, or landscape significance, or are designated for the conservation of urban green spaces. First-grade zones, which become the major habitats, ecological networks, or ecological corridors for endangered wildlife, require the protection and restoration of natural environments. Second-grade zones, which are areas outside first-grade zones, are required to protect first-grade zones, and use of the natural environment within these zones is permitted for conservation activities or controlled development activities. Third-grade zones comprise all other areas and development activities are permitted in these zones.

We used these maps to verify the reliability of habitat quality assessment results in pixel units (30 × 30 m). We derived Spearman correlation coefficients between model assessment results and these verification maps issued officially by the central government using R software (version 4.1.0). Based on these coefficients, we analyzed the extent to which the conservation zones and the management zones corresponded spatially with the first-, second-, and third-grade zones and with the fourth- and fifth-grade zones, respectively, on the environmental conservation value assessment map. For the additional verification of the habitat quality of the conservation zones, a comparative analysis was conducted based on the separately managed zones and the first- and second-grade zones of the ecological and natural map.

3. Results

3.1. Land-Cover Change between 1989 and 2019

Human-induced change, including land use change and exploitation of biological resources, has had the largest negative impact on biodiversity since 1989. Between 1989 and 2019, overall changes in the area of each land-cover type showed that urban land and agricultural land had increased 1.7- and 0.6-fold, respectively, and that grassland and forest land had decreased 0.3- and 0.2-fold, respectively (Figure 2). In the mountainous zone, forest land constituted 89.6% of the land in 2019, and there were no significant changes to land-cover types over the last 30 years, in terms of legal management based on the Natural Parks Acts. In the semi-mountainous area, the proportion of forest land in 2019 was 51.2%, which was similar to the proportion recorded in 1989. However, urban and agricultural land had increased 6.0- and 0.2-fold, respectively, while grassland had decreased 0.2-fold. In the low-lying zone, there had been a rapid change in land-cover due to productive, economic, and recreational activities. Urban land and agricultural land showed a large increase (1.5- and 0.8-fold, respectively), whereas grassland and forest land showed a marked decline (0.4- and 0.6-fold, respectively). In the coastal zone, urban land had increased 1.3-fold, while grassland and forest land had decreased 0.1- and 0.5-fold, respectively.

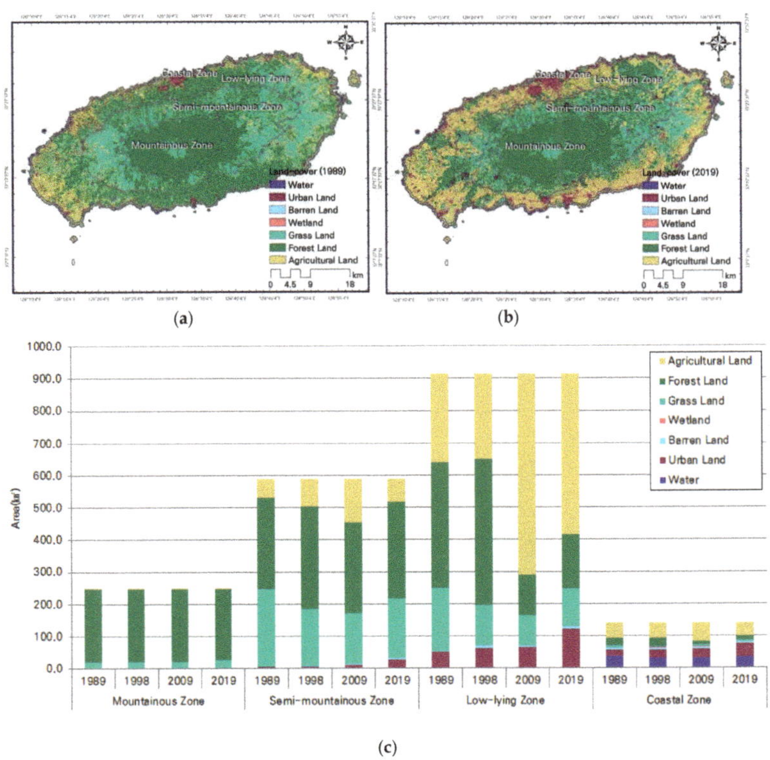

(c)

Figure 2. Changes in land-cover on Jeju, South Korea. (**a**) Distribution of land-cover types in 1989. (**b**) Distribution of land-cover types in 2019. (**c**) Changes in land-cover.

3.2. Habitat Quality Change between 1989 and 2019

The negative trend in land-cover change has contributed to worsening habitat quality. The average habitat quality on Jeju was 0.60 in 2019 and declined 0.2-fold from 1989 to 2019. The habitat quality of the conservation zones also decreased 0.2-fold, and 21.0% of the island changed from a conservation zone to a management zone (Figure 3). In the mountainous zone, average habitat quality was maintained at 0.82 from 1989 to 2019. The conservation zones underwent little change, and the management zones underwent a 0.3-fold increase. In the mountainous zone, 51.6% of the area showed a decline in habitat quality. In the semi-mountainous zone, the average habitat quality declined from 0.62 in 1989 to 0.61 in 2019, impacting 70.9% of the area in this zone. The conservation and management zones showed a 0.1-fold increase and a 0.1-fold decrease in habitat quality, respectively. In the low-lying zone, the average habitat quality declined from 0.55 in 1989 to 0.38 in 2019, impacting 83.0% of the area in this zone. The conservation zones underwent a 0.6-fold decrease in cover, while the management zone cover increased 0.4-fold. Over this time, 7.5% of the area was upgraded from a management zone to a conservation zone, whereas 31.3% was downgraded from a conservation zone to a management zone. In the coastal zone, the average habitat quality continuously decreased 0.3-fold from 0.46 in 1989 to 0.33 in 2019, impacting 64.3% of the area in this zone. The conservation and management zones underwent a 0.2-fold decrease and a 0.2-fold increase, respectively. In the coastal zone, 5.0% of the area was upgraded from a management zone to a conservation zone, whereas 16.5% was downgraded from a conservation zone to a management zone.

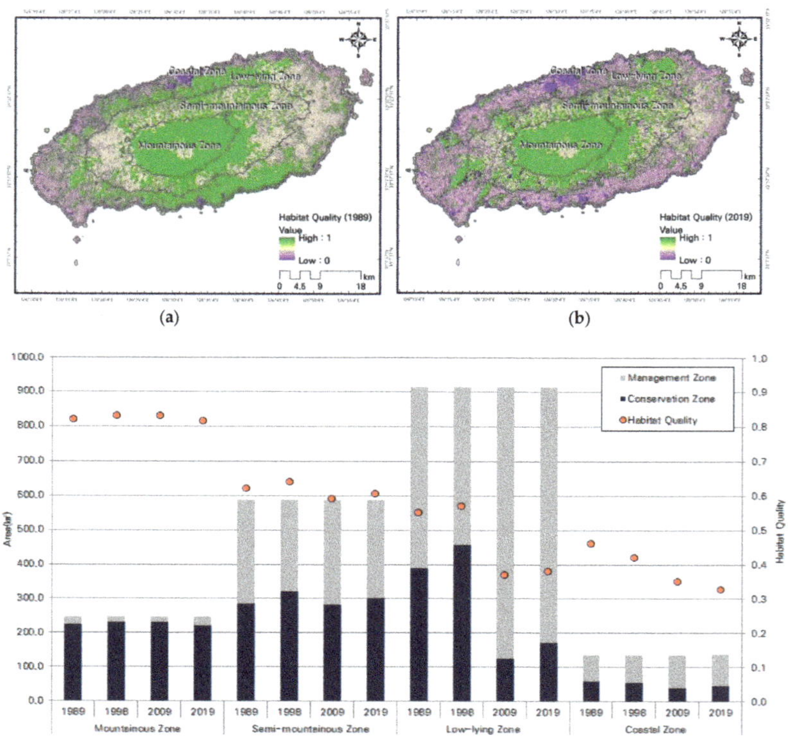

Figure 3. Changes in habitat quality on Jeju, South Korea. (**a**) Distribution of habitat quality in 1989. (**b**) Distribution of habitat quality in 2019. (**c**) Changes in habitat quality.

3.3. Reliability of Habitat Quality Assessments

Statistical analysis showed that the results of habitat quality assessment for 2019 correlated with the environmental conservation value assessment map value of 0.45, as well as the ecological and natural map value of 0.59 (Table 2). Conservation zones recorded for 2019 showed 56.9% consistency with the first-, second-, and third-grade zones of the environmental conservation value assessment map. However, the conservation zones recorded for 2019 were 76.5% consistent with the first-grade, second-grade, and separately managed zones of the ecological and natural map, supporting the validity of our findings. Management zones showed 86.1% consistency with the fourth- and fifth-grade zones of the environmental conservation value assessment map.

Table 2. Spearman correlation coefficient between the results of habitat quality assessment and verification materials of Jeju, South Korea.

	Habitat Quality	ECVAM *	Ecological and Natural Map
Habitat quality	1.00	0.45	0.59
ECVAM *	0.45	1.00	0.53
Ecological and natural map	0.59	0.53	1.00

Number of observations = 2,098,074; p-value < 0.0001. * ECVAM: environmental conservation value assessment map.

4. Discussion

Jeju was found to have a higher habitat quality, compared to other spots where wildlife has been found. This insight triggered rapid land-use changes related to development and tourism (Figure 2). The presence of man-made habitats, including agricultural land, barren land, and urban land, has rapidly increased in extent over the last 30 years, whereas that of natural habitats, including forest land, grassland, wetland, and water, has rapidly decreased. These might be the primary drivers associated with damage to biodiversity on the island (Figure 3). Local residents are also wary of the increasing scale of the aforementioned social and environmental effects due to economic development and support the need for the sustainable management of biodiversity. In a survey conducted among Jeju's residents (September 2015; 120 respondents), 60.2% of the respondents agreed that natural resources need to be protected by immediately implementing appropriate measures, rather than maintaining the status quo through the continuation of existing policies (29.6%), or development for the sake of economic growth (10.2%) [44]. The local government is devising and executing various policies to enhance biodiversity, reflecting the local residents' demands; survey, protection, conservation, expansion, restoration, control, restriction, nationalization, monitoring, research, education, ecotourism, and collaboration [45] (Figure 4).

Despite these efforts by the local government, policy effect has not been achieved due to the disconnection between policy alternatives and their application target zones based on scientific assessments and monitoring. International societies have encouraged policy decision-making based on scientific assessments for habitat, landscape, ecosystem, and biodiversity management [6,7]. However, this has yet to be realized at the local level in South Korea, owing to the lack of personnel, budget, skill, and experience. The identification of excellent or vulnerable spaces for biodiversity [46,47] and the establishment of countermeasures to improve biodiversity [45,48] have been conducted individually. The only habitats to have been properly linked with policy, considering spatial characteristics, are the specific habitats for natural monuments (cactuses) or endangered species (*Azolla japonica*) [49,50], or other specific protected areas [51–53]. Our study findings can be incorporated systematically in policy decision-making by scientifically quantifying the status of, and change in, habitat quality as a proxy for biodiversity, as well as by spatially connecting these results with policy reform. We suggest the following approaches as potential solutions for the biodiversity management of Jeju, based on habitat quality monitoring (Figure 4).

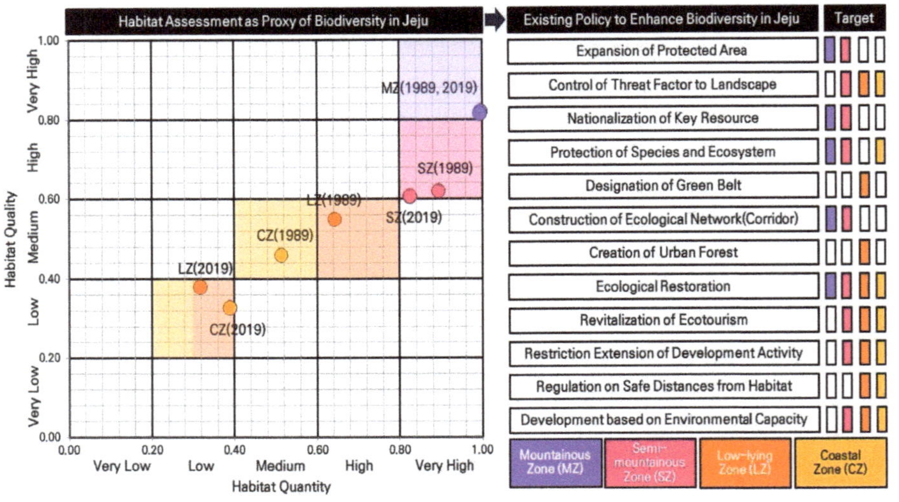

Figure 4. Connection between existing policy alternatives and application target zones based on habitat quality assessment and monitoring for the improvement of biodiversity in Jeju, South Korea.

The habitat quantity (defined for natural habitats in this; forest land, grassland, wetland, and water) and the habitat quality of the mountainous zone were very high and had not undergone much change, owing to legislation-based strict management. However, over the 30-year period, the habitat quality in the zone had declined to half. Thus, the expansion of protected areas, nationalization of key resources, protection of species and ecosystems, construction of ecological networks, and restoration of ecosystems should be selected as areas requiring stronger conservation policies for biodiversity. Both the habitat quantity and the habitat quality of the semi-mountainous zone were very high and had undergone minimal change. However, the habitat quality had declined over more than 70% of this zone due to the large-scale investment into tourism-related land development and the construction of tourist complexes, golf courses, and amusement parks. Policies must be urgently implemented to consider the expansion of protected area, control of threats to the landscape, nationalization of key resources, protection of species and ecosystems, construction of ecological networks, restoration of ecosystems, revitalization of ecotourism, restriction extension of development activities, and development based on environmental capacity. The habitat quantity and the habitat quality of the low-lying zone were low and had undergone major degradation, with the habitat quality having declined across more than 80% of this zone, owing to intensive developmental activities and rapid increase in land area allocation for residential development, tourism, and economic activity. Policies that can be used to mitigate the effects of increasing developmental activities are as follows: control of threat factors to landscape; designating a green belt; creation of urban forests; restoration of ecosystems; revitalization of ecotourism; restriction extension of development activities; regulation on safe distances from habitat; development based on environmental capacity. The habitat quantity and the habitat quality of the coastal zone were low and had undergone moderate decline, with habitat quality having declined over more than 60% of this zone, owing to the construction of coastal roads and encroachment of the shoreline. Primary policy interventions should focus on the control of threat factors to the landscape, protection of species and ecosystems, restoration of ecosystems, revitalization of ecotourism, restriction extension of development activities, regulation on safe distances from habitat, and development based on the environmental capacity.

The mountainous zone is rich in tropical, temperate, cold temperate, and cold species. Particularly, Halla Mountain National Park is an important habitat of the mountainous zone for protection of its notably rich biodiversity, undamaged natural ecosystem, and spectacular scenery. At this park, only minimal measures are allowed for scientific research or the conservation of nature, as per the Natural Parks Act. Additionally, the construction of park facilities, in accordance with limited standards, is permitted at this park. However, with the abolishment of a park entrance fee and designation of the park as an international protected area, there has been a rapid increase in the number of visitors. Consequently, a portion of it was released from protected status and has been allocated for construction and development, and thus, the biodiversity of the mountain is under growing pressure. Based on the results of the present study, significant evidence supporting critical degradation of the habitat can be provided. Habitat quantity of the park decreased from 99.89% in 1989 to 99.62% in 2019, showing negligible change, while the habitat quality of the park decreased from 0.82 in 1989 to 0.80 in 2019, impacting 8.5% of the park. Given this situation, we propose the following strategies for protecting, conserving, restoring, controlling, and utilizing biodiversity of Halla Mountain Park, based on habitat quality monitoring (Figure 5).

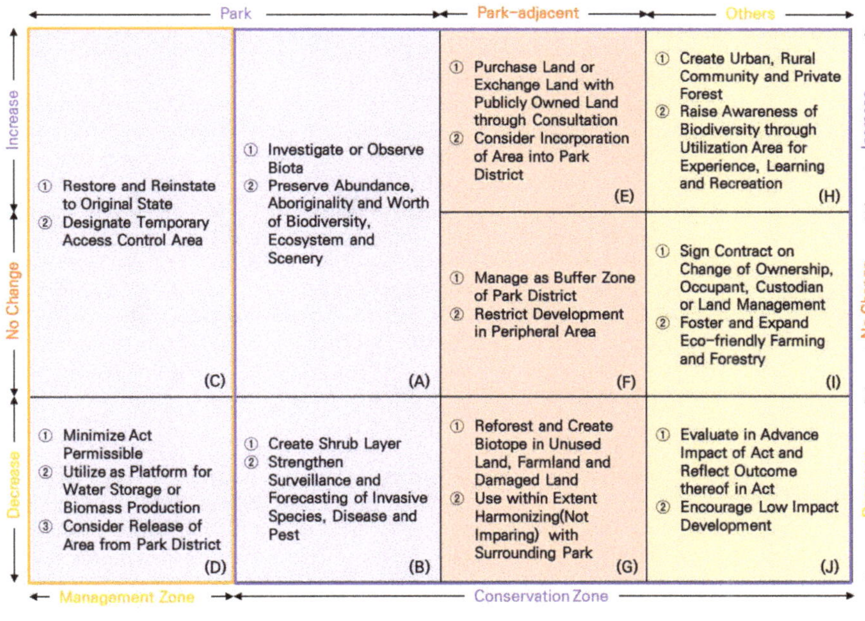

Figure 5. Biodiversity conservation and management strategies for Halla Mountain National Park, based on habitat quality assessment and monitoring in Jeju, South Korea.

The management strategies for the Halla Mountain National Park to preserve and promote the sustainable utilization of biodiversity are as follows. First, a scientific and specialized investigation on the status and characteristics of the topography, ecosystem, biodiversity, and scenery should be conducted over a long term in the areas where habitat quality has improved or has subsequently been maintained as conservation zones within the park district (A). The preservation of the abundance, aboriginality, and worth of the ecosystem, biodiversity, and scenery should be implemented for the long term, based on the aforementioned investigations in this area (A). Second, habitat threats should be mitigated by creating a shrub layer and strengthening surveillance for invasive species, disease, and pests, particularly in the conservation zones where habitat quality has diminished within the park district (B). Third, a special protection area or temporary access control area should be designated to prohibit the access of visitors or the passage of vehicles over the area for a fixed period, and the destroyed biodiversity should be restored to its original state in areas where habitat quality has been maintained or increased as management zones within the park district (C). Fourth, the activities permissible in these areas, including the construction of facilities, should be minimized, with retaining activities required for national defense, public interest, and safety of visitors, where habitat quality has diminished, as management zones, including areas bordering the boundary the park (D). In addition, the utilization as a platform for water storage or biomass production, or the release from the park district should be considered in these areas (D).

Potential management strategies for non-park districts or areas neighboring park districts are as follows. First, conservation zones where the habitat quality has improved should be purchased or exchanged with publicly owned land through consultations and land transactions with owners (E). Such areas should be incorporated into a park district when validity is ascertained through examination of the feasibility of the park district (E). Second, in conservation zones where the habitat quality has been maintained, a park buffer zone should be created to restrict development activities in peripheral areas (F). Third, in conservation zones where habitat quality has diminished, the unused, arable, and damaged lands should be reforested to create a designated biotope (G). This area should be used in union with the surrounding park, without impairment of the scenery and the ecological and cultural environments of the park (G).

Management strategies suggested for locations outside of these areas are as follows. First, in conservation zones where habitat quality has improved, urban eco-villages, private forests, and field programs should be created to enhance public awareness on the conservation of biodiversity and on the adoption of environmental-friendly practices (H). Second, conservation zones where habitat quality has been maintained should be protected as wildlife habitats, if necessary, through contracts detailing regulations related to change in the owner, occupant, and custodian, or in the land management methods (the cultivation of land, use of chemical materials, and creation of wetlands) (I). Eco-friendly farming and forestry practices should be instituted and expanded upon in this area (I). Third, in conservation zones where habitat quality has declined, regulations should be implemented to evaluate, in advance, the impact of land use on biodiversity and to only permit low-impact development (J).

5. Conclusions

There has been a significant increase in public awareness on biodiversity, as well as on its importance to the quality of life of human beings. However, biodiversity is still being lost and degraded, and many of nature's contributions to humans are being compromised [4–6]. Policy-making and implementation, based on scientific and systematic assessment of biodiversity, could lead to adoption of best practices, including the conservation, restoration, and sustainable use of biodiversity. This would, in turn, contribute to the achievement of the Sustainable Development Goals and the Aichi Biodiversity Targets, in addition to compliance with the New Deal for Nature and People. Accordingly, international societies intend to provide scientifically credible and independent up-to-date assessments of biodiversity using indices to realize evidence-informed policy decisions and actions [6,7]. However, prevalent indices used to assess biodiversity, such as the Mean Species Abundance, Living Planet Index, and National Biodiversity Index, which are used in formulating biodiversity strategies, are not sufficient for the qualitative consideration of biodiversity. Habitat quality assessment allows for complex factors that contribute to long-term biodiversity degradation to be identified and adequately informs policy making, as the habitat quality is defined as the ability of a habitat to provide conditions that facilitate the persistence of a species, a population, or a community with available resources within a given ecosystem [11]. Therefore, we assessed the status and trends of habitat quality with land-cover change as a proxy for biodiversity, and proposed strategies that should be considered to conserve biodiversity comprehensively and systematically and to use biological resources sustainably.

Jeju, South Korea, where various protected areas are located, includes a national park, a wetland protection area, and a specified island, among others, thereby encompassing a region with both rich biodiversity and rapid development pressures. A habitat-focused scientific study is urgently needed to identify viable solutions and to formulate appropriate biodiversity management schemes. We monitored the change in habitat quality with land-cover change between 1989 and 2019 in Jeju. We provided detailed evidence regarding critical degradation of the habitat over a long period. We found that natural habitats throughout Jeju had rapidly decreased by 24.9%, from 1989 to 2019, and this change contributed to the degradation of the habitat quality by 15.8%. Using this information, we elucidated significant evidence on the critical degradation of biodiversity over a long period and highlighted the urgent need for policies and behaviors to enhance biodiversity. We proposed appropriate policy measures that could prompt people to conserve better, restore effectively, and use biodiversity sustainably, based on the accumulated knowledge gathered through the aforementioned assessment. These findings can also be used to trigger a shift in focus from the quantitative to qualitative aspects, with respect to decision-making on biodiversity management, which could result in improvement of biodiversity.

Detailed analyses of how biodiversity interacts holistically with pressure/status/trends/resilience in habitats, conducted in line with the advice of interdisciplinary expert panels, working groups, and committees, are required for each habitat type. Professional reviews, between habitat quality and factors affecting habitat quality, should be conducted based on the significance test, and the reliability of habitat quality assessments should be enhanced based on various verification data. The prediction of changes to habitat quality based on various scenarios that reflect past, current, and future conditions can assist policy and decision makers in devising more strategic and timely policy-based interventions to sustain biodiversity, which safeguards socio-economic welfare. Based on the aforementioned contributions, we believe that the results of our study can improve international indices used for the quantitative assessment of biodiversity, provide scientific and evidence-based guidance for policy-making on biodiversity enhancement, and support the achievement of the global vision, which is living in harmony with nature.

Author Contributions: Conceptualization, H.-J.H. and C.-K.K.; methodology, H.-J.H., C.K-K., and W.-K.L.; software, H.-J.H. and C.-K.K.; validation, H.-J.H., C.-K.K., and W.-K.L.; formal analysis, H.-J.H., C.-K.K., and H.-W.L.; investigation, H.-J.H., C.-K.K., and H.-W.L.; resources, H.-J.H. and C.-K.K.; data curation, H.-J.H. and C.-K.K.; writing—original draft preparation, H.-J.H.; writing—review and editing, H.-J.H., C.-K.K., H.-W.L., and W.-K.L.; visualization, H.-J.H.; supervision, C.-K.K. and H.-W.L.; project administration, H.-W.L.; funding acquisition, H.-W.L. All authors have read and agreed to the published version of the manuscript.

Funding: This study was funded by "Development of Decision Supporting Framework to Enhance Natural Capital Sustainability (RR2015-11)" and "An Integrated Assessment to Environmental Valuation via Impact Pathway Analysis (GP2021-09)", supported by the Korea Environment Institute.

Institutional Review Board Statement: Not applicable.

Informed Consent Statement: Not applicable.

Conflicts of Interest: The authors declare no conflict of interest.

References

1. United Nations Convention on Biological Diversity (1992): Article 2. Use of Terms. 1992. Available online: https://www.cbd.int/doc/legal/cbd-en.pdf (accessed on 25 March 2020).
2. Millennium Ecosystem Assessment. *Ecosystems and Human Well-Being: Synthesis*; Island Press: Washington, DC, USA, 2005.
3. Díaz, S.; Fargione, J.; Chapin, F.S.; Tilman, D. Biodiversity Loss Threatens Human Well-Being. *PLoS Biol.* **2006**, *4*, e277. [CrossRef]

4. OECD. *OECD Environmental Outlook to 2050*; OECD Publishing: Paris, France, 2012. [CrossRef]
5. UN Environment. *Glob. Environ. Outlook—GEO-6: Healthy Planet, Healthy People*; UN Environment: Nairobi, Kenya, 2019. [CrossRef]
6. IPBES. *Summary for Policymakers of the IPBES Global Assessment Report on Biodiversity and Ecosystem Services*; IPBES Secretariat: Born, Germany, 2019.
7. Secretariat of the Convention on Biological Diversity. *Global Biodiversity Outlook 2*; Convention on Biological Diversity: Montreal, QC, Canada, 2006.
8. Primack, R. *A Primer of Conservation Biology*, 2nd ed.; Sinauer Associates: Sunderland, UK, 2000.
9. Walther, G.R.; Post, E.; Convey, P.; Menzel, A.; Parmesan, C.; Beebee, T.J.; Fromentin, J.M.; Hoegh-Guldberg, O.; Bairlein, F. Ecological Responses to Recent Climate Change. *Nature* **2002**, *416*, 389–395. [CrossRef] [PubMed]
10. Yuk, G.; Kang, M.; Kang, W.; Go, I.; Bae, S.; Lee, M.; Choi, K.; Heo, J.; Lee, D. Biodiversity and Threat Factors, the Foundation of Ecosystem Services and Human Culture. *J. Environ.* **2010**, *49*, 1–25.
11. Hall, L.S.; Krausman, P.R.; Morrison, M.L. The Habitat Concept and a Plea for Standard Terminology. *Wildl. Soc. Bull.* **1997**, *25*, 173–182.
12. Wilcove, D.S.; Rothstein, D.; Dubow, J.; Phillips, A.; Losos, E. Quantifying Threats to Imperiled Species in the United States. *BioScience* **1998**, *48*, 607–615. [CrossRef]
13. Myers, N.; Mittermeier, R.A.; Mittermeier, C.G.; Da Fonseca, G.A.B.; Kent, J. Biodiversity Hotspots for Conservation Priorities. *Nature* **2000**, *403*, 853–858. [CrossRef] [PubMed]
14. OECD. *Environmental Indicators for Agriculture: Methods and Results*; OECD Publications: Paris, France, 2001.
15. Primack, R. *A Primer of Conservation Biology*, 3rd ed.; Sinauer Associates: Sunderland, UK, 2004.
16. Ferrier, S. Mapping Spatial Pattern in Biodiversity for Regional Conservation Planning: Where to From Here? *Syst. Biol.* **2002**, *51*, 331–363. [CrossRef]
17. World Resources Institute. *World Resources 1994–1995: A Guide to the Global Environment*; Oxford University Press: New York, NY, USA, 1994.
18. Beazley, K.; Smandych, L.; Snaith, T.; MacKinnon, F.; Austen-Smith, P.; Duinker, P. Biodiversity Considerations in Conservation System Planning: Map-Based Approach for Nova Scotia, Canada. *Ecol. Appl.* **2005**, *15*, 2192–2208. [CrossRef]
19. Hayward, L.S.; Busch, D.S. Conservation Behavior and Endocrinology. In *Encyclopedia of Animal Behavior*; Academic Press: London, UK, 2010.
20. Ah-King, M. Flexible Mate Choice. In *Encyclopedia of Animal Behavior*; Academic Press: London, UK, 2010.
21. Jeltsch, F.; Moloney, K.A.; Schwager, M.; Körner, K.; Blaum, N. Consequences of Correlations between Habitat Modifications and Negative Impact of Climate Change for Regional Species Survival. *Agric. Ecosyst. Environ.* **2011**, *145*, 49–58. [CrossRef]
22. Nelson, E.; Mendoza, G.; Regetz, J.; Polasky, S.; Tallis, H.; Cameron, D.R.; Chan, K.M.; Daily, G.C.; Goldstein, J.; Kareiva, P.M.; et al. Modeling Multiple Ecosystem Services, Biodiversity Conservation, Commodity Production, and Tradeoffs at Landscape Scales. *Front. Ecol. Environ.* **2009**, *7*, 4–11. [CrossRef]
23. Polasky, S.; Nelson, E.; Pennington, D.; Johnson, K.A. The Impact of Land-Use Change on Ecosystem Services, Biodiversity and Returns to Landowners: A Case Study in the State of Minnesota. *Environ. Resour. Econ.* **2011**, *48*, 219–242. [CrossRef]
24. Terrado, M.; Sabater, S.; Chaplin-Kramer, B.; Mandle, L.; Ziv, G.; Acuña, V. Model Development for the Assessment of Terrestrial and Aquatic Habitat Quality in Conservation Planning. *Sci. Total Environ.* **2016**, *540*, 63–70. [CrossRef]
25. Ntshane, B.S.; Gambiza, J. Habitat Assessment for Ecosystem Services in South Africa. *Int. J. Biodivers. Sci. Ecosyst. Serv. Manag.* **2016**, *12*, 242–254. [CrossRef]
26. Salata, S.; Ronchi, S.; Arcidiacono, A.; Ghirardelli, F. Mapping Habitat Quality in the Lombardy Region, Italy. *One Ecosyst.* **2017**, *2*, e11402. [CrossRef]
27. Xu, L.; Chen, S.; Xu, Y.; Li, G.; Su, W. Impacts of Land-Use Change on Habitat Quality during 1985–2015 in the Taihu Lake Basin. *Sustainability* **2019**, *11*, 3513. [CrossRef]
28. Terrado, M.; Sabater, S.; Acuña, V. Identifying Regions Vulnerable to Habitat Degradation under Future Irrigation Scenarios. *Environ. Res. Lett.* **2016**, *11*, 114025. [CrossRef]
29. Nematollahi, S.; Fakheran, S.; Kienast, F.; Jafari, A. Application of InVEST Habitat Quality Module in Spatially Vulnerability Assessment of Natural Habitats (Case Study: Chaharmahal and Bakhtiari Province, Iran). *Environ. Monit. Assess.* **2020**, *192*, 1–17. [CrossRef] [PubMed]
30. Wang, Y.; Fu, B.; Colvin, C.; Ennaanay, D.; McKenzie, E.; Chen, M. *Mapping Ecosystem Function Conservation Areas to Integrate Ecosystem Services into Land Use Plans in Baoxing County, China*. 2010. Available online: http://www.teebweb.org/wp-content/uploads/2013/01/Mapping-conservation-areas-for-ecosystem-services-in-land-use-planning-China.pdf (accessed on 17 March 2021).
31. Duarte, G.T.; Ribeiro, M.C.; Paglia, A.P. Ecosystem Services Modeling as a Tool for Defining Priority Areas for Conservation. *PLoS ONE* **2016**, *11*, e0154573. [CrossRef]
32. Zhu, C.; Zhang, X.; Zhou, M.; He, S.; Gan, M.; Yang, L.; Wang, K. Impacts of Urbanization and Landscape Pattern on Habitat Quality Using OLS and GWR Models in Hangzhou, China. *Ecol. Indic.* **2020**, *117*, 106654. [CrossRef]
33. Gibson, D.M.; Quinn, J.E. Application of Anthromes to Frame Scenario Planning for Landscape-Scale Conservation Decision Making. *Land* **2017**, *6*, 33. [CrossRef]
34. Hong, H.; Lee, H.; Yun, I. *Foundational Study on Biosafety Legislation*; Korea Environment Institute: Sejong, Korea, 2015.

35. Hallasan Mountain Research Institute. *Halla Mountain National Park Natural Resource Survey*; Jeju Special Self-Government Province: Jeju, Korea, 2012.
36. Jeon, S.; Hong, H.; Kang, S. Simulation of Urban Growth and Urban Living Environment with Release of the Green Belt. *Sustainability* **2018**, *10*, 3260. [CrossRef]
37. Sharp, R.; Tallis, H.T.; Ricketts, T.; Guerry, A.D.; Wood, S.A.; Chaplin-Kramer, R.; Nelson, E.; Ennaanay, D.; Wolny, S.; Olwero, N.; et al. *InVEST 3.9.0. User's Guide*. 2021. Available online: https://invest-userguide.readthedocs.io/_/downloads/en/3.9.0/pdf/ (accessed on 17 March 2021).
38. Zhong, L.; Wang, J. Evaluation on Effect of Land Consolidation on Habitat Quality Based on InVEST Model. *Trans. Chin. Soc. Agric. Eng.* **2017**, *33*, 250–255. [CrossRef]
39. Gao, Y.; Ma, L.; Liu, J.; Zhuang, Z.; Huang, Q.; Li, M. Constructing Ecological Networks Based on Habitat Quality Assessment: A Case Study of Changzhou, China. *Sci. Rep.* **2017**, *7*, 46073. [CrossRef] [PubMed]
40. Park, C.; Oh, G.; Oh, C.; Han, B. *Development Guidelines for the Construction of an Urban Ecological Axis*; Korea Environment Institute: Seoul, Korea, 2007.
41. Jang, K. A Study on the Construction of the Keumgang West Sea Basin Green Belt Network Using Nearest Feature Model. *Kor. J. Land Arch.* **2007**, *35*, 56–63.
42. Kellner, C.J.; Brawn, J.D.; Karr, J.R. What Is Habitat Suitability and How Should It Be Measured? In *Wildlife 2001: Populations*; Springer: Berlin/Heidelberg, Germany, 2001; pp. 476–488.
43. National Inventory of Natural Heritage. Available online: https://inpn.mnhn.fr/programme/sensibilite-ecologique?lg=en (accessed on 12 December 2020).
44. Lee, H.; Kim, C.; Hong, H.; Noh, Y.; Kang, S.; Kim, J.; Shin, S.; Lee, S.; Kim, T.; Kang, J. *Development of Decision Supporting Framework to Enhance Natural Capital Sustainability: Focusing on Ecosystem Service Analysis*; Korea Environment Institute: Sejong, Korea, 2015.
45. Jeju Green Environment. *Mid-Term Basic Plan for Environment Conservation of Jeju Special Self-Government Province*; Jeju Special Self-Government Province: Jeju, Korea, 2016.
46. Koo, K.; Park, S. Prioritizing Ecologically Important Areas Under Land-Use Changes in Jeju Island, Jeju, Korea. *J. Korean Geogr. Soc.* **2020**, *55*, 253–264.
47. Lee, D.; Jeon, S.W. Estimating Changes in Habitat Quality through Land-Use Predictions: Case Study of Roe Deer (Capreolus pygargus Tianschanicus) in Jeju Island. *Sustainability* **2020**, *12*, 10123. [CrossRef]
48. Kang, H. Jeju Vision for the Future and Special Law on Jeju Island. *World Environ. Isl. Stud.* **2017**, *7*, 237–256.
49. Lee, C.; Jang, G.; Ryu, T.; Choi, B. A Study on Vascular Plants, Distribution Status and Management Plans of the Cactus Habitat (No. 429 Natural Monument) in Wolryung-ri, Jeju Island. *J. Korean Inst. Tradit. Landsc. Archit.* **2018**, *36*, 55–66. [CrossRef]
50. Oh, S.; Kim, B.W.; Kwak, S.; Park, H.; Rim, H.; Song, U. Spatial Distribution, Growth Conditions and Local Utilization for Conservation Strategy of an Endangered Species Azolla japonica. *J. Plant Biol.* **2021**. [CrossRef]
51. Kang, H.; Kim, C.; Kim, E. Human Influence, Regeneration, and Conservation of the Gotjawal Forests in Jeju Island, Korea. *J. Mar. Isl. Cultures* **2013**, *2*, 85–92. [CrossRef]
52. Keith, B.; Kim, E.; Yang, Y.; Lee, S. A Study on Restoration Plans of Jeju Hanon Maar Crater. *World Environ. Isl. Stud.* **2014**, *4*, 45–82.
53. Byun, K.H.; Kang, E.J.; Kim, K.H. Environment Management for Sustainability of Hallasan National Park in Jeju Island, Korea. *Adv. Mater. Res.* **2014**, *905*, 334–338. [CrossRef]

land

Article

China's Wildlife Management Policy Framework: Preferences, Coordination and Optimization

Lu Feng [1], Qiyi Cai [1], Yang Bai [1] and Wenjie Liao [2,*]

1 Law School, Sichuan University, Chengdu 610065, China; fenglu@scu.edu.cn (L.F.); caiqiyi@stu.scu.edu.cn (Q.C.); 2019141460383@stu.scu.edu.cn (Y.B.)
2 Institute of New Energy and Low-Carbon Technology, Sichuan University, Chengdu 610065, China
* Correspondence: wenjieliao@scu.edu.cn or wenjieliao@outlook.com; Tel.: +86-28-6213-8375

Abstract: The coronavirus disease 2019 pandemic led to global concerns about the delicate relationship between humans and wildlife. However, quantitative research on the elements of a wildlife management policy framework in a certain country is lacking. In this study, we try to close this research gap by analyzing the formulation preferences of key elements in the wildlife management policy framework, as well as the coordination between them, in China, which is generally regarded as a main wildlife consumption country. Based on the content analysis of China's wildlife management policy documents, with a three-dimensional analytical framework, we find that: China's wildlife management policy framework prefers the use of compulsory tools, while voluntary and mixed tools are not fully used; adequate attention is paid to the biodiversity conservation objectives and attention is paid to the objectives of public health protection and wildlife welfare, while the utilization objective is restricted to some extent; government sectors, industry, citizens, and non-governmental organizations are involved in wildlife management policies and the degrees of participation of citizens and non-governmental organizations are relatively low. In conclusion, we draw wider implications for China's wildlife management policy formulation, arguing for a more coordinated and participatory policy framework.

Keywords: wildlife protection; biodiversity conservation; policy framework; policy formulation; China

Citation: Feng, L.; Cai, Q.; Bai, Y.; Liao, W. China's Wildlife Management Policy Framework: Preferences, Coordination and Optimization. *Land* **2021**, *10*, 909. https://doi.org/10.3390/land10090909

Academic Editors: Juan F. Beltrán, Pedro Abellán and John Litvaitis

Received: 31 July 2021
Accepted: 27 August 2021
Published: 28 August 2021

1. Introduction

The coronavirus disease 2019 (COVID-19) pandemic has severely disrupted the global society and economy. As COVID-19 is a zoonotic disease and the earlier cases were found in Huanan Seafood Wholesale Market in Wuhan, the wildlife consumption in China led to global concerns about the relationship between wildlife and humans. Actually, issues of wildlife protection in China have drawn long-time attention by the international society, since China has some of the world's richest biodiversity which, at the same time, is highly threatened [1]. For instance, wildlife in the Himalayan region, which accounts for 70% of total area of natural reserves in China, is suffering from illegal hunting and trade [2]. According to the data by TRAFFIC, the wildlife trade monitoring network, a total of 1 million pangolins were captured in the period from 2000 to 2013, and China is one of the largest markets of trafficked pangolins [3]. As analysts, policymakers, and scholars have recognized, given that China has some of the richest biodiversity, and a large amount of wildlife consumption, in the world, any strategy to address wildlife conservation and human health risk should take account of the situation in China.

The problems of wildlife protection, especially overconsumption of wildlife in China, are mainly due to, as pointed out by many scholars [4–7], the lack of a suitable regulatory framework, administrative interference, local protectionism, and issues of public acceptance related to wildlife protection. The above-mentioned problems are closely related to the defects of wildlife-related legislation and policy design [5]. In response to the risk of zoonotic diseases such as COVID-19, China has adopted a legally binding decision to ban consumption of terrestrial wildlife on 24 February 2020. However, scholars believed that this ban is not strong enough in the long run [6,7].

A review of existing wildlife-protection-related studies shows a large number of literature in this area, but the challenges related to wildlife management policies are still under investigated. The limited studies on China's wildlife management policies are mainly divided into three kinds: (1) normative studies of the challenges and measures of the Wildlife Protection Law of China and the recent wildlife consumption ban at a macro level [5,6,8,9], (2) empirical studies of the effects of specific wildlife management policies in China at the micro level [10–12], and (3) comparative studies of discussing wildlife management policies in China and other countries, as well as the political and institutional contexts that influence policymaking [13–15]. However, no scholars have studied the elements of a wildlife management policy framework in a certain country using quantitative methods, resulting in a vague understanding of the key elements of a wildlife management policy framework, such as policy tools, policy objectives, and policy subjects, and a lack of analysis of the relationship between these key policy elements. In this study, we try to close this research gap by analyzing the formulation preferences of the key elements in the wildlife management policy framework, as well as the coordination between them in China, which is generally regarded as a significant actor in worldwide wildlife conservation. Based on the content analysis of China's wildlife policy documents with a three-dimensional analytical framework, we have some new findings about the formulation preferences and coordination of China's wildlife management policy framework, which can provide the academia with enlightening information from China. Besides, our research results provide insights into the optimization of China's wildlife management policy formulation.

To clarify how the formulation preferences of policy tools, objectives, and subjects as key policy elements impacts the coordination of China's wildlife management policy framework, this paper is organized as follows. After the introduction, Section 2 sets up a three-dimensional framework to analyze the key elements of China's wildlife management policy framework. In Section 3, the texts of wildlife management policy documents are encoded and quantitatively analyzed based on the three-dimensional framework. Section 4 discusses the preferences of China's wildlife management policy formulation in terms of policy tools, policy objectives, and policy subjects and the coordination between them. Section 5 draws wider implications, arguing for a more coordinated and participatory wildlife management policy framework.

2. Research Method, Materials and Analytical Framework

2.1. Research Method and Materials

The content analysis method is based on grounded theory and conducts systematic analysis of text content by combining qualitative and quantitative analyses [16], which can be better implemented when there is limited discussion on a phenomenon in existing research literature [17]. The literature review above shows that studies on China's wildlife management policies are relatively limited. Therefore, it is appropriate to adopt the content analysis method, which can, to a certain extent, overcome the subjectivity, uncertainty and ambiguity in the existing qualitative studies on wildlife management policies [18]. This paper follows the usual carry-out steps of the content analysis method, including determination of the research question, selection of research samples, establishment of analytical dimensions, encoding of text units, and quantitative processing and analysis [19].

The analysis materials in this paper are the texts of China's wildlife management policy documents. The policy documents were mainly selected from the "China Law Retrieval System" (that is, "vip.chinalawinfo.com"), which is the most mature and professional retrieval system of laws and regulations in China. Based on the characteristics of wildlife management policy expressions, "wildlife", "terrestrial wildlife", "aquatic", "wildlife protection", "natural reserves", and "habitat" were used as key words individually and in pairs, and full-text retrieval was carried out within the system through a fuzzy search, with the period from 1949 to 2021. Based on the above retrieval conditions, 219 policy documents were finally selected and a text table was made in accordance with the chronological order of policy promulgation for the next-step analysis (Table 1 and Appendix A). These 219 policy documents are all currently valid. If a certain policy has been revised, the text of the policy used for content analysis is the latest revised version, and the latest revision time is shown in Appendix A (Table A1).

Table 1. China's Wildlife Management Policies (Excerpts).

No.	Name of Policy Document	Year of Promulgation
1	Instruction on the Active Protection and Reasonable Utilization of Wildlife Resources	1962
2	Notice on Approving the Report on the Development of Hunting Production	1971
3	Notice on Strengthening Bird Protection and Implementing the China-Japan Migratory Bird Protection Agreement	1981
...	......	...
217	Opinions on Strictly Punishing Illegal Activities during the Prevention and Control of COVID-19 Epidemic in Accordance with the Law	2020
218	Opinions on Punishing Illegal Fishing and Other Crimes in the Yangtze River Basin in Accordance with the Law	2020
219	Norms of Fishery Administrative Enforcement	2021

In terms of the selection of policy documents, the following issues need to be explained. Firstly, the term "policy" is used in a broad sense in this paper, including laws, plans, and policies in a narrow sense [20]. Although China has joined some international wildlife-protection-related conventions, such as the *Convention on International Trade in Endangered Species of Wild Fauna and Flora* (1975) and its Annexes I and II, the *Convention on Wetlands of International Importance Especially as Waterfowl Habitat* (1975), and the *Convention on Biological Diversity* (1993), these international conventions need to be translated and applied through domestic legislation before they can actually be binding on China; therefore, they are not analyzed in this paper. Secondly, in terms of the political system, China is a country featuring a centralized leadership system, so central policies are thus generally applied throughout the country. So far, local governments at different levels of China have issued fewer policies on wildlife protection, and most of them are duplicates of central policies, so the selected documents do not include local policies. Thirdly, policies are selected based on whether they are pertinent to the rights and obligations of wildlife protection stakeholders, regardless of the name of the policy document, such as laws, actions, plans, catalogues, and notices.

2.2. *Analytical Framework*

In order to study the preferences of China's wildlife management policy formulation, the key elements of policy framework, namely policy tools, policy subjects, and policy objectives, are first identified, and a three-dimensional framework composed of the three elements is constructed to analyze China's wildlife management policies.

2.2.1. Dimension X: Policy Tools

The policy tool is the management behavior of a policy subject to achieve a policy objective, and a governance choice for the government's available resources. In view of their complexity, scholars have divided policy tools in many ways according to various standards [21–23]. By observing the application of different policy tools, we can analyze the degree of government intervention in wildlife management and the degree of regulation of actors, to judge the government's preferences in the selection of policy tools. Therefore, in this paper, in terms of the dimension of wildlife management policy tools, Howlett's classification method is adopted and policy tools are divided into compulsory, voluntary, and mixed tools according to the degree of state power intervention in social affairs [24]. Compulsory policy tools mean that the government requires, or prohibits, certain acts of the actor on the basis of its own authority, and forces the actor to comply with regulations, such as the compulsory supervision on public health risks caused by the wildlife industry, and the legal liability for violation of wildlife management regulations. Voluntary policy tools refer to soft and indirect policy means that guide actors to engage in wildlife-management-related activities, including market access, publicity and education, public opinion guidance, information tools, and international exchanges. Mixed policy tools refer to policy means that indirectly encourage and support actors to carry out wildlife management activities through external conditions, including financial subsidies, tax reduction, and science-based species protection standards.

2.2.2. Dimension Y: Policy Objectives

The policy objective is the purpose and effect of policy tools. The same policy tool can be used for different policy objectives; it is not comprehensive enough to use policy tools alone to conduct policy analysis. Therefore, a second dimension, the policy objectives, is usually involved in policy analysis. Based on the existing literature and official reports [15,25,26], and after serval rounds of discussions with scholars in wildlife protection and officials of wildlife-management-related government sectors, wildlife management policy objectives are divided into four categories in this paper: biodiversity conservation, public health protection, utilization, and wildlife welfare. Scholars generally believe that the primary goal of wildlife management policies is to strengthen and improve biodiversity and habitat conservation to maintain the integrity of ecosystems [6,25,26]. Thus, in this paper, biodiversity conservation is identified as the primary objective of China's wildlife management policies. The COVID-19 pandemic created a strong incentive for the Chinese government to prevent the future outbreak of zoonotic disease through policy tools, public health protection is also included in wildlife management policy objectives. Wildlife is considered as a natural resource in the existing Chinese wildlife legislation. However, as societal awareness of wildlife protection is rising, wildlife welfare and animal right issues have attracted public attention and the Chinese government has provided guidance on health and welfare standards for wildlife farming [13,15]. Utilization and wildlife welfare are thus also indispensable to wildlife management policy objectives.

2.2.3. Dimension Z: Policy Subjects

The distinction among policy subjects should be fully considered in the realization process "from policy tools to policy objectives". Therefore, a third dimension is introduced, namely, policy subjects. In addition to be a policymaker, the government sector is also the main implementer of wildlife management policies. The wildlife breeding industry has grown to provide significant economic and social value to China [27,28]. Although the wildlife farming and trading industry does not directly participate in policy formulation, it affects the formulation and implementation of wildlife management policies and is an important policy subject. Non-governmental organizations (NGOs), facilitates communications among governmental sectors, industry, and citizens with their neutrality and professionalism. NGOs in this paper are used in a broad sense, that is, research institutions where scientists or researchers work are included. In addition, citizens' participation also plays an important role in the policymaking and implementation of wildlife management policies in China and other countries [29–31]. Thus, citizens should be analyzed as a separate policy subject. As for landowners, they are included in industry if they engage in activities related to wildlife utilization activities; they are classified as citizens if they are just landowners. Accordingly, in this paper, the policy subjects are divided into four categories: the government sectors, industry, NGOs, and citizens.

By combining policy subjects with policy tools and objectives, a three-dimensional analytical framework for China's wildlife management policy documents is finally formed (Figure 1).

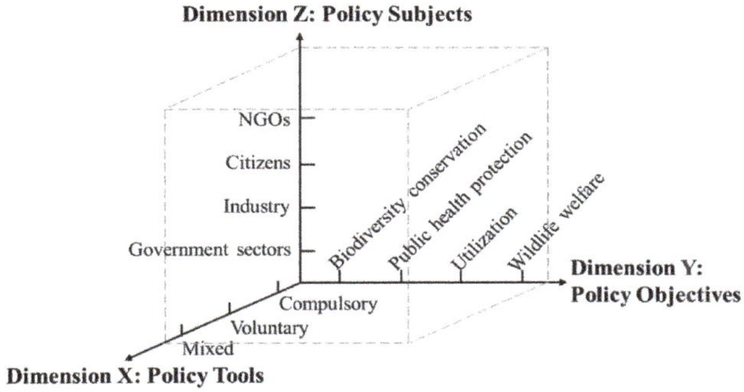

Figure 1. Three-dimensional Analytical Framework of China's Wildlife Management Policy Documents.

3. Content Analysis of China's Wildlife Management Policy Framework

3.1. Text Encoding and Reliability Test

In this paper, the text encoding of China's wildlife management policies was specifically conducted as follows. Firstly, the full text of the 219 policy documents was carefully read through to determine the content analysis unit, which was encoded and classified according to the format of "policy text number-item number". The computer-aided qualitative analysis software NVivo12.0 was used to automate document labeling and coding [32]. Then, the specific content of all the 2652 analysis units that were formed and numbered was classified according to the three-dimensional analytical framework. As for the determination of the analysis unit, chapter, paragraph, subsection, and even sentence can be identified as the appropriate analysis unit [33]. Because the objects of content analysis in this paper are standardized policy texts, policy clause is identified as the content analysis unit. Finally, based on the frequency statistics, the content of all analysis units was further classified into 11 second-level categories under 3 first-level categories, with a total of 4115 reference points (Table 2).

Table 2. Encoding of Content Analysis Units of Wildlife Management Policy Documents.

First-Level Category	Second-Level Category	Encoded Fragment (Excerpts)
Policy tools	Compulsory	In violation of the provisions of these regulations, if the construction unit causes damage to surrounding scenery, water bodies, forests and vegetation, wildlife resources, and topography during the construction process, the scenic area management agency shall order it to stop the illegal act, restore the original state within a time limit, or take other remedies measures and impose a fine from RMB 20,000 to RMB 100,000; if the original status is not restored or effective measures are not taken within the time limit, the scenic area management agency shall order the construction to stop. (81-1)
	Voluntary	The zoo management agency shall formulate a plan for popularizing wild animal science education and set up a dedicated person to be responsible for the popularization of science and use various methods to publicize and educate the public, especially the youth. (40-1)
	Mixed	During the "Ninth Five-Year Plan" period, it is necessary to further improve the national standards for laboratory animals, implement a unified laboratory animal quality certification and notification system throughout the country, and select qualified units as the nationally recognized laboratory animal quality inspection agency to undertake the task of laboratory animal quality inspection. (48-4)

Table 2. *Cont.*

First-Level Category	Second-Level Category	Encoded Fragment (Excerpts)
Policy objectives	Biodiversity conservation	The state encourages zoos to actively carry out scientific research and ex-situ protection of rare and endangered wildlife. (40-2)
	Public health protection	The zoo management agency shall have facilities for sanitation and epidemic prevention, medical rescue, and anesthesia, and shall conduct epidemic prevention and disinfection at regular intervals. Qualified zoos must have animal disease quarantine stations. (40-4)
	Utilization	The governments at all levels shall incorporate fishery production into their national economic development plans and take measures to strengthen the unified planning and comprehensive utilization of waters. (12-2)
	Wildlife welfare	Artificial breeding of wildlife under national key protection shall be conducive to species protection and scientific research and shall not damage wild population resources. According to the habits of wild animals, it should ensure that they have the necessary space for activities, survival and reproduction, sanitation and health conditions, and have the use, type, development site, facilities and technology that are compatible with the scale of wildlife. It should comply with relevant technical standards and epidemic prevention requirements. Do not abuse wildlife. (17-25)
Policy subjects	Government sectors	The National Endpoint Management Office can focus on the on-site supervision and inspection of the import and export activities of bulk trade species and sensitive species, as well as the import and export activities of endangered wildlife. (135-4)
	Industry	In order to guide the healthy development of the domestication and breeding of musk deer and bear, the local area should study and establish an incentive mechanism of "who invests, who owns, and who benefits" according to local situation, and guide and encourage the relevant Chinese medicine enterprises that use the above-mentioned raw materials to actively participate in the artificial breeding of musk deer and improve the technical conditions of bear bile farming. (68-2)
	Citizens	Residents and tourists in the scenic area shall protect the scenery, water bodies, forest and grass vegetation, wildlife and various facilities in the scenic area. (81-1)
	NGOs	Any organization or individual who finds that wild animals need to be sheltered and rescued due to injuries, traps, etc., shall promptly report to the local forestry authority and its wildlife shelter and rescue agency. (164-3)

It is necessary to verify the accuracy of the encoding and avoid subjectivity after all wildlife management policy texts are encoded. Formula (1) is used to calculate the level of agreement between two coders, and Formula (2) is used to calculate the reliability of all codes.

$$A = \frac{2 \times M}{N_1 + N_2} \tag{1}$$

$$R = n \times \frac{A}{1 + [(n-1) \times A]} \tag{2}$$

where, M is the number of items on which both coders agree completely; N_1 is the number of items agreed on by the first coder; N_2 is the number of items agreed on by the second coder; n is the number of coders; A is the level of agreement that both coders agree on, that is, the same level of agreement; R is the code reliability. In this research, another coder (that is, the second coder) was invited to use the "text search" function in the software NVivo12.0 to perform a second round of recoding of the 219 wildlife management policy documents according to the node system established by the first coder. The results showed that the two coders totally agreed on 2900 items, the first coder agreed on 4115 items, and the second coder agreed on 3900 items. The reliability (R) obtained according to the formula was 0.84.

Generally, if the calculation result is greater than 0.8, the reliability test criterion is met [34]. Thus, the reliability of codes in this research conforms to the established criterion.

3.2. One-Dimensional Analysis on the Preferences of China's Wildlife Management Policy Formulation

As shown in Table 3, China's wildlife management policy formulation takes into account compulsory, voluntary, and mixed policy tools, covers the objectives of biodiversity conservation, public health protection, utilization, and wildlife welfare, and involves the subjects of government sectors, industry, citizens, and NGOs, which provides an institutional and regulatory framework for wildlife protection in China. In this part, the formulation preferences of policy tools, objectives, and subjects will be analyzed according to the distribution of 4115 reference points.

Table 3. Node Distribution of China's Wildlife Management Policy Framework among Different Policy Tools, Objectives, and Subjects.

First-Level Category	Second-Level Category	Number of Ref. Points	Total Ref. Points	Proportion
Policy tools	Compulsory	853	1432	59.57%
	Voluntary	266		18.58%
	Mixed	313		21.86%
Policy objectives	Biodiversity conservation	743	1350	55.04%
	Public health protection	292		21. 63%
	Utilization	190		14.07%
	Wildlife welfare	125		9.26%
Policy subjects	Government sectors	744	1333	55.82%
	Industry	226		16.95%
	Citizens	225		16.88%
	NGOs	138		10.35%

From the distribution of the three policy tools, compulsory policy tools account for 59.57%, followed by mixed policy tools (21.86%) and voluntary policy tools (18.58%). Traditionally China was a totalitarian state [35]. Thereby compulsory policy tools were most frequently used, and so were the wildlife management policy tools. However, greater compulsion does not mean higher possibility of achieving the expected policy objectives. Specifically, to a large extent, successful wildlife protection requires cooperation rather than compulsion [7,36]. Since the reform and opening up, China has been shifted from the totalitarian era to the "post-totalitarian era", with the characteristics changing from authoritarianism to bureaucratism [37]. The state has changed its governance strategy and relied on technical governance to a certain extent. Accordingly, the preference of wildlife management policy tools has also changed, as reflected by the proportion of voluntary and mixed policy tools (40.43%).

From the distribution of the four policy objectives, biodiversity conservation is the most frequently referred objective of China's wildlife management policies, accounting for 55.04%, followed by public health protection (21.63%), utilization (14.07%) and wildlife welfare (9.26%). The proportion of each objective of wildlife management policies is relatively reasonable. For a long time, scholars have fiercely criticized that China's wildlife management policies adopted a utilization approach through legalizing and regulating wildlife farming and trading [4,7,27,38]. Nevertheless, according to the statistical results in Table 3, among the policy objectives, the objective of biodiversity conservation accounts for a high proportion while the objective of utilization accounts for a relatively small proportion. In addition, scholars generally hold that the objective of China's wildlife management policies ignores reducing the risk of zoonotic diseases, while the coded data in Table 3 shows that the policy design has paid attention to public health protection to some extent. Furthermore, the welfare of wild animals is also a concern of China's wildlife management policies.

From the distribution of the four policy subjects, China's wildlife management policies mainly involve government sectors, accounting for 55.82%, followed by industry (16.95%), citizens (16.88%), and NGOs (10.35%) China's wildlife management policy framework prefers a government-centric and top-down path, thereby government sectors play a guiding role in wildlife management activities. However, public policy is shaped by various actors, including industry and civil society, who influence policymaking through bottom-up pressure. Public attitude, especially, plays a very important role in the formulation and implementation of wildlife management policies [39]. The statistical results in Table 3 show that, in the framework of China's wildlife management policy, attention to citizens and NGOs is apparently weaker than that to government sectors.

3.3. *Two-Dimensional Analysis on the Preferences of China's Wildlife Management Policy Formulation*

With the help of the three-dimensional analytical framework of wildlife management policy documents in Figure 1, policy tools, policy objectives, and policy subjects are paired for two-dimensional matrix encoding, so as to analyze the coordination between the policy tools, objectives, and subjects of China's wildlife management policy framework in a more intuitive and comprehensive way.

3.3.1. Two-Dimensional Analysis of Policy Subjects and Policy Objectives

According to the results of the matrix coding of policy subjects and policy objectives (Table 4), in terms of biodiversity conservation and public health protection, government sectors bear much higher responsibilities than the other three subjects, accounting for 71.02% and 86.03%, respectively; government sectors and industry are mainly responsible for the realization of utilization, accounting for 67.92% and 18.87%, respectively; in terms of wildlife welfare, government sectors and non-governmental actors (including industry, citizens and NGOs) bear almost equal responsibilities, accounting for 49.06% and 50.94%, respectively. According to the results of the matrix coding of policy subjects and policy objectives (Table 4), in terms of biodiversity conservation and public health protection, government sectors bear much higher responsibilities than the other three subjects, accounting for 71.02% and 86.03%, respectively; government sectors and industry are mainly responsible for the realization of utilization, accounting for 67.92% and 18.87%, respectively; in terms of wildlife welfare, government sectors and non-governmental actors (including industry, citizens, and NGOs) bear almost equal responsibilities, accounting for 49.06% and 50.94%, respectively.

Table 4. Matrix Coding of Policy Subjects and Policy Objectives.

	Biodiversity Conservation	Public Health Protection	Utilization	Wildlife Welfare	Total Ref. Points	Proportion
Government sectors	326	117	36	26	505	72.04%
Industry	47	2	10	11	70	9.99%
Citizens	56	13	5	9	83	11.84%
NGOs	30	4	2	7	43	6.13%
Total ref. points	459	136	53	53	701	100.00%
Proportion	65.48%	19.40%	7.56%	7.56%	100.00%	-

It can be seen that, first of all, government sectors play a vital role in the realization of all policy objects. This is consistent with the conclusion drawn from the aforementioned one-dimensional analysis, that is, government sectors are the main subjects of China's wildlife management policies. On the one hand, biodiversity conservation and public health protection are the policy objectives that are most closely coupled with government sectors, which indicates a lot of government-led supervision of the delicate relationship between humans and wildlife in policy formulation. On the other hand, in the relatively new field of wildlife welfare, policy formulation has considered the joint efforts of the government and civil society. Secondly, all policy objectives have a low degree of coupling with citizens and NGOs, which is consistent with the conclusion in the aforementioned one-dimensional analysis that China's wildlife management policies have not paid enough attention to citizens and NGOs. Thirdly, further statistics show that the coupling degree between biodiversity conservation and industry (67.14%) is even much higher than that between utilization and industry (14.28%). Regarding the wildlife breeding and trading industry, policy formulation should focus on regulating its utilization, rather than imposing excessive protection requirements.

3.3.2. Two-Dimensional Analysis of Policy Objectives and Policy Tools

According to the matrix coding of policy objectives and policy tools (Table 5), compulsory tools are primarily provided to achieve the objective of biodiversity conservation (54.61%), followed by the objectives of utilization (20.65%) and public health protection (17.92%). Voluntary tools are mainly used for the objective of biodiversity conservation (61.04%). For mixed tools, biodiversity conservation is also the main objective, accounting for 67.04%.

Table 5. Matrix Coding of Policy Objectives and Policy Tools.

	Compulsory	Voluntary	Mixed	Total Ref. Points	Proportion
Biodiversity conservation	320	141	181	642	59.06%
Public health protection	105	49	52	206	18.95%
Utilization	121	22	12	155	14.26%
Wildlife welfare	40	19	25	84	7.73%
Total ref. points	586	231	270	1087	100.00%
Proportion	53.91%	21.25%	24.84%	100.00%	-

On the positive side, the design of China's wildlife policy tools gives targeted response to the realization of different policy objectives. In most cases, the objectives of biodiversity conservation, public health protection, and wildlife welfare are achieved through a combination of relatively equivalent compulsory and non-compulsory tools, namely voluntary tools and mixed tools. However, just as it was found by the aforementioned one-dimensional analysis that more compulsory tools are used in China's wildlife management policies, the two-dimensional analysis further shows that, on the one hand, the objective of biodiversity conservation is mainly achieved through compulsory tools, which does not meet the practical needs of wildlife protection [7,36]; on the other hand, the objective of utilization is also mainly achieved through compulsory tools, with the coding coupling degree of 78.06%, which is much higher than that between the utilization objective and non-compulsory tools (21.94%). This reflects that China's wildlife policy tools have not yet created enough social space to guide other policy subjects to participate in wildlife utilization activities in accordance with their own mechanisms and relationships.

3.3.3. Two-Dimensional Analysis of Policy Tools and Policy Subjects

According to the matrix coding of the policy tools and policy subjects (Table 6), most of the policy tools allocated to government sectors are compulsory, accounting for 58.83%, followed by mixed tools (25.11%); the policy tools allocated to industry, citizens and NGOs are also mostly compulsory tools, accounting for 59.15%, 59.69%, and 54.26%, respectively.

Table 6. Matrix Coding of Policy Tools and Policy Subjects.

	Government Sectors	Industry	Citizens	NGOs	Total Ref. Points	Proportion
Compulsory	663	168	154	102	1087	58.54%
Voluntary	181	85	87	73	426	22.94%
Mixed	283	31	17	13	344	18.52%
Total ref. points	1127	284	258	188	1857	100.0%
Proportion	60.69%	15.29%	13.89%	10.12%	100.00%	-

From the above data, it can be seen that China's wildlife management policy subjects and policy tools are relatively matched. Wildlife protection is an area that requires strict government supervision. Therefore, government sectors use more compulsory tools than mixed and voluntary tools. However, just as found by the aforementioned one-dimensional analysis that voluntary tools are least used among China's wildlife management policy tools, the two-dimensional matrix coding further shows that among the policy tools allocated to citizens, voluntary tools account for only 33.73%, which is much lower than compulsory tools (59.59%). For citizens, it is advisable to raise their awareness of wildlife protection through voluntary means such as publicity, education, and public opinion guidance. In addition, among the policy tools allocated to industry, mixed tools account for the smallest proportion (10.92%), which is lower than both compulsory tools (59.15%) and voluntary tools (29.93%). Over-deployment of compulsory tools may hinder industry from playing a self-regulatory role in wildlife breeding and trading activities. At the same time, completely voluntary tools may not be able to effectively control its profit-seeking tendency. Therefore, mixed tools, such as financial subsidies, tax reduction, and industry standards, are more suitable for the industry.

4. Discussion

Based on the one-dimensional and two-dimensional analyses of China's wildlife management policy formulation, we can find the design preferences in terms of policy tools, policy objectives and policy subjects, as well as the coordination between them, and then explore some defects in policy formulation.

Firstly, in terms of policy tools, preferences are given to compulsory tools with the intervention of state power, while voluntary and mixed tools are limitedly used. This is consistent with other scholars' research results [7,36]; the content analysis further indicates that government sectors rely excessively on compulsory tools to achieve all policy objectives, especially biodiversity conservation. As a traditional totalitarian state, China preferred to use compulsory tools to implement policies through a top-down approach, which was regarded as an effective form of command-and-control regulation [40]. Although some changes have taken place with reform and opening up, as well as the introduction of market economy, preferences are still given to compulsory policy tools since the system of administrative appointments and political accountability remains centralized [41]. However, one of the most prominent problems in the implementation of China's wildlife management policies is that the implementation responsibility is distributed across different government sectors, including the forestry department, agriculture department, and market supervision department. These departments are responsible for overseeing wildlife conservation, public health, food safety, and animal health. In the process of policy implementation, insufficient cross-sector collaboration has weakened market supervision, judicial evidence collection, and law enforcement [7]; overlapping management has led to inefficiency and waste of administrative resources [36]. Correspondingly, excessive use of compulsory tools by government sectors will not help resolve the problems of separated or overlapping management but may increase the obstacles to effective wildlife protection in practice given the inflexibility of compulsory tools.

Secondly, in terms of policy objectives, the allocation of four policy objectives are generally reasonable, reflecting that policymakers are fully concerned about biodiversity conservation and public health protection, restricting utilization, and are beginning to pay attention to wildlife welfare. Many scholars have pointed out that China's wildlife management policy texts have ignored biodiversity conservation and paid more attention to utilization through qualitative research [4,7,27,38], but the results of content analysis show that this is not the case. However, there are still some shortcomings in the coordination of policy elements based on the aforementioned content analysis: the excessive protection obligations imposed on the wildlife breeding and trading industry and excessive compulsory tools for achieving the utilization objective have probably increased the operating cost of the industry. Scholars found that, if the cost of illegal wildlife farming and trade is lower than that of legal ones, breeders and traders may circumvent the mandatory provisions of policies, such as circumventing license management, and laundering illegally sourced animals into legally bred ones; on the contrary, if there is no economic advantage to illegal utilization, they may decrease the illegal acts [42–44]. It is in indicated that the vast majority of wild animals in the Chinese market are locally poached and internationally smuggled wild animals [15,45]. This status quo can be attributed not only to weak law enforcement, but also to the inherent logical contradiction in policy design, that is, the wildlife farming and trading industry, who seeks profit, has been imposed with excessive obligations of wildlife protection and allocated more compulsory tools.

Finally, in terms of policy subjects, government sectors, industry, citizens, and NGOs participate in China's wildlife management policies as multiple subjects. Among them, government sectors play a vital role, while citizens and NGOs play a smaller role. Although researchers have pointed out that there is a lack of sufficient public participation in environmental protection, including wildlife protection, in China [6,46], the reflection of this lack of public participation in wildlife management policy framework is still vague. Wildlife farming was expected to help achieve the goal of poverty alleviation [47]. However, studies showed that few wet markets provide low-income citizens with an important source of protein, and wildlife consumers have turned to young, highly-educated, and white-collar citizens; wildlife consumption is seen as a delicacy that demonstrates status and hospitality [11,48]. The shift of wildlife consumers from the poor to the upper-middle class is a serious problem, which indicates the urgency and importance of improving citizens' awareness of public health and biodiversity protection by minimizing the consumption of wildlife. The aforementioned content analysis shows that the neglect of guidance and regulation on citizens' acts, pertinent to wildlife protection in China's wildlife management policy framework, cannot effectively respond to this serious concern. In addition, many people have difficulty in obtaining detailed information on environmental issues and often rely on social media, which can be biased and lead to public misunderstanding [49,50]. NGOs, as a crucial part of civil society, are a bridge for creating an interactive relationship among other policy subjects. Accordingly, the neglect of NGOs in China's wildlife management policy framework may hinder NGOs from playing their due roles of communication and education.

5. Conclusions

In the context of the global COVID-19 pandemic, it is the right and critical time to review the delicate relationship between humans and nature. As far as China's wildlife management policy framework is concerned, it should be formulated in a more reasonable and optimized way among various policy tools, policy objectives, and policy subjects. For this reason, some optimization suggestions are proposed to improve the policy integration towards a more coordinated and participatory wildlife management policy framework.

Firstly, more voluntary and mixed policy tools rather than compulsory ones should be used to achieve the policy objectives of biodiversity conservation and utilization. During the past two decades, the form of governance has undergone great changes, from top-down into a more diversified and participatory process [51,52]. From management to governance, the construction of a service-oriented government has triggered a challenge of the relationship between government, market, and society. As China's population continues to grow, the pressure on the society and economy cannot be avoided. When other job options are exhausted, rural and indigenous communities are more likely to turn to natural resource extraction and wildlife-related shadow economy [53,54]. Accordingly, it is not advisable to excessively adopt compulsory tools for wildlife protection and utilization. More consideration should be given to mixed and voluntary policy tools; for instance, redesigning wet market access mechanisms, guiding wildlife farmers to quit from wildlife industry with the incentive of financial subsidies or taxes, and formulating science-based standards on which species can be utilized.

Secondly, in terms of policy objectives, in addition to further restricting the objective of utilization, the protection obligations for industry should also be reduced in policy design to resolve the aforementioned logical contradiction. Unfortunately, absolute protection currently dominates public opinion and sympathy, with scientific discussion and rational thinking often discounted [55]. Absolute protection includes the random release of animals, which has led to the invasion of alien species weakening and hindering the process of wildlife conservation itself [56]. Similarly, it is not rational to require the wildlife farming and trading industry to bear many responsibilities of wildlife conservation and animal welfare. A more feasible way is not only to ban farming, trading, and consumption of all terrestrial wildlife for food consumption, which was stipulated in the legally binding decision by Chinese legislative body in February 2020, but also to ban various non-food use of wildlife products, gradually changing from commercial utilization to the utilization for public welfare.

Thirdly, the public participation in China's wildlife management policy framework should be strengthened. There is general agreement that public participation in policy formulation has many benefits, including enhancing democratic capacity, increasing justice, empowering citizens, harnessing local information and knowledge, and lessening potential conflicts [57–59]. The contemporary good governance also emphasizes the participation and collaboration of multiple actors in governance. Accordingly, in the field of wildlife management in China, more emphasis should be put on public participation, particularly the participation of citizens and NGOs. A promising example is that, compared with other cities in China, residents of Beijing and Shanghai have significantly increased their support for wildlife protection. This was related to continuous public awareness education campaigns led by various government sectors and NGOs [11]. Recently, some NGOs have brought litigation to court to stop development projects that convert wildlife habitats in China [60]. Therefore, the next policy formulation should enhance continuous science-based popularization and education campaigns of biodiversity conservation, mainly led by NGOs, to ensure long-term behavioral change.

This paper is focused on the discussion of the design and coupling of key elements such as policy tools, objectives, and subjects through the quantitative research of China's wildlife management policy documents. Because China's wildlife management policy framework is broad, with more than 200 policy documents, existing studies' focus on the Wildlife Protection Law of China—although it is the most representative policy—and ignorance of the analysis of most other policies may lead to vague or inaccurate conclusions. In future studies of this field, policy texts and practice should be further effectively connected, and the international cooperation and exchange of policymaking, as well as the effect of policy implementation, should be deeply explored through an effective combination of qualitative research (such as theoretical discussion, case study, and interviews) and quantitative research.

Author Contributions: Conceptualization, L.F. and W.L.; Investigation, Q.C. and L.F.; Methodology, L.F. and Q.C.; Supervision, W.L.; Writing—original draft, Q.C. and L.F.; Writing—review and editing, L.F., W.L. and Y.B. All authors have read and agreed to the published version of the manuscript.

Funding: This work was funded by the National Science & Technology Major Project of China (2016ZX05040001), Soft Science Project of Sichuan Provincial Department of Science and Technology (2020JDR0098), and Special Research Projects of Prevention and Control on COVID-19 Pandemic of Law School, Sichuan University by the Fundamental Research Funds for the Central Universities (2020fxzy-08).

Acknowledgments: The authors appreciate three anonymous reviewers and the academic editor for their constructive critiques on improving the manuscript.

Conflicts of Interest: The authors declare no conflict of interest.

Appendix A

Table A1. China's Wildlife Management Policies.

No.	Name of Policy Document	Year of Promulgation
1	Instruction on the Active Protection and Reasonable Utilization of Wildlife Resources	1962
2	Notice on Approving the Report on the Development of Hunting Production	1971
3	Notice on Strengthening Bird Protection and Implementing the China-Japan Migratory Bird Protection Agreement	1981
4	Marine Environment Protection Law of the People's Republic of China (2017 Amendment)	1983
5	Interim Regulations on the Management of Laboratory Animals in the Health System	1983
6	Rules for the Implementation of the Regulations of the People's Republic of China on Import and Export Animal and Plant Quarantine	1983
7	Circular on Strictly Protecting Precious and Rare Wildlife	1983
8	Notice on Strengthening the Management and Scientific Research of Macaque Resources	1983
9	Grassland Law of the People's Republic of China (2013 Amendment)	1985
10	Forest Law of the People's Republic of China (2019 Amendment)	1985
11	Measures for the Administration of Forest and Wildlife Nature Reserves	1985
12	Fishery Law of the People's Republic of China (2013 Amendment)	1986
13	Urgent Notice on Resolutely Preventing Arbitrary Hunting, Reselling, and Smuggling of Rare Wildlife	1987
14	Regulation on the Administration of Laboratory Animals (2017 Amendment)	1988
15	Notice on Issuing the Interim Management Measures for "Draining Bear Gallbladder"	1988
16	Environmental Protection Law of the People's Republic of China (2014 Amendment)	1989
17	Wildlife Conservation Law of the People's Republic of China (2018 Amendment)	1989
18	Law of the People's Republic of China on the Prevention and Treatment of Infectious Diseases (2013 Amendment)	1989
19	Notice on Strengthening the Administration of the Entry and Exit of Antelope Horns	1990
20	Circular of the General Office of the State Council on the Current Situation of Illegal Hunting, Killing, Purchase and Resale of Rare Wildlife	1990
21	Notice on Strengthening the Administration of Snake Exports	1990
22	Notice on Severely Cracking Down on Illegal Hunting, Purchasing, Reselling, and Smuggling of Wildlife	1990
23	Measures for the Administration of the Domestication and Breeding Licenses of Wildlife under Special State Protection (2015 Amendment)	1991
24	Notice of the National Tourism Administration on the Implementation of the "Emergency Notice of the State Council on Strengthening the Protection of Wildlife and Severely Cracking down on Illegal and Criminal Activities"	1991
25	Notice on Strengthening the Work Report of Wildlife Protection and Management	1991
26	Urgent Notice on Strengthening the Protection of Wildlife and Severely Cracking down on Illegal and Criminal Activities	1991
27	Reply on Clarifying the Attributes of Endangered and Precious Animal Skins, Feathers, Metacarpal Bones and Organs	1991
28	Notice on Strengthening the Protection and Management of Marine Aquatic Wildlife	1991
29	Measures for the Implementation of the Law of the People's Republic of China on the Prevention and Control of Infectious Diseases	1991
30	Regulation of the People's Republic of China on the Protection of Terrestrial Wildlife (2016 Amendment)	1992
31	Notice on the Current Situation of Deforestation, Indiscriminate Hunting, and Comprehensive Control Measures	1992
32	Regulations of the People's Republic of China on the Protection of Aquatic Wildlife (2013 Amendment)	1993
33	Notice on Issuing the "Measures for Charges for the Protection and Management of Terrestrial Wildlife Resources"	1993
34	Notice on Strengthening Enforcement of Environmental and Resource Protection Law Enforcement	1993
35	Notice on Strengthening Environmental Protection Law Enforcement Inspection and Severely Cracking Down on Illegal Activities	1993
36	Notice on Strengthening the Protection of Wildlife in Zoos	1993
37	Notice on Vigorously Strengthening the Protection of Wildlife and Propaganda for Prohibiting the Trade of Endangered Species and Their Products According to Law	1993
38	Agriculture Law of the People's Republic of China (2012 Amendment)	1993
39	Regulations of the People's Republic of China on Nature Reserves (2017 Amendment)	1994
40	Provisions for Administration of City Zoo (2004 Amendment)	1994
41	Notice on Strengthening the Protection and Management of Forest Resources	1994
42	Letter on Issues Concerning the Execution of the Power of Punishment by the Administrative Department of Terrestrial Wildlife	1994
43	Yangtze River Fishery Resources Management Regulations (2004 Amendment)	1995
44	Notice on Strengthening Environmental Protection Work in Tourist Areas	1995
45	Notice on Strengthening the Administration of the Import and Export of Live Wildlife and Plants	1996
46	Criminal Law of the People's Republic of China (2020 Amendment)	1997
47	Notice on Issuing the "Outline of China's Nature Reserve Development Plan (1996–2010)"	1997
48	Several Opinions on the Development of Laboratory Animals during the "Ninth Five-Year Plan" Period	1997
49	Animal Epidemic Prevention Law of the People's Republic of China (2015 Amendment)	1998
50	Fishery Administrative Penalty Provisions	1998

Table A1. *Cont.*

No.	Name of Policy Document	Year of Promulgation
51	Notice on Several Issues Concerning Doing a Good Job in Agriculture and Rural Work in the Autumn and Winter of 1998	1998
52	Notice on launching a Special Struggle against Illegal and Criminal Activities That Destroy Forest Resources	1998
53	Regulations of the People's Republic of China on Concession for Utilization of Aquatic Wildlife (Revised in 2019)	1999
54	Notice on the Implementation of the Simultaneous Monitoring Activities of the Baiji, Finless Porpoise and the Yangtze River Environment	1999
55	Interpretation on Several Issues Concerning the Specific Application of Law in the Trial of Criminal Cases of Destroying Wildlife Resources	2000
56	Notice on Severely Cracking Down on Illegal and Criminal Activities That Destroy Wildlife Resources	2000
57	Urgent Notice on Strengthening the Management of Crocodile	2001
58	Announcement on the No Longer Approving Health Foods Produced from Bear Bile Powder and Creatine	2001
59	Notice on Restricting the Production of Health Food from Wildlife, Plants and Their Products	2001
60	Notice on Prohibiting the Selling of Rare Wildlife and Rare Plants and Their Products in Stores in Isolated Areas at Ports	2002
61	Regulations on the Management of Ocean Fisheries (2020)	2003
62	Notice on Establishing the National Environmental Protection Center for the Protection of Key Aquatic Wildlife of the Yangtze River	2003
63	Notice on the implementation of the "Notice on the Prohibition of Illegal Hunting and Management of Terrestrial Wildlife in Adapting to the Needs of the Situation"	2003
64	Urgent Notice on Cracking Down on Illegal Capture, Management and Utilization of Aquatic Wildlife	2003
65	Urgent Notice on the Prevention and Control of Atypical Pneumonia in Public Places and Food Production and Business Units	2003
66	Notice on Strengthening Food Safety and Market Supervision and Administration during the New Year's Day and Spring Festival	2004
67	Opinions on implementing the "Notice on Strengthening the Protection and Management of Biological Species Resources" of the General Office of the State Council	2004
68	Notice on Further Strengthening the Protection of Musk Deer and Bear Resources and the Management of Their Products as Medicines	2004
69	Notice on Issuing the "Guiding Opinions on Promoting the Sustainable Development of Wildlife and Plants"	2004
70	Notice on Strengthening the Protection and Management of Biological Species Resources	2004
71	Urgent Notice on Strictly Preventing the Spread of Highly Pathogenic Avian Influenza Caused By the Migration of Migratory Birds	2004
72	Notice on Issuing the Work Plan for the Prevention and Control of Infectious SARS and Influenza in the Winter and Spring of 2004-2005	2004
73	Regulation on Handing Major Animal Epidemics Emergencies (2017 Amendment)	2005
74	Measures for the Administration of Examination and Approval of the Species and Quantity of Introduced Terrestrial Wildlife and Alien Species (2016 Amendment)	2005
75	Urgent Notice on Strengthening New Year's Day and Spring Festival Market Supervision to Ensure Consumer Safety	2005
76	Opinions on Continuing to Implement the "Decision of the Central Committee of the Communist Party of China and the State Council on Accelerating Forestry Development"	2005
77	Several Opinions on Accelerating the Construction of the Project of Fast-growing and High-yielding Timber Forest Base	2005
78	Opinions on Strengthening the Construction and Management of Nature Reserves	2005
79	Notice on Issuing the Main Points of Work in 2005	2005
80	Urgent Notice on Strengthening the Research and Management of Highly Pathogenic Microorganisms	2005
81	Regulations on Scenic and Historic Areas (2016 Amendment)	2006
82	Regulations of the People's Republic of China on the Administration of Import and Export of Endangered Wildlife and Plants (2019 Amendment)	2006
83	Regulations on the Prevention and Control of Schistosomiasis (2019 Amendment)	2006
84	Notice on 16 Practical Events for Promoting the Construction of a New Socialist Countryside in 2006	2006
85	Opinions on the Implementation of the "Several Opinions of the Central Committee of the Communist Party of China and the State Council on Promoting the Construction of a New Socialist Countryside"	2006
86	Notice on Further Strengthening the Management of Natural Resources in Nature Reserves	2006
87	Notice on Printing and Distributing the "Main Points of Work of the State Forestry Administration in 2006"	2006
88	Opinions on the implementation of the "Outline of Action for the Conservation of Aquatic Biological Resources in China" to do a good job in the current key fishery work	2006
89	Notice on Issuing the "Eleventh Five-Year Plan for National Fishery Development (2006–2010)"	2006
90	Notice on Implementing Label Management on Tiger Skins, Leopard Skins and Their Products and Further Regulating Their Business and Utilization Activities	2007
91	Urgent Notice on Strengthening the Prevention and Control of Petit Ruminants	2007
92	Notice on Issuing the "Strategy for Prevention and Control of Sudden Acute Infectious Diseases"	2007
93	Notice on Issuing the Guiding Opinions on Promoting the Recovery of Wildlife and Plant Resources and Ecosystems in Nature Reserves after Disasters	2008
94	Urgent Notice on Doing a Good Job in the Prevention and Mitigation of Heavy Rainfalls	2008
95	Administrative Measures for Practicing Veterinarians (2013 Amendment)	2009

Table A1. *Cont.*

No.	Name of Policy Document	Year of Promulgation
96	Notice to Jointly Launch a Special Rectification Action Plan for the Live and Live Bird Market	2009
97	Notice on Printing and Distributing the "Key Points of the State Forestry Administration's 2009 Forestry Publicity and Ideological and Cultural Work"	2009
98	Notice on Printing and Distributing the "Main Points of Work of the State Forestry Administration in 2009"	2009
99	Regulations on Propagation and Release of Aquatic Organisms	2009
100	Notice on Strengthening the Surveillance and Control of Wild Animal Epidemic Diseases This Winter and Next Spring	2010
101	Notice on Issuing the "China Biodiversity Conservation Strategy and Action Plan" (2011–2030)	2010
102	Notice on Strengthening the Management of Crocodile	2010
103	Notice on Strengthening the Management of Aquatic Wildlife Domestication Exhibition Activities in Aquariums and Aquariums	2010
104	Opinions on Further Strengthening Zoo Management	2010
105	Measures for the Administration of National Forest Parks	2011
106	Notice on Printing and Distributing the Action Points for Forestry's Response to Climate Change in the Twelfth Five-Year Plan	2011
107	Notice on Printing and Distributing the "Main Points of Work of the State Forestry Administration in 2011"	2011
108	Notice on Issuing the "Measures for the Management of State-owned Forest Farms"	2011
109	Notice on Strengthening the Work Safety of Oceanic Fisheries	2011
110	Notice on Issuing the Twelfth Five-Year Plan for the Development of National Fisheries	2011
111	Urgent Notice on Effectively Strengthening Law Enforcement of Wildlife Protection	2012
112	Urgent notice on Strictly Preventing Illegal Hunting of Migratory Birds and Other Wildlife	2012
113	Notice on Issuing the National Medium and Long-term Animal Disease Prevention and Control Plan (2012–2020)	2012
114	Notice on the Special Law Enforcement Inspection of Illegal Development and Construction Projects in Aquatic Biological Nature Reserves Above the Provincial Level	2012
115	Administrative Measures for Monitoring and Control of Epidemics and Epidemic Sources for Terrestrial Wildlife	2013
116	Management Regulations on Wetland Protection (2017 Amendment)	2013
117	Notice on Effectively Doing a Good Job in Monitoring and Controlling Wild Animal Epidemic Diseases in Spring	2013
118	Notice on Printing and Distributing the "Main Points of National Forestry Propaganda and Ideological and Cultural Work in 2013"	2013
119	Opinions on the Implementation of the "Agricultural Technology Extension Law of the People's Republic of China"	2013
120	Notice on Printing and Distributing the "Main Points of Work of the State Forestry Administration in 2013"	2013
121	Notice on Further Strengthening the Management of Chinese Medicinal Materials	2013
122	Notice on Issuing the Tasks of the Fishery Resources Protection Project in 2013	2013
123	Notice on Doing a Good Job in the 2013 Yangtze River Basin Fishery Law Enforcement and Fishery Protection Action	2013
124	Notice on Doing a Good Job in the 2013 Pearl River Basin Fishery Administration, Law Enforcement and Fishery Protection Action	2013
125	Notice on Issuing the National Zoo Development Outline	2013
126	Urgent Notice on Actively Responding to Human Infection with H7N9 Avian Influenza	2014
127	Emergency Notice on Effectively Strengthening the Protection of Migratory Birds in Autumn and Winter and the Monitoring and Control of Epidemic Diseases	2014
128	Notice on Issuing the Tasks of the Fishery Resources Protection Project in 2014	2014
129	Notice on Doing a Good Job in the 2014 Yangtze River and Pearl River Basin Fishery Administration, Law Enforcement and Fishery Protection Operations	2014
130	Notice on further strengthening the protection and management of the Yangtze finless porpoise	2014
131	Notice on Further Strengthening the Protection of Marine Wildlife	2015
132	Notice on Doing a Good Job in the Law Enforcement of Fishery Administration in Related Waters in 2015	2015
133	Notice on Strengthening the Protection and Standardization of Operation and Utilization Management of Giant Salamander Resources	2015
134	Provisions on Several Issues Regarding the Trial of Related Cases Occurring in my country's Waters Under jurisdiction (1)	2016
135	Announcement No. 1, 2016 of the Office of Import and Export Administration of Endangered Species-Measures for the Supervision and Inspection of Licensees for the Administrative License of Import and Export Certificates of the National Office of Endangered Species Measures for the Supervision and Inspection of Licensees (2016 Amendment)	2016
136	Guiding Opinions on the use of government and social capital cooperation models to promote forestry construction	2016
137	Notice on Effectively Doing a Good Job in the Protection and Publicity Activities of Spring Migratory Birds and Other Wildlife	2016
138	Notice on Doing a Good Job in Rehabilitation and Reconstruction of Forestry after Flood Disaster	2016
139	Notice on Printing and Distributing China's National Plan for Implementing the 2030 Agenda for Sustainable Development-Forestry Action Plan	2016
140	Notice on Doing a Good Job in the Law Enforcement of the Protection of Migratory Birds and Other Wildlife in Autumn and the Monitoring and Control of Epidemic Diseases	2016

Table A1. *Cont.*

No.	Name of Policy Document	Year of Promulgation
141	Announcement on the Results of the National Special Action against Reclamation and Occupation of Lakes and Wetlands	2016
142	Guiding Opinions on Promoting the Development of China's Forestry Internet of Things	2016
143	Notice on Printing and Distributing the "Main Points of Work of the State Forestry Administration in 2016"	2016
144	Notice on Issuing the National Comprehensive Disaster Prevention and Mitigation Plan (2016–2020)	2016
145	Notice on Issuing the Wetland Protection and Restoration System Plan	2016
146	Notice on Increasing Fishery Poverty Alleviation and Winning the Battle of Poor Fishermen in Poverty Alleviation	2016
147	Notice on Doing a Good Job in the Management of Related Waters and Fisheries in 2016	2016
148	Guiding Opinions on Accelerating the Promotion of Fishery Conversion Mode and Structure Adjustment	2016
149	Notice on Issuing the National Veterinary Health Service Development Plan (2016–2020)	2016
150	Notice on Issuing the "Thirteenth Five-Year Plan for the Development of National Fisheries"	2016
151	Standards for Quality Management of Non-clinical Drug Research (2017)	2017
152	Notice on the Implementation of the "Wild Animal Protection Law"	2017
153	Emergency Notice on Further Strengthening the Law Enforcement of the Protection of Migratory Birds and Other Wildlife in Autumn and Winter and the Monitoring and Control of Epidemic Diseases	2017
154	Notice on Doing a Good Job in the Protection of Rare and Endangered Wildlife and Forest Resources such as Siberian Tigers, Giant Pandas, Snow Leopards	2017
155	Notice on Issuing the 13th Five-Year Plan for the Construction of the National Emergency Response System	2017
156	Notice on Jointly Launching Special Law Enforcement Actions to Combat the Violation of Sea Turtle Resources	2017
157	Notice on Printing and Distributing the Key Points of Fishery and Fishery Administration in 2017	2017
158	Notice on Doing a Good Job in the Management of Related Waters and Fishery Administration in 2017	2017
159	Notice on implementing the "Wildlife Protection Law" and strengthening the management of aquatic wildlife protection	2017
160	Notice on Further Strengthening the Management and Control of Domestic Fishing Vessels and Implementing the Total Management of Marine Fishery Resources	2017
161	Notice on the Investigation and Research on the Reform of the Collection and Use System of Fishery Resources Proliferation Protection Fees	2017
162	Notice on organizing the national "fish release day" on June 6th to promote and release activities simultaneously	2017
163	Overall Plan for Establishing a National Park System	2017
164	Measures for the Sheltering and Rescue of Wildlife	2018
165	Key Points of Fishery and Fishery Administration in 2018	2018
166	Notice on Issuing the Measures for the Administration of National Wetland Parks (2017 Amendment)	2018
167	Notice on Issuing the "Technical Specifications for Black Bear Breeding and Utilization" and other standards	2018
168	Opinions on Strengthening the Protection of the Yangtze River's Aquatic Organisms	2018
169	Notice on Strictly Controlling the Operation and Utilization of Rhino and Tiger and Their Products	2018
170	Measures for the Supervision and Administration of the Inspection and Quarantine of Imported and Exported Meat Products(2018 Amendment)	2018
171	Key Points of Fishery and Fishery Administration in 2018	2018
172	Notice on Issuing the Action Plan for the Protection of Sea Turtles (2019–2033)	2018
173	Aquatic Biodiversity Conservation Plan in Key River Basins	2018
174	Notice of the General Office of the CPC Central Committee and the General Office of the State Council on Issuing the Provisions on the Functions, Structure and Staffing of the State Forestry and Grassland Administration	2018
175	Implementation of the "Guiding Opinions of the State Forestry and Grassland Administration on Promoting the High-quality Development of the Forest and Grass Industry" Task Division Plan	2019
176	Guiding Opinions on Promoting the High-quality Development of Forest and Grass Industry	2019
177	Urgent Notice on Strengthening the Protection and Management of Wildlife and Combating the Illegal Hunting, Killing, Operation and Utilization of Wildlife	2019
178	Notice on Effectively Strengthening the Protection of Migratory Birds in Autumn and Winter	2019
179	Main Points of Work of the State Forestry and Grassland Administration in 2019	2019
180	Key Points of Fishery and Fishery Administration in 2019	2019
181	"China Fishery Policy Action 2019" Series of Special Law Enforcement Action Plans	2019
182	Notice on Jointly Carrying out Special Rectification Actions for Wildlife Protection	2019
183	Notice on the issuance of the 2019 network market supervision special action (Internet Sword Action) plan	2019
184	National Ecological Civilization Pilot Zone (Hainan) Implementation Plan	2019
185	Decision of the Standing Committee of the National People's Congress to Comprehensively Prohibit the Illegal Trade of Wildlife, Break the Bad Habit of Excessive Consumption of Wildlife, and Effectively Secure the Life and Health of the People	2020
186	Notice of Issuing the Opinions on Punishing Criminal and Illegal Activities that Hinder the Prevention and Control of Novel Coronavirus Pneumonia	2020
187	Proposal to Mobilize the Masses to Actively Participate in the Patriotic Health Campaign	2020
188	Notice on Doing a Good Job in Monitoring and Controlling Wild Animal Epidemic Diseases	2020

Table A1. *Cont.*

No.	Name of Policy Document	Year of Promulgation
189	Notice on the implementation of the "Decision of the Standing Committee of the National People's Congress on the Comprehensive Banning of Illegal Wild Animal Trade, Eliminating the Bad Habits of Wildlife, and Effectively Protecting the People's Life, Health and Safety"	2020
190	Notice on Regulating the Scope of Classified Management of Fasting Wildlife	2020
191	Notice on Effectively Strengthening the Protection of Birds	2020
192	Notice on Effectively Strengthening the Protection of Birds and Other Wildlife in Autumn and Winter	2020
193	Notice on the follow-up work of fasting Wildlife	2020
194	Letter on Organizing and Implementing the "Technical Guidelines for Properly Handling Wildlife in Raising"	2020
195	Notice on Effectively Doing a Good Job in Relevant Work of Ban on Arrest in the Yangtze River	2020
196	Opinions on Implementing the Division of Labor of Key Work Departments of the "Government Work Report" (2020)	2020
197	Notice on Further Strengthening the Administration of Wildlife Transportation in Accordance with the Law	2020
198	Notice on Further Regulating the Protection and Management of Frogs	2020
199	Opinions on Further Strengthening the Management of Fishing in the Yangtze River Basin	2020
200	Key Points for the Green Development of Agriculture and Rural Areas in 2020	2020
201	Key Points of Fishery and Fishery Administration in 2020	2020
202	Notice on organizing the national "fish release day" in 2020 to synchronize reproduction and release	2020
203	Notice on the implementation of the "Decision of the Standing Committee of the National People's Congress on the Comprehensive Banning of Illegal Wildlife Trade, Eliminating the Bad Habits of Wildlife, and Effectively Protecting the People's Life, Health and Safety" to further strengthen the protection and management of aquatic wildlife	2020
204	Notice on Further Strengthening Animal Quarantine Work	2020
205	Special Law Enforcement Action Plan of "China Fishery Policy Liangjian 2020" Series	2020
206	Key Points of Work of the Yangtze River Basin Fishery Administration Office of the Ministry of Agriculture and Rural Affairs in 2020	2020
207	Guiding Opinions on the Normalization of the Prevention and Control of the New Crown Pneumonia Epidemic in Catering Services	2020
208	Notice on Issuing the Guidelines for the Prevention and Control of Operation and Service of Retail and Catering Enterprises During the New Coronavirus Epidemic	2020
209	Emergency Notice on Strengthening the Supervision of the Wildlife Market and Actively Doing a Good Job in Epidemic Prevention and Control	2020
210	Announcement on the Prohibition of Wildlife Trading	2020
211	Notice on the Issuance of the 2020 Online Market Supervision Special Action (Internet Sword Action) Plan	2020
212	Notice on the Joint Implementation of Special Law Enforcement Actions against Illegal Wildlife Trade	2020
213	Notice on Further Strengthening Food Safety Supervision during the Period of Epidemic Prevention and Control	2020
214	Action Plan for Improving Food and Beverage Quality and Safety	2020
215	Announcement on Launching the Special Action of "The Yangtze River Banning Catching, Cracking Down on Non-chain Breaking"	2020
216	Notice on Doing a Good Job during the Spring Festival and New Year's Day in 2021	2020
217	Opinions on Strictly Punishing Illegal Activities during the Prevention and Control of COVID-19 Epidemic in Accordance with the Law	2020
218	Opinions on Punishing Illegal Fishing and Other Crimes in the Yangtze River Basin	2020
219	Norms of Fishery Administrative Enforcement	2021

References

1. Jiang, Z.; Ma, K. *The Principles of Conservation Biology*; Science Press: Beijing, China, 2014; pp. 27–34. (In Chinese)
2. Li, Y.; Gao, Z.; Li, X.; Wang, S.; Jari, N. Illegal Wildlife Trade in the Himalayan Region of China. *Biodivers. Conserv.* **2008**, *9*, 901–918.
3. TRAFFIC. Pangolins—Species We Work with at TRAFFIC. 2015. Available online: https://www.traffic.org/what-we-do/species/pangolins/ (accessed on 24 April 2021).
4. Li, P. Enforcing Wildlife Protection in China: The Legislative and Political Solutions. *China Inf.* **2007**, *21*, 71–107. [CrossRef]
5. Feng, L.; Liao, W.; Hu, J. Towards a More Sustainable Human–Animal Relationship: The Legal Protection of Wildlife in China. *Sustainability* **2019**, *11*, 3112. [CrossRef]
6. Huang, Q.; Wang, F.; Yang, H.; Valitutto, M.; Songer, M. Will the COVID-19 Outbreak be a Turning Point for China's Wildlife Protection: New Developments and Challenges of Wildlife Conservation in China. *Biodivers. Conserv.* **2021**, *254*, 108937.
7. Xiao, L.; Lu, Z.; Li, Y.; Zhao, X.; Li, V. Why Do We Need a Wildlife Consumption Ban in China? *Curr. Biol.* **2021**, *31*, R161–R185. [CrossRef]
8. Harris, R.B. Chinese Legal Institutions and Wildlife. In *Wildlife Conservation in China: Preserving the Habitat of China's Wild West*; Routledge: New York, NY, USA, 2015; pp. 92–104.
9. Park, E.; Choi, Y.; Hee, J.; Hong, I. A Review of China's Endangered Species Protection Act and Suggestions for Improvement. *J. Korean Soc. Mar. Environ. Saf.* **2018**, *24*, 112–118. [CrossRef]

10. Zhang, L.; Hua, N.; Sun, S. Wildlife Trade, Consumption and Conservation Awareness in Southwest China. *Biodivers. Conserv.* **2008**, *17*, 1493–1516. [CrossRef]
11. Zhang, L.; Yin, F. Wildlife Consumption and Conservation Awareness in China: A Long Way to Go. *Biodivers. Conserv.* **2014**, *23*, 2371–2381. [CrossRef]
12. Miao, Z.; Wang, Q.; Chen, D.; Song, Z.; Zhang, W.; Zhou, X.; MacMillan, D.C. Current Societal Views about Sustainable Wildlife Management and Conservation: A Survey of College Students in China. *Animals* **2020**, *10*, 1821. [CrossRef]
13. Lu, J.; Bayne, K.; Wang, J. Current Status of Animal Welfare and Animal Rights in China. *Altern. Lab. Anim.* **2013**, *41*, 351–357. [CrossRef] [PubMed]
14. Chang, J. China's Legal Response to Trafficking in Wild Animals: The Relationship between International Treaties and Chinese Law. *Am. J. Int. Law* **2017**, *11*, 408–412. [CrossRef]
15. Whitfort, A. COVID-19 and Wildlife Farming in China: Legislating to Protect Wild Animal Health and Welfare in the Wake of a Global Pandemic. *J. Environ. Law* **2021**, *33*, 57–84. [CrossRef]
16. Qiu, J.; Zou, F. Research on Content Analysis. *J. Libr. Sci. China* **2002**, *2*, 12–17. (In Chinese)
17. Hsieh, H.F.; Shannon, S.E. Three Approaches to Qualitative Content Analysis. *Qual. Health Res.* **2005**, *15*, 1277–1288. [CrossRef]
18. Wu, S. Outline of Content Analysis Methodology. *Inf. Docum. Serv.* **1991**, *2*, 37–39. (In Chinese)
19. Liu, W. The Application of Content Analysis Method in Research. *Chin. Public Adm.* **2014**, *6*, 93–98. (In Chinese)
20. Feng, L.; Liao, W. Legislation, Plans, and Policies for Prevention and Control of Air Pollution in China: Achievements, Chal-lenges, and Improvements. *J. Clean. Prod.* **2016**, *112*, 1549–1558. [CrossRef]
21. Dahl, R.A.; Lindblom, C.E. *Politics, Economics, and Welfare: Planning and Politico-Economic Systems Resolved into Basic Social Processes*; Harper& Brothers: New York, NY, USA, 1953.
22. McDonnell, L.M.; Elmore, R.F. Getting the Job Done: Alternative Policy Instruments. *Educ. Eval. Policy Anal.* **1987**, *9*, 133–152. [CrossRef]
23. Howlett, M. Policy Instruments, Policy Styles, and Policy Implementation: National Approaches to Theories of Instrument Choice. *Policy Stud. J.* **1991**, *19*, 1–21. [CrossRef]
24. Howlett, M. *Designing Public Policies: Principles and Instruments*, 2nd ed.; Routledge: London, UK, 2019.
25. Xu, H.; Wang, S.; Xue, D. Biodiversity Conservation in China: Legislation, Plans and Measures. *Biodivers. Conserv.* **1999**, *8*, 819–837. [CrossRef]
26. McBeath, J.; McBeath, J.H. Biodiversity Conservation in China: Policies and Practice. *J. Int. Wildl. Law Pol.* **2006**, *9*, 293–317. [CrossRef]
27. Yang, N.; Liu, P.; Li, W.; Zhang, L. Permanently Ban Wildlife Consumption. *Science* **2020**, *367*, 1434. [PubMed]
28. Foley, K.-E.; Stengel, C.; Shepherd, C. Pills, Powders, Vials and Flakes: The Bear Bile Trade in Asia (TRAFFIC 2011). Available online: https://www.traffic.org/publications/reports/pills-powders-vials-flakes-the-bear-biletrade-in-asia/ (accessed on 18 February 2021).
29. Zhou, Q.; Grumbine, E. National Parks in China Experiments with Protecting Nature and Human Livelihoods in Yunnan Province, Peoples' Republic of China (PRC). *Biol. Conserv.* **2011**, *144*, 1314–1321. [CrossRef]
30. Freyaa, V.J.; Gareth, E.; Julia, P.G.J. Opinions of the Public, Conservationists and Magistrates on Sentencing Wildlife Trade Crimes in the UK. *Environ. Conserv.* **2012**, *39*, 154–161.
31. Clifton, J. Refocusing Conservation through a Cultural Lens: Improving Governance in the Wakatobi National Park. *Indones. Mar. Pol.* **2013**, *41*, 80–86. [CrossRef]
32. McLafferty, E.; Farley, A. Analyzing Qualitative Research Data Using Computer Software. *Computer* **2006**, *14*, 111–117.
33. Li, G.; Lan, S. *Content Analysis Method of Public Policy: Theory and Application*; Chongqing University Press: Chongqing, China, 2007; p. 10. (In Chinese)
34. Qu, H. *Social and Behavioral Science Research Method: Qualitative Research Method. Social Sciences*; Academic Press: Beijing, China, 2013. (In Chinese)
35. Zou, D. *China's Politics in 20th Century*; Oxford University Press: Hong Kong, China, 1994; p. 69. (In Chinese)
36. Gong, S.; Wu, J.; Gao, Y.; Fong, J.J.; Parham, F.J.; Shi, H. Integrating and Updating Wildlife Conservation in China. *Curr. Biol.* **2020**, *30*, R915–R919. [CrossRef]
37. Mann, M. *States, War and Capitalism: Studies in Political Sociology*; Oxford University Press: New York, NY, USA, 1988; pp. 6–7.
38. Wang, W.; Yang, L.; Wronski, T.; Chen, S.; Hu, Y.; Huang, S. Captive Breeding of Wildlife Resources—China's Revised Sup-ply-side Approach to Conservation. *Wildl. Soc. Bull.* **2019**, *43*, 425–435. [CrossRef]
39. Zhou, X.; Wan, X.; Jin, Y.; Zang, W. Concept of Scientific Wildlife Conservation and Its Dissemination. *Zool. Res.* **2016**, *37*, 270–274.
40. Sinclair, D. Self-Regulation Versus Command and Control? Beyond False Dichotomies. *Law Policy* **1997**, *19*, 529–559. [CrossRef]
41. Liu, M.; Tao, R. Local Governance and Fiscal Reform. In *Paying for Progress in China*; Shue, V., Wong, C., Eds.; Routledge: London, UK, 2007; pp. 166–189.
42. Gratwicke, B.; Mills, J.; Dutton, A.; Gabriel, G.; Long, B.; Seidensticker, J.; Wright, B.; You, W.; Zhang, L. Attitudes Towards Consumption and Conservation of Tigers in China. *PLoS ONE* **2008**, *3*, e2544. [CrossRef]
43. Lyons, A.J.; Natusch, J.D.D. Wildlife Laundering Through Breeding Farms: Illegal Harvest, Population Declines and a Means of Regulating the Trade of Green Pythons (*Morelia viridis*) from Indonesia. *Biol. Conserv.* **2011**, *144*, 3073. [CrossRef]

44. Phelps, J.; Carrasco, L.R.; Webb, L.E. A Framework for Assessing Supply-side Wildlife Conservation. *Conserv. Biol.* **2013**, *28*, 244. [CrossRef]
45. Abbott, B.; Kooten, V.G.C. Can Domestication of Wildlife lead to Conservation? The Economics of Tiger Framing in China. *Ecol. Econ.* **2011**, *70*, 721. [CrossRef]
46. Feng, L.; Wu, Q.; Wu, W.; Liao, W. Decision-Maker-Oriented VS. Collaboration: China's Public Participation in Environmental Decision-Making. *Sustainability* **2020**, *12*, 1334. [CrossRef]
47. Ministry of Agriculture and Rural affairs of the People's Republic of China. *Central File No.1 2018*; Ministry of Agriculture and Rural Affairs of the People's Republic of China: Beijing, China, 2018.
48. Yuan, J.; Lu, Y.; Cao, X.; Cui, H. Regulating wildlife conservation and food safety to prevent human exposure to novel virus. *Ecosyst. Health Sustain.* **2020**, *6*, 1741325. [CrossRef]
49. Coyle, K. *Environmental Literacy in America*; The National Environmental & Training Foundation: Washington, DC, USA, 2005.
50. Ladle, R.J.; Jepson, P.; Whittaker, R.J. Scientists and the Media: The Struggle for Legitimacy in Climate change and Conservation Science. *Interdiscip. Sci. Rev.* **2005**, *30*, 231–240. [CrossRef]
51. Swyngedouw, E. Governance Innovation and the Citizen: The Janus Face of Governance-Beyond-the-State. *Urban Stud.* **2005**, *42*, 1991–2006. [CrossRef]
52. Mirjana, D. The Development of the Right to Public Participation in Environmental Matters as a New Concept of Adminis-trative Decision Making in Serbia. *Transylv. Rev. Adm. Sci.* **2015**, *40*, 74–90.
53. McNamara, J.; Robinson, E.J.Z.; Abernethy, K.; Iponga, D.M.; Sackey, H.N.K.; Wright, J.H.; Milner-Gulland, E.J. COVID-19, Systemic Crisis, and Possible Implications for the Wild Meat Trade in Sub-Saharan Africa. *Environ. Resour. Econ.* **2020**, *76*, 1045–1066. [CrossRef] [PubMed]
54. Roe, D.; Dickman, A.; Kock, R.; Milner-Gulland, E.J.; Rihoy, E.; T'Sas-Rolfes, M. Beyond Banning Wildlife Trade: COVID-19, Conservation and Development. *World Dev.* **2020**, *1*, 5121.
55. Zhang, W.; Zhou, X.; Li, Q.; Zhao, X.; Xu, Y. Background and Methods on Scientific Wildlife Protection Education of Under-graduates under the Construction of Ecological Civilization. *Sichuan J. Zool.* **2015**, *34*, 141–144. (In Chinese)
56. Karanth, K.K.; Kramer, R.A.; Qian, S.S. Examining Conservation Attitudes, Perspectives, and Challenges in India. *Biol. Conserv.* **2008**, *141*, 2357–2367. [CrossRef]
57. Fiorino, D. Citizen Participation and Environmental Risk: A Survey of Institutional Mechanisms. *Sci. Technol. Hum. Values* **1990**, *15*, 226–243. [CrossRef]
58. Haajer, M. A Frame in the Fields: Policymaking and Reinvention of Politics. In *Deliberative Policy Analysis—Understanding Governance in the Network Society*; Haajer, M., Wagenaar, H., Eds.; Cambridge University Press: New York, NY, USA, 2003; pp. 88–112.
59. Stewart, J.M.; Sinclair, A.J. Meaningful Public Participation in Environmental Assessment: Perspectives from Canadian Participants, Proponents, and Government. *J. Environ. Assess. Policy Manag.* **2007**, *9*, 161–183. [CrossRef]
60. Tang, W.; Wang, X.; Yan, M.; Zeng, G.; Liang, J. China's Dams Threaten Green Peafowl. *Science* **2019**, *364*, 943.

land

MDPI

Article

Amphibian Metacommunity Responses to Agricultural Intensification in a Mediterranean Landscape

Luis Albero [1], Íñigo Martínez-Solano [2,*], Ana Arias [3], Miguel Lizana [3] and Eloy Bécares [1]

[1] Área de Ecología, Departamento de Biodiversidad y Gestión Ambiental, Facultad de Ciencias Biológicas, Callejón Campus Vegazana s/n, Universidad de León (ULE), 24071 León, Spain; lalbm@unileon.es (L.A.); ebecm@unileon.es (E.B.)

[2] Departamento de Biodiversidad y Biología Evolutiva, Museo Nacional de Ciencias Naturales (MNCN-CSIC), c/ José Gutiérrez Abascal, 28006 Madrid, Spain

[3] Departamento de Biología Animal (Zoología), Universidad de Salamanca, 37007 Salamanca, Spain; aniarias@hotmail.es (A.A.); lizana@usal.es (M.L.)

* Correspondence: inigomsolano@mncn.csic.es; Tel.: +34-91-411-1328 (ext. 988968)

Citation: Albero, L.; Martínez-Solano, Í.; Arias, A.; Lizana, M.; Bécares, E. Amphibian Metacommunity Responses to Agricultural Intensification in a Mediterranean Landscape. *Land* **2021**, *10*, 924. https://doi.org/10.3390/land10090924

Academic Editors: Juan F. Beltrán, Pedro Abellán and John Litvaitis

Received: 31 July 2021
Accepted: 29 August 2021
Published: 2 September 2021

Abstract: Agricultural intensification has been associated with biodiversity declines, habitat fragmentation and loss in a number of organisms. Given the prevalence of this process, there is a need for studies clarifying the effects of changes in agricultural practices on local biological communities; for instance, the transformation of traditional rainfed agriculture into intensively irrigated agriculture. We focused on pond-breeding amphibians as model organisms to assess the ecological effects of agricultural intensification because they are sensitive to changes in habitat quality at both local and landscape scales. We applied a metacommunity approach to characterize amphibian communities breeding in a network of ponds embedded in a terrestrial habitat matrix that was partly converted from rainfed crops to intensive irrigated agriculture in the 1990s. Specifically, we compared alpha and beta diversity, species occupancy and abundance, and metacommunity structure between irrigated and rainfed areas. We found strong differences in patterns of species occurrence, community structure and pairwise beta diversity between agricultural management groups, with a marked community structure in rainfed ponds associated with local features and the presence of some rare species that were nearly absent in the irrigated area, which was characterized by a random community structure. Natural vegetation cover at the landscape scale, significantly lower on the irrigated area, was an important predictor of species occurrences. Our results suggest that maintaining both local and landscape heterogeneity is key to preserving diverse amphibian communities in Mediterranean agricultural landscapes.

Keywords: agricultural management; amphibians; beta diversity; community ecology; metacommunities

1. Introduction

Biodiversity loss due to anthropogenic activities is a global concern, which calls for practices favoring coexistence between human needs and the viability of natural communities. In the current sixth mass extinction [1,2], we need to understand how human activities affect biodiversity and find ways to mitigate their negative effects [3]. One of the main threats to biodiversity is the increasing replacement of natural habitats with human settlements and farmland [4]. The amount of land cover devoted to agricultural practices has increased for centuries in a continuous trend, and after the "green revolution" of the 20th century, higher crop yields were achieved with the use of heavy machinery, agrochemicals, irrigation and a shift to extensive monocultures [4–6]. This intensified agriculture has been associated with biodiversity declines and deleterious effects for farmland animal and plant communities [7–10].

Traditional extensive agriculture, characterized by rainfed (supplied primarily with rainfall) heterogeneous crops in smaller plot sizes, is usually associated with landscape heterogeneity. The terrestrial habitat matrix includes patches of natural vegetation separating different crops and can potentially host more species than homogeneous habitats [11,12]. The transformation of traditional extensive agriculture to intensive (irrigated) agriculture implies the replacement of patches of natural habitats by crops and an increasing isolation of the remaining patches. The effect of this process of landscape homogenization on farmland biodiversity is an important research topic with applied implications [13–15].

Water bodies, such as ponds, represent key ecosystems in agricultural landscapes [16], hosting sets of biological communities linked by the dispersal of multiple, potentially interacting species (i.e., metacommunities [17]). Metacommunity ecology is a powerful framework to assess the impact of human activities on entire biotic assemblages, explicitly addressing the role of factors acting at different spatial scales, like environmental conditions, predation, competition, or dispersal [17]. Most metacommunity studies have focused on discrete water bodies that are considered local populations, especially in temperate and cold regions [18,19]. Mediterranean ponds in agricultural landscapes have been comparatively less studied [20] (but see [21,22]). Mediterranean ponds typically differ from the comparatively stable environments of temperate ponds in their marked seasonality, with strong and largely unpredictable changes in hydroperiod across seasons and years [23,24]. From a metacommunity perspective, this unpredictability can result in stronger environmental sorting (*sensu* [17]) in Mediterranean communities, with increasing probability of local extinction events following extended dry periods.

Since water is a major limiting factor in Mediterranean agro-ecosystems, irrigation is often used to achieve higher crop yields, and is in fact one of the main drivers of agricultural intensification in the region [25]. Negative effects of the conversion of rainfed to irrigated agriculture on biodiversity have been reported in terrestrial organisms like farmland and steppe birds [26,27] and insects [28], but few studies have focused on aquatic or semi-aquatic organisms, such as amphibians. Amphibians are key components of Mediterranean biotas, with many regional endemics (N = 68) [29], but threats like habitat destruction, alteration and fragmentation are causing population declines and extinctions [30,31]. However, artificial habitats can be used successfully by some species [32,33]. Most community studies of Mediterranean pond-breeding amphibians have been carried out in natural rather than in agricultural areas and have focused mainly on the influence of local wetland features on individual species [34,35]. Negative effects of irrigation, mediated by the increased use of fertilizers and pesticides alongside strong changes in landscape transformation, have been described in biotic communities associated with Mediterranean ponds [36]. However, higher humidity as a consequence of irrigation, and the construction of water channels could also have a positive effect, creating humid terrestrial habitats and improving connectivity [37,38].

Therefore, our knowledge about the effects of agricultural management practices on Mediterranean amphibian communities is incomplete. Further studies are needed to better characterize the effects of the transformation of rainfed into irrigated agriculture on amphibians and other associated biotic communities. Investigating the effects of agricultural intensification and irrigation on amphibian communities is required to identify and mitigate the negative effects of particular management regimes and to provide guidelines for the conservation of diverse amphibian communities in agricultural areas.

We used a metacommunity approach to compare the amphibian communities of two contiguous agricultural areas in northern Iberia. Both areas were managed as rainfed crops (cereal) until the 1990s, when part of the area was transformed into intensively managed irrigated crops (corn), and have thus been subject to different agricultural management regimes for several amphibian generations. We explore the effects of this process of agricultural intensification on local amphibian community structure, beta diversity and patterns of species occurrence and abundance. Our expectation is that amphibian communities in the irrigated area will be altered as a result of habitat homogenization at the landscape scale, and this will be reflected in differences in patterns of amphibian occupancy, abundance and community structure across differentially managed areas. Our a priori hypothesis is that irrigated ponds will host less diverse amphibian communities with an altered community structure.

2. Materials and Methods

2.1. Study Area

The study area is located in SE León Province (Castilla y León, Spain), in the Iberian North Plateau, near the Esla river, part of the Duero basin (Figure 1). The climate is Mediterranean with continental influence, with cold winters and dry and hot summers. Rainfall concentrates in the spring and autumn. Until the mid-1990s, the landscape was dominated by a traditional agro-ecosystem, a mixture of rainfed crops (mostly cereal), oak forest patches (*Quercus pyrenaica*) and meadows used for pasture. A considerable number of natural ponds occur in this area, favored by local lithological features, mostly comprising thin post-Tertiary deposits from the nearby Cantabrian Mountains [39]. All sampled ponds are located in a formerly homogeneous area, part of which was transformed into intensive irrigated agriculture around 1998, following the construction of the Payuelos irrigation channel. Studies carried out in 1994 and 1995, a few years before the shift to irrigation, showed that all ponds were homogeneous in their hydrochemistry [40] and in their aquatic macrophyte and invertebrate communities [41], with eutrophication and hydroperiod mainly shaping biotic differences among ponds. This is the first study about amphibian communities in this region.

We selected two groups of 16 ponds with similar characteristics in each agricultural management regime; that is, 16 ponds in the rainfed area and 16 in the irrigated area (Table S1 in Supplementary Materials, Figure 1). These ponds are mostly shallow and temporary, drying up during the summer months (July to September). Mean pond area was 8551 m^2 (range: 416–42,201 m^2), and mean depth was 74 cm (range: 35–140 cm). Only natural ponds were sampled; other wetlands, such as excavated ponds, inundated gravel pits and irrigation reservoirs, were excluded from analyses. Surveys in these artificial sites and in other natural ponds around our study area show they share a common amphibian species pool with ponds in our study area (unpublished data).

2.2. Amphibian Sampling

Amphibians were sampled at night, during their breeding period (February to May, 2019). Each pond was sampled monthly during this period, resulting in four visits per pond, covering the breeding phenology of all species expected to occur in the area. We selected nights with optimal conditions for amphibian activity (temperature > 5 °C, humidity > 70% or raining) for sampling. Sampling took place from sunset until 4–5 h later. On each visit, we first conducted an acoustic monitoring survey for five to ten minutes, with no artificial lights, in order to detect and, if possible, quantify all male anurans of the different species calling. Then, we surveyed through the entire surface of each pond, except in the largest pond (>40,000 m^2), where we surveyed a 5 m band along the full perimeter of the pond, where most breeding individuals concentrate. Amphibians were located directly with a flashlight or through their eye shine. For capture, when necessary, we used a dip-net.

We recorded amphibian species as "present" in a sampled pond when we were able to confirm reproduction of that species in the pond. In the case of newts, reproduction was confirmed with the observation of adults in their aquatic phase, larvae or eggs, and in anurans, by the observation of adults with nuptial pads, calling males, amplectant pairs, larvae or eggs. Individual counts were performed for all species except for *Pelophylax perezi*, due to the high abundance of this aquatic frog in some ponds. As a surrogate of amphibian abundance in each pond we took the maximum count across all visits, considering only adults. Most Iberian amphibians are explosive breeders, so the maximum number of individuals recorded in a pond is expected to reflect the breeding peak of each species, thus representing a good surrogate for population size. We standardized this value across all ponds relative to pond area, obtaining a general value of abundance for each species and pond in units of individuals/1000 m^2.

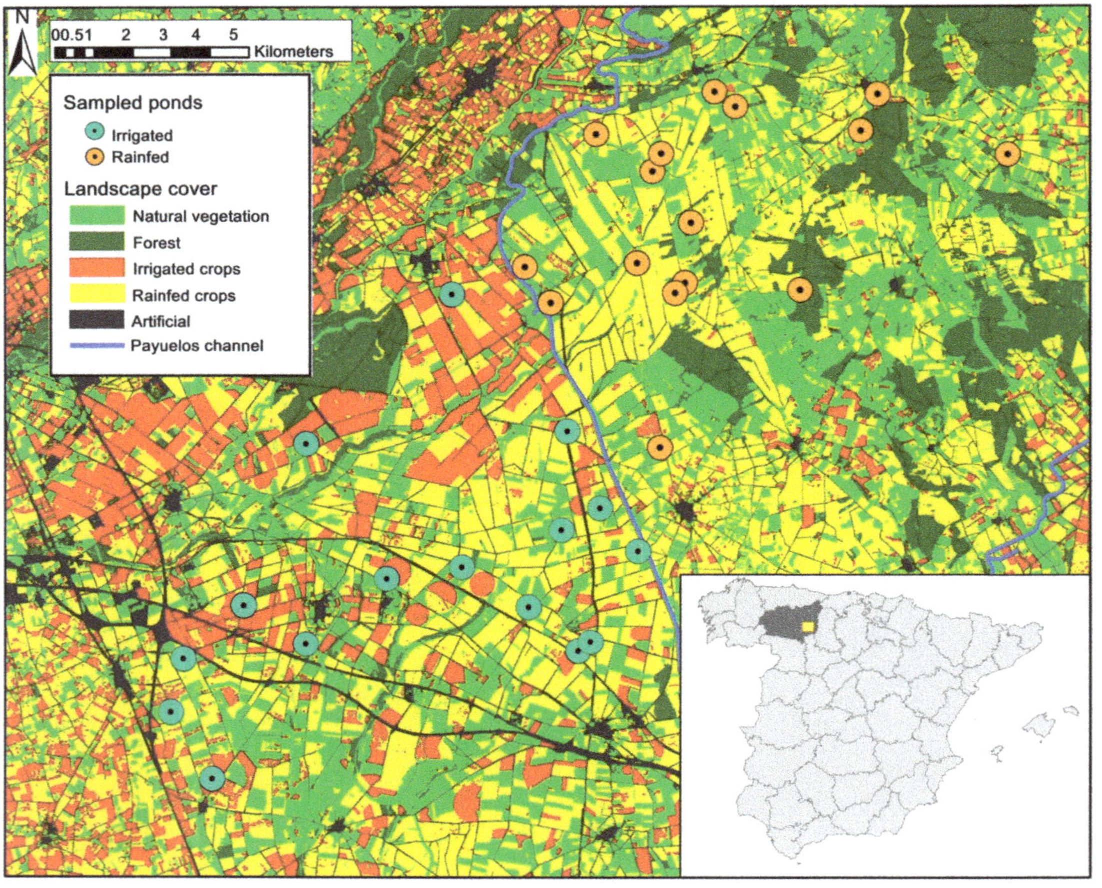

Figure 1. Map of the study area showing its location in León Province (Spain, inset), landscape cover types and sampled ponds in each agricultural management group. The blue line represents Payuelos Channel. "Natural vegetation" includes scrub, grassland and bare soil; "Artificial" includes roads and urban areas.

2.3. Environmental Variables

Following the ecological metacommunity approach [42], we considered environmental predictors at the local and landscape scales for the study of amphibian community patterns on both agricultural management groups. At the local scale, we selected variables affecting amphibian dynamics in temperate regions as reported in previous studies [43–46], including pond area (m^2), medium (average across visits) and maximum depth (cm), and helophyte vegetation cover (%). Pond area and helophyte cover were measured directly from satellite images (the more recent images available from the Spanish "Centro Nacional de Información Geográfica", CNIG) using ArcGIS 10.7 (®ESRI). Pond depth was recorded on each visit using a graduated stick, at the point of maximum depth detected. Landscape variables were considered at two spatial scales, 500 m and 1000 m from each pond, accounting for recorded dispersal distances of amphibian species in the area [47,48]. We recorded the following landscape types: natural forest, forest plantation, grassland, scrub, bare soil, wetlands, irrigated crops and non-irrigated (i.e., rainfed) crops. The importance of each landscape type on a pond was calculated as the proportion cover (%) of a circular buffer of 500 m or 1000 m radio centered in the pond centroid. Land cover data was extracted from the database of "Mapa de Cultivos y Superficies Naturales de Castilla y León" (MCSNCyL, Junta de Castilla y León). Proportion cover was calculated with ArcGIS.

For subsequent analyses, environmental variables were filtered to account for multicollinearity using Pearson's pairwise correlation as calculated with R package *stats* [49] (Table S2). We found high correlation between all landscape variables at the 500 and 1000 m spatial scales (Pearson's Correlation Coefficient, $r > 0.7$), so we retained only one value for each landscape spatial category, selecting the one less correlated on average with the rest of environmental predictors. Retained variables included natural forest, artificial surfaces, bare soil, scrub, irrigated and non-irrigated crops for the 500 m scale, and plantation forest, grassland and wetland cover for the 1000 m scale (Table S3). We also found a high correlation between maximum and medium pond depth ($r = 0.93$) so we retained only medium depth as the predictor less correlated with the remaining variables.

2.4. Statistical Analysis

In a first level of inference, we tested for differences in species richness, abundance and patterns of co-occurrence between the amphibian communities of the two agricultural management groups. We first tested for environmental differences using local and landscape environmental variables. Levene's normality test was significant for almost all variables, so we used the non-parametric Mann–Whitney test. To compare alpha diversity (richness) and species abundance estimates between ponds in the two management groups we used an Analysis of Similarities (ANOSIM). If significant, we looked for the species with the highest contribution to inter-group differences using a Similarity Percentages (SIMPER) test with 9999 permutations. Finally, we tested for differences in species by species paired occurrences in irrigated and rainfed ponds using Fisher's exact test. All analyses were performed with packages *stats* and *vegan* in R.

For the second level of inference, we quantified differences in community composition among and within agricultural management groups. We calculated beta diversity following the additive partition approach [50], which decomposes total beta diversity (calculated by the Sørensen index, βsor) in two components explaining different measures of inter-site differences: spatial turnover (measured by Simpson's index, βsim) and nestedness (βnes). Differences due to turnover are caused by the replacement of some species by others between two sites, and nestedness occurs when the species composition of a community is a subset of that of another richer community. We used presence/absence matrices to calculate the three components of beta diversity for all ponds and separately for each group using the function 'beta.multi' implemented in R package *betapart* [51]. Then, we calculated the three beta components for each pair of ponds within the groups using the function 'beta.pair' in *betapart* to test for differences in beta diversity between irrigated and rainfed ponds. Pairwise approaches have been used in many ecological comparisons of multi-site data because they provide a good surrogate of average differences between paired samples [52,53]. Pairwise values were pooled for each group using the function 'dist2list' in R library *spaa* [54]. Since pairwise beta value distributions did not fit a normal distribution, we used Mann–Whitney's test to assess differences between irrigated and rainfed amphibian communities in all three beta diversity components.

The third inference level focused on differences in amphibian community structure between ponds and agricultural management groups. We followed the "Elements of Metacommunity Structure" (EMS) approach [55], which uses analyses of coherence, turnover and boundary clumping to distinguish several idealized metacommunity patterns (Evenly Spaced, Gleasonian, Clementsian, Nested, Checkerboard and Random). Prior to this analysis, species by site matrices were ordered by reciprocal averaging following the main axis of community variation. Coherence is assessed by counting the number of embedded absences, and significance is assessed by comparing against a null distribution created by random reordering of the community matrix. Then, turnover is calculated with the number of replacements between species among sites and Morisita's Index is used for the evaluation of Boundary Clumping [56]. The distinct metacommunity structure patterns represent a gradient from metacommunities driven by environmental species sorting (Clementsian, Checkerboard) to other patterns driven by species dispersal (Nested) or intermediate situations (Evenly Spaced, Gleasonian, Random). This analysis was performed with package *metacom* in R [57].

Finally, we explored the main drivers of individual species preferences using both an ordination analysis via redundancy analysis (RDA) and a binomial regression analysis, with species composition per site as the response variable and all environmental predictors as factors. For the regression analysis we summarized environmental variables via Principal Component Analysis (PCA), which reduced the set of predictor variables to six principal components reflecting the main gradients of environmental variation among ponds (Table S4). Then, we conducted logistic regression analyses with these principal components as the new environmental variables. Using this approach decreases the complexity of fitting regression models [37] and also reduces problems of perfect separation (which could be detected and thus excluded from the final models). Model selection was based on Akaike's Information Criterion (AIC) using a stepwise procedure. Analyses were performed with package *vegan* in R. For the RDA analyses we used software Canoco 4.5 CanoDraw4 to analyze and plot results [58].

3. Results

The environmental comparison between ponds in the two agricultural areas showed significant differences for many landscape variables, but not for the local ones (Table S3). The overall landscape was dominated by grasslands (≈20% cover), and rainfed crops (>40%). Natural vegetation cover, excluding grasslands, was significantly higher in the traditionally managed landscape (>12% versus <1%), with more proportion of natural and planted forest, bare ground and scrub. As expected, the landscape surrounding ponds in the irrigated area showed a significantly higher proportion cover of irrigated crops and artificial surfaces. Wetland cover, reflecting pond density, was not different between both agricultural areas. The irrigated area was mainly characterized by a greater diversity of crops, both irrigated and rainfed, but with less proportion of natural cover (forest and pastures). Therefore, the rainfed area has a greater level of landscape heterogeneity, with a mixture of crops and natural areas, whereas the irrigated area is more homogenous, containing mostly crops.

We found eight amphibian species in the study area: marbled newt (*Triturus marmoratus*), Iberian ribbed newt (*Pleurodeles waltl*), western spadefoot toad (*Pelobates cultripes*), Iberian painted frog (*Discoglossus galganoi*), spiny toad (*Bufo spinosus*), natterjack toad (*Epidalea calamita*), Iberian treefrog (*Hyla molleri*) and Iberian green frog (*Pelophylax perezi*). All species were found at least once in both irrigated and rainfed areas except for the painted frog, which was only found in five ponds in the rainfed area. All species were more frequent in the rainfed area, except for the marbled newt and natterjack toad, which occurred in the same number of ponds in the two areas. Fisher's test showed significant differences in the occurrences of *P. cultripes* ($p = 0.015$) and *D. galganoi* ($p = 0.043$) between the two pond groups, with both species being more frequent in the rainfed area (Table S5).

Species richness ranged from one to seven amphibian species across ponds. Mean species richness by pond also differed between the two areas, with ponds in the rainfed area having significantly more amphibian species on average (ANOSIM R: 0.097, $p < 0.02$; Figure 2). The SIMPER analysis showed *P. cultripes*, *E. calamita* and *P. waltl* to be responsible for more than 50% of inter-group differences in species richness (Table S6). Overall, amphibian abundance was more similar between agricultural management groups than species richness (Figure 2), and abundance estimates for the different species were not significantly different between irrigated and rainfed ponds (Table S7), although *P. cultripes* and *P. waltl* were considerably more abundant in rainfed ponds (>40% more).

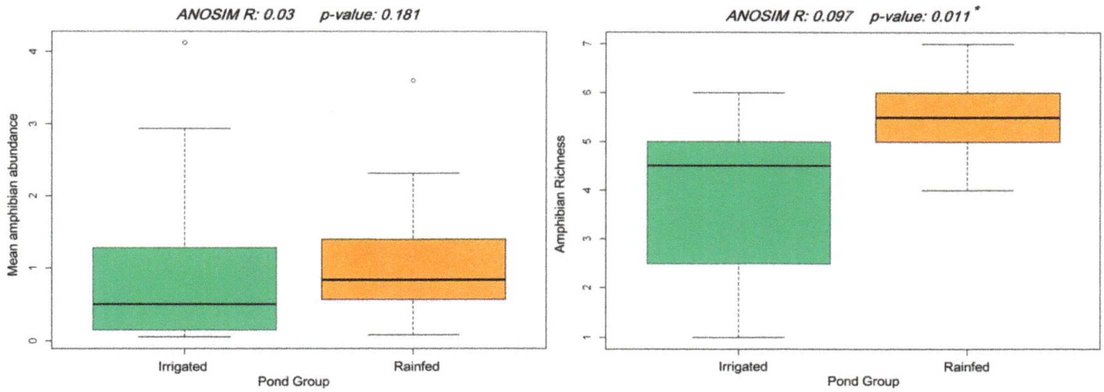

Figure 2. Comparison of mean amphibian abundance (**left**) and species richness (**right**) between irrigated and rainfed areas. * Significant under $\alpha = 0.05$.

Partition of beta diversity (Table 1) showed higher total beta diversity in rainfed ponds, with higher values of the nestedness and turnover components in comparison with the irrigated area. Turnover was more important than nestedness in both areas, but the partition also showed stronger differences between the two components for irrigated than for rainfed ponds. Rainfed ponds hosted more amphibian species on average, with some of them being present in only a few ponds (*E. calamita*, *B. spinosus*, *D. galganoi*). Thus, they are expected to have higher total beta diversity values than irrigated ponds, which have fewer species and similar occurrences.

Table 1. Partition of total and pairwise beta diversity in irrigated and rainfed ponds, following the additive partition approach. The statistical significance of differences between groups was assessed with Mann–Whitney tests. * Significant under $\alpha = 0.05$.

	Sorensen	Simpson	Nestedness
All ponds Rainfed	0.75	0.5	0.25
All ponds Irrigated	0.6	0.42	0.18
All ponds Rainfed + Irrigated	0.81	0.61	0.2
Mean Pairwise Rainfed	0.177	0.092	0.084
Mean Pairwise Irrigated	0.368	0.166	0.202
Pairwise U Mann-Whitney	16,996	30,014	21,802
Pairwise *p*-value	2.20×10^{-16} *	0.06	3.478×10^{-11} *

The density distribution of pairwise values is shown in Figure 3. Beta diversity among all ponds was higher in the rainfed area, but pairwise beta diversity reached significantly higher values for irrigated ponds both in total beta diversity and in the nestedness component (Table 1), highlighting the existence of some pairs of well-differentiated ponds in their community composition (Figure 3). Based on pairwise comparisons, mean nestedness was higher than turnover in the irrigated group. Rainfed ponds showed less difference between nestedness and turnover beta diversity on the mean pairwise values than on the overall beta value. In spite of the major importance of turnover for total beta diversity, pairwise measures showed some pairs of ponds with high nestedness in the irrigated group, with these nestedness pairwise values being significantly higher than in the rainfed group (Figure 3, Table 1). Thus, in the irrigated group there are some species-poor ponds whose species composition is a subset of that found in other ponds.

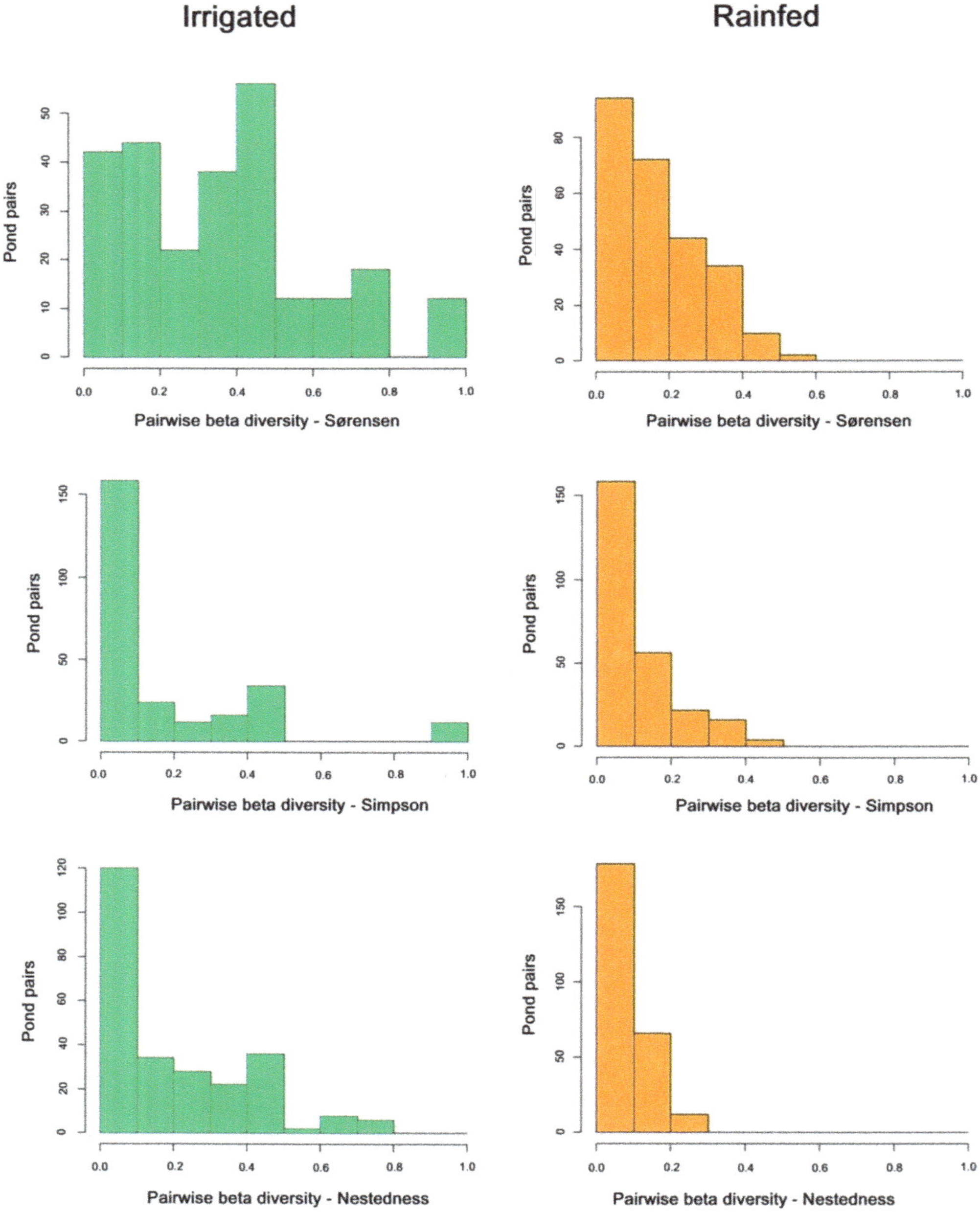

Figure 3. Distribution of pairwise beta diversity values for each component of beta diversity in irrigated (**left**) and rainfed (**right**) ponds, following the additive partition approach.

The results of the EMS analysis are summarized in Table 2. EMS showed a random pattern for the community structure of ponds in the irrigated area and also when ponds of the two areas were considered jointly. Rainfed ponds showed a significant coherence pattern and a Clementsian-type community, characterized by positive turnover and significant boundary clumping. Species by site matrices ordered by reciprocal averaging for each management group are shown in Figure 4.

Table 2. Results of the Elements of Metacommunity Structure Analysis. * Significant under α = 0.05.

	Coherence z	Coherence *p*-Value	Turnover z	Turnover *p*-Value	Morisita Index	Morisita *p*-Value	Structure
Rainfed	2.05	0.04 *	1.78	0.075	2	0.0012 *	Clementsian
Irrigated	0.62	0.53	1.7	0.09	NA	NA	Random
All ponds	0.96	0.34	2.41	0.015 *	NA	NA	Random

Figure 4. Occurrence matrices of all amphibian species in irrigated (**left**) and rainfed (**right**) ponds, ordered by maximum coherence following EMS analysis. Black boxes indicate species presence (columns) in each pond (rows). EC: *Epidalea calamita*, PC: *Pelobates cultripes*, PW: *Pleurodeles waltl*, HM: *Hyla molleri*, PP: *Pelophylax perezi*, TM: *Triturus marmoratus*, DG: *Discoglossus galganoi*, BS: *Bufo spinosus*.

Regarding the environmental preferences of individual species, regression analyses (Table S8) mostly showed weak and non-significant responses of amphibian species to the main gradients of environmental variation. *Discoglossus galganoi* and *P. cultripes* showed preference for large ponds in heterogeneous landscapes with low irrigation, in accordance with their lower abundance and occurrence in the irrigated area.

On the other hand, the RDA analysis combining landscape and local variables showed clear differences between rainfed and irrigated ponds (Figure 5). The main two axes of the redundancy analysis explained up to 61.9% of amphibian community variation, with similar importance (Axis 1: 32.6%; Axis 2: 29.3%). Both axes were driven by a combination of local and landscape features and thus represent landscape heterogeneity. Axis 1 can be interpreted as a gradient from vegetated ponds in intensive agricultural sites (high axis values), to more insolated ponds in heterogeneous landscapes with a greater proportion of natural forest. Axis 2 mostly represents pond size and landscape heterogeneity, with negative values correlating with the amount of irrigated and non-irrigated crops. Overall, the RDA plot (Figure 5) shows separation between ponds in the two agricultural management regimes, with landscape variables being most discriminative.

Irrigated ponds are associated with high scores on axis 1 and markedly low scores on axis 2 (corresponding to smaller ponds in agricultural landscapes).

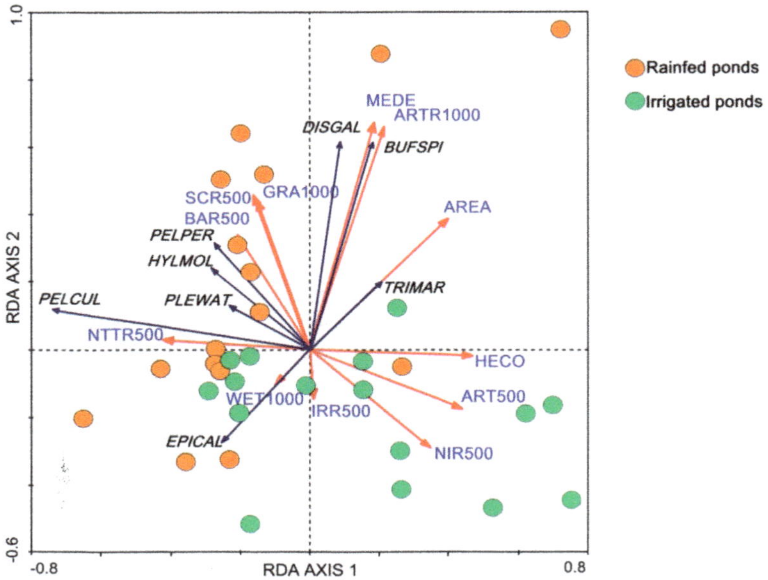

Figure 5. RDA ordination plot showing the main two redundancy axes and their relationships with local and landscape environmental variables (red arrows) and patterns of amphibian species occurrence (blue arrows). PLEWAT: *Pleurodeles waltl*; TRIMAR: *Triturus marmoratus*; PELCUL: *Pelobates cultripes*; DISGAL: *Discoglossus galganoi*; BUFSPI: *Bufo spinosus*; EPICAL: *Epidalea calamita*; HYLMOL: *Hyla molleri*; PELPER: *Pelophylax perezi*. Environmental variables at the local scale: pond area in m^2 (AREA); medium pond depth (MEDE); helophyte cover in % pond area (HECO). Landscape variables in % cover at a 1000 m scale from the pond: man-made forest, poplar or pine (ARTR1000); grasslands (GRA1000); wetlands (WET1000). In % cover at a 500 m scale from the pond: natural oak forest (NTTR500), artificial surfaces (ART500), bare ground (BAR500), scrub (SCR500).

Species responses were in general positively correlated with natural vegetation and pond size, except for *E. calamita*, which preferred smaller ponds close to other wetlands (Figure 5). RDA also offers an environmental explanation for the community turnover pattern among rainfed ponds. The responses to environmental variables were similar for *B. spinosus* and *D. galganoi*, both showing a strong preference for large and deep ponds. On the other hand, the presence of *E. calamita* decreased with pond area and depth and increased with the vicinity of wetlands. The presence of *P. cultripes* was highly correlated with natural forests and low agricultural cover at the landscape scale, preferring insolated ponds (with low helophyte cover). *Triturus marmoratus* was more frequent in large ponds with aquatic vegetation. Finally, the responses of *P. perezi*, *H. molleri* and *P. waltl* were similar and positively correlated with the presence of natural vegetation at the landscape scale and negatively affected by both types of croplands.

RDA found no clear differences in the effects of irrigated and rainfed crops on amphibian presence. Both types of crops showed negative effects on all species, except on *E. calamita*, which seems to be slightly favored by irrigated crops. The effect of rainfed crop cover on species occurrences seems to be stronger than that of irrigated crops, but this could be a consequence of the higher abundance of rainfed crops in the study area. The species which seemed more negatively affected by crop cover were also the scarcest in the intensively managed area: *D. galganoi* and *B. spinosus*. Other species, like *H. molleri*,

P. perezi, *P. cultripes*, and *P. waltl* were also negatively correlated with crop cover, as the amount of crop cover was also correlated with lower cover of natural vegetation.

4. Discussion

Our analyses showed significant differences between the two agricultural management groups in the terrestrial landscape matrix surrounding sampled ponds, and in their associated amphibian communities, which differed in structure and distribution patterns. These differences are linked to the effect of landscape (e.g., higher natural vegetation in rainfed area), but not local variables across management groups. This suggests that the transformation of rainfed crops into irrigated agriculture has resulted in an overall homogenization of the landscape with negative consequences for amphibian communities.

As we predicted, amphibian species richness was significantly higher in rainfed than in irrigated ponds (Figure 2), in accordance with previous studies supporting a positive relationship between landscape heterogeneity and amphibian richness [43,59]. Species composition was similar in irrigated and rainfed areas, with the exception of *D. galganoi*, which was only present in the latter, albeit in a small number of ponds. Overall, total species abundance was not significantly different in the two agro-systems, but some species were markedly less abundant in the irrigated area, including *B. spinosus*, *P. cultripes* and *P. waltl*. *Pelobates* showed the most contrasting patterns of occupancy between the two areas, as shown by our SIMPER analysis. This species has low dispersal capacity compared to other sympatric amphibian species [44], and is strongly dependent on non-compacted soils, where they burrow for refuge [60]. Traditionally managed areas could favor the presence of the species by maintaining a larger extent of bare soil and fallow. Negative effects of intensive agriculture on this species were also reported by [33], who found a strong negative correlation between agricultural cover and larval abundance. On the other hand, *Discoglossus* usually breeds in shallow and ephemeral puddles [61]; the loss of inundated grassland habitats surrounding ponds in the irrigated area may have negatively affected this species. Deleterious effects of agrochemicals have been reported for both *Pelobates* and *Discoglossus* in their larval stages [62,63], suggesting that reduced abundance in intensified areas could be associated with the increased use of fertilizers and pesticides around irrigated ponds. A previous study [64] found important concentrations of pesticides and fertilizers in the study area, with a positive relationship between the intensity of agricultural management and the amount of polycyclic aromatic hydrocarbons (PAHs, a family of pollutants) in the ponds.

Beta diversity also differed across management groups, while the turnover component dominated on both irrigated and rainfed pond groups, as in other amphibian studies [65,66] (but see [67]). The latter showed higher beta diversity on all three components. The pairwise analysis showed a "harmonic" pattern on the rainfed side, with a majority of pond pairs showing small differences in community composition, in contrast with irrigated ponds, where we found a non-normal pattern, with more pronounced community differences among ponds. This may be the result of local extinction processes in the irrigated pond metacommunity, with some irrigated ponds having only three or less species. Local extinctions can increase pairwise beta diversity due to turnover among impoverished communities (subtractive heterogenization [53]). Pairwise beta diversity due to nestedness is also expected to increase among pairs of ponds with different levels of degradation, as the most degraded will have lost a higher number of species. *Pleurodeles waltl*, *B. spinosus*, *H. molleri*, *P. perezi* and especially *P. cultripes* occur less frequently in irrigated than in rainfed ponds, and could thus be considered to be more sensitive towards intensification, driving observed inter-pond differences. On the other hand, *E. calamita* and *T. marmoratus* show an equal number of occurrences on both irrigated and rainfed areas, and can therefore be regarded as more tolerant to changes in agricultural practices. This decrease in beta diversity associated with agricultural intensification has not been described in amphibians or in Mediterranean environments so far, but it has been reported in other taxonomic groups in different geographic regions [68–70].

To our knowledge, this is the first study to apply the EMS approach for the characterization of amphibian metacommunities. Our results suggest that amphibian communities in areas affected by agricultural intensification are more randomly structured, whereas in traditionally managed areas, species arrange in communities formed by species with similar environmental preferences. The analysis shows rainfed ponds display a Clementsian metacommunity pattern, whereas no clear pattern could be associated with irrigated ponds, as their incidence matrices were not significantly coherent [56]. Clementsian patterns are named after the classic text by Clements [71], and represent discrete species assemblages that replace each other along an environmental gradient [55]. Each assemblage can be referred to as a compartment [72]; based on ordered matrices we can distinguish three assemblages in rainfed ponds: ponds with *D. galganoi* and *B. spinosus*, ponds with *E. calamita*, and ponds hosting only the most generalist species. This result is in accordance with the higher importance of the turnover beta diversity component on the rainfed side, both for total beta and for pairwise values. The absence of *D. galganoi* and the extreme rarity of *B. spinosus* in the irrigated area, along with the greater scarcity of species that are common in the rainfed area, probably explain the random metacommunity structure of irrigated ponds. Differences in community structure could thus represent the disassembly of a recognizable metacommunity pattern (Clementsian, Gleasonian or Nested) into a random pattern following anthropic disturbances [73,74].

Our analysis for individual amphibian species showed lower potential to detect possible effects of agricultural management regimes compared to analyses at the community-level. This shows the advantage of metacommunity approaches when addressing large scale ecological perturbations associated with human activities. Specifically, logistic regressions were not very informative, probably because many species were present in most of the ponds, while others appeared in a small number of ponds. This reduces the power of this type of analysis to relate the presence of each species to environmental variables. On the other hand, RDA provided further insights on the role of different factors in shaping community differences across management groups. Specifically, these analyses revealed a positive association of all amphibian species with the amount of natural vegetation, and negative responses to crop cover. The exception is *E. calamita*, which seems to be more dependent on local variables, especially pond area and depth [75], and thus more indifferent to agricultural management regimes. A positive relationship between amphibian richness and natural vegetation cover has been found in previous studies [45,76–78]. Amphibian species in our study area showed different responses to anthropogenic alterations driven by agricultural intensification, as found in previous studies [40,79,80], and while some do not appear to be sensitive to landscape homogenization, those most affected are Iberian endemics of conservation concern (*D. galganoi*, *P. cultripes*), included in the EU Habitats Directive.

Our results have important implications for amphibian conservation in Mediterranean croplands. First, local variables seem to be more important than landscape variables in driving the observed turnover (i.e., Clementsian) pattern among rainfed ponds. In this context, conservation planning is recommended to target multiple ponds, not only those with higher species richness [53]. In contrast, conservation efforts should focus on the richest sites at the expense of species-poorer communities when nested patterns are found, as in some ponds in the irrigated area. In Mediterranean pond-breeding amphibian communities, which are characterized by strong rainfall variation and unpredictability and thus with high probability of local extinctions due to stochastic factors, these general guidelines should be complemented with explicit assessments of inter-pond connectivity to account for possible rescue effects [81].

Agricultural intensification has been shown to promote population fragmentation in amphibians, but crop compositional heterogeneity can have positive effects for some species [59]. In our study area, however, diversification of crop types with the introduction of irrigated crops seems to have negatively affected amphibian communities. Moreover, the increase of the water content of soils and the presence of channels with excess water from irrigation in the irrigated area did not have a positive effect on amphibian communities, in contrast with the significant negative effect of reduced natural vegetation patches (forest, scrub and grassland). Our study supports the application of environmental measures for biodiversity conservation in European agricultural policies, including preserving strips of forest or scrub across crop plots to favor the persistence of diverse amphibian communities. Patches of natural grassland and bare soil, especially near water bodies, can also provide adequate foraging and terrestrial refugial habitats for some species. Regulations like the Common Agriculture Policy should stress the role of patches of natural vegetation and promote the construction and maintenance of small ponds and wetlands, providing breeding habitat, terrestrial corridors and refugial areas for amphibians.

Supplementary Materials: The following are available online at https://www.mdpi.com/article/10.3390/land10090924/s1, Table S1: Location of the sampled 32 ponds and agricultural management groups (Datum: WGS84); Table S2: Pearson's correlation between environmental variables from sampled ponds. In brown, removed variables due to high correlation. Local variables are AREA: pond area (m^2), MXDE: maximum pond depth on all visits (cm), MEDE: medium pond depth on all visits (cm), HECO: proportion of helophyte cover over pond area (0–1). Landscape variables (proportion of surface on a 500 m or 1000 m buffer around the pond) are ARTR: man-made forest, poplar or pine, GRA: grasslands, WET: wetlands other than the sampled pond, NTTR: natural oak forest, ART: artificial surfaces, BAR: bare ground, SCR: scrublands; Table S3: Means and range of values for environmental variables recorded in ponds in the two agricultural management areas, with statistical significance of differences between irrigated and rainfed groups assessed with Mann-Whitney tests. Variable names as in Table S2; Table S4: Environmental variables summarized via Principal Components Analysis (PCA) showing the scores for each principal component (PC) and environmental variable; Table S5: Occurrences of amphibian species (number and proportion of occupied ponds) in irrigated vs. rainfed ponds. Statistical significance of differences was assessed by means of Fisher's test; Table S6: Results of the SIMPER analysis between irrigated and rainfed pond communities; Table S7: Abundances of amphibian species (individuals/1000 m^2) in irrigated vs. rainfed ponds. Statistical significance of differences between the two agricultural areas was assessed with ANOSIM test.

Author Contributions: Conceptualization, L.A., Í.M.-S. and E.B.; methodology, L.A., A.A.; formal analysis, L.A.; resources, M.L., E.B.; writing—original draft preparation, L.A.; writing—review and editing, all authors; supervision, Í.M.-S., E.B.; project administration, E.B.; funding acquisition, E.B. All authors have read and agreed to the published version of the manuscript.

Funding: This research was funded by the Spanish Ministerio de Economía, Industria y Competitividad, project Metaponds, (ref: CGL2017-84176-R); and by Fundación Biodiversidad, Ministerio para la Transición Ecológica y el Reto Demográfico (project BT-2019). L. Albero is funded by a Phd grant from Universidad de León (ULE).

Institutional Review Board Statement: All experimental protocols and sampling permits were approved by the regional authority (Consejería de Medio Ambiente, Junta de Castilla y León, reference: AUES/CYL/693/2019). We followed ethical guidelines provided by Junta de Castilla y León and León University (Universidad de León). No amphibians were harmed during sampling; and handling, when necessary, involved the minimum time required for species identification. We disinfected field equipment between ponds to prevent transmission of infectious diseases.

Acknowledgments: We wish to thank all people who helped during field sampling: O. Rodríguez, A. Casanova, M. Borrego, V. Ezquerra, P. Oviedo, A. Cachorro, R. de Prado and specially A. Benito. We also thank Fundación SCAYLE for their support with bioinformatics.

Conflicts of Interest: The authors declare no conflict of interest. The funders had no role in the design of the study; in the collection, analyses, or interpretation of data; in the writing of the manuscript, or in the decision to publish the results.

References

1. Ceballos, G.; Ehrlich, P.R. The misunderstood sixth mass extinction. *Science* **2018**, *360*, 1080–1081. [CrossRef]
2. Pimm, S.L.; Jenkins, C.N.; Abell, R.; Brooks, T.M.; Gittleman, J.L.; Joppa, L.N.; Raven, P.H.; Roberts, C.M.; Sexton, J.O. The biodiversity of species and their rates of extinction, distribution, and protection. *Science* **2014**, *344*. [CrossRef] [PubMed]

3. Grant, E.H.C.; Muths, E.; Schmidt, B.R.; Petrovan, S.O. Amphibian conservation in the Anthropocene. *Biol. Conserv.* **2019**, *236*, 543–547. [CrossRef]
4. Foley, J.A.; DeFries, R.; Asner, G.P.; Barford, C.; Bonan, G.; Carpenter, S.R.; Chapin, F.S.; Coe, M.T.; Daily, G.C.; Gibbs, H.K.; et al. Global consequences of land use. *Science* **2005**, *309*, 570–574. [CrossRef]
5. Martin-Guay, M.O.; Paquette, A.; Dupras, J.; Rivest, D. The new Green Revolution: Sustainable intensification of agriculture by intercropping. *Sci. Total Environ.* **2018**, *615*, 767–772. [CrossRef] [PubMed]
6. Pingali, P.L. Green revolution: Impacts, limits, and the path ahead. *Proc. Natl. Acad. Sci. USA* **2012**, *109*, 12302–12308. [CrossRef] [PubMed]
7. Balestrieri, A.; Gazzola, A.; Formenton, G.; Canova, L. Long-term impact of agricultural practices on the diversity of small mammal communities: A case study based on owl pellets. *Environ. Monit. Assess.* **2019**, *191*. [CrossRef]
8. Habel, J.C.; Ulrich, W.; Biburger, N.; Seibold, S.; Schmitt, T. Agricultural intensification drives butterfly decline. *Insect Conserv. Divers.* **2019**, *12*, 289–295. [CrossRef]
9. Morgado, R.; Santana, J.; Porto, M.; Sánchez-Oliver, J.S.; Reino, L.; Herrera, J.M.; Rego, F.; Beja, P.; Moreira, F. A Mediterranean silent spring? The effects of olive farming intensification on breeding bird communities. *Agric. Ecosyst. Environ.* **2020**, *288*, 106694. [CrossRef]
10. Uchida, K.; Ushimaru, A. Biodiversity declines due to abandonment and intensification of agricultural lands: Patterns and mechanisms. *Ecol. Monogr.* **2014**, *84*, 637–658. [CrossRef]
11. Betts, M.G.; Wolf, C.; Pfeifer, M.; Banks-Leite, C.; Arroyo-Rodríguez, V.; Bandini Ribeiro, D.; Barlow, J.; Eigenbrod, F.; Faria, D.; Fletcher, R.J., Jr.; et al. Extinction filters mediate the global effects of habitat fragmentation on animals. *Science* **2019**, *366*, 1236–1239. [CrossRef]
12. Fahrig, L. Ecological Responses to Habitat Fragmentation Per Se. *Annu. Rev. Ecol. Evol. Syst.* **2017**, *48*, 1–23. [CrossRef]
13. Benton, T.G.; Vickery, J.A.; Wilson, J.D. Farmland biodiversity: Is habitat heterogeneity the key? *Trends Ecol. Evol.* **2003**, *18*, 182–188. [CrossRef]
14. Fahrig, L.; Baudry, J.; Brotons, L.; Burel, F.G.; Crist, T.O.; Fuller, R.J.; Sirami, C.; Siriwardena, G.M.; Martin, J.L. Functional landscape heterogeneity and animal biodiversity in agricultural landscapes. *Ecol. Lett.* **2011**, *14*, 101–112. [CrossRef]
15. Córdova-Lepe, F.; Del Valle, R.; Ramos-Jiliberto, R. The process of connectivity loss during habitat fragmentation and their consequences on population dynamics. *Ecol. Model.* **2018**, *376*, 68–75. [CrossRef]
16. Lewis-Phillips, J.; Brooks, S.J.; Sayer, C.D.; Patmore, I.R.; Hilton, G.M.; Harrison, A.; Robson, H.; Axmacher, J.C. Ponds as insect chimneys: Restoring overgrown farmland ponds benefits birds through elevated productivity of emerging aquatic insects. *Biol. Conserv.* **2020**, *241*, 108253. [CrossRef]
17. Leibold, M.A.; Holyoak, M.; Mouquet, N.; Amarasekare, P.; Chase, J.M.; Hoopes, M.F.; Holt, R.D.; Shurin, J.B.; Law, R.; Tilman, D.; et al. The metacommunity concept: A framework for multi-scale community ecology. *Ecol. Lett.* **2004**, *7*, 601–613. [CrossRef]
18. Heino, J.; Melo, A.S.; Siqueira, T.; Soininen, J.; Valanko, S.; Bini, L.M. Metacommunity organisation, spatial extent and dispersal in aquatic systems: Patterns, processes and prospects. *Freshw. Biol.* **2015**, *60*, 845–869. [CrossRef]
19. Chaparro, G.; Horváth, Z.; O'Farrell, I.; Ptacnik, R.; Hein, T. Plankton metacommunities in floodplain wetlands under contrasting hydrological conditions. *Freshw. Biol.* **2018**, *60*, 380–391. [CrossRef] [PubMed]
20. Álvarez Cobelas, M.; Rojo, C.; Angeler, D.C. Mediterranean Limnology: Current status, gaps and the future. *J. Limnol.* **2005**, *64*, 13–29. [CrossRef]
21. Ferreira, M.; Beja, P. Mediterranean amphibians and the loss of temporary ponds: Are there alternative breeding habitats? *Biol. Conserv.* **2013**, *165*, 179–186. [CrossRef]
22. Ruhí, A.; Sebastian, O.S.; Feo, C.; Franch, M.; Gascón, S.; Richter-Boix, À.; Boix, D.; Llorente, G. Man-made Mediterranean temporary ponds as a tool for amphibian conservation. *Ann. Limnol. Int. J. Limnol.* **2012**, *48*, 81–93. [CrossRef]
23. Bolle, H.J. (Ed.) *Mediterranean Climate. Variability and Trends*; Springer: Berlin, Germany, 2003; 320p.
24. Beklioglu, M.; Romo, S.; Kagalou, I.; Quintana, X.; Becares, E. State of the art in the functioning of shallow Mediterranean lakes: Workshop conclusions. *Hydrobiologia* **2007**, *584*, 317–326. [CrossRef]
25. Harmanny, K.S.; Malek, Ž. Adaptations in irrigated agriculture in the Mediterranean region: An overview and spatial analysis of implemented strategies. *Reg. Environ. Change* **2019**, *19*, 1401–1416. [CrossRef]
26. Giralt, D.; Pantoja, J.; Morales, M.B.; Traba, J.; Bota, G. Landscape-Scale Effects of Irrigation on a Dry Cereal Farmland Bird Community. *Front. Ecol. Evol.* **2021**, *9*, 1–8. [CrossRef]
27. De Frutos, A.; Olea, P.P.; Mateo-Tomás, P. Responses of medium- and large-sized bird diversity to irrigation in dry cereal agroecosystems across spatial scales. *Agric. Ecosyst. Environ.* **2015**, *207*, 141–152. [CrossRef]
28. Clemente-Orta, G.; Madeira, F.; Batuecas, I.; Sossai, I.; Juárez-Escario, A.; Albajes, R. Changes in landscape composition influence the abundance of insects on maize: The role of fruit orchards and alfalfa crops. *Agric. Ecosyst. Environ.* **2020**, *291*, 106805. [CrossRef]
29. Cox, N.; Chanson, J.; Stuart, S. *The Status and Distribution of Reptiles and Amphibians of the Mediterranean Basin*; IUCN: Gland, Switzerland; Cambridge, UK, 2006; 42p.
30. Beja, P.; Alcazar, R. Conservation of Mediterranean temporary ponds under agricultural intensification: An evaluation using amphibians. *Biol. Conserv.* **2003**, *114*, 317–326. [CrossRef]

31. Fortuna, M.A.; Gómez-Rodríguez, C.; Bascompte, J. Spatial network structure and amphibian persistence in stochastic environments. *Proc. R. Soc. Lond. B Biol. Sci.* **2005**, *273*, 1429–1434. [CrossRef]
32. Caballero-Díaz, C.; Sánchez-Montes, G.; Butler, H.M.; Vredenburg, V.T.; Martínez-Solano, Í. The role of artificial breeding sites in amphibian conservation: A case study in rural areas in central Spain. *Herpetol. Conserv. Biol.* **2020**, *15*, 87–104.
33. Valdez, J.W.; Gould, J.; Garnham, J.I. Global assessment of artificial habitat use by amphibian species. *Biol. Conserv.* **2021**, *257*, 109129. [CrossRef]
34. Jakob, C.; Poizat, G.; Veith, M.; Seitz, A.; Crivelli, A.J. Breeding phenology and larval distribution of amphibians in a Mediterranean pond network with unpredictable hydrology. *Hydrobiologia* **2003**, *499*, 51–61. [CrossRef]
35. Richter-Boix, A.; Llorente, G.A.; Montori, A. Structure and dynamics of an amphibian metacommunity in two regions. *J. Anim. Ecol.* **2007**, *76*, 607–618. [CrossRef] [PubMed]
36. Fuentes-Rodríguez, F.; Juan, M.; Gallego, I.; Lusi, M.; Fenoy, E.; León, D.; Peñalver, P.; Toja, J.; Casas, J.J. Diversity in Mediterranean farm ponds: Trade-offs and synergies between irrigation modernisation and biodiversity conservation. *Freshw. Biol.* **2003**, *58*, 63–78. [CrossRef]
37. Aspe, C.; Gilles, A.; Jacqué, M. Irrigation canals as tools for climate change adaptation and fish biodiversity management in Southern France. *Reg. Environ. Change* **2014**, *16*, 1975–1984. [CrossRef]
38. Riedener, E.; Rusterholz, H.P.; Baur, B. Land-use abandonment owing to irrigation cessation affects the biodiversity of hay meadows in an arid mountain region. *Agric. Ecosyst. Environ.* **2016**, *185*, 144–152. [CrossRef]
39. Fernández Aláez, M.; Fernández Aláez, C.; Rodríguez, S.; Bécares, E. Evaluation of the state of conservation of shallow ponds in the province of Leon (Northwest Spain) using botanical criteria. *Limnetica* **1999**, *17*, 107–117.
40. Fernandez-Aláez, C.; Fernández-Aláez, M.; Trigal, C.; Luis, B. Hydrochemistry of northwest Spain ponds and its relationships to groundwaters. *Limnetica* **2006**, *25*, 433–452.
41. Pozo, R.; Fernandez-Aláez, M.; Fernández-Aláez, C. Composición de las comunidades de macrófitos y establecimiento del estado de conservación de charcas y lagunas de la Depresión del Duero (noroeste de España) en base a criterios botánicos. *Limnetica* **2012**, *31*, 47–58.
42. Logue, J.B.; Mouquet, N.; Peter, H.; Hillebrand, H. Empirical approaches to metacommunities: A review and comparison with theory. *Trends Ecol. Evol.* **2011**, *26*, 482–491. [CrossRef]
43. Boissinot, A.; Besnard, A.; Lourdais, O. Amphibian diversity in farmlands: Combined influences of breeding-site and landscape attributes in western France. *Agric. Ecosyst. Environ.* **2019**, *269*, 51–61. [CrossRef]
44. Hecnar, S.J.; M'Closkey, R.T. Species richness patterns of amphibians in southwestern Ontario ponds. *J. Biogeogr.* **1998**, *25*, 763–772. [CrossRef]
45. Piha, H.; Luoto, M.; Merila, J. Amphibian Occurrence Is Influenced by Current and Historic Landscape Characteristics. *Ecol. Appl.* **2007**, *17*, 2298–2309. [CrossRef] [PubMed]
46. Indermaur, L.; Schaub, M.; Jokela, J.; Tockner, K.; Schmidt, B.R. Differential response to abiotic conditions and predation risk rather than competition avoidance determine breeding site selection by anurans. *Ecography* **2010**, *33*, 887–895. [CrossRef]
47. Cayuela, H.; Valenzuela-Sánchez, A.; Teulier, L.; Martínez-Solano, Í.; Léna, J.P.; Merilä, J.; Muths, E.; Shine, R.; Quay, L.; Denoël, M.; et al. Determinants and consequences of dispersal in vertebrates with complex life cycles: A review of pond-breeding amphibians. *Quart. Rev. Biol.* **2020**, *95*, 1–36. [CrossRef]
48. Sánchez-Montes, G.; Wang, J.; Ariño, A.H.; Martínez-Solano, Í. Mountains as barriers to gene flow in amphibians: Quantifying the differential effect of a major mountain ridge on the genetic structure of four sympatric species with different life history traits. *J. Biogeogr.* **2018**, *45*, 318–331. [CrossRef]
49. R Core Team. *R: A Language and Environment for Statistical Computing*; R Foundation for Statistical Computing: Vienna, Austria, 2014; Available online: http://www.R-project.org/ (accessed on 30 January 2021).
50. Baselga, A. Partitioning the turnover and nestedness components of beta diversity. *Glob. Ecol. Biogeogr.* **2010**, *19*, 134–143. [CrossRef]
51. Baselga, A.; Orme, C.D.L. Betapart: An R package for the study of beta diversity. *Methods Ecol. Evol.* **2012**, *3*, 808–812. [CrossRef]
52. Marion, Z.H.; Fordyce, J.A.; Fitzpatrick, B.M. Pairwise beta diversity resolves an underappreciated source of confusion in calculating species turnover. *Ecology* **2017**, *98*, 933–939. [CrossRef] [PubMed]
53. Socolar, J.B.; Gilroy, J.J.; Kunin, W.E.; Edwards, D.P. How Should Beta-Diversity Inform Biodiversity Conservation? *Trends Ecol. Evol.* **2016**, *31*, 67–80. [CrossRef]
54. Zhang, J. Species Association Analysis. 2016. Available online: https://cran.r-project.org/web/packages/spaa/spaa.pdf (accessed on 30 January 2021).
55. Leibold, M.A.; Mikkelson, G.M. Coherence, species turnover, and boundary clumping: Elements of meta-community structure. *Oikos* **2002**, *97*, 237–250. [CrossRef]
56. Presley, S.J.; Higgins, C.L.; Willig, M.R. A comprehensive framework for the evaluation of metacommunity structure. *Oikos* **2010**, *119*, 908–917. [CrossRef]
57. Dallas, T. Metacom: An R package for the analysis of metacommunity structure. *Ecography* **2014**, *37*, 402–405. [CrossRef]
58. Ter Braak, C.J.F.; Smilauer, P. *CANOCO Reference Manual and CanoDraw for Windows User's Guide: Software for Canonical Community Ordination (Version 4.5)*; Microcomputer Power: Ithaca, NY, USA, 2002; 500p.

59. Collins, S.J.; Fahrig, L. Responses of anurans to composition and configuration of agricultural landscapes. *Agric. Ecosyst. Environ.* **2017**, *239*, 399–409. [CrossRef]
60. Recuero, E. Sapo de Espuelas—*Pelobates cultripes*. In *Enciclopedia Virtual de los Vertebrados Españoles*; Salvador, A., Martínez Solano, I., Eds.; Museo Nacional de Ciencias Naturales: Madrid, Spain, 2014. Available online: http://www.vertebradosibericos.org/ (accessed on 30 January 2021).
61. Martínez-Solano, I. Sapillo pintojo ibérico—*Discoglossus galganoi*. In *Enciclopedia Virtual de los Vertebrados Españoles*; Salvador, A., Martínez-Solano, I., Eds.; Museo Nacional de Ciencias Naturales: Madrid, Spain, 2014. Available online: http://www.vertebradosibericos.org/ (accessed on 30 January 2021).
62. Ortiz, M.E.; Marco, A.; Saiz, N.; Lizana, M. Impact of ammonium nitrate on growth and survival of six European amphibians. *Arch. Environ. Contam. Toxicol.* **2004**, *47*, 234–239. [CrossRef]
63. Polo-Cavia, N.; Burraco, P.; Gomez-Mestre, I. Low levels of chemical anthropogenic pollution may threaten amphibians by impairing predator recognition. *Aquat. Toxicol.* **2016**, *172*, 30–35. [CrossRef]
64. Hijosa-Valsero, M.; Bécares, E.; Fernández-Aláez, C.; Fernández-Aláez, M.; Mayo, R.; Jiménez, J.J. Chemical pollution in inland shallow ponds in the Mediterranean region (NW Spain): PAHs, insecticides and herbicides in water and sediments. *Sci. Total Environ.* **2016**, *544*, 797–810. [CrossRef]
65. Tavares, H.N.; Da Silva, F.R. Species turnover drives the spatial distribution of frog beta diversity in farmland ponds. *J. Trop. Ecol.* **2019**, *35*, 199–202. [CrossRef]
66. Boix, D.; Caria, M.C.; Gascón, S.; Mariani, M.A.; Sala, J.; Ruhí, A.; Compte, J.; Bagella, S. Contrasting intra-annual patterns of six biotic groups with different dispersal mode and ability in Mediterranean temporary ponds. *Mar. Freshw. Res.* **2017**, *68*, 1044–1060. [CrossRef]
67. Ficetola, G.F.; De Bernardi, F. Amphibians in a human-dominated landscape: The community structure is related to habitat features and isolation. *Biol. Conserv.* **2004**, *119*, 219–230. [CrossRef]
68. Santana, J.; Porto, M.; Reino, L.; Moreira, F.; Ribeiro, P.F.; Santos, J.L.; Rotenberry, J.T.; Beja, P. Using beta diversity to inform agricultural policies and conservation actions on Mediterranean farmland. *J. Appl. Ecol.* **2017**, *54*, 1825–1835. [CrossRef]
69. Sueyoshi, M.; Ishiyama, N.; Nakamura, F. β-diversity decline of aquatic insects at the microhabitat scale associated with agricultural land use. *Landsc. Ecol. Eng.* **2016**, *12*, 187–196. [CrossRef]
70. Johnson, R.K.; Angeler, D.G. Effects of agricultural land use on stream assemblages: Taxon-specific responses of alpha and beta diversity. *Ecol. Indic.* **2014**, *45*, 386–393. [CrossRef]
71. Clements, F.E. *Plant Succession. An Analysis of the Development of Vegetation*; Carnegie Institution of Washington: Washington, DC, USA, 1917. [CrossRef]
72. Lewinsohn, T.M.; Inácio Prado, P.; Jordano, P.; Bascompte, J.; Olesen, J.M. Structure in plant-animal interaction assemblages. *Oikos* **2006**, *113*, 174–184. [CrossRef]
73. Brasil, L.S.; Vieira, T.B.; de Oliveira-Junior, J.M.B.; Dias-Silva, K.; Juen, L. Elements of metacommunity structure in Amazonian Zygoptera among streams under different spatial scales and environmental conditions. *Ecol. Evol.* **2019**, *7*, 3190–3200. [CrossRef] [PubMed]
74. Leavitt, D.J.; Fitzgerald, L.A. Disassembly of a dune-dwelling lizard community due to landscape fragmentation. *Ecosphere* **2013**, *4*, 1–15. [CrossRef]
75. Gómez-Mestre, I. Sapo corredor—*Epidalea calamita*. In *Enciclopedia Virtual de los Vertebrados Españoles*; Salvador, A., Martínez-Solano, I., Eds.; Museo Nacional de Ciencias Naturales: Madrid, Spain, 2014. Available online: http://www.vertebradosibericos.org/ (accessed on 30 January 2021).
76. Price, S.J.; Marks, D.R.; Howe, R.W.; Hanowski, J.A.M.; Niemi, G.J. The importance of spatial scale for conservation and assessment of anuran populations in coastal wetlands of the western Great Ponds, USA. *Landsc. Ecol.* **2005**, *20*, 441–454. [CrossRef]
77. Rubbo, M.J.; Kiesecker, J.M. Amphibian Breeding Distribution in an Urbanized Landscape. *Conserv. Biol.* **2005**, *19*, 504–511. [CrossRef]
78. Couto, A.P.; Ferreira, E.; Torres, R.T.; Fonseca, C. Local and landscape drivers of pond-breeding amphibian diversity at the northern edge of the Mediterranean. *Herpetologica* **2017**, *73*, 10–17. [CrossRef]
79. Pyron, R.A. Global amphibian declines have winners and losers. *Proc. Natl. Acad. Sci. USA* **2018**, *115*, 3739–3741. [CrossRef]
80. Nowakowski, A.J.; Thompson, M.E.; Donnelly, M.A.; Todd, B.D. Amphibian sensitivity to habitat modification is associated with population trends and species traits. *Glob. Ecol. Biogeogr.* **2017**, *26*, 700–712. [CrossRef]
81. Capellà-Marzo, B.; Sánchez-Montes, G.; Martínez-Solano, I. Contrasting demographic trends and asymmetric migration rates in a spatially structured amphibian population. *Integr. Zool.* **2020**, *15*, 482–497. [CrossRef] [PubMed]

land

Article

Four Years Continuous Monitoring Reveals Different Effects of Urban Constructed Wetlands on Bats

Han Li [1,*], Radmila Petric [1,2], Zinah Alazzawi [1], Jake Kauzlarich [1], Rania H. Mahmoud [1], Rasheed McFadden [1], Niklas Perslow [1], Andrea Rodriguez Flores [1], Hadi Soufi [1], Kristina Morales [1], Matina C. Kalcounis-Rueppell [3], Malcolm D. Schug [1] and Lindsey A. Zarecky [4]

1 Department of Biology, University of North Carolina Greensboro, Greensboro, NC 27412, USA; r_petric@uncg.edu (R.P.); z_alazza@uncg.edu or zinah.alazzawi97@gmail.com (Z.A.); jrkauzla@uncg.edu or jkauzlarichenvbio@gmail.com (J.K.); r_mahmou@uncg.edu or rmahmoud.medicine@gmail.com (R.H.M.); remcfad2@uncg.edu or rasheedmcfad@gmail.com (R.M.); ngperslow@uncg.edu or niklasperslow.biotech@gmail.com (N.P.); asrodrig@uncg.edu or andreasrflores@gmail.com (A.R.F.); h_soufi@uncg.edu or soufihadi1@gmail.com (H.S.); k_morale@uncg.edu (K.M.); mdschug@uncg.edu (M.D.S.)
2 Institute for the Environment, University of North Carolina Chapel Hill, Chapel Hill, NC 27517, USA
3 Department of Biological Sciences, University of Alberta, Edmonton, AB T6G 2E1, Canada; kalcounis.rueppell@ualberta.ca
4 Greensboro Science Center, Greensboro, NC 27455, USA; lzarecky@greensboroscience.org
* Correspondence: hanli.urbanecology@gmail.com or h_li6@uncg.edu; Tel.: +1-336-256-1043

Citation: Li, H.; Petric, R.; Alazzawi, Z.; Kauzlarich, J.; Mahmoud, R.H.; McFadden, R.; Perslow, N.; Rodriguez Flores, A.; Soufi, H.; Morales, K.; et al. Four Years Continuous Monitoring Reveals Different Effects of Urban Constructed Wetlands on Bats. *Land* **2021**, *10*, 1087. https://doi.org/10.3390/land10101087

Academic Editors: Juan F. Beltrán, Pedro Abellán and John Litvaitis

Received: 14 September 2021
Accepted: 9 October 2021
Published: 14 October 2021

Abstract: Proactive artificial wetland constructions have been implemented to mitigate the loss of wetlands and their ecosystem services. As wetlands are habitats for bats, short-term (one or two years) studies find that constructed wetlands can immediately increase local bat activity and diversity. However, it is not clear how constructed wetlands affect bats through time while the wetlands are aging. We collected four years of continuous bat acoustic monitoring data at two constructed wetlands in an urban park in Greensboro, NC, USA. We examined bat activity and community composition patterns at these wetlands and compared them with reference sites in the city. With four years of data, we found that the effects of constructed wetlands were both habitat- and species-specific. The wetland in forests significantly increased bat activity, while the wetland in the open grass altered bat community composition. Specifically, in terms of species, we found that over time, constructed wetlands no longer attracted more big brown, silver-haired, or evening bats than control sites while the wetlands aged, highlighting the need to study broadly how each bat species uses natural and artificial wetlands. We emphasize the importance of long-term monitoring and the periodical evaluation of wildlife conservation actions.

Keywords: constructed wetlands; bats; urban ecology; biodiversity; long-term monitoring; acoustics; city parks; community dynamics; conservation evaluation

1. Introduction

Wetlands represent a continuum between both aquatic and terrestrial ecosystems [1]. Despite covering only 6–7% of the Earth's surface, wetlands are one of the most biologically productive ecosystems [2–4]. These species-rich ecosystems provide invaluable services including protection from ecological disturbances such as hurricanes and floods, water filtration, food chain support, and carbon sequestration [3,5]. Ecosystem services provided by wetlands are also fundamental to local economies [6–8]. Natural wetlands usually form an interconnective channel network of water and land infrastructure, providing an important habitat for local wildlife [9,10].

Bats, the second most diverse mammal group, also provide invaluable ecosystem services [11]. Many bat species are insectivores and consume large amounts of nocturnal insects. Thus, bats constitute an important pest control service in forest and agricultural systems [12,13]. Bat predation also limits vector-borne diseases that spread via insects [14].

Due to a wide range of interactions between omnivorous bats and flora, bats serve as pollinators and seed dispersers and increase the yield of plants in their environments [11,15]. The influence of bats on ecosystems is large in geographical scope because of their mobility and migration behaviors [16]. Furthermore, bats are bio-indicators for habitat degradation, pollution, and climate change [17–19].

Worldwide, wetlands are important habitats for bats and support a wide range of species due to the high abundance of native insects to support foraging [20–23]. Wetlands are also a source of drinking water for bats [24–26]. In addition to the foraging opportunities, high-quality riparian vegetation in wetlands offers roosting structures [27,28]. However, due to climate change, pollution, and the ever-increasing need for land conversion, wetlands are predicted to continue to decline [29]. Pollution and fragmentation of natural wetlands negatively impact wetlands' ability to support bats, which is particularly severe in urbanized areas [30–33]. Proactive artificial wetland construction has been implemented to mitigate the increasing losses of wetlands [10,22,34].

Constructed wetlands can be beneficial to wildlife by providing essential habitats for many taxa [10,35]. Constructed wetlands are known to create safe havens for bats, especially in areas that are heavily modified by humans [21,22,34,36]. Menzel et al. (2005) and Parker et al. (2018) demonstrated that the short-term benefits of constructed wetlands to bat communities were observed immediately after the construction or restoration of wetlands within a year or two [34,36]. Yet, there is limited knowledge about natural wetlands and their mechanisms to support wildlife, which in turn limits our understanding of the benefits constructed wetlands provide to wildlife [37–39]. The primary purposes of wetland construction in many cases may not be to protect wildlife and consequently may pose risks to specific wildlife or even form an ecological trap [39–42].

The ecological trap scenario occurs when animals prefer a low-quality habitat over other available higher quality habitats following rapid environmental changes induced by humans [43]. The hypothesis is that environmental changes pose as false cues of high-quality habitats and confuse animals during habitat selection, which eventually leads to lower fitness of individuals [44]. There is evidence that constructed wetlands serve as ecological traps for many species, especially those with limited mobility [40,45]. For example, urban wetlands are often constructed to mitigate stormwater runoff that contains contaminants such as heavy metals, pesticides, and other harmful materials. Frogs living in these constructed wetlands showed lower survival and lacked responses to predator cues [40,46]. Similarly, bats that forage for insects over polluted water are likely to accumulate pollutants over time, even though insects may be more abundant locally [31,47,48].

Pollutant accumulation in constructed wetlands has been well studied. Existing literature shows that pollutants accumulate fastest in the first one or two years after the wetland's construction and that older wetlands contain comparable high levels of pollutants regardless of how artificial wetlands are constructed [49–51]. However, the monitoring and evaluation of biodiversity at constructed wetlands through time is lacking [37,39]. The existing literature tends to present conflicting findings depending on wetland type. For example, in constructed wetlands for wastewater treatment, scientists found decreasing plant diversity over time due to competition [52]. In contrast, mitigation bank wetlands and small restored wetlands on farms showed increasing native plant diversity as wetlands aged [53,54]. A study of amphibian communities in constructed ridge-top wetlands showed that the age of the wetland did not affect the community as a whole or any individual species [55]. However, in a series of constructed urban floodway wetlands, scientists found a higher diversity of macroinvertebrates in older wetlands [56]. To our knowledge, there has not been any multiple-year study on how constructed wetlands affect bats through time.

The objective of our study was to examine bat activity and community composition patterns at two constructed wetlands, through time, as the wetlands aged. We used four years of continuous acoustic monitoring data to investigate whether short-term benefits of constructed wetlands to bats would persist or whether the bat response would attenuate over time. Previous research compared these constructed wetlands with nearby control sites and documented an immediate increase in overall bat activity after wetland construction within a year [34]. Several species including big brown bats (*Eptesicus fuscus*), silver-haired bats (*Lasionycteris noctivagans*), evening bats (*Nycticeius humeralis*), and Mexican free-tailed bats (*Tadarida brasiliensis*) were attracted to the constructed wetlands and consequently increased overall bat diversity at the wetlands. We hypothesized that overall bat activity and certain species' activity would continue to be higher at wetland sites than at the nearby control sites over time. Regarding the community, we hypothesized that bat diversity at the wetlands would be higher than the control sites and that bat community composition at the wetland would be different from the control sites through time. We also hypothesized that bat community composition would be similar across years but vary among seasons at each site. Furthermore, we compared the wetland bat community composition with long-term monitoring sites in a large city park in our study area to explore how the constructed wetlands could alter bat community composition. Previous studies have shown that larger urban parks would have higher bat diversity and more evenly distributed communities [57–59]. Our wetlands were constructed in a small urban park. The previous study at our wetlands already demonstrated increased bat diversity within a year of the construction [34]. Therefore, we hypothesized that constructed wetlands altered bat community composition and made bat communities in a small urban park similar to bat communities in a large urban park immediately after the construction. We also hypothesized that the community similarity would persist over time as the wetlands aged.

2. Materials and Methods

2.1. Study Sites

In March 2017, the University of North Carolina at Greensboro (UNCG) constructed two small wetlands (less than 1000 m^2 each) on its campus in the Peabody Park (a small downtown park, 0.14 km^2) near tributaries to North Buffalo Creek, part of the Cape Fear River water basin, in Greensboro (36°4′48″ N 79°49′10″), NC, USA. The wetlands were constructed at two locations representing two different habitats in the park: one in a wooded forest (named the UNCG woody wetland) and one in an open grass field (named the UNCG open wetland). The goals of constructed wetlands include improving runoff water quality, promoting local biodiversity, enhancing campus aesthetics, and providing educational opportunities. Parker et al. (2018) described the design and construction details of the wetlands [34].

To understand how constructed wetlands affect local biodiversity, a paired monitoring design was implemented. Near each constructed wetland, we identified a matching control site. The control sites have similar vegetation structures comparable to the corresponding wetland with one in the forest (named the UNCG woody control) and one in the grass field (named the UNCG open control). At all four sites, a series of non-invasive biodiversity monitoring equipment was installed prior to the wetland construction and continues through to the present. Bat acoustic monitoring was part of the long-term urban wildlife monitoring effort at these sites.

In addition to the wetland vs. control site pairs, we also included three sites from the Greensboro Science Center (GSC), which is located in a 2.2 km^2 forested park complex approximately 7.5 km northwest of UNCG, representing a large urban greenspace. The three sites at GSC were selected at different heights to specifically monitor bat acoustic activities below, within, and above the park forest canopy for a complete acoustic profile of bats in the large urban park [60]. Monitoring at those sites started between April and July 2017 and continues through to the present.

2.2. Bat Monitoring and Acoustic Analysis

In total, we assayed seven sites in this study using bat acoustic monitoring. We used Song Meter SM4BAT-FS ultrasonic detectors with the SMM-U2 omnidirectional microphone (Wildlife Acoustics Inc., Maynard, MA, USA) at all sites. All detectors were set to record continuously from sunset to sunrise nightly throughout the year and powered by D cell batteries. The specific detector and microphone settings have been previously described [34,61]. All detectors were checked and maintained every two to four weeks throughout the year.

At all UNCG sites and the below canopy site at GSC (named the GSC ground level), the bat detector was strapped to a tree. The microphone and the connecting cable were run through PVC pipes that were strapped to the tree as well. . The microphone was cantilevered away from the tree with a 1 m PVC pipe facing the open space. At these five sites, microphones were approximately 8 m above the ground. At the within canopy site at GSC (named the GSC canopy level), a similar microphone setup was used on a recreational tower with the detector and PVC pipes strapped to the tower pillar, resulting in the microphone facing the open space in the forest canopy approximately 11 m above the ground. The final site at GSC was on the rooftop of a building (named the GSC rooftop level). The microphone was projected above the forest canopy by a 15 m tall weighted station. The specific setup and photos of this site were presented by Li et al. (2020) [62].

We analyzed bat acoustic recordings from April 2017 to December 2020 at all sites except for the GSC rooftop level, which started monitoring in July 2017. For this site, we analyzed bat acoustic recordings from July 2017 to December 2020. We used Kaleidoscope (version 4.5, Wildlife Acoustics Inc., Maynard, MA, USA) to process acoustic recording files and assign species identification. Each recording file had to contain at least three complete bat echolocation calls within 0.5 s to be classified as a bat pass. Others were classified as noise. To assign species identification to a bat pass, we selected big brown bats (EPFU), eastern red bats (*Lasiurus borealis*, LABO), hoary bats (*Lasiurus cinereus*, LACI), silver-haired bats (LANO), evening bats (NYHU), tricolored bats (*Perimyotis subflavus*, PESU), and Mexican free-tailed bats (TABR) in the Kaleidoscope reference library as the only candidate species with the neutral auto-identification setting. This is because previous studies in the area only found these species [63,64]. After the automatic processing, we used the match ratio generated by Kaleidoscope for each bat pass to determine whether we accepted a species identification. We only considered a bat pass identified to species if the match ratio was greater than 0.60, which was a value necessary to be accurate in our study area after comparing Kaleidoscope automatic identification and manual identification by a bat acoustics expert [18,64]. The remaining bat passes were identified as "no ID". Acoustic analysis yielded the total bat passes (including bat passes identified to a species and no ID) and species-specific bat passes for each recording night at each site. We also counted how many species were recorded each night at each site as nightly species richness.

2.3. Statistical Analysis

We used R (version 4.1.0, [65]) for all statistical analyses and data visualization. Since we recorded bats nightly throughout the year and night length varies through the year in Greensboro, we standardized bat passes by night length. We used R package "suncalc" [66] to extract night length in hours and divided bat passes by night length. For each recording night at a site, we had total bat passes per hour and species-specific bat passes per hour. We also assigned seasons to each recording night using meteorological seasons, as follows: spring (March–May), summer (June–August), fall (September–November), and winter (December–February of the following year). We presented our results in a chronological way with seasons as the blocking factor for statistical analyses because life-history events in different seasons could significantly alter bat acoustic activities and communities [67].

To test whether bat activity levels at the constructed wetlands were higher than at the control sites, we compared the dependent variables, total bat passes, and species-specific bat passes at each site pair. We compared the open pair separately from the woody pair because the physical environmental differences (due to vegetation and other obstacles) could significantly affect the probability to detect and record bats and thus the amount of bat passes recorded [62,68,69]. With the pair design, both the wetland and the control site had similar physical conditions for sound transmission and the same weather condition. We could assume the probability to detect and record a bat was similar for paired sites for a species. We did not make any cross-species comparisons for bat activity as the detection probability could be species-specific. We first checked the normality of dependent variables using the Kolmogorov–Smirnov test for large sample sizes [70] and found that all dependent variables were not normally distributed. Therefore, we used the nonparametric Wilcoxon rank-sum test to compare the medians between the wetlands and their matching control site. We conducted Wilcoxon tests by season. For the spring of 2017, we only included April and May data (wetlands constructed in late March). For the winter of 2020, we only included December data due to logistic reasons. To visualize whether the wetland or the control site had higher bat activities during each season, we extracted the Wilcoxon test results for each season and plotted a tile graph for each wetland vs. its control pair using R package "ggplot2" [71]. Each tile in the graph represented a specific season and showed which site had significantly higher bat activities indicated by the Wilcoxon test.

To understand how constructed wetlands affected the bat community, we first examined bat species richness at all seven sites. Previous research in our area suggested that weather conditions including temperature, wind, and precipitation could affect bat acoustic activities [62,64,72]. Thus, we extracted daily weather data (temperature in °C, wind in km/h, and precipitation in cm/h) from the source as described in Li et al. (2020) [62]. To incorporate weather data into analyses, we constructed generalized linear models with nightly species richness as the dependent variable and site as the independent variable. For the dependent variable, we used the Poisson distribution link for generalized linear models as this variable is a count of species recorded on each night [70]. We constructed generalized linear models for each season separately and used the UNCG open control site as the reference level. We used a backward approach for the covariates and included all three weather variables as regression covariates in an initial model and eliminated nonsignificant covariates. All weather covariate results are reported in Supplementary Material Table S1. For each final model, we plotted residuals to visually examine the model fit. We compared site pairs using Tukey HSD tests in the post hoc analysis. To visualize the generalized linear model results, we plotted box plots via "ggplot2" and used different colors to indicate sites that had significantly different levels of species richness within a season. It is important to note that differences in bat species richness among sites could be affected by the physical conditions near a site. Sites with more vegetation coverage could usually detect bats in a smaller range [62,64,72] and possibly fewer species, given that we did not have forest interior specialist species in our study area [63,64].

Next, we conducted Mantel's tests to compare community composition between sites. Mantel's test compares two matrices for correlation, based on multivariate relative distances [73]. We used the month as the sample unit to describe bat communities at each site by calculating mean species-specific bat activities in a month and compared sites in pairs in each year. We binned nightly data together for monthly comparison to reduce the uncertainty among nights caused by different detection probabilities across species. When comparing other sites with the GSC rooftop level for 2017, we only used data between July to December in 2017 from other sites as the Mantel's test requires two matrices to be the same size, and the GSC rooftop level was set up in July 2017. It is important to note that the bat community was represented by the acoustic activity level instead of number of individual bats. We used the Bray–Curtis distance to calculate the dissimilarity matrix due to there being many zeros in our data. For all Mantel's tests, we ran 9999 permutations per test and used Spearman's methods to calculate correlations. To visualize relationships among sites based on the Mantel test, we generated a correlogram indicating if any site pair showed a significant correlation based on monthly mean bat activity matrices via "ggplot2".

Lastly, we used nonmetric multidimensional scaling (NMDS) to describe how community composition changed over time at each site. Nonmetric multidimensional scaling is an ordination technique that graphically presents community relationships by projecting each community from a multidimensional space into a lower-dimensional plot [73]. In our NMDS analysis, we used the same method as with the Mantel test to describe bat communities at each site monthly. Each data point, representing a site at each month, was quantified in a seven-dimensional space, where each dimension represented a bat species. This analysis described whether one or a few species dominated the acoustic space at each site during each month and whether dominance patterns changed over time. The dominance was indicated by the relative amount of bat acoustic activity in comparison among species at a site and did not reflect the absolute amount of bat acoustic activity among sites. We conducted 500 runs with random starts to search for the best two-dimensional solution with the lowest stress using R package "Vegan" [74]. For each NMDS solution at each site, we extracted NMDS scores for each axis and plotted the graph by "ggplot2" and reported stress value for each graph. The interpretation of NMDS plots should focus on the spatial proximity patterns instead of the NMDS scores on each axis [73]. The physical conditions at each site could affect the probability to detect and record bats and thus affect the NMDS scores.

3. Results

From the spring of 2017 to the winter of 2020, we conducted bat acoustic monitoring through 16 seasons at seven sites. In total, we recorded 744,286 bat passes and identified 444,916 passes to species (Table 1, Supplementary Material Table S2). The UNCG open wetland and the UNCG open control had the highest numbers of bat passes across all sites. The UNCG woody control or the GSC ground level tended to have the lowest numbers of bat passes. Summer was the season with the highest number of bat passes, whereas winter had the lowest bat passes for all sites. Interestingly, at UNCG sites, the spring usually had more bat passes than during the fall. In contrast, at GSC sites, the fall had more bat passes than during the spring (Table 1).

Table 1. Total bat passes recorded at each study site in each season in Greensboro, NC, USA.

	UNCG Open Wetland	UNCG Open Control	UNCG Woody Wetland	UNCG Woody Control	GSC Rooftop Level	GSC Canopy Level	GSC Ground Level
2017 Spring	22,481	23,379	8824	3559	N/A	1394	262
2017 Summer	36,428	36,284	5565	1131	8792	7725	1148
2017 Fall	7917	5938	2152	2054	5514	2514	1635
2017 Winter	4794	2445	1526	1561	309	281	280
2018 Spring	19,952	19,578	9588	3964	4765	1909	497
2018 Summer	35,358	19,219	5879	398	11,350	4972	1482
2018 Fall	2621	8898	1481	1898	7554	1104	909
2018 Winter	2728	1460	844	546	131	205	115
2019 Spring	18,481	33,980	10,098	5217	2783	852	824
2019 Summer	32,177	30,544	3906	873	13,172	4789	738
2019 Fall	3662	5744	558	1436	11,869	1045	1183
2019 Winter	2154	1806	812	887	225	153	117
2020 Spring	20,712	28,819	12,160	10,979	1235	1177	498
2020 Summer	35,598	26,851	2512	1227	10,072	8586	2274
2020 Fall	6775	3020	1858	435	9066	3122	756
2020 Winter	417	271	81	83	242	19	59

3.1. Wetland vs. Control Bat Activity Comparison by Wilcoxon Tests

For the wetland versus control comparison on bat activity, we found varying results by wetland type, season, and species. For total bat activity at the open sites, there was no difference between the wetland and the control sites in the first five seasons after construction (Figure 1). In the subsequent eleven seasons, there were six with statistical differences, three with higher total bat activity at the wetland and three at the control, without a consistent pattern (Figure 1, Supplementary Material Table S2). In contrast, at the woody sites, total bat activity was significantly higher at the wetland for ten seasons, including every summer and most springs (Figure 2, Supplementary Material Table S2). At both wetlands, the effect of constructed wetlands on total bat activity did not change through time. Similar to total bat activity, the effect of constructed wetlands on the eastern red bat and the tricolored bat was generally consistent over time at both open and woody wetlands. For the eastern red bat, bat activity was higher at both wetlands than the corresponding control sites for all summers and most springs (Figures 1 and 2, Supplementary Material Table S2). For the tricolored bat, wetland construction generally resulted in lower activity at wetland sites as compared to the control sites across most seasons (Figures 1 and 2, Supplementary Material Table S2).

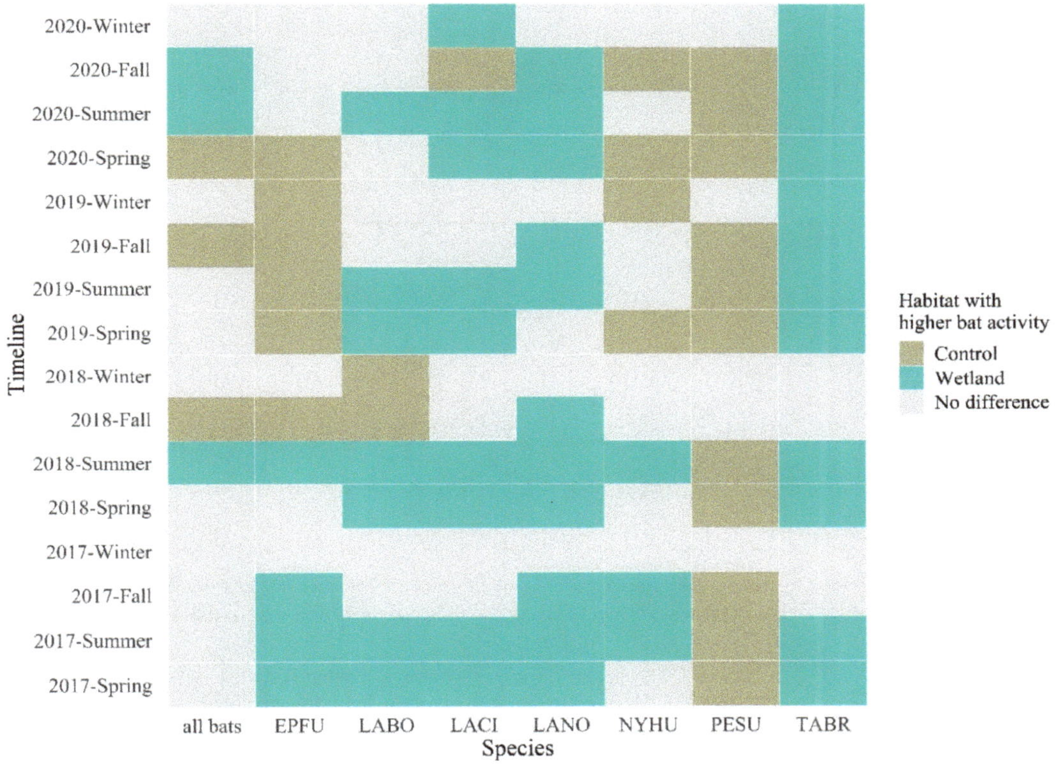

Figure 1. Tile plot showing in different seasons whether the open wetland or the open control had higher total and species-specific bat activity in the Peabody Park in Greensboro, NC, USA. Statistical significance was determined by Wilcoxon tests. Species abbreviations in all figures: EPFU, *Eptesicus fuscus*; LABO, *Lasiurus borealis*; LACI, *L. cinereus*; LANO, *Lasionycteris noctivagans*; NYHU, *Nycticeius humeralis*; PESU, *Perimyotis subflavus*; TABR, *Tadarida brasiliensis*.

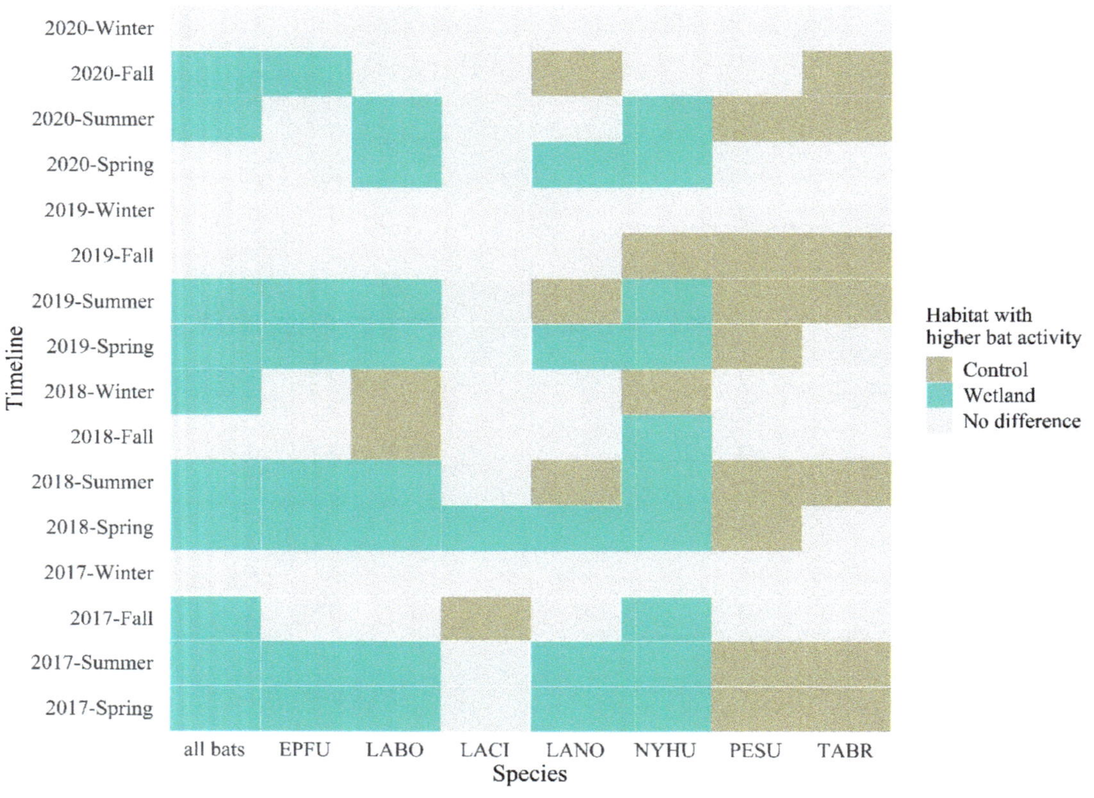

Figure 2. Tile plot showing in different seasons whether the woody wetland or the woody control had higher total and species-specific bat activity in the Peabody Park in Greensboro, NC, USA. Statistical significance was determined by Wilcoxon tests.

For some species, the effect of constructed wetlands varied, depending on whether the wetland was in the open grass or the woods. Hoary bat activity was significantly higher at the open wetland than the open control for ten seasons and no difference was found for the remaining six seasons (Figure 1, Supplementary Material Table S2). However, in the woods, there was no difference between the wetland and the control for most of the seasons (Figure 2, Supplementary Material Table S2). The open wetland had significantly higher Mexican free-tailed bat activity than the open control in 12 seasons (Figure 1, Supplementary Material Table S2). However, there were seven seasons when the woody wetland had significantly lower Mexican free-tailed bat activity than the woody control (Figure 2, Supplementary Material Table S2).

We found that the effect of constructed wetlands on bat activity changed over time for the silver-haired bat, the big brown bat, and the evening bat at certain wetlands. Generally, both the open wetland and the woody wetland had higher bat activity compared to control sites during the first year for these species (Figures 1 and 2, Supplementary Material Table S2). However, starting in the spring or summer of 2018, roughly a year to fifteen months after the wetland construction, the activity patterns changed. For the silver-haired bat, the open wetland continued having higher activity than the open control. However, the woody wetland and woody control alternately had higher activity among seasons (Figures 1 and 2, Supplementary Material Table S2). For the big brown bat and the evening bat, the woody wetland continued to have higher activity than the woody control.

However, the open wetland started having significantly lower bat activity than the control (Figures 1 and 2, Supplementary Material Table S2).

3.2. Species Richness Comparison by Generalized Linear Models

Both season and site affected whether the wetland had higher species richness than the control. In all summers and three springs (except spring 2020), the woody wetland had higher species richness than the woody control (Figure 3, Supplementary Material Table S1). However, for most falls and winters, there was no difference for this pair. In three falls (except fall 2018) and two summers (2018 and 2019), the open wetland had higher species richness than the open control (Figure 3, Supplementary Material Table S1). In two seasons, fall 2020 for the woody pair and spring 2018 for the open pair, the control site had higher species richness than the wetland. Generally, species richness was higher at the open pair than the woody pair. When comparing sites on the UNCG campus with three sites at GSC, the open wetland generally had the same level of species richness as the GSC rooftop site, both having the highest species count within a season. Species richness at the woody wetland was more often lower than the GSC canopy level or ground level. In eleven seasons, the woody control site had the lowest species richness (Figure 3).

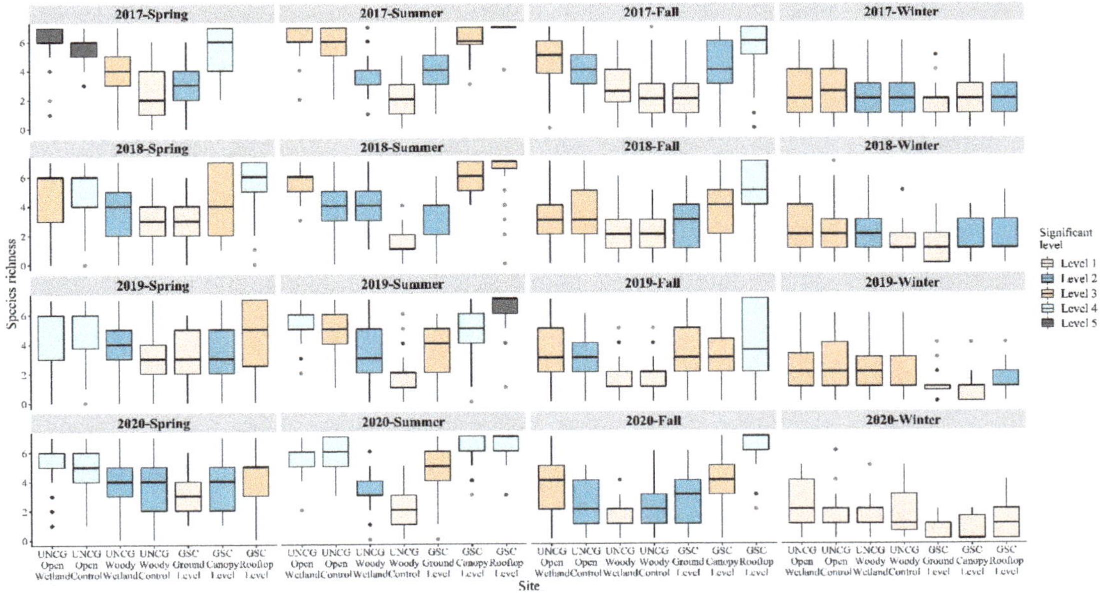

Figure 3. Boxplot (25–75% quantiles) for nightly species richness against sites within each season in Greensboro, NC, USA. Post hoc Tukey HSD tests were used to determine species richness difference between sites, indicated by different colors. When two sites are shown in different colors/significant levels, they have different species richness.

3.3. Community Composition Comparison by Mantel's Tests

Both wetlands had different community compositions from control sites in 2017 immediately after the wetland construction (both $p > 0.05$, Figure 4, Supplementary Material Table S3). However, over time, the community composition difference between a wetland and its control site disappeared (Figure 4). Between the open wetland and the open control, only the first year was different. The woody wetland and the woody control were different in 2017 and 2018 but showed no difference in 2019 and 2020. In 2017, after the construction, the community composition matrices were correlated between the two wetlands ($p < 0.05$, Supplementary Material Table S3), suggesting similar compositions. Interestingly, when comparing the open wetland with three GSC sites, there were strong community matrices correlations (all $p < 0.005$, Figure 4), indicating the open wetland had a community composition similar to sites in a large urban park. As the open wetland aged, the composition matrices were still correlated. In contrast, the woody wetland community composition was never similar to sites at GSC except for 2018, when the woody wetland had a composition similar to the GSC canopy level. At the woody control site, the community composition was always different from any site at GSC. All three sites at GSC did not show any compositional difference among them (Figure 4).

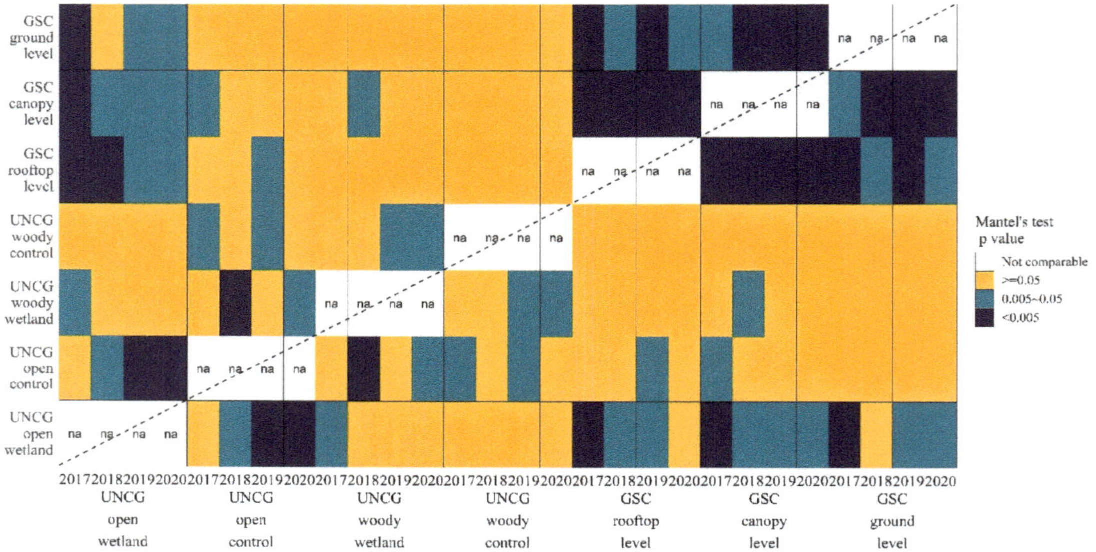

Figure 4. Correlogram showing community composition comparison among study sites in Greensboro, NC, USA. Mantel's tests were used to compare sites. A p value smaller than 0.05 (shown in light and dark blue) indicated two sites were significantly correlated and had no compositional difference whereas a p value larger than 0.05 (shown in yellow) indicated two communities being different.

3.4. Community Composition Changes over Time by NMDS

Among the four sites on the UNCG campus, the open wetland showed the most obvious seasonal community composition variation (Figure 5a). In the winter, the community at the open wetland mostly associated with the hoary bat. Spring seemed to be the transition season with most species present. In the summer of 2017, the bat community at the open wetland had associations with the tricolored bat or the evening bat and the big brown bat. Later in 2019 or 2020, the dominant species in the summer shifted to the big brown bat, suggesting a changing summer community since the construction of the wetland. In the fall, the community associated with either the evening bat or the tricolored bat.

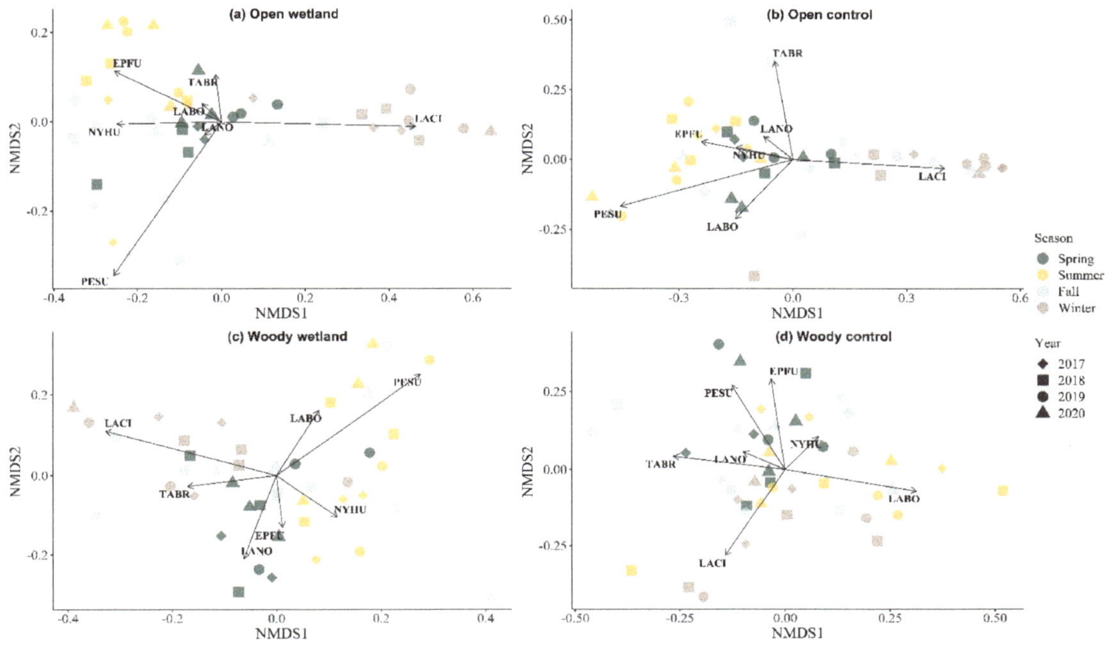

Figure 5. NMSD plots showing monthly bat community composition varied among seasons and across years at four sites in the Peabody Park in Greensboro, NC, USA. Each panel for a site: (**a**) open wetland (stress 0.099); (**b**) open control (stress 0.101); (**c**) woody wetland (stress 0.1554); (**d**) woody control (stress 0.163).

The community composition pattern at the open control sites was similar to the open wetland, with the hoary bat being the dominant species for the winter and the big brown bat for the summer (Figure 5b). However, there were a few summer and fall months in 2019 and 2020 when the tricolored bat became more prevalent at this site. In the fall, the community composition varied and did not form a clear grouping pattern. At the woody wetland, the winter and summer community compositions separated distinctively. However, neither spring nor fall had a clear pattern associating the site with certain species (Figure 5c). At the woody control site, the separation among seasons was weaker than the other three sites, suggesting no particular species used this site more often than others at any time of the year (Figure 5d). No clear separation by year was found at both woody wetland and woody control.

Among three sites at GSC, the rooftop level site had the clearest seasonal separation. Similar to the open wetland at UNCG, in the winter the community was dominated by the hoary bat (Figure 6a). However, different from the open wetland, the association with the evening or tricolored bat at this site was found in the summer instead of the fall. The big brown bat was the dominant species at the rooftop level site in the fall. Both the canopy and ground-level sites only had the summer separated from other seasons (Figure 6b,c). The ground level site at GSC was dominated by the big brown bat in the summer, except for the summer of 2020 when the evening bat was the most common species in the community (Figure 6c). In fact, at all three sites, the evening bat was prevalent in the summer of 2020, making the community different from other summer seasons.

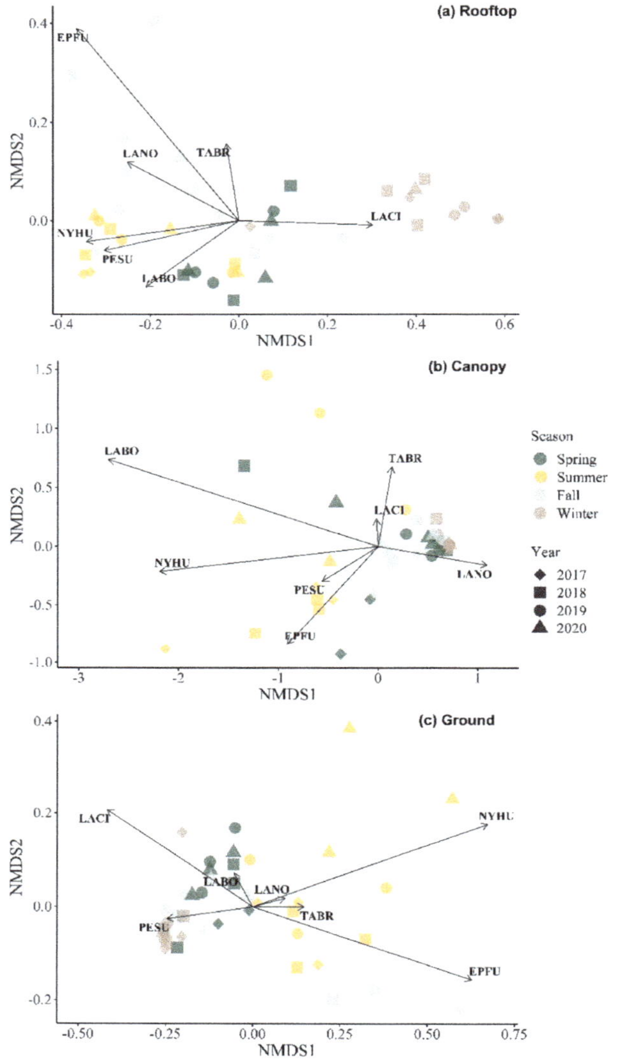

Figure 6. NMSD plots showing monthly bat community composition varied among seasons and across years at three sites at the Greensboro Science Center in Greensboro, NC, USA. Each panel for a site: (**a**) rooftop level (stress 0.042); (**b**) canopy level (stress 0.144); (**c**) ground level (stress 0.048).

4. Discussion

Four years of continuous monitoring at two constructed wetlands revealed different outcomes. How bats responded to constructed wetlands could change over time. We found that total bat activity was higher at the woody wetland than the control, whereas generally, total bat activity was not different for the wetland and its control in the open grass. Constructed wetlands in forests have not been as well studied as wetlands in the open. However, it is known that water sources within forests provide important drinking and foraging habitats for bats [24,75]. For the wetland in the open grass, our results are consistent with a previous study in the Southeastern US, which found no difference between wetlands and controls after the wetland restoration at a much broader scale [36].

However, two studies in South Africa and Germany showed that constructed wetlands in open areas had higher total bat activity than other land covers [21,22], suggesting there could be species-specific responses to constructed wetlands.

Among the seven species we studied, we found that only four species showed a consistent response between open and woody wetlands over time. Eastern red bat activity was higher at wetlands compared to controls, consistent with the short-term one-year response that we found at these sites [34]. We also found that tricolored bat activity was consistently lower at wetlands compared to controls, suggesting constructed wetlands might repel this species. Previous studies have demonstrated that the tricolored bat prefers relatively low-quality eutrophicated water at both local and regional scales, likely due to emerging aquatic insects associated with eutrophicated water [18,76]. Wetlands have the ability to filter water and improve water quality [3,5], and water quality at these wetlands may not have been suitable for tricolored bats to forage preferred insect preys. Interestingly, in a broad scale analysis in the same region, the tricolored bat was found positively correlated to woody wetlands [77], suggesting there should be more studies on how the tricolored bat responds to both vegetation and water. For the hoary bat and the Mexican free-tailed bat, only the open wetland showed an effect. This is likely because these species usually fly over the canopy in open space and are less suitable for maneuvering through forests [60,78,79]. This finding suggests that the location of small constructed urban wetlands is important for attracting bats.

For three species, we found that constructed wetlands had higher bat activity than controls immediately after construction but the difference disappeared over time. In the short term, within a year, the big brown bat, the silver-haired bat, and the evening bat all showed increases with construction wetlands [34]. However, starting in the second or third year, some wetlands had lower bat activity than controls for these species. This is a result that can only be found in multiple-year studies and has not been reported previously. Why would there be a discrepancy between short and long term? Our wetlands changed over time as planted aquatic vegetation became mature and might have altered the water surface area available for drinking. It is likely that water quality also changed, which could alter insect prey availability for bats at the wetlands. Studies of the relationship between aquatic vegetation structure and bat activity deserve future attention. We also propose future studies to examine constructed wetlands with different vegetation management schemes. Notwithstanding, it is still puzzling why only certain species of bats changed their preference. It is likely that different species of bats benefit from wetlands in unique and different ways and we need to better understand what aspects of wetlands each species is responding to and, more broadly, how each bat species uses wetlands. How natural wetlands support wildlife is not completely understood [37–39] and our results at constructed wetlands underscore this.

Recognizing that each bat species has specific needs and responses to wetlands, it is also important to consider the bat community and how interspecific interactions might affect the effects of constructed wetlands on bats. Our analysis on species richness showed that wetlands generally have more species than the control for most seasons, which was consistent with previous studies [22,34]. In a behavioral study conducted at these wetlands, we found increased territory defense calls when multiple species were present [61]. It is likely that constructed wetlands attracted more species and increased interspecific interactions. Consequently, increased interspecific interactions might shape bat communities over time [80,81].

Our community-level analyses also showed seasonal variations. The big brown bat was associated with most seasons except for the winter, when the hoary bat was the dominant species. All seven species have been documented to be residents in the Piedmont of North Carolina for the winter [34,72]. However, we found that the big brown bat and the silver-haired bat were the most active species in the winter in non-urban settings instead of the hoary bat [72]. We suspect that there might be local scale seasonal migrations, similar to a study on big brown bats in Colorado where urban big brown bats left the city during the winter season [82]. Community differences among seasons were always stronger at the wetlands than at the control sites, suggesting that wetlands might increase interspecific interactions and certain species used the wetlands more often than others. Over time, we observed the open wetland bat community losing evening bats and becoming more big brown bat-dominant. The construction of a wetland could be considered as a change in the environment. A previous study showed that the big brown bat was more adaptive to anthropogenic changes than evening bats [83], which might explain why evening bats stopped using the wetland site once the wetland was constructed.

Combining results that the open wetland became more big brown bat-dominant, and that big brown bat activity was higher at the control than the wetland, we suggest that the wetland attracted big brown bats to Peabody Park on the UNCG campus. This is supported by yearly comparisons of community composition between sites where we found that the composition was different between the wetland and their matching controls in the first year after wetland construction but not thereafter. We speculate that wetland construction initially attracted more bats to Peabody Park. Over time, these bats could have explored the entire park and found more preferable habitats within it. Peabody Park is a small urban park near the city center. Studies have shown that the size of urban green spaces and their relative location in the city could affect bat activity and community composition [58–60], likely because of vegetation, water, noise level, and other environmental characteristics [62,84,85]. Generally, larger parks tend to have higher bat activity and higher diversity. Interestingly, our community-level analyses found that the wetland in the open grass had a bat community composition similar to sites in GSC, a large urban park. This result further demonstrates the potential benefits of small constructed wetlands in urban areas.

5. Conclusions

Our analysis of four years of continuous bat monitoring data in constructed wetlands found that how bats responded to constructed wetlands was both habitat- and species-specific. A constructed wetland's ability to attract bats depended on the wetland's location. Constructed wetlands in forests could significantly increase bat activity, while constructed wetlands in the open grass played a bigger role in altering bat community composition. Interspecific interactions were likely increased by the constructed wetlands. Therefore, it is important to study how each species uses wetlands specifically regarding drinking and foraging activities in the future. We suggest that the location of a constructed wetland must be carefully planned based on what bat species the constructed wetland is intended to conserve to ensure its goal of conserving bats. Overall, constructed wetlands in a small urban park in our study were beneficial for bat diversity and community composition on a scale that is typically seen only in large urban parks. However, as time passes, constructed wetlands may no longer attract more bats due to other environmental changes. Therefore, we emphasize the importance of long-term monitoring and the periodical evaluation of wildlife conservation actions which may have unanticipated effects over time.

Supplementary Materials: The following are available online at https://www.mdpi.com/article/10.3390/land10101087/s1, Table S1: Generalized linear model for species richness analysis results (including weather covariates results for each model) and post hoc paired Tukey comparison results for bat species richness difference among sites within each season, Table S2: bat passes recorded and bat passes identified to species by site and by season, median total bat activity and species-specific bat activity and Wilcoxon test p value for paired wetland versus control sites, Table S3: Mantel's test p values comparing bat communities among sites year by year.

Author Contributions: Conceptualization, H.L., R.P., Z.A., J.K., R.H.M., R.M., N.P., A.R.F., H.S., K.M., M.C.K.-R., M.D.S. and L.A.Z.; methodology, H.L., R.P., Z.A., J.K., R.H.M., R.M., N.P., A.R.F., H.S., K.M., M.C.K.-R., M.D.S. and L.A.Z.; software, H.L., R.P. and L.A.Z.; validation, H.L., R.P. and L.A.Z.; formal analysis, H.L.; investigation, R.P., K.M. and L.A.Z.; resources, M.D.S. and L.A.Z.; data curation, H.L., R.P., K.M. and L.A.Z.; writing—original draft preparation, H.L., R.P., Z.A., J.K., R.H.M., R.M., N.P., A.R.F., H.S. and L.A.Z.; writing—review and editing, H.L., R.P., Z.A., J.K., R.H.M., R.M., N.P.,

A.R.F., H.S., K.M., M.C.K.-R., M.D.S. and L.A.Z.; visualization, H.L.; supervision, H.L., R.P., M.D.S. and L.A.Z.; project administration, H.L., M.D.S. and L.A.Z.; funding acquisition, M.C.K.-R., M.D.S. and L.A.Z. All authors have read and agreed to the published version of the manuscript.

Funding: The bat monitoring on the University of North Carolina Greensboro campus was funded by grants from the Duke Energy Water Resources Fund, the University of North Carolina Greensboro Green Fund and the University of North Carolina Greensboro Provost Office. The bat monitoring at Greensboro Science Center was partially funded by the North Carolina Science Museums Grant Program. The funding bodies were not involved in the design of the research question, field data collection, analysis and interpretation of data, or writing the manuscript.

Institutional Review Board Statement: Not applicable.

Informed Consent Statement: Not applicable.

Data Availability Statement: The data presented in this study are openly available in FigShare at https://doi.org/10.6084/m9.figshare.16574999.v1 (accessed on 9 June 2021).

Acknowledgments: The bat monitoring at the University of North Carolina Greensboro campus was supported by the following: The City of Greensboro, UNCG Department of Chemistry, UNCG Department of Geography, Environment, and Sustainability, Peabody Park Preservation Committee, UNCG Office of Sustainability, UNCG Undergraduate Research, Scholarship, and Creativity Office, UNCG Facilities Design and Construction, UNCG Facilities Grounds and Garage, and the community of volunteers. The bat monitoring at the Greensboro Science Center was supported by its staff, interns, and patrons. This paper was written as part of a special scientific reading and writing course supported by the UNCG Department of Biology motivated by participation in the Council of Undergraduate Research (CUR) Transformation Project, for which we thank gratefully. Comments by journal editors and four anonymous reviewers greatly improved earlier versions of this manuscript.

Conflicts of Interest: The authors declare no conflict of interest.

References

1. Brinson, M.M. *A Hydrogeomorphic Clasification for Wetlands*; U.S. Environmental Laboratory: Vicksburg, MS, USA; Engineer Research and Development Center: Vicksburg, MS, USA, 1993; Available online: https://hdl.handle.net/11681/6483 (accessed on 10 September 2021).
2. Davidson, N.C. How much wetland has the world lost? Long-term and recent trends in global wetland area. *Mar. Freshw. Res.* **2014**, *65*, 934–941. [CrossRef]
3. Gardner, R.C.; Finlayson, C. *Global Wetland Outlook: State of the World's Wetlands and Their Services to People*; Social Science Research Network: Rochester, NY, USA, 2018; Available online: https://ssrn.com/abstract=3261606 (accessed on 10 September 2021).
4. Finlayson, C.M.; Davidson, N.C.; Spiers, A.G.; Stevenson, N.J. Global wetland inventory—Current status and future priorities. *Mar. Freshw. Res.* **1999**, *50*, 717–727. [CrossRef]
5. Verhoeven, J.T.A.; Setter, T.L. Agricultural use of wetlands: Opportunities and limitations. *Ann. Bot.* **2010**, *105*, 155–163. [CrossRef] [PubMed]
6. Ingraham, M.W.; Foster, S.G. The value of ecosystem services provided by the U.S. National Wildlife Refuge System in the contiguous U.S. *Ecol. Econ.* **2008**, *67*, 608–618. [CrossRef]
7. Woodward, R.T.; Wui, Y.-S. The economic value of wetland services: A meta-analysis. *Ecol. Econ.* **2001**, *37*, 257–270. [CrossRef]
8. Mitsch, W.J.; Gosselink, J.G. The value of wetlands: Importance of scale and landscape setting. *Ecol. Econ.* **2000**, *35*, 25–33. [CrossRef]
9. Russi, D.; Brink ten, P.; Farmer, A.; Badura, T.; Coates, D.; Forster, J.; Kumar, R.; Davidson, N. The Economics of Ecosystems and Biodiversity for Water and Wetlands. 2013; Available online: http://www.teebweb.org/wp-content/uploads/2013/04/TEEB_WaterWetlands_Report_2013.pdf (accessed on 10 September 2021).
10. Zhang, C.; Wen, L.; Wang, Y.; Liu, C.; Zhou, Y.; Lei, G. Can Constructed Wetlands be Wildlife Refuges? A Review of Their Potential Biodiversity Conservation Value. *Sustainability* **2020**, *12*, 1442. [CrossRef]
11. Kunz, T.H.; De Torrez, E.B.; Bauer, D.; Lobova, T.; Fleming, T.H. Ecosystem services provided by bats. *Ann. N. Y. Acad. Sci.* **2011**, *1223*, 1–38. [CrossRef]
12. Boyles, J.G.; Cryan, P.M.; McCracken, G.F.; Kunz, T.H. Economic Importance of Bats in Agriculture. *Science* **2011**, *332*, 41–42. [CrossRef]
13. Rodríguez-San Pedro, A.; Allendes, J.L.; Beltrán, C.A.; Chaperon, P.N.; Saldarriaga-Córdoba, M.M.; Silva, A.X.; Grez, A.A. Quantifying ecological and economic value of pest control services provided by bats in a vineyard landscape of central Chile. *Agric. Ecosyst. Environ.* **2020**, *302*, 107063. [CrossRef]
14. de Jesús Crespo, R.; Méndez Lázaro, P.; Yee, S.H. Linking Wetland Ecosystem Services to Vector-borne Disease: Dengue Fever in the San Juan Bay Estuary, Puerto Rico. *Wetlands* **2019**, *39*, 1281–1293. [CrossRef]

15. Tremlett, C.J.; Moore, M.; Chapman, M.A.; Zamora-Gutierrez, V.; Peh, K.S.-H. Pollination by bats enhances both quality and yield of a major cash crop in Mexico. *J. Appl. Ecol.* **2020**, *57*, 450–459. [CrossRef]
16. Medellin, R.A.; Wiederholt, R.; Lopez-Hoffman, L. Conservation relevance of bat caves for biodiversity and ecosystem services. *Biol. Conserv.* **2017**, *211*, 45–50. [CrossRef]
17. Jones, G.; Jacobs, D.; Kunz, T.; Willig, M.; Racey, P. Carpe noctem: The importance of bats as bioindicators. *Endanger. Species Res.* **2009**, *8*, 93–115. [CrossRef]
18. Li, H.; Kalcounis-Rueppell, M. Separating the effects of water quality and urbanization on temperate insectivorous bats at the landscape scale. *Ecol. Evol.* **2018**, *8*, 667–678. [CrossRef]
19. Russo, D.; Salinas-Ramos, V.B.; Cistrone, L.; Smeraldo, S.; Bosso, L.; Ancillotto, L. Do We Need to Use Bats as Bioindicators? *Biology* **2021**, *10*, 693. [CrossRef] [PubMed]
20. Mickleburgh, S.P.; Hutson, A.M.; Racey, P.A. A review of the global conservation status of bats. *Oryx* **2002**, *36*, 18–34. [CrossRef]
21. Stahlschmidt, P.; Pätzold, A.; Ressl, L.; Schulz, R.; Brühl, C.A. Constructed wetlands support bats in agricultural landscapes. *Basic Appl. Ecol.* **2012**, *13*, 196–203. [CrossRef]
22. Sirami, C.; Jacobs, D.S.; Cumming, G.S. Artificial wetlands and surrounding habitats provide important foraging habitat for bats in agricultural landscapes in the Western Cape, South Africa. *Biol. Conserv.* **2013**, *164*, 30–38. [CrossRef]
23. Fukui, D.; Murakami, M.; Nakano, S.; Aoi, T. Effect of emergent aquatic insects on bat foraging in a riparian forest. *J. Anim. Ecol.* **2006**, *75*, 1252–1258. [CrossRef] [PubMed]
24. Salvarina, I. Bats and aquatic habitats: A review of habitat use and anthropogenic impacts. *Mammal Rev.* **2016**, *46*, 131–143. [CrossRef]
25. Blakey, R.V.; Law, B.S.; Straka, T.M.; Kingsford, R.T.; Milne, D.J. Importance of wetlands to bats on a dry continent: A review and meta-analysis. *Hystrix Ital. J. Mammal.* **2018**, *12*, 41–52. [CrossRef]
26. Seibold, S.; Buchner, J.; Bässler, C.; Muller, J. Ponds in acidic mountains are more important for bats in providing drinking water than insect prey. *J. Zool.* **2013**, *290*, 302–308. [CrossRef]
27. Lintott, P.R.; Bunnefeld, N.; Park, K.J. Opportunities for improving the foraging potential of urban waterways for bats. *Biol. Conserv.* **2015**, *191*, 224–233. [CrossRef]
28. Ober, H.K.; Hayes, J.P. Influence of Vegetation on Bat Use of Riparian Areas at Multiple Spatial Scales. *J. Wildl. Manag.* **2008**, *72*, 396–404. [CrossRef]
29. Sofaer, H.R.; Skagen, S.K.; Barsugli, J.J.; Rashford, B.S.; Reese, G.C.; Hoeting, J.A.; Wood, A.W.; Noon, B.R. Projected Wetland Densities under Climate Change: Habitat Loss but Little Geographic Shift in Conservation Strategy. *Ecol. Appl.* **2016**, *26*, 1677–1692. [CrossRef]
30. Straka, T.M.; Lentini, P.E.; Lumsden, L.F.; Wintle, B.A.; van der Ree, R. Urban bat communities are affected by wetland size, quality, and pollution levels. *Ecol. Evol.* **2016**, *6*, 4761–4774. [CrossRef]
31. Naidoo, S.; Vosloo, D.; Schoeman, M.C. Foraging at Wastewater Treatment Works Increases the Potential for Metal Accumulation in an Urban Adapter, the Banana Bat (*Neoromicia nana*). *Afr. Zool.* **2013**, *48*, 39–55. [CrossRef]
32. Mendes, E.S.; Pereira, M.J.R.; Marques, S.F.; Fonseca, C. A Mosaic of Opportunities? Spatio-Temporal Patterns of Bat Diversity and Activity in a Strongly Humanized Mediterranean Wetland. *Eur. J. Wildl. Res.* **2014**, *60*, 651–664. [CrossRef]
33. Lookingbill, T.R.; Elmore, A.J.; Engelhardt, K.A.; Churchill, J.B.; Gates, J.E.; Johnson, J.B. Influence of wetland networks on bat activity in mixed-use landscapes. *Biol. Conserv.* **2010**, *143*, 974–983. [CrossRef]
34. Parker, K.A.; Springall, B.T.; Garshong, R.A.; Malachi, A.N.; Dorn, L.E.; Costa-Terryll, A.; Mathis, R.A.; Lewis, A.N.; MacCheyne, C.L.; Davis, T.T.; et al. Rapid Increases in Bat Activity and Diversity after Wetland Construction in an Urban Ecosystem. *Wetlands* **2019**, *39*, 717–727. [CrossRef]
35. Giosa, E.; Mammides, C.; Zotos, S. The importance of artificial wetlands for birds: A case study from Cyprus. *PLoS ONE* **2018**, *13*, e0197286. [CrossRef] [PubMed]
36. Menzel, J.M.; Menzel, M.A.; Kilgo, J.C.; Ford, W.M.; Edwards, J.W. Bat response to Carolina bays and wetland restoration in the southeastern U.S. Coastal Plain. *Wetlands* **2005**, *25*, 542–550. [CrossRef]
37. Whigham, D.F. Ecological issues related to wetland preservation, restoration, creation and assessment. *Sci. Total Environ.* **1999**, *240*, 31–40. [CrossRef]
38. Gopal, B. Natural and Constructed Wetlands for Wastewater Treatment: Potentials and Problems. *Water Sci. Technol.* **1999**, *40*, 27–35. [CrossRef]
39. Wiegleb, G.; Dahms, H.-U.; Byeon, W.; Choi, G. To What Extent Can Constructed Wetlands Enhance Biodiversity? *Int. J. Environ. Sci. Technol.* **2017**, *8*, 561–569. [CrossRef]
40. Sievers, M.; Parris, K.M.; Swearer, S.E.; Hale, R. Stormwater wetlands can function as ecological traps for urban frogs. *Ecol. Appl.* **2018**, *28*, 1106–1115. [CrossRef] [PubMed]
41. Haberl, R.; Perfler, R.; Mayer, H. Constructed Wetlands in Europe. *Water Water Sci. Technol.* **1995**, *32*, 305–315. [CrossRef]
42. Sievers, M.; Hale, R.; Parris, K.M.; Swearer, S.E. Impacts of Human-Induced Environmental Change in Wetlands on Aquatic Animals. *Biol. Rev. Camb. Philos. Soc.* **2018**, *93*, 529–554. [CrossRef]
43. Robertson, B.A.; Hutto, R.L. A Framework for Understanding Ecological Traps and an Evaluation of Existing Evidence. *Ecology* **2006**, *87*, 1075–1085. [CrossRef]

44. Hale, R.; Swearer, S.E. Ecological traps: Current evidence and future directions. *Proc. Biol. Sci.* **2016**, *283*, 20152647. [CrossRef] [PubMed]
45. Severns, P.M. Habitat restoration facilitates an ecological trap for a locally rare, wetland-restricted butterfly. *Insect Conserv. Divers.* **2011**, *4*, 184–191. [CrossRef]
46. Sievers, M.; Hale, R.; Swearer, S.E.; Parris, K.M. Contaminant mixtures interact to impair predator-avoidance behaviours and survival in a larval amphibian. *Ecotoxicol. Environ. Saf.* **2018**, *161*, 482–488. [CrossRef]
47. Clarke-Wood, B.K.; Jenkins, K.M.; Law, B.S.; Blakey, R.V. The ecological response of insectivorous bats to coastal lagoon degradation. *Biol. Conserv.* **2016**, *202*, 10–19. [CrossRef]
48. Laverty, T.M.; Berger, J. Do bats seek clean water? A perspective on biodiversity from the Namib Desert. *Biol. Conserv.* **2020**, *248*, 108686. [CrossRef]
49. Sharley, D.J.; Sharp, S.M.; Marshall, S.; Jeppe, K.; Pettigrove, V.J. Linking urban land use to pollutants in constructed wetlands: Implications for stormwater and urban planning. *Landsc. Urban Plan.* **2017**, *162*, 80–91. [CrossRef]
50. Paing, J.; Guilbert, A.; Gagnon, V.; Chazarenc, F. Effect of climate, wastewater composition, loading rates, system age and design on performances of French vertical flow constructed wetlands: A survey based on 169 full scale systems. *Ecol. Eng.* **2015**, *80*, 46–52. [CrossRef]
51. Hijosa-Valsero, M.; Sidrach-Cardona, R.; Becares, E. Comparison of interannual removal variation of various constructed wetland types. *Sci. Total Environ.* **2012**, *430*, 174–183. [CrossRef]
52. Saggaï, M.M.; Ainouche, A.; Nelson, M.; Cattin, F.; El Amrani, A. Long-term investigation of constructed wetland wastewater treatment and reuse: Selection of adapted plant species for metaremediation. *J. Environ. Manag.* **2017**, *201*, 120–128. [CrossRef] [PubMed]
53. Reinartz, J.A.; Warne, E.L. Development of vegetation in small created wetlands in southeastern Wisconsin. *Wetlands* **1993**, *13*, 153–164. [CrossRef]
54. Spieles, D.J.; Coneybeer, M.; Horn, J. Community Structure and Quality after 10 Years in Two Central Ohio Mitigation Bank Wetlands. *Environ. Manag.* **2006**, *38*, 837–852. [CrossRef] [PubMed]
55. Denton, R.D.; Richter, S.C. Amphibian communities in natural and constructed ridge top wetlands with implications for wetland construction. *J. Wildl. Manag.* **2013**, *77*, 886–896. [CrossRef]
56. Schad, A.N.; Kennedy, J.H.; Dick, G.O.; Dodd, L. Aquatic macroinvertebrate richness and diversity associated with native submerged aquatic vegetation plantings increases in longer-managed and wetland-channeled effluent constructed urban wetlands. *Wetl. Ecol. Manag.* **2020**, *28*, 461–477. [CrossRef]
57. Silva De Araújo, M.L.V.; Bernard, E. Green remnants are hotspots for bat activity in a large Brazilian urban area. *Urban Ecosyst.* **2016**, *19*, 287–296. [CrossRef]
58. Tena, E.; Fandos, G.; De Paz, Ó.; De La Peña, R.; Tellería, J.L. Size does matter: Passive sampling in urban parks of a regional bat assemblage. *Urban Ecosyst.* **2020**, *23*, 227–234. [CrossRef]
59. Krauel, J.J.; Lebuhn, G. Patterns of Bat Distribution and Foraging Activity in a Highly Urbanized Temperate Environment. *PLoS ONE* **2016**, *11*, e0168927. [CrossRef]
60. Kalcounis, M.C.; Hobson, K.A.; Brigham, R.M.; Hecker, K.R. Bat Activity in the Boreal Forest: Importance of Stand Type and Vertical Strata. *J. Mammal.* **1999**, *80*, 673–682. [CrossRef]
61. Springall, B.T.; Li, H.; Kalcounis-Rueppell, M.C. The In-Flight Social Calls of Insectivorous Bats: Species Specific Behaviors and Contexts of Social Call Production. *Front. Ecol. Evol.* **2019**, *7*, 441. [CrossRef]
62. Li, H.; Crihfield, C.; Feng, Y.; Gaje, G.; Guzman, E.; Heckman, T.; Mellis, A.; Moore, L.; Romo Bechara, N.; Sanchez, S.; et al. The Weekend Effect on Urban Bat Activity Suggests Fine Scale Human-Induced Bat Movements. *Animals* **2020**, *10*, 1636. [CrossRef]
63. Loeb, S.C.; Post, C.J.; Hall, S.T. Relationship between urbanization and bat community structure in national parks of the southeastern U.S. *Urban Ecosyst.* **2009**, *12*, 197–214. [CrossRef]
64. Schimpp, S.A.; Li, H.; Kalcounis-Rueppell, M.C. Determining species specific nightly bat activity in sites with varying urban intensity. *Urban Ecosyst.* **2018**, *21*, 541–550. [CrossRef]
65. R Development Core Team. *R: A Language and Environment for Statistical Computing*; R Foundation for Statistical Computing: Vienna, Austria, 2018; ISBN 3-900051-07-0; Available online: https://www.r-project.org/ (accessed on 10 September 2021).
66. Thieurmel, B.; Elmarhraoui, A. Suncalc: Compute Sun Position, Sunlight Phases, Moon Position and Lunar Phase. 2019; Available online: https://cran.r-project.org/web/packages/suncalc/index.html (accessed on 10 September 2021).
67. Reichert, B.E.; Bayless, M.; Cheng, T.L.; Coleman, J.T.H.; Francis, C.M.; Frick, W.F.; Gotthold, B.S.; Irvine, K.M.; Lausen, C.; Li, H.; et al. NABat: A top-down, bottom-up solution to collaborative continental-scale monitoring. *Ambio* **2021**, *50*, 901–913. [CrossRef]
68. Loeb, S.C.; Hines, B.A.; Armstrong, M.P.; Zarnoch, S.J. Effects of Omnidirectional Microphone Placement and Survey Period on Bat Echolocation Call Quality and Detection Probabilities. *Acta Chiropterol.* **2020**, *21*, 453–464. [CrossRef]
69. Findlay, S.V.; Barclay, R.M.R. Acoustic Surveys for Bats are Improved by Taking Habitat Type into Account. *Wildl. Soc. Bull.* **2020**, *44*, 86–93. [CrossRef]
70. Quinn, G.P.; Keough, M.J. *Experimental Design and Data Analysis for Biologists*; Cambridge University Press U.S.: New York, NY, USA, 2002; ISBN 0-521-00976-6.
71. Wickham, H.; Chang, W. Ggplot2: An Implementation of the Grammar of Graphics. R Package Version 0.7; Available online: http://CRAN.+R-project.+org/package=+ggplot2+2008 (accessed on 10 September 2021).

72. Parker, K.A., Jr.; Li, H.; Kalcounis-Rueppell, M.C. Species-Specific Environmental Conditions for Winter Bat Acoustic Activity in North Carolina, United States. *J. Mammal.* **2020**, *101*, 1502–1512. [CrossRef]
73. McCune, B.; Grace, J.B.; Urban, D.L. *Analysis of Ecological Communities*; MJM Software Design: Gleneden Beach, OR, USA, 2002; ISBN 978-0-9721290-0-8.
74. Oksanen, J.; Blanchet, F.G.; Friendly, M.; Kindt, R.; Legendre, P.; McGlinn, D.; Minchin, P.R.; O'Hara, R.B.; Simpson, G.L.; Solymos, P.; et al. Vegan: Community Ecology (R Package). 2018; Available online: https://cran.r-project.org/web/packages/vegan/index.html (accessed on 10 September 2021).
75. Vindigni, M.A.; Morris, A.D.; Miller, D.A.; Kalcounis-Rueppell, M.C. Use of modified water sources by bats in a managed pine landscape. *For. Ecol. Manag.* **2009**, *258*, 2056–2061. [CrossRef]
76. Kalcounis-Rueppell, M.C.; Payne, V.H.; Huff, S.R.; Boyko, A.L. Effects of wastewater treatment plant effluent on bat foraging ecology in an urban stream system. *Biol. Conserv.* **2007**, *138*, 120–130. [CrossRef]
77. Li, H.; Parker, K.A.; Kalcounis-Rueppell, M.C. The luxury effect beyond cities: Bats respond to socioeconomic variation across landscapes. *BMC Ecol.* **2019**, *19*, 46. [CrossRef] [PubMed]
78. Li, H.; Wilkins, K.T. Patch or mosaic: Bat activity responds to fine-scale urban heterogeneity in a medium-sized city in the United States. *Urban Ecosyst.* **2014**, *17*, 1013–1031. [CrossRef]
79. Li, H.; Wilkins, K.T. Selection of Building Roosts by Mexican Free-Tailed Bats (*Tadarida brasiliensis*) in an Urban Area. *Acta Chiropterol.* **2015**, *17*, 321–330. [CrossRef]
80. Lewanzik, D.; Sundaramurthy, A.K.; Goerlitz, H.R.; Derryberry, E. Insectivorous bats integrate social information about species identity, conspecific activity and prey abundance to estimate cost–benefit ratio of interactions. *J. Anim. Ecol.* **2019**, *88*, 1462–1473. [CrossRef]
81. Culina, A.; Garroway, C.J. Bats use social information within and across species. *J. Anim. Ecol.* **2019**, *88*, 1444–1446. [CrossRef] [PubMed]
82. Neubaum, D.J.; O'Shea, T.J.; Wilson, K.R. Autumn Migration and Selection of Rock Crevices as Hibernacula by Big Brown Bats in Colorado. *J. Mammal.* **2006**, *87*, 470–479. [CrossRef]
83. Duchamp, J.E.; Sparks, D.W.; Whitaker, J.O. Foraging-habitat selection by bats at an urban–rural interface: Comparison between a successful and a less successful species. *Can. J. Zool.* **2004**, *82*, 1157–1164. [CrossRef]
84. Lehrer, E.W.; Gallo, T.; Fidino, M.; Kilgour, R.J.; Wolff, P.J.; Magle, S.B. Urban bat occupancy is highly influenced by noise and the location of water: Considerations for nature-based urban planning. *Landsc. Urban Plan.* **2021**, *210*, 104063. [CrossRef]
85. Moretto, L.; Francis, C.M. What factors limit bat abundance and diversity in temperate, North American urban environments? *J. Urban Ecol.* **2017**, *3*, jux016. [CrossRef]

land MDPI

Review

Addressing the Early-Successional Habitat Needs of At-Risk Species on Privately Owned Lands in the Eastern United States

John A. Litvaitis [1,*], Jeffery L. Larkin [2,3], Darin J. McNeil [4], Don Keirstead [5] and Bridgett Costanzo [6]

1 Department of Natural Resources and the Environment, University of New Hampshire, Durham, NH 03824, USA
2 Department of Biology, Indiana University of Pennsylvania, Indiana, PA 15705, USA; larkin@iup.edu
3 American Bird Conservancy, The Plains, VA 20198, USA
4 Department of Environmental Sciences, University of North Carolina—Wilmington, Wilmington, NC 28403, USA; mcneild@uncw.edu
5 Natural Resources Conservation Service, Dover, NH 03820, USA; donald.keirstead@usda.gov
6 Working Lands for Wildlife, Natural Resources Conservation Service, Washington, DC 20250, USA; bridgett.costanzo@usda.gov
* Correspondence: john@unh.edu

Abstract: Public lands alone are insufficient to address the needs of most at-risk wildlife species in the U.S. As a result, a variety of voluntary incentive programs have emerged to recruit private landowners into conservation efforts that restore and manage the habitats needed by specific species. We review the role of one such effort, Working Lands for Wildlife (WLFW), initiated by the Natural Resources Conservation Service in partnership with the U.S. Fish and Wildlife Service. Using two at-risk species in the eastern U.S. (where private lands dominate), we show the substantial potential that WLFW has for restoring and maintaining needed habitats. Monitoring how effective these efforts are on populations of the target species has been challenging, and both monitoring and implementation are being modified in response to new information. Identifying landowner motivations is essential for developing long-term relationships and conservation success. As WLFW projects develop, they are moving toward a more holistic ecosystem approach, within which the conservation goals of at-risk species are embedded.

Keywords: at-risk species; golden-winged warbler; landowner incentives; New England cottontail; Natural Resources Conservation Service; U.S. Department of Agriculture; Working Lands for Wildlife

Citation: Litvaitis, J.A.; Larkin, J.L.; McNeil, D.J.; Keirstead, D.; Costanzo, B. Addressing the Early-Successional Habitat Needs of At-Risk Species on Privately Owned Lands in the Eastern United States. *Land* **2021**, *10*, 1116. https://doi.org/10.3390/land10111116

Academic Editors: Ioannis N. Vogiatzakis and Robert Baldwin

Received: 15 July 2021
Accepted: 19 October 2021
Published: 21 October 2021

1. Introduction

As human populations and their influence expand, the challenge of maintaining adequate habitat for species that are threatened with extinction has become urgent. In the United States, just over 13% of the terrestrial land area is protected (e.g., designated as national parks, wilderness areas, permanent conservation easements, state parks, national wildlife refuges, and national monuments) [1]. As a result, privately owned lands are especially important when addressing the needs of at-risk taxa because these lands support populations of more than two-thirds of the species listed under the U.S. Endangered Species Act, with 10% of the listed species occurring only on private lands [2]. Additionally, hundreds of species that are in documented declines occur on private lands [2]. Because rates of habitat destruction within the range of imperiled species are greater on private lands than protected lands [3], it is clear that efforts to maintain at-risk taxa require working on both public and privately owned lands [4–7].

Although some landowners consider their conservation responsibilities as a priority [8,9], others may perceive wildlife as a liability [10]. Because economic concerns affect decisions made by private landowners, incentive programs have been developed by state and federal agencies or non-governmental organizations to encourage landowner participation in conservation actions. These programs are intended to benefit a range of taxa from popular game species to at-risk plants and animals. They include monetary grants, cost sharing, incentive payments, rental contracts, and conservation easement purchases [2]. Grants or cost-share programs pay all or part of the costs associated with restoration or enhancement of habitats for specific species or communities. For example, a cost-share program in the state of Wisconsin provides funds to landowners to manage, restore, and preserve woodlands, savannah, wetlands, and prairie. That program provides funds for the cost of labor for prescribed burning, as well as in-kind materials, such as burning equipment and grass seed. A 10-year commitment is made by participating landowners, and the cost-share funds come from the sale of turkey and pheasant hunting permits purchased by hunters [11].

At the national level, several agencies are involved with the conservation of important vegetation communities on private lands. The Natural Resources Conservation Service (NRCS) within the U.S. Department of Agriculture (USDA) was established over 80 years ago as the Soil Conservation Service to address soil conservation needs in response to the Dust Bowl of the 1930s. Today, the NRCS works with private landowners to conserve soil, water, air, plants, and animals that contribute toward productive lands and healthy ecosystems [12]. In 2012, NRCS and the U.S. Fish and Wildlife Service (USFWS) developed a partnership to provide long-term predictability in regulation of the Endangered Species Act (ESA) for farmers, ranchers, and forest landowners who voluntarily participate in Working Lands for Wildlife (WLFW) projects [12]. Specifically, participating landowners in WLFW are in compliance with ESA regulations as long as they follow their NRCS-approved conservation plans.

A substantial portion of the needed funds is provided through the U.S. Farm Bill, legislation that covers most federal government policies related to agriculture in the United States. Conservation programs within the Farm Bill are the largest single federal source of funding for private land conservation. It is renewed approximately every 5 years [13], and support for conservation efforts on private lands has grown. The 1985 Farm Bill included the Conservation Reserve Program (CRP), which provided rental payments and cost-share assistance to establish grass or tree cover on environmentally sensitive croplands. Following the passage of the 2002 Farm Bill, the Conservation Effects Assessment Project (CEAP) was created by multiple agencies within the USDA to document the benefits of conservation practices and programs and to provide the science and education base needed for effective planning, implementation, management decisions, and policy [14]. In the 2018 Farm Bill, funding for the Environmental Quality Incentives Program (EQIP), the primary program for funding conservation practices on working lands, increased to $9.2 billion for the years 2019 to 2023, with the expressed goal of maximizing the environmental benefits of conservation funding [15]. In the 2018 Farm Bill, WLFW was codified by the U.S. Congress as a permanent mechanism of the NRCS for directing EQIP and other Farm Bill program funds toward strategic conservation initiatives. WLFW is not a funded program itself; instead, it is an approach used to target and measure both outputs (e.g., area affected) and outcomes (e.g., threats mitigated or species recovered) across landscapes using Farm Bill funds and NRCS staff expertise. Initial efforts were targeted to benefit specific at-risk species [2]. So far, WLFW projects have affected more than 4 million hectares in 48 states [16]. Herein, we review WLFW efforts to aid in the recovery of two at-risk species in the eastern United States where private lands dominate the region and partnerships with landowners are essential for achieving habitat goals. Both species were among the first to be included in WLFW efforts and therefore may provide insights into the effectiveness of this approach.

2. Case Studies

2.1. *New England Cottontails*

2.1.1. Status and Habitat Needs

The New England cottontail (*Sylvilagus transitionalis*, here abbreviated as NEC) is a medium-sized lagomorph (1–1.4 kg, Figure 1) and is the only rabbit native to the New

England region of the United States [17]. NECs are restricted to sites with dense understory vegetation, including native shrublands, beaver (*Castor canadensis*) flowages, old fields, and early-successional forests [18–20]. Its historic range extended throughout much of the six New England states and eastern New York (Figure 2), a region that has experienced substantial land-use changes since European colonization [21].

Figure 1. New England cottontails are dependent on dense understory vegetation that provides food and cover. Such vegetation has become rare in recent decades. Photo courtesy of Linda Cullivan.

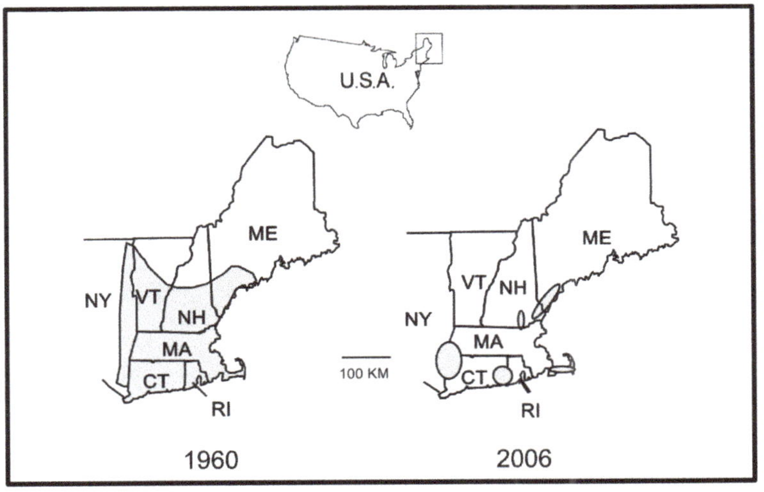

Figure 2. Historic and estimated range of New England cottontails in seven states within the northeastern United States. Populations in Vermont (VT) have been extirpated [22].

Initial subsistence agriculture expanded to widespread clearing of forests until the mid-1800s, when transportation corridors made more productive farmlands in the midwestern states available to eastern markets [21]. Widespread farm abandonment throughout New England during the late eighteenth and early twentieth centuries then resulted in a large increase in early-successional forests, with a concurrent increase in NEC populations [21]. However, this vegetation is ephemeral and is only occupied by cottontails until second-growth forests mature and the understory cover thins [23].

Since the 1960s, NEC populations have been declining. Although a number of factors have been implicated in the decline, including competition and hybridization with non-native eastern cottontails (*S. floridanus*), habitat loss via forest maturation and fragmentation seems to be the most parsimonious explanation for the range-wide decline [24]. The remaining habitats are often small, isolated patches of shrublands or young forests. Combined, populations of NEC are now restricted to five geographically disjunct locations that collectively occupy <15% of the historic range (Figure 2) [22]. Small patches of habitat support few rabbits that are exposed to high rates of predation [18], and these habitats are embedded in substantially modified landscapes [25,26]. The abundance of early-successional forests and shrublands in the region continues to decline [27,28], so without intervention, the long-term viability of NEC populations is unlikely [29–31]. As a result, the NEC was listed in 1989 as a candidate species for threatened or endangered status [32].

2.1.2. Conservation Strategy

Rather than delay recovery until a listing decision was made, several governmental (USFWS, NRCS, and state fish and wildlife agencies within the current range of the NEC) and non-governmental organizations initiated efforts to restore and expand habitats for NECs [33] and prepared a conservation strategy [34]. These efforts included plans to systematically develop and maintain habitat for NECs on public and private lands [34], and were considered sufficient enough that in 2015, the USFWS decided not to list the NEC as threatened or endangered under the ESA [35].

The strategy for NECs has specific goals for habitat and populations [34]. Initially, the USFWS established a range-wide restoration goal of approximately 11,000 ha of young forests or shrublands to support 13,500 rabbits. Subsequently, states within the current range of NECs collectively established a more ambitious goal of 17,200 ha of habitat for 21,650 rabbits [34]. To achieve these targets, the NEC Technical Committee (a group of wildlife biologists from all the states within the current range of the NEC and federal agencies) delineated focus areas for restoration activities (Figure 3). Focus areas were based on specific criteria (e.g., soils capable of supporting dense woody vegetation) within landscapes that supported or recently supported the NEC [34]. Each focus area has 11 or more planned or existing patches of suitable habitat with a combined capacity to support 80 metapopulations of NECs and is considered sufficient for long-term viability of the NEC [34]. Of the approximately 940 planned management operations, half were identified as suitable for creating patches greater than 10 ha (minimum habitat patch size) and over 40% of these are planned to be on private lands [34].

To initiate the recruitment of private landowners, NRCS staff met with biologists from state and federal agencies, university researchers, and other non-governmental organizations familiar with the status of the NEC within specific focus areas. Efforts were localized to landscapes where ownership patterns and land-use activities were amenable to conservation actions. These landscapes presented many complexities regarding the size of land parcels, socioeconomic conditions, and land ownership. Challenges exist in much of the range of the NEC, especially where the parcel size is small, land values are high, and there is a general unwillingness among landowners to engage in government programs and perceived regulation risk. In these instances, suitable patches were assembled into anchor parcels and steppingstones. Anchor parcels are large and instrumental to NEC persistence in a landscape and are considered "source habitats" within a metapopulation [36]. From our experience, the owners of such parcels are landowners with legacies in a particular town and their land has been handed down through generations. These landowners have an appreciation of working the land, and WLFW participation places a premium on "keeping working lands working." As a result, it is important that managing for the NEC will not impede any future decisions on the uses of their land. Candidate conservation agreements [37] aid in assuring the landowner that they maintain control of their land. Building trust is a main focus during the development of management plans, so that the landowner is confident that other goals and objectives are considered.

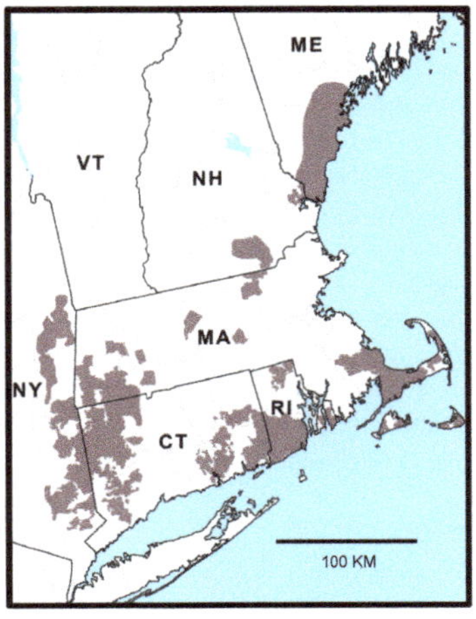

Figure 3. Focal areas where managed habitats for New England cottontails are prioritized within the species' historic range [34].

Landowner recruitment is challenging. Often, the most successful approach is to identify a local conservation commission member or other resident that is familiar with key landowners and can reach out and initiate engagement. Less personal approaches, such as mailings or phone calls, have mixed results and are less effective unless carefully managed. It is especially important to avoid the threat of future regulation as a motivator for the landowners to participate.

Once a landowner's goals and objectives are clear, next comes the discussion of funding. Although landowners are willing to host a project on their property, they are generally unwilling to spend their own money on implementation. The NRCS provides financial assistance based on various metrics, especially the size of the area being managed. Cost-share payments from the NRCS to landowners enrolled in NEC projects are typically 75% or 90% of the project costs and cover actions that promote early-successional vegetation (e.g., brush mowing, tree removal, and herbicide treatments). Where landowners are unwilling to pay for the costs, matching grants are often necessary and are brought in by a third party.

2.1.3. Positive Outcomes and Potential Concerns

By 2020, nearly 6500 ha had been managed for NECs within focal areas [38]. CEAP-funded projects produced a suitability model to monitor progress in generating and maintaining habitats for the NEC [39] and was subsequently modified for rapid assessments [40]. A sample of managed sites was evaluated in 2017 and 2018, and included sites in all six states in the current range of the NEC (J. A. Litvaitis, unpublished data). This evaluation was based on features associated with NEC occupancy [39], especially understory density. The majority of sites evaluated were enrolled in WLFW and were visited once every 1–5 growing seasons after a management action (e.g., removal of overstory trees or brush mowing). The intent of the rapid assessment was to gauge progress toward achieving NEC suitability and identify any limitations that could be addressed with additional actions [40]. Of the 55 sites visited, 11 were considered "rabbit ready" (considered suitable for NECs), 12 were "moving toward suitability", 18 were "too soon to project suitability", 13 were "unlikely to develop essential features required by NEC", and one was not designated because of varied conditions. Sites designated as "unlikely to develop essential features" did not seem capable of supporting dense understory vegetation and were often characterized by poor, thin soils. Such sites, however, may function as steppingstones between source habitats [29]. However, it is important to acknowledge that the inclusion of small or low-quality parcels is, at least partly, a consequence of the difficulty of recruiting landowners in some areas.

In southwestern focus areas, the creation of open early-successional shrublands may be inadvertently promoting non-native eastern cottontails. Management to benefit NEC populations has primarily focused on clearcutting forest stands, mowing, or burning sites to promote dense understory vegetation. Cheeseman et al. [41] suggested that these management actions may result in vegetation that is not structurally equivalent to what is generated by natural disturbances, such as beaver activity, wind-generated canopy gaps, or wildfires. Additionally, human disturbances can facilitate the spread of invasive plants [42]. Some invasive shrubs have been shown to support higher tick abundances than native shrub cover [43], and tick burden may affect the survival of NECs, especially juveniles [41]. Partners acknowledge that challenges remain [38]. A recent range-wide survey revealed a 50% decline over the last decade in the number of sites known to be occupied by NECs [44]. As a result, modifications of the conservation strategy may be needed.

2.2. Golden-Winged Warblers

2.2.1. Status and Habitat Needs

The golden-winged warbler (*Vermivora chrysoptera*, here abbreviated as GWW) is a migratory songbird that breeds in young forests and shrublands of eastern North America (Figure 4; [45]). In its breeding grounds, the species' distribution is largely restricted to the Great Lakes and Appalachian Mountains [46]. There, GWWs nest within early-successional communities situated within otherwise heavily forested landscapes [47]. Within appropriate landscapes, GWWs nest on the ground, where their nesting cycle is completed in about 25 days [45]. After the young leave the nest, adults provide parental care for an additional 4 weeks within densely stocked stands of woody vegetation, such as regenerating forests or older forests with taller canopies and well-developed understories [48]. In most of their range, the nesting habitat is ephemeral (e.g., regenerating forests that occur following disturbance events), rarely remaining suitable for nesting beyond 15 growing seasons [49]. Like the NEC, the GWW initially benefitted from widespread abandonment of farmland across the northeastern United States [50]. However, regenerating forests matured and natural disturbances (e.g., beavers and wildfire) were not sufficient to reverse the decline of GWW populations [50,51]. Although more persistent nesting habitats exist (e.g., shrub wetlands), the availability of these communities is restricted mostly to the Great Lakes Region or very locally in the Appalachian Mountains (e.g., Pocono Plateau) [45,52].

Figure 4. Golden-winged warblers (**A**) are dependent on early-successional communities with a mix of trees, shrubs, saplings, and herbaceous plants such as those provided by regeneration after timber harvests (**B**), and shrublands (**C**) that provide nesting and foraging resources. (Photos by D.J. McNeil).

GWW populations have been declining for at least the past 50 years (2.57%/year) and these declines have been more pronounced in the Appalachian Mountains (7.82%/year) than in the Great Lakes region (e.g., state of Wisconsin: 2.57%/year [53]; Figure 5). Several factors are likely to contribute to the decline, including the range expansion of and hybridization with blue-winged warblers (*Vermivora cyanoptera*), nest parasitism by brown-headed cowbirds (*Molothrus ater*), and loss of non-breeding habitats [45]. However, the loss of breeding habitat is considered the most important threat [47]. In 2010, the species was petitioned for listing under the ESA, and the USFWS found that the petition was substantial enough to warrant further review [54].

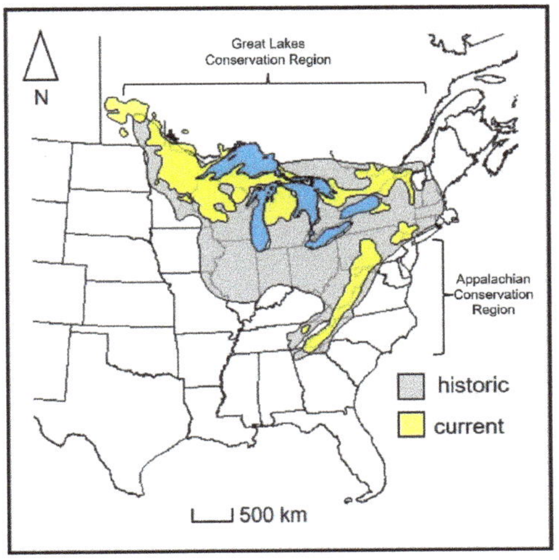

Figure 5. Historic and current estimated range of golden-winged warblers within eastern North America. The range has contracted by approximately 66% over the past 50 years.

2.2.2. Conservation Strategy

To understand the drivers of population decline and develop a conservation strategy, a group of government agencies, conservation organizations, and academics formed the Golden-winged Warbler Working Group in 2004. This group prepared a status review and conservation plan [47] that included three primary goals: (a) increase the range-wide breeding habitat by 400,000 hectares, (b) stabilize the Appalachian Mountains population by doubling the number of breeding adults, and (c) grow the range-wide population by 50% by 2050. The plan also identified focal areas for implementing vegetation management. Focal areas are defined as places where the maintenance of core breeding populations will be important for sustaining and expanding the species' current distribution, and their boundaries were delineated based on expert opinion, remote sensing data (elevation and percent forest cover), and distance to blue-winged warbler breeding populations. At the same time, habitat management guidelines were developed to provide landowners and managers with descriptions of actions for creating and enhancing habitats for GWW [47,49,55].

In 2012, the inclusion of the GWW by the NRCS as one of nine target species for the WLFW partnership added considerable funding and momentum toward efforts on private lands in several Appalachian states (Figure 6). Landowners interested in participating in the WLFW GWW program first contacted their local NRCS office to determine if their property met the general requirements for enrollment in the initiative (i.e., it fell within the initiative's boundary and was within a local landscape dominated by forest cover). If a property met these initial screening criteria, a partner forester and/or biologist conducted a site visit to discuss the landowner's stewardship objectives and to identify areas that have potential for habitat management. If it was determined that a property was a good fit for the WLFW GWW program, the landowner completed an NRCS application which included a conservation plan that was prepared by the partner staff. All applications for a given fiscal year are ranked based on a set of criteria that considers each application's potential for success. The NRCS provides cost-share funding to the highest ranked applications until all available funding has been obligated.

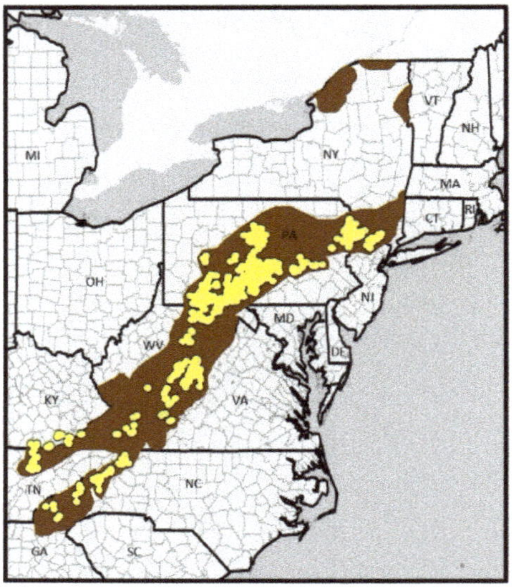

Figure 6. The Working Lands for Wildlife golden-winged warbler initiative's boundaries (brown shading) and priority areas for conservation (PACs; yellow shading). All private landowners within the initiative's boundaries are eligible to receive financial and technical assistance to create or enhance nesting habitats for this declining species, but properties within a PAC are ranked higher. Note: PAC boundaries were not delineated for NY, as they joined this effort in a later year.

To guide delivery and set benchmarks for success, NRCS staff completed a conservation strategy that set a goal of creating approximately 6400 hectares of nesting habitat on private lands across several Appalachian states over a 5-year period (2017–2021). The goal was reached a year early and a new 5-year goal will be set in 2022. Outreach activities directed toward private landowners including targeted mailings, seminars, webinars, and field tours proved effective in attracting participants. Additional efforts were made to stress economic advantages, forest health, ecosystem services, and game management as co-benefits. Important to achieving WLFW goals was the availability of funds to employ conservation planners, biologists, and foresters to assist the NRCS field offices with various components of program delivery, including landowner outreach, assistance with the application process, conducting site visits, drafting conservation plans, and site preparation (i.e., marking project boundaries and residual trees).

After private landowners enroll in WLFW and management actions are completed, do warblers respond to the habitat created? From 2012 to 2014, a CEAP-funded monitoring team evaluated the GWW's responses (i.e., density, nest success, fledgling survival) to management actions throughout the Appalachian breeding range [48,52,56–59]. In 2015, CEAP again funded a multi-year effort to monitor GWW occurrence across hundreds of properties enrolled in WLFW [60]. Collectively, the monitoring results from CEAP were used to assess program success [60], revise species-specific vegetation management guidelines [47,57], and to further inform where to target the delivery of WLFW efforts [61].

2.2.3. Positive Outcomes and Potential Concerns

More than 450 private landowners across 10 Appalachian states have enrolled >9400 ha in WLFW during 2012–2020. These hectares were administered through 580 contracts totaling $14.4 million in financial assistance to participating landowners. Most of these landowners have or will implement conservation practices that result in young forests, with the remaining implementing practices associated with old-field management and prescribed grazing.

During the 2015–2017 breeding seasons, monitoring teams conducted 1145 point count surveys across 459 managed sites (n = 267 on WLFW lands and n = 192 on comparable nearby public lands). GWWs were detected on 26% (121/459) of all sites monitored, with GWWs detected on 17% and 39% of the WLFW and public land sites, respectively [60]. Across all 459 sites, the mean probability of GWW occupancy was 0.22 [60]. Sites that had at least eight growing seasons after management had the highest probability of occupancy (0.77; 95% CI 0.66-0.85). In addition to within-stand characteristics, occupancy was strongly driven by metapopulation dynamics and local landscape composition, whereby managed sites with low proportions of mixed and coniferous forest cover that were <15 km from other known GWW subpopulations were most likely to be occupied. Colonization of restored habitats was especially pronounced within areas of moderate to high breeding output (>1.6 juveniles/pair/year) [62]. Although a 22% mean occupancy rate in the managed sites is not an overwhelming success, it is substantial, given the current status of the GWW in the Appalachians. Moreover, it is promising that GWWs were detected on 39% (75/192) of public land sites during CEAP monitoring (2015–2017), which is nearly double the rate (22.5%; 50/222 sites) for public land sites monitored in this region prior to WLFW in 2010–2011 [49]. It is also important to note that monitoring is essential to guide future management actions to increase occupancy rates as conservation efforts advance.

Boundaries for the Appalachian conservation region and focal areas associated with the Golden-winged Warbler Working Group's status review and conservation plan [47] were used for targeting implementation efforts during the initial years of the WLFW GWW partnership. However, CEAP-generated monitoring data combined with those from other agency-led efforts and a citizen science program [46] enabled the NRCS and its partners to identify portions of the GWW Appalachian Conservation Region where responses to management (i.e., successful occupancy) were rare. Using this data-driven approach, the NRCS refined its boundaries, delineating where best to prioritize program delivery. Specifically, Priority Areas for Conservation (PACs) were identified by using GWW occurrence data and information on the associated vegetation to generate spatially explicit models of GWW occupancy [61]. As a result of the increased spatial focus resulting from the PACs, program success (occupancy) is expected to increase as resources are concentrated within sites in landscapes with high-ranked suitability (Figure 6), thus better contributing to the population goals outlined in the GWW status review and conservation plan. PAC boundaries are re-evaluated periodically, which is important for data-driven expansion of PAC footprints as GWWs colonize restored habitat in the peripheral portions of PACs, thus increasing the likelihood of colonizing sites beyond the original PAC footprints.

To this end, while sites managed outside the PACs have a much lower likelihood of becoming occupied by GWWs due to the strong metapopulation structure of the species' Appalachian range, implementation of management practices within the larger WLFW boundary supports overall forest health and provides much needed early-successional communities used by many other declining taxa [63]. Moreover, as GWW subpopulations grow in response to conservation actions inside PACs, the availability of sites outside the PACs will help facilitate population spread by providing areas to host dispersing pairs. In the meantime, all landowners within the larger project area are eligible to participate in WLFW, but properties within PACs are ranked highest for receiving finite financial and technical assistance for implementing conservation practices.

3. Discussion

Given the distribution of at-risk taxa and the limitations of relying exclusively on public lands, engagement with private landowners is essential for creating adequate amounts of the needed habitat. Opportunities exist in other countries to increase participation by private landowners. For example, landholders in Australia are invited to submit a bid to carry out conservation work on their property. Bids are then ranked according to the best financial value for conservation benefit [64].

3.1. Not Waiting for Endangered Species Act Listing

Although several species aided by WLFW projects have been or are currently being considered for listing under the ESA (including the NEC and GWW), this is not a requirement for WLFW support. Listed species have waited a median of 12.1 years to receive ESA protection [65]. Notably, NEC were first listed as a candidate for listing in 1989 and it was not until 2015 that the USFWS decided not to list them as threatened or endangered, largely because of the recovery efforts that were initiated by the partners and WLFW several years earlier.

Listing under the ESA triggers protection from take and trade, designation of critical habitats, development of a recovery plan, and specific requirements on federal agencies to avoid jeopardizing listed species' persistence or adversely modifying their critical habitat, as well as making species eligible for recovery funding. ESA listings can also have economic effects that are detrimental to the public's competing priorities and create conflict with private landowners, which often dampens willingness to participate in conservation actions. Prior to listing, at-risk species may experience further declines, thus increasing both the cost of recovery and ultimately extinction risk [66–68]. The advantages of WLFW include responding to population declines before federal listing is necessary to avoid both extinction risks and adverse economic impacts.

3.2. Examining Outcomes

WLFW was originally established to focus on large-scale conservation challenges based on a suite of target species that either already had ESA status or were at some risk of being listed as threatened or endangered. Over time, WLFW has expanded to include other species (e.g., the northern bobwhite quail, *Colinus virginianus*) with well-documented habitats and population declines but no ESA implications, and has shifted its emphasis from single target species to a greater emphasis on restoring at-risk ecosystems such as native grasslands and the wildlife communities at large therein. As a result, monitoring and outcome assessments include tracking single-species responses as well as landscape-wide effects.

There are a number of challenges in monitoring the outcomes of WLFW activities. Perhaps the most obvious is the need to communicate individually with hundreds of landowners who expect a level of privacy and coordination. In comparison, monitoring efforts on public lands may require conversations with a small group of individuals. When a species either has or is being considered for ESA listing, private landowners may be reluctant or unwilling to participate in conservation practices or, secondarily, in monitoring due to fear of regulation. The difficulty practitioners experience in contacting and gaining access to private lands for management activities similarly affects monitoring goals. It can take several months for NRCS staff and partners to contact individual landowners to obtain permission for monitoring. Additionally, lands enrolled in WLFW projects are often sparsely spread across large areas and thus require considerable time traveling between sites.

Aligning the timing of monitoring with the target species' response to management is another challenge for assessing WLFW outcomes. For some target species, specific seral stages represent critical habitats, and there may be a lag between management action and when the habitat develops. For example, NECs may not occupy a site until 10 years after management [23] and GWW occupancy was highest 8 years after management [60]. As a result, it may not be clear for some time if efforts have been successful in increasing the size of imperiled populations. Such relationships can muddle an evaluation of WLFW activities if they are based on responses by a single target species. Therefore, a broader evaluation of the community or ecosystem response is also relevant for evaluating the outcomes of WLFW efforts [69], as well as a proactive approach toward identifying the limitations of management actions [70].

NRCS staff evaluate their efforts on responses by target species, the larger wildlife community, ecosystem health, and economic outcomes for producers and rural communities. Conservation objectives set by WLFW are often a subset of the larger habitat and population goals set by partners, and reaching these is only one marker for success. WLFW effectiveness has, at times, been gauged by the outcomes of ESA listing decisions; certainly, these milestones should be celebrated. However, threats such as disease, climate change, and cultural and socio-economic drivers can all interact to undermine this measure of WLFW outcomes. Often, managers define their goals as returning a species to a stable or common occurrence within historic habitats; but as landscapes change and human populations increase, these goals may not be easily achieved. As a result, examining the responses by other taxa affiliated with the habitats managed for individual target species has become an important approach for understanding the impact of actions taken on private lands. For instance, among a sample of sites managed for or occupied by the NEC, 11 shrubland-affiliated species of birds were detected [71]. These included prairie warblers (*Setophaga discolor*), a species of regional conservation concern [71]. Additional associations between other shrubland birds and herbaceous vegetation and low shrubs indicated that land managed for NECs but not yet suitable for cottontails could benefit an additional suite of birds [71]. Further, abundant and diverse flowering plants within sites managed for NECs and GWWs were also found to attract a substantial group of native bees [72–74]. These observations indicate that species requiring early-successional conditions may benefit from conservation practices, but monitoring is needed to evaluate the effectiveness of the practices for other species.

Nonetheless, quantifying the responses by target species remains a valuable metric beyond amount of land enrolled, as it provides the conservation community an understanding of the extent to which WLFW contributes to achieving population goals and addressing regulatory considerations. Acknowledging the difficulties of monitoring does not dismiss the need for improvement. Collaborative monitoring of management activities and their outcomes among landowners, NRCS personnel, and research scientists could establish information feedback loops between actions taken and conservation outcomes, and subsequently improve outcomes [75]. Although it was not the focus of this study, WLFW also conducts outcomes assessments for the economic impacts of the initiatives on landowners and communities, and this dual focus is key to the conservation of working lands conducted by the NRCS and its partners.

3.3. Landowner Retention

Given the challenges associated with enrolling private landowners in WLFW efforts, it can be advantageous to have previously engaged landowners re-enroll and encourage others to participate in these efforts to maintain positive conservation gains. Recent surveys of landowners participating in GWW initiatives indicated that they did not respond uniformly after their individual contracts ended [76–78]. Several factors seemed to influence their enthusiasm for continued post-program management, including forest health, future timber value, wildlife, and recreational opportunities [77]. Therefore, the degree to which WLFW contributes to achieving GWW habitat goals will likely be driven by outreach and a technical assistance approach that continues to appeal to a broad set of landowner motivations rather than an approach that solely stresses habitats for a single species, and landowner outreach has been adapted accordingly [79].

Landowners' experiences with conservation programs are important in affecting management outcomes [80,81]. Among the landowners involved in the GWW initiative, those who interacted with monitoring technicians in the field showed a greater level of agency trust than those landowners who did not interact with monitoring technicians [76], suggesting that personnel interactions could bolster program enrollment. Surprisingly, the presence of GWWs had a negative effect on continued management, suggesting that results for the target species may have been outweighed by broader landowner priorities for participation in conservation programs. This is not unusual or to be lamented, as developing a shared vision with landowners is not dependent on shared motivations. Beyond wildlife, other benefits (e.g., enhancing forest health and scenery) could affect landowners' behavior [78].

4. Conclusions

The contributions of WLFW projects for developing and protecting habitats for at-risk species have been substantial, and these efforts are usually nested within larger partnerships with agencies that track population trends as part of their mission. The NRCS itself does not set population goals or track population trends. Instead, the NRCS conducts broader assessments that document priority species' use of implemented projects to meet basic habitat needs, measures and tracks ecosystem health, and assesses local economic benefits to gauge WLFW's effectiveness. Recognizing and appealing to landowner motivations are essential toward developing good relationships. Having a shared vision with private landowners should aid in ensuring the longevity of conservation actions in agricultural or timbered landscapes.

WLFW is built upon a foundational philosophy of encouraging win–win solutions for producers ("Working Lands") and target species ("For Wildlife"). The NRCS develops implementation plans based on threats, conservation actions, and habitat and population goals identified by integrated partnerships of state and federal agencies collaborating with non-government conservation organizations, university experts, and private landowners. WLFW initiative partners strive to incorporate principles from existing conservation frameworks designed to achieve multiple objectives for wildlife, natural resources, and humans (e.g., [82]). In our examples, the conservation strategy for the New England cottontail and the golden-winged warbler's status assessment and conservation plan were developed by technical committees representing each target species' recovery needs. These existing conservation strategies were enhanced by monitoring and modeling to guide the delivery of WLFW [83] using many components of effective conservation planning (e.g., [82]). It cannot be overstated that well-funded conservation efforts such as WLFW have great potential for addressing resource concerns (i.e., forest health, water quality) and recovering declining wildlife populations, but the degree to which such programs are impactful, efficient, and sustained will largely be dependent upon the use of proven conservation frameworks and adaptive management.

Author Contributions: Conceptualization, J.A.L.; methodology, J.A.L., J.L.L., D.J.M., D.K. and B.C.; writing—original draft preparation, J.A.L., J.L.L., D.J.M., D.K. and B.C.; writing—review and editing, J.A.L., J.L.L., D.J.M., D.K. and B.C. All authors have read and agreed to the published version of the manuscript.

Funding: This review received no external funding.

Acknowledgments: We thank the hundreds of private landowners that have enrolled in WLFW projects. C. Rewa, S. Weber, and three anonymous reviewers made useful recommendations on drafts of this manuscript.

Conflicts of Interest: The authors declare no conflict of interest.

References

1. Protected Planet. Discover the world's Protected Areas—United States of America. Available online: https://www.protectedplanet.net/country/USA (accessed on 18 August 2021).
2. Baier, L.E. *Saving Species on Private Lands—Unlocking Incentives to Conserve Wildlife and Their Habitats*; Rowman & Littlefield: Lantham, MD, USA, 2020.

3. Eichenwald, A.J.; Evans, M.J.; Malcom, J.W. US imperiled species are most vulnerable to habitat loss on private lands. *Front. Ecol. Environ.* **2020**, *18*, 439–446. [CrossRef]
4. Knight, R.L. Private lands: The neglected geography. *Conserv. Biol.* **1999**, *13*, 223–224.
5. Kremen, C.; Merenlender, A.M. Landscapes that work for biodiversity and people. *Science* **2018**, *362*. [CrossRef] [PubMed]
6. Robles, M.D.; Flather, C.H.; Stein, S.M.; Nelson, M.D.; Cutko, A. The geography of private forests that support at-risk species in the conterminous United States. *Front. Ecol. Environ.* **2008**, *6*, 301–307. [CrossRef]
7. Clancy, N.G.; Draper, J.P.; Wolf, J.M.; Abdulwahab, U.A.; Pendleton, M.C.; Brothers, S.; Brahney, J.; Weathered, J.; Hammill, E.; Atwood, T.B. Protecting endangered species in the USA requires both public and private land conservation. *Sci. Rep.* **2020**, *10*, 11925. [CrossRef] [PubMed]
8. Alexander, L.; Kellert, S.R. Forest landowners' perspectives on wildlife management in New England. *Trans. N. Am. Wildl. Nat. Resour. Conf.* **1984**, *49*, 164–173.
9. Daley, S.S.; Cobb, D.T.; Bromley, P.T.; Sorenson, C.E. Landowner attitudes regarding wildlife management on private land in North Carolina. *Wildl. Soc. Bull.* **2004**, *32*, 209–219. [CrossRef]
10. Noonan, P.F.; Zagata, M.D. Wildlife in the market place: Using the profit motive to maintain wildlife habitat. *Wildl. Soc. Bull.* **1982**, *10*, 46–49.
11. Defenders of Wildlife. Conservation in America: State Government Incentives for Habitat Conservation: A Status Report. 2002. Available online: https://www.defenders.org/sites/default/files/publications/conservation_in_america.pdf (accessed on 11 March 2021).
12. Martinez, M. Working lands for wildlife: Targeted landscape-scale wildlife habitat conservation. *Nat. Resour. Environ.* **2015**, *29*, 36–39.
13. Land Trust Alliance. Farm Bill Conservation Programs. 2018. Available online: https://www.landtrustalliance.org/topics/federal-programs/farm-bill-conservation-programs (accessed on 20 August 2021).
14. Duriancik, L.; Bucks, D.; Dobrowolski, J.; Drewes, T.; Eckles, S.; Jolley, L.; Kellogg, R.; Lund, D.D.; Makuch, J.R.; O'Neill, M.; et al. The first five years of the Conservation Effects Assessment Project. *J. Soil Water Conserv.* **2008**, *63*, 185A–197A. [CrossRef]
15. Natural Resources Conservation Service. Environmental Quality Incentives Program. *Fed. Reg.* **2020**, *85*, 67637–67648.
16. Natural Resources Conservation Services. Supporting America's Working Lands. USDA Natural Resources Conservation Service. Available online: https://www.nrcs.usda.gov/wps/portal/nrcs/detail/national/programs/initiatives/?cid=stelprdb1046975 (accessed on 29 March 2021).
17. Chapman, J.A.; Litvaitis, J.A. Eastern cottontails and their allies. In *Wild Mammals of North America: Biology, Management, and Conservation*, 2nd ed.; Feldhammer, G.A., Thompson, B.C., Chapman, J.A., Eds.; John Hopkins Press: Baltimore, MD, USA, 2003; pp. 101–125.
18. Barbour, M.S.; Litvaitis, J.A. Niche dimensions of New England cottontails in relation to habitat patch size. *Oecologia* **1993**, *95*, 321–327. [CrossRef]
19. Litvaitis, J.A.; Johnson, B.; Jakubas, W.; Morris, K. Distribution and habitat features associated with remnant populations of New England cottontails in Maine. *Can. J. Zool.* **2003**, *81*, 877–887. [CrossRef]
20. Tash, J.P.; Litvaitis, J.A. Characteristics of occupied habitats and identification of sites for restoration and translocation of New England cottontail populations. *Biol. Conserv.* **2007**, *137*, 584–598. [CrossRef]
21. Litvaitis, J.A. Response of early successional vertebrates to historic changes in land use. *Conserv. Biol.* **1993**, *7*, 866–873. [CrossRef]
22. Litvaitis, J.A.; Tash, J.P.; Litvaitis, M.K.; Marchand, M.N.; Kovach, A.I.; Jenkins, R. A range-wide survey to determine the current distribution of New England cottontails. *Wildl. Soc. Bull.* **2006**, *34*, 1190–1197. [CrossRef]
23. Litvaitis, J.A. Importance of early-successional habitats to mammals in eastern forests. *Wildl. Soc. Bull.* **2001**, *29*, 466–473.
24. Litvaitis, J.A.; Barbour, M.S.; Brown, A.L.; Kovach, A.I.; Oehler, J.D.; Probert, B.L.; Smith, D.F.; Tash, J.P.; Villafuerte, R.; Litvaitis, M.K. Testing multiple hypotheses to identify the causes of the decline of a lagomorph species: The New England cottontail as a case study. In *Biology of Lagomorphs—Evolution, Ecology and Conservation*; Alves, P.C., Ferrand, N., Hackländer, K., Eds.; Springer-Verlag: New York, NY, USA, 2007; pp. 167–185.
25. Brown, A.L.; Litvaitis, J.A. Habitat features associated with predation of New England cottontails: What scale is appropriate? *Can. J. Zool.* **1995**, *73*, 1005–1011. [CrossRef]
26. Litvaitis, J.A. Are pre-Columbian conditions relevant baselines in managed forests of the northeastern United States? *For. Ecol. Manag.* **2003**, *185*, 113–126. [CrossRef]
27. Litvaitis, J.A.; Wagner, D.L.; Confer, J.L.; Tarr, M.D.; Snyder, E.J. Early successional forests and shrub-dominated habitats: Land-use artifact or critical community in the northeastern United States? *Northeast Wildl.* **1999**, *54*, 101–118.
28. Brooks, R.T. Abundance, distribution, trends, and ownership patterns of early successional forests and native shrublands in the northeastern United States. *For. Ecol. Manag.* **2003**, *185*, 65–74. [CrossRef]
29. Litvaitis, J.A.; Villafuerte, R. Factors affecting the persistence of New England cottontail metapopulations: The role of habitat management. *Wildl. Soc. Bull.* **1996**, *24*, 686–693.

30. Fenderson, L.E.; Kovach, A.I.; Litvaitis, J.A.; Litvaitis, M.K. Population genetic structure and history of fragmented remnant populations of the New England cottontails (*Sylvilagus transitionalis*). *Conserv. Gen.* **2011**, *12*, 943–958. [CrossRef]
31. Cheeseman, A.E.; Cohen, J.B.; Whipps, C.M.; Kovach, A.I.; Ryan, S.J. Hierarchical population structure of a rare lagomorph indicates recent fragmentation has disrupted metapopulation function. *Conserv. Gen.* **2019**, *20*, 1–13. [CrossRef]
32. U.S. Fish and Wildlife Service. Animal notice of review. *Fed. Reg.* **1989**, *54*, 554–579.
33. Arbuthnot, M. A Landowner's Guide to New England Cottontail Habitat Management. Environmental Defense Fund. Available online: http://apps.edf.org/documents/8828_New-England-Cottontail-Guide.pdf (accessed on 6 January 2013).
34. Fuller, S.; Tur, A. Conservation Strategy for the New England Cottontail (Sylvilagus Transitionalis). Available online: http://www.newenglandcottontail.org/sites/default/files/research_documents/conservation_strategy_final_12-3-12.pdf (accessed on 9 January 2013).
35. U.S. Fish and Wildlife Service. Endangered and threatened wildlife and plants; 12-month finding on a petition to list the New England cottontail as an endangered or threatened species. *Fed. Reg.* **2015**, *80*, 55286–55304.
36. Harrison, S. Local extinction in a metapopulation context: An empirical evaluation. *Biol. J. Linn. Soc.* **1991**, *42*, 73–88. [CrossRef]
37. U.S. Fish and Wildlife Service. Candidate Conservation Agreements. Available online: https://www.fws.gov/endangered/esa-library/pdf/CCAs.pdf (accessed on 25 April 2021).
38. New England Cottontail Regional Initiative Performance Report 2020. Available online: https://newenglandcottontail.org/resource/new-england-cottontail-conservation-progress-2020 (accessed on 28 April 2021).
39. Warren, A.; Litvaitis, J.A.; Keirstead, D. Developing a habitat suitability index to guide management of New England cottontail habitats. *Wildl. Soc. Bull.* **2016**, *40*, 69–77. [CrossRef]
40. Working Lands for Wildlife. A Rapid Assessment for Tracking Success of New England Cottontail Conservation Efforts. Available online: https://www.nrcs.usda.gov/wps/PA_NRCSConsumption/download?cid=nrcseprd1503016&ext=pdf (accessed on 25 April 2021).
41. Cheeseman, A.E.; Cohen, J.B.; Ryan, S.; Whipps, C.M. Is conservation based on best available science creating an ecological trap for an imperiled lagomorph? *Ecol. Evol.* **2021**, *11*, 912–930. [CrossRef]
42. Johnson, V.S.; Litvaitis, J.A.; Lee, T.D.; Frey, S.D. The role of spatial and temporal scale in colonization and spread of exotic shrubs in early-successional habitats. *For. Ecol. Manag.* **2006**, *228*, 124–134. [CrossRef]
43. Elias, S.P.; Lubelczyk, C.B.; Rand, P.W.; Lacombe, E.H.; Holman, M.S.; Smith, R.P., Jr. Deer browse resistant exotic-invasive understory: An indicator of elevated human risk of exposure to *Ixodes scapularis* (Acari: Ixodidae) in southern coastal Maine woodlands. *J. Med. Ent.* **2006**, *43*, 1142–1152. [CrossRef]
44. Rittenhouse, C.D.; Kovach, A.I. Assessment of alternative sampling designs for range-wide monitoring of New England cottontail. *Wildl. Soc. Bull.* **2020**, *44*, 798–806. [CrossRef]
45. Confer, J.L.; Hartman, P.; Roth, A.M. Golden-winged Warbler (*Vermivora chrysoptera*), version 1.0. In *Birds of the World*; Poole, A.F., Ed.; Cornell Lab of Ornithology: Ithaca, NY, USA, 2020.
46. Fink, D.; Auer, T.; Johnston, A.; Strimas-Mackey, M.; Robinson, O.; Ligocki, S.; Hochachka, W.; Wood, C.; Davies, I.; Iliff, M.; et al. *eBird Status and Trends, Data Version: 2019*; Cornell Lab of Ornithology: Ithaca, NY, USA, 2020.
47. Roth, A.M.; Rohrbaugh, R.W.; Will, T.; Swarthout, S.B.; Buehler, D.A. (Eds.) Golden-Winged Warbler Status Review and Conservation Plan, 2nd ed. 2019. Available online: www.gwwa.org//wp-content/uploads/2020/06/GWWA_Conservation-Plan_191007_low-res.pdf (accessed on 15 August 2021).
48. Fiss, C.J.; McNeil, D.J.; Duchamp, J.; Rodewald, A.D.; Larkin, J.L. Post-fledging Golden-winged Warblers Require Forests with Multiple Stand Developmental Stages. *Condor Ornithol. Appl.* **2020**, *122*, 1–13. [CrossRef]
49. Bakermans, M.H.; Smith, B.W.; Jones, B.C.; Larkin, J.L. Stand and within-stand factors influencing Golden-winged Warbler use of regenerating stands in the central Appalachian Mountains. *Avian Conserv. Ecol.* **2015**, *10*, 10. [CrossRef]
50. Rosenberg, K.V.; Will, T.; Buehler, D.A.; Swarthout, S.B.; Thogmartin, W.E.; Bennett, R.E.; Chandler, R. Dynamic distributions and population declines of Golden-winged Warblers. In *Studies in Avian Biology 49—Golden-Winged Warbler Ecology, Conservation, and Habitat Management*; Streby, H.M., Andersen, D.E., Buehler, D.A., Eds.; CRC Press: Boca Raton, FL, USA, 2016; pp. 3–28.
51. Rohrbaugh, R.W.; Buehler, D.A.; Swarthout, S.B.; King, D.I.; Larkin, J.L.; Rosenberg, K.V.; Roth, A.M.; Vallender, R.; Will, T. Conservation perspectives: Review of new science and primary threats to Golden-winged Warblers. In *Studies in Avian Biology 49—Golden-Winged Warbler Ecology, Conservation, and Habitat Management*; Streby, H.M., Andersen, D.E., Buehler, D.A., Eds.; CRC Press: Boca Raton, FL, USA, 2016; pp. 207–215.
52. McNeil, D.J.; Fiss, C.J.; Wood, E.M.; Duchamp, J.E.; Bakermans, M.J.; Larkin, J.L. Using a Natural Reference System to Evaluate Songbird Habitat Restoration. *Avian Conserv. Ecol.* **2018**, *13*, 22. [CrossRef]
53. Sauer, J.R.; Niven, D.K.; Hines, J.E.; Ziolkowski, D.J.; Pardieck, K.L.; Fallon, J.E.; Link, W.A. *The North American Breeding Bird Survey, Results and Analysis 1966–2015. Version 2.07*; USGS Patuxent Wildlife Research Center: Laurel, MD, USA, 2017.
54. Sewell, A. Petition to List the Golden-Winged Warbler (*Vermivora chrysoptera*) as a Threatened or Endangered Species under the U.S. Endangered Species Act. 2010. Available online: www.fws.gov/midwest/es/soc/birds/GoldenWingedWarbler/GoldenWingedWarblerPetition.pdf (accessed on 15 August 2021).
55. Bakermans, M.H.; Larkin, J.L.; Smith, B.W.; Fearer, T.M.; Jones, B.C. *Golden-Winged Warbler Habitat Best Management Practices for Forestlands in Maryland and Pennsylvania*; American Bird Conservancy: The Plains, Virginia, 2011; 26p.

56. McNeil, D.J.; Aldinger, K.R.; Bakermans, M.H.; Lehman, J.A.; Tisdale, A.C.; Jones, J.A.; Wood, P.B.; Buehler, D.; Smalling, C.G.; Siefferman, L.; et al. An Evaluation and Comparison of Conservation Guidelines for an At-Risk Migratory Songbird. *Glob. Ecol. Conserv.* **2017**, *9*, 90–103. [CrossRef]
57. Frantz, M.W.; Aldinger, K.R.; Wood, P.B.; Duchamp, J.; Nuttle, T.; Vitz, A.; Larkin, J.L. Space and habitat use of breeding Golden-winged Warblers in the central Appalachian Mountains. In *Golden-Winged Warbler Ecology, Conservation, and Habitat Management. Studies in Avian Biology (no. 49)*; Streby, H.M., Buehler, D., Andersen, D.E., Eds.; CRC Press: Boca Raton, FL, USA, 2016; pp. 81–94.
58. Aldinger, K.R.; Wood, P.B. Reproductive success and habitat characteristics of Golden-winged Warblers in high-elevation pasturelands. *Wil. J. Ornith.* **2014**, *126*, 279–287. [CrossRef]
59. Fiss, C.J.; McNeil, D.J.; Rodewald, A.D.; Hegganstaller, D.; Larkin, J.L. Cross-Scale Habitat Selection Reveals Within-Stand Structural Requirements for Fledgling Golden-Winged Warblers. *Avian Conserv. Ecol.* **2021**, *16*, 16. [CrossRef]
60. McNeil, D.J.; Rodewald, A.D.; Ruiz-Gutierrez, V.; Johnson, K.; Strimas-Mackey, M.; Petzinger, S.; Robinson, O.J.; Soto, G.E.; Dhondt, A.; Larkin, J.L. Multi-scale Drivers of Restoration Outcomes for an Imperiled Songbird. *Rest. Ecol.* **2020**, *28*, 880–891. [CrossRef]
61. Lott, C.A.; Larkin, J.L.; McNeil, D.J.; Fiss, C.J.; Costanzo, B.E. Mapping areas with suitable landscape context for Golden-winged Warbler nesting habitat restoration and management in the Appalachian Mountains. In *Wildlife Management and Landscapes: Principles and Applications*; Porter, W.F., Parent, C.J., Stewart, R.A., Williams, D.M., Eds.; Johns Hopkins University Press in Affiliation with The Wildlife Society: Baltimore, MD, USA, 2021.
62. McNeil, D.J.; Rodewald, A.D.; Robinson, O.J.; Fiss, C.J.; Rosenberg, K.V.; Ruiz-Gutierrez, V.; Aldinger, K.; Petzinger, S.; Dhondt, A.; Larkin, J.L. Regional Abundance and Local Breeding Productivity Explain Occupancy of Restored Habitats in a Migratory Songbird. *Biol. Conserv.* **2020**, *245*, 108463. [CrossRef]
63. Gilbart, M. *Under Cover: Wildlife of Shrublands and Young Forest*; Wildlife Management Institute: Cabot, VT, USA, 2012.
64. Fitzsimons, J.; Cooke, B. Key questions for conservation tenders as a means for delivering biodiversity benefits on private land. *Ecol. Manag. Restor.* **2021**, *22*, 110–114. [CrossRef]
65. Puckett, E.E.; Kesler, D.C.; Greenwald, D.N. Taxa, petitioning agency, and lawsuits affect time spent awaiting listing under the US Endangered Species Act. *Biol. Conserv.* **2016**, *201*, 220–229. [CrossRef]
66. Wilcove, D.S.; McMillan, M.; Winston, K.C. What exactly is an endangered species? An analysis of the United States Endangered Species List: 1985–1991. *Conserv. Biol.* **1993**, *7*, 87–93. [CrossRef]
67. Suckling, K.; Slack, R.; Nowicki, B. Extinction and the Endangered Species Act. Center for Biological Diversity, 2004. Available online: https://www.biologicaldiversity.org/publications/papers/ExtinctAndESA.pdf (accessed on 4 July 2021).
68. Neel, M.C.; Leidner, A.K.; Haines, A.; Goble, D.D.; Scott, J.M. By the numbers: How is recovery defined by the Endangered Species Act? *BioScience* **2012**, *62*, 646–657. [CrossRef]
69. Maresch, W.; Walbridge, M.R.; Kugler, D. Enhancing conservation on agricultural landscapes: A new direction for the Conservation Effects Assessment Project. *J. Soil Water Conserv.* **2008**, *63*, 198A–203A. [CrossRef]
70. Runge, M.C. An introduction to adaptive management for threatened and endangered species. *J. Fish Wildl. Manag.* **2011**, *2*, 220–233. [CrossRef]
71. Bauer, M.L. Assessing the Effects of Habitat Restoration on Shrubland Specialists: Case Study on the New England Cottontail and Shrubland Birds. Ph.D. Thesis, University of New Hampshire, Durham, NH, USA, 2018.
72. Milam, J.C.; Litvaitis, J.A.; Warren, A.; Keirstead, D. Bee assemblages in managed early-successional habitats in southeastern New Hampshire. *Northeast. Nat.* **2018**, *25*, 437–459. [CrossRef]
73. Mathis, C.L.; McNeil, D.J.; Lee, M.R.; King, D.I.; Grozinger, C.M.; Otto, C.R.V.; Larkin, J.L. Pollinator communities vary with vegetation structure and time since management within regenerating timber harvests of the Central Appalachian Mountains. *For. Ecol. Manag.* **2021**, *496*, 119373. [CrossRef]
74. Lee, M.; McNeil, D.J.; Mathis, C.L.; Grozinger, C.M.; Larkin, J.L. Microhabitats Created by Log Landings Support Abundant Flowers and Insect Pollinators within Regenerating Mixed-Oak Stands in the Central Appalachian Mountains. *For. Ecol. Manag.* **2021**, *497*, 119472. [CrossRef]
75. Briske, D.D.; Bestlemeyer, B.T.; Brown, J.R.; Brunson, M.W.; Thurow, T.L.; Tanka, J.A. Assessment of USDA-NRCS rangeland conservation programs: Recommendation for evidence-based conservation platform. *Ecol. Appl.* **2017**, *27*, 94–104. [CrossRef]
76. Lutter, S.H.; Dayer, A.A.; Heggenstaller, E.; Larkin, J.L. Effects of biological monitoring and results outreach on private landowner conservation management. *PLoS ONE* **2018**, *13*, e0194740. [CrossRef]
77. Lutter, S.H.; Dayer, A.A.; Larkin, J.L. Young forest conservation incentive programs: Explaining re-enrollment and post-program persistence. *Environ. Manag.* **2018**, *63*, 270–281. [CrossRef]
78. Lutter, S.H.; Dayer, A.A.; Rodewald, A.D.; McNeil, D.J.; Larkin, J.L. Early successional forest management on private lands as a coupled human and natural system. *Forest* **2019**, *10*, 499. [CrossRef]
79. Williams, B.K.; Brown, E.D. Adaptive management: From more talk to real action. *Environ. Manag.* **2014**, *53*, 465–479. [CrossRef] [PubMed]
80. Selinske, M.J.; Coetzee, J.; Purnell, K.; Knight, A.T. Understanding the motivations, satisfaction, and retention of landowners in private land conservation programs. *Conserv. Lett.* **2015**, *8*, 282–289. [CrossRef]

81. Farmer, J.R.; Ma, Z.; Drescher, M.; Knackmuhs, E.G.; Dickinson, S.L. Private landowners, voluntary conservation programs, and implementation of conservation friendly land management practices. *Conserv. Lett.* **2017**, *10*, 58–66. [CrossRef]
82. Schwartz, M.W.; Cook, C.N.; Pressey, R.L.; Pullin, A.S.; Runge, M.C.; Salafsky, N.; Sutherland, W.J.; Williamson, M.A. Decision support frameworks and tools for conservation. *Conserv. Lett.* **2018**, *11*, e12385. [CrossRef]
83. Naugle, D.E.; Maestas, J.D.; Allred, B.W.; Hagen, C.A.; Jones, M.O.; Falkowski, M.J.; Randall, B.; Rewa, C.A. CEAP quantifies conservation outcomes for wildlilfe and people on western grazing lands. *Rangelands* **2019**, *41*, 211–217. [CrossRef]

land

Article

Occurrence and Abundance of an Apex Predator and a Sympatric Mesopredator in Rural Areas of the Coastal Range of Southern Chile

Fernando García-Solís [1,2,*], Jaime R. Rau [2] and Edwin J. Niklitschek [3]

1 Programa de Doctorado en Ciencias, Mención Conservación y Manejo de Recursos Naturales, Universidad de Los Lagos, Avda, Fuchslocher 1305, Osorno 5290000, Chile
2 Laboratorio de Ecología, Departamento de Ciencias Biológicas y Biodiversidad, Universidad de Los Lagos, Osorno, Castilla 933, Osorno 5290000, Chile; jrau@ulagos.cl
3 Centro i-mar, Universidad de Los Lagos, Camino Chinquihue Km 6, Puerto Montt 5480000, Chile; edwin.niklitschek@ulagos.cl
* Correspondence: fernando.garcia-solis@alumnos.ulagos.cl; Tel.: +34-690-884-935

Abstract: The two mammalian carnivores, puma (*Puma concolor*) and South American grey fox (*Lycalopex griseus*) were studied, in a remote area located in the humid temperate forest of the coastal range of southern Chile. A total of six locations were selected in three landscapes: pre-mountain range, mountain range, and coast. The chosen study locations are relevant because they correspond to threatened areas with different levels of human intervention., so they offer the ideal setting for studying how different species of carnivores respond to both human presence and activities. A dataset was collected for 24 months during 2016–2018 through photo-trapping (13 camera traps placed along 50 photo-trap stations). Wes estimated the apparent occurrence and relative abundance index (RAI) of the fauna registered, by means of generalized linear models to contrast those of an apex predator, such as the puma and a sympatric mesopredator, the South American grey fox, across the three landscapes. The ecological variables assessed were the RAI of the other carnivore considered, exotic carnivores such as dogs and cats, human intervention, farmland effect, prey availability, and habitat quality. The primary hypothesis was that the apparent occurrence and RAI of puma and fox would be positively associated with the RAI of prey and livestock and negatively with human intervention. On the other hand, the secondary hypothesis dealt with the interactions between puma and fox faced with different degrees of human intervention. The results showed that the apparent occurrence of the puma was statistically explained by location only, and it was highest at the mountain range. The apparent occurrence of foxes was explained by both puma apparent occurrence and relative integrated anthropization index (INRA), being highest in the pre-mountain range. Concerning the RAI of pumas, high values were yielded by location and fox RAI. For the RAI of foxes, they were location, puma RAI, and INRA. It can be suggested that eucalyptus plantations from the pre-mountain range could offer an adequate habitat for the puma and the fox, but not the coastal range, as the mountain range could be acting as a biological barrier. Due to the nature of the data, it was not possible to detect any relevant effect between the two carnivores' considered, between their respective preys, or the very abundant presence of dogs.

Keywords: camera-trapping; conservation puma; relative integrated anthropization index; INRA; South American grey fox

Citation: García-Solís, F.; Rau, J.R.; Niklitschek, E.J. Occurrence and Abundance of an Apex Predator and a Sympatric Mesopredator in Rural Areas of the Coastal Range of Southern Chile. *Land* **2022**, *11*, 40. https://doi.org/10.3390/land11010040

Academic Editor: Juan F. Beltrán

Received: 5 December 2021
Accepted: 15 December 2021
Published: 28 December 2021

Publisher's Note: MDPI stays neutral with regard to jurisdictional claims in published maps and institutional affiliations.

1. Introduction

Mammalian carnivores tend to have large home ranges, low densities, and slow growth rates, making them especially vulnerable to extinction [1–3]. Because of the lack of protection, habitat loss, and human action; most wild carnivores have undergone significant decreases in their abundance and diversity [1,2,4–8]. The conflict with humans is the leading cause of the decline in carnivore populations [9,10]. These conflicts happen mainly because of suspected predation on livestock and on some wild species with trophy hunting interests [11]. These human carnivore conflicts are a worldwide problem [10,12] with plenty of examples of carnivores killing livestock or even attacking humans. Carnivores have an essential role in the community of which they are part of, primarily by regulating it through trophic cascades. Their effects can be produced by consumption or by behavior [13]. The consumption function is also called lethal and can directly regulate prey population size [8,14] and mesopredators (in the case of apex predators) [15,16]; or indirectly by providing carrion [17,18], promoting higher biodiversity levels [19], or even influencing soil composition [20]. Their effects by behavior can be direct and indirect as well: Directly by influencing prey behavior and habitat use [21,22], prey pack size [23], reproductive physiology [24], and natural selection [25]. Indirectly by modulating prey population dynamics [26,27], limiting herbivory, or maintaining plant diversity [28,29]. Therefore, carnivore protection is one of the priorities in biological conservation using the top-down approach [10].

Most ecological ecosystems are human-modified environments [4,30] due to urban development or exploitation of natural resources. Carnivores are affected by human activities in many different ways: by habitat fragmentation, physical barriers limiting gene flow, road death tolls, behavioral changes, dispersal, disease spreading, and exposure to poisons [3,31,32].

An essential aspect in carnivore conservation and management is based on their interactions in sympatry. It is crucial to understand the structure of the ecological community in which they are inserted [33], as it may influence the distribution, activity patterns, and or diet of the carnivores involved. The competitive exclusion principle proposes that two species with identical niches cannot coexist indefinitely; therefore, some degree of partitioning must materialize in the realized niche of coexisting species [33–36]. Such partitioning is commonly observed across time, space, and trophic axes. In addition, the particular association of coexistence established between apex predators and mesopredators should be considered. The latter being defined as those at intermediate trophic levels, where the former control the populations of the latter [15,16,33,37].

The present study is focused on the apparent occurrences, relative abundance indexes (RAI) [38], and connections of an apex predator, the puma (*Puma concolor*) and one mesopredator, the South American grey fox (*Lycalopex griseus*), considering ecological variables such as the abundance of others wild and exotic carnivores, human intervention, farmland effect, prey availability, and habitat quality. Consideration of dogs is also important as well because we observed abundant free-roaming individuals were observed, which may influence both native species (by predation, competition, disease transmission) [39–42], and livestock [43,44]. The chosen study locations are relevant, because they are threatened areas with different levels of human intervention. These features offer the ideal setting for studying of how different carnivore species respond to human presence and activities. The primary hypothesis was that the apparent occurrence and RAI of puma and fox would be positively associated with RAI of prey and livestock and negatively related to human intervention. On the other hand, the secondary hypotheses were related to the interactions between puma and fox faced with different degrees of human intervention.

In this work, the puma and fox apparent occurrence and RAI between three contrasting landscapes were compared, characterized by considerable differences in human population and intervention. Under this central hypothesis, lower RAI and apparent occurrence of both predators in the pre-mountain landscape are expected, which was more anthropized. Besides this, the authors were interested in assessing several secondary hypotheses that might explain the variability observed between localities. They included a negative relationship between puma and fox, positive effects of prey and livestock apparent abundance, and negative effects of humans and free-roaming dogs on the apparent occurrence and RAI of both carnivores. Nonetheless, the large collinearity between most of these explanatory variables and the small number of localities where they were tested precluded proper isolation of their effects, leading to shape the current assessment as an exploratory analysis.

2. Materials and Methods

The methodology chosen was camera-trapping, which is non-invasive, considers the wellbeing of the animals by minimizing the disturbance of their activities, and guarantee their safety. These methods have been increasingly used over the last 30 years [45], especially in the study of carnivores [46,47]. As shown, camera trapping is an effective method for wildlife surveys, is easy to use, allows to obtain information remotely, is resilient to weather conditions, and has a reasonable cost [45].

2.1. Study Site

The study area is located in the Valdivian Eco-region (40°–42° S) in the humid temperate forest of the coastal range [48]. Six locations from the Purranque Commune in Osorno Province of Los Lagos Region were selected across three landscapes: pre-mountain range (locations of Hueyusca and Los Riscos), mountain range (locations: slope and the peak), and coast (locations of San Pedro Bay and Manquemapu) (Figure 1). The latter two belong to the Lafken Mapu Lahual Indigenous Protected Area [49]. The Valdivian Rainforest is one of the top conservation priorities worldwide due to its high levels of endemism and biodiversity. The three landscapes studied have very different degrees of human intervention, which offers the chance to study and compare the effects of such intervention upon carnivore RAI and apparent occurrence. Due to a limited number of cameras available and logistic difficulties to access the study area, only two locations per landscape were chosen.

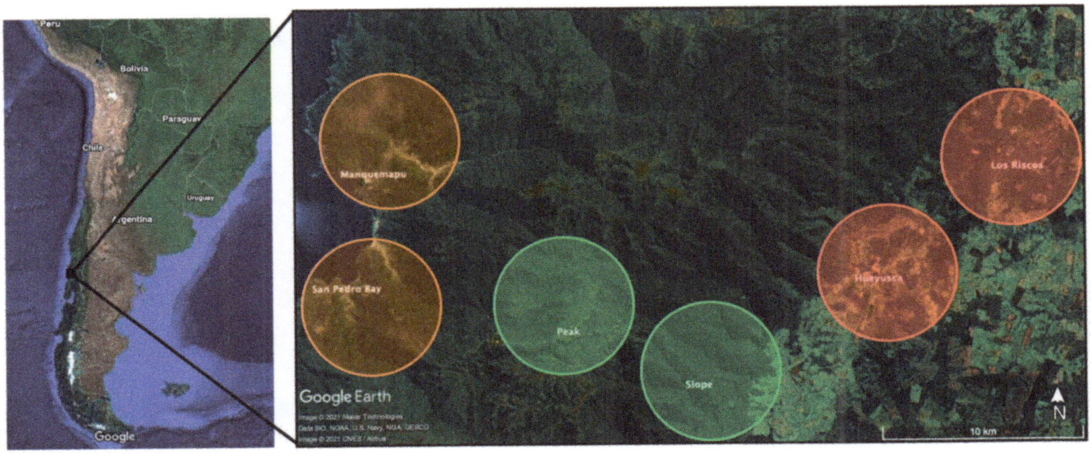

Figure 1. Study area (800 km^2 approximately). Coast locations: Manquemapu and San Pedro Bay (Orange). Mountain range locations: peak and slope (Green). Pre-mountain range locations: Hueyusca and Los Riscos (red).

The climate is rainy temperate, characterized by moderate temperatures (average of the coldest month is 7.5 °C, of the warmest month is 22 °C, with a yearly average over 10 °C [50,51]). Rains occur throughout the year, lacking a dry season [50,51]. During 2017, the rainiest month was August (289.4 mm) and the lowest precipitation was during November (22.8 mm), averaging 112 mm yearly [52].

The pre-mountain range is a human-dominated landscape, with small-family livestock owners and large patches of exotic plantations of eucalyptus (mainly Eucalyptus nitens and Eucalyptus globulus [53]) and pines (*Pinus* spp.). The location of Hueyusca has 399 inhabitants [54] whose main activity is related to small-scale livestock raising and agriculture. The location of Los Riscos has 130 inhabitants [54], mainly related to eucalyptus forestry practices. There are still fragments of deciduous forest of Patagonian oak (*Nothofagus obliqua* now *Lophozonia heterocarpa* [55]), and Chilean laurel (*Laurelia sempervirens*), coigüe

(*Nothofagus dombeyi*), and ulmo (*Eucryphia cordifolia*) mixed forest closer to the mountain range [56].

The mountain range is a more pristine landscape with a low human population and intervention. The location of slope has 89 inhabitants [54] and the peak has no official population records, five inhabited houses were observed nearby the sampling site. The vegetation is dominated by a mixed forest of coigüe with ulmo in the east slope, a narrow strip of Patagonian cypress (*Fitzroya cupressoides*) at the top, and tineo (*Weinmannia trichosperma*) with tepa (*Laureliopsis philippiana*) on the west slope [56].

The coast landscape has a few small indigenous settlements, whose main activity is fishing, complemented by the collection and handwork of local wood [57]. The vegetation surrounding these settlements is dominated by tineo and tepa [56]. No official population records are available for San Pedro Bay or Manquemapu, though the local government estimates they have about 40 and 100 inhabitants, respectively.

An essential feature of our study area is their inhabitants, as there are several native communities of Huilliche natives (people from the south) they are one of the several Mapuche ethnic groups, whose lives are linked to nature and its resources, especially the Patagonian cypress. Wood handicraft is one of their main activities, but they also work the land, raise livestock and crops, or do fishing if they live close to the coast, all in a traditional fashion [57,58].

2.2. Study Design

A total of 13 camera traps (Bushnell 8MP Trophy Cam HD Hybrid Trail Camera with Night Vision) were individually placed along 50 photo-trap stations to maximize the number of cameras: 14 in the coast (6 in Manquemapu and 8 in San Pedro Bay), 16 in the mountain range (8 in the slope and 8 at the peak), and 20 in the pre-mountain range (9 in Hueyusca and 11 in Los Riscos). The photo-trap period lasted from April 2016 to March 2018, with a survey period of 5772 camera days. The cameras were placed between 50–70 cm high [59–62] along secondary paths, randomly within the specific location [59,63] and with a minimum separation distance of 3 km among them [46,59,61,64], to promote the spatial independence in detections. To further optimize the use of cameras, specific attractants for carnivores were applied [62,65–67], chiefly commercial Eurasian lynx (*Lynx lynx*) urine.

The camera traps were placed on large-diameter trees to prevent or hinder their removal by humans, which was a problem during our study. Another measure to prevent these events was the use of two locks and one chain per camera. Once a month the status of the cameras, battery levels, and memory cards were checked, and their contents were transferred if it necessary for the research design. The camera settings corresponded to the following: mode: camera, image size: 5M pixel, image format: full screen, capture number: 3 photo, led control: medium, camera name: input, interval: 5 s, sensor level: low, NV shutter: low, camera mode: 24 h, format: execute, time stamp: on, and field scan: off. Those photographic records with animals were considered as independent events when images contained species within a 60-minute period. If another animal of the same species was captured in this time window, it was not registered unless it could be recognized as a different individual [60,64,68].

To identify which of the three Chilean fox species known were recorded, the photos were reviewed by the study team, determining the species positively as South American grey fox (*Lycalopex griseus*), which agreed with the bibliography labelling it as more of a lowland animal than culpeo fox (*Lycalopex culpaeus*), the latter being more of a mountain dweller of the Andes range [69,70].

2.3. Data and Statistical Analysis

The data were analyzed with the statistical program R [71]. RAI by species, location, season, and camera was computed as the total number of independent and recognizable pictures of each species recorded by a single camera placed at a particular location within a single season. As the number of deployment days was variable between cameras, locations, and seasons, RAI was standardized to a fixed 100-day period [38,72]. The apparent occurrence was computed as a dichotomic variable indicating presence for all RAI values > 0 and absence otherwise. Apparent occurrence and RAI of puma and fox were then analyzed using a generalized linear models (GLMs) framework [73] (Table 1). On the other hand, a binomial distribution was inherent for apparent occurrence data, a zero-inflated negative binomial distribution [74] was used to analyze RAI responses, respectively. Model assumptions were assessed using a simulated residuals approach [75] as implemented in the R package DHARMa [76]. Dichotomic uses of *p*-values were purposely avoided following recommendations made by the American Statistical Association [77] and a growing number of scientists worldwide [78].

Table 1. Model used to estimate the apparent occurrence and RAI of puma and fox. AO = Apparent occurrence.

Puma	Model in R
Puma AO	gpuma.bin = glmer(formula = Puma.bin~Habitat + Dog + Fox + PumaPrey + Livestock + HumanPresence + Inra + (1 ǀ Locality), family = "binomial", data = data3, na.action = "na.pass")
Fox AO	gzorro.bin2 = glmer(formula = Zorro.bin~Habitat + Dog + Puma + FoxPrey + Livestock + HumanPresence + Inra + (1 ǀ Locality), family = "binomial", data = data3, na.action = "na.pass")
Puma RAI	gpuma.ab = glmmadmb(Puma~ Habitat + Dog + Fox + PumaPrey + Livestock + HumanPresence + Inra + (1 ǀ Locality), data = data3, zeroInflation = TRUE, family = "nbinom")
For RAI	gzorro.ab = glmmadmb(Zorro~Habitat + Dog + Puma + FoxPrey + Livestock + HumanPresence + Inra + (1 ǀ Locality), data = data3, zeroInflation = TRUE, family = "nbinom")

In accordance with the primary hypothesis, landscape effects upon RAI and apparent occurrence of puma and fox were assessed by means of marginal likelihood ratio-tests [79]. To properly isolate landscape effects, GLMs used for this purpose also included season and location effects. This sampling design-based analysis was followed by an exploratory analysis of the secondary hypotheses, where locality and seasonal effects were replaced by six quantitative variables: dog (*Canis familiaris*), competitor, prey and livestock apparent relative abundances (records/100 camera-days), human presence and degree of anthropization. Competitor RAI corresponded to either puma or fox standardized records, while apparent prey abundances summed over European hare (*Lepus europaeus*) and pudu (*Pudu puda*) records for both predators, plus red deer (*Cervus elaphus*) records for puma. Livestock apparent abundance summed over apparent abundances of horse (*Equus caballus*), cattle (*Bos taurus*), sheep (*Ovis aries*) and pig (*Sus scrofa domestica*). Human presence was indexed summing over people, vehicles, and machinery records.

Anthropization was indexed through the relative integrated anthropization index (INRA) [80,81], computed after assigning intervention values (0.000–1.000) reflecting the use or land cover by subunits of analysis (SUA) [80,82], 0 being no-intervention level and 1 maximum intervention level. In this case, 0.1×0.1 km quadrants were used (aerial images from Google Earth). The categories assigned for land use or land cover were the following: native vegetation (0.000), native vegetation + clearing (0.125), native vegetation + clearing + crops (0.250), native vegetation + crops (0.375), clearing (0.5000), clearing + crops (0.625), cultivation (0.750), rural population (0.850), and urban nucleus (1.000). Once the SUA values were obtained, the INRA value of the analysis units (UA) was obtained as:

$$\text{INRA} = (\sum \text{SUA}' / \text{n}) \cdot 100 \quad (1)$$

where $\sum$SUA′ = the sum of the partial anthropization value of all SUA and n = total number of SUAs.

The exploratory analysis was performed following a two-steps approach. First, a deviance analysis followed by marginal likelihood ratio-tests was used to assess all main effects of habitat and the six quantitative variables considered at once [79]. Second, a multi-model comparison approach based on second-order Akaike's information criterion [82] was used to compare and rank all possible combinations of habitat and quantitative variables using the second-order Akaike's information criterion and Akaike's weight, which was interpreted as the probability of being the most parsimonious model within the set of candidate models being compared [82].

3. Results

A total of 3611 records (Tables A1 and A2) were obtained with an average of 55 per camera, location, and year. Because of the low number, data from the coastal range were removed from further analyses. Among carnivores, it was possible to detect: kodkod (*Leopardus guigna*), Molina's hog-nosed skunk (*Conepatus chinga*), puma (*Puma concolor*), South American grey fox (*Lycalopex griseus*), and two exotic species, dog (*Canis familiaris*), and cat (*Felis catus*).

The relative integrated anthropization Index (INRA) showed the general trend of highest values appearing in the pre-mountain range, followed by the coast, and ended by the mountain range (Table 2).

Table 2. Relative integrated anthropization index (INRA) with respective SUAs (subunits of analysis) by location in the coastal range of southern Chile. (C: coast, MR: mountain range, PMR: pre-mountain range).

Units of Analysis	Landscape	SUA1	SUA2	SUA3	SUA4	SUA5	SUA6	SUA7	SUA8	SUA9	INRA
Manquemapu	C	0.125	0	0.125	0	0	0.125	0	0	0	4.167
San Pedro	C	0	0.5	0	0	0	0.5	0	0.125	0	12.500
Peak	MR	0	0.125	0	0	0.5	0.125	0	0	0	8.333
Slope	MR	0	0	0	0	0	0	0	0	0.125	1.389
Hueyusca	PMR	0.25	0.125	0.75	0	0.25	0.625	0.125	0.375	0.125	29.167
Los Riscos	PMR	0.125	0.125	0.125	0.25	0.375	0.75	0.375	0.25	0.75	34.722

3.1. Exploratory Analysis of Primary Hypothesis

Significant differences in RAI and apparent puma occurrence were observed between landscapes and locations (Figures 2 and 3). Although evident landscape effects (Table 3) resulted from the complete absence of puma records in the coast (Figure 2), no apparent differences were found when locality means were compared between the pre-mountain and mountain landscapes. Thus, the highest puma apparent occurrence and RAI means were not consistently observed in the mountain landscape, although their maximum values did occur in the less anthropized location of the slope (Figures 2 and 3, Table 2).

Table 3. Marginal likelihood ratio tests for the effects of landscape, season, landscape: location, and landscape: Season on the apparent occurrence and RAI of puma (*Puma concolor*) and South American grey fox (*Lycalopex griseus*) in six locations of the Southern Chile coastal range. p-values < 0.1 highlighted in bold.

	Degrees of Freedom	Puma				Fox			
		Apparent Occurrence		RAI		Apparent Occurrence		RAI	
		LR X^2	p (>X^2)	LR X^2	p (>X^2)	LR X^2	p (>X^2)	LR X^2	p (>X^2)
Landscape	2	19.08	**<0.001**	24.51	**<0.001**	8.32	**0.016**	12.10	**0.002**
Season	3	2.58	0.461	4.14	0.247	3.74	0.291	3.68	0.298
Landscape: Location	3	1.64	0.651	6.90	**0.075**	10.33	**0.016**	5.79	0.123
Landscape: Season	6	3.36	0.762	6.36	0.384	6.92	0.329	8.70	0.191

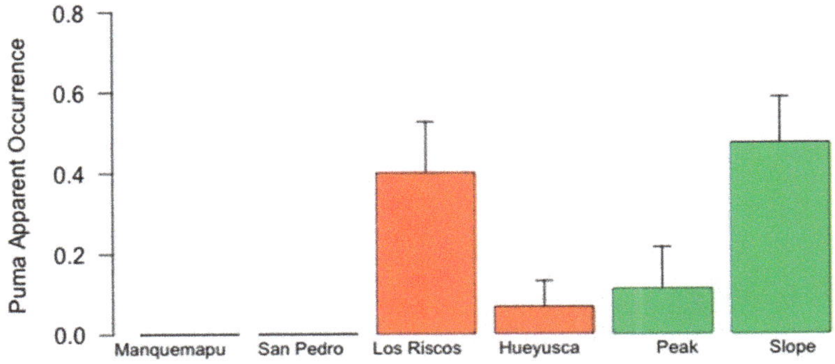

Figure 2. Puma apparent occurrence in six locations of the coastal range of southern Chile (whiskers represent 1 standard error). White, red and green bars identify coast, pre-mountain range and mountain range landscapes, characterized by intermediate, high and low degrees of human intervention, respectively.

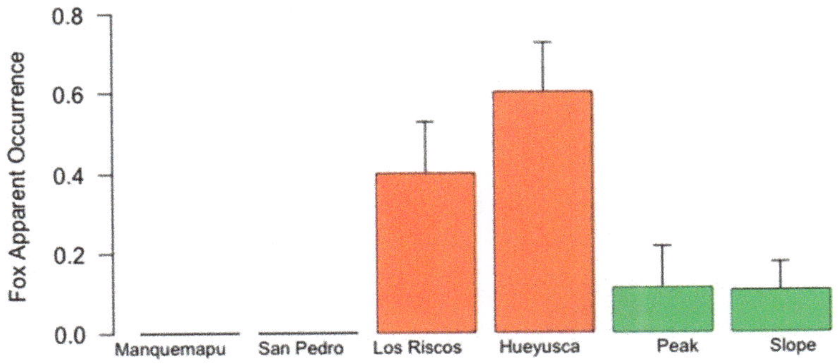

Figure 3. Fox apparent occurrence in six locations of the coastal range of southern Chile (whiskers represent 1 standard error). White, red and green bars identify coast, pre-mountain range and mountain range landscapes, characterized by intermediate, high and low degrees of human intervention, respectively.

Variability in fox apparent occurrence appeared more clearly linked to landscape features (Table 2, Figure 4), with higher mean values found in the pre-mountain range and no evidence of fox presence in the coast. Variability in fox RAI was inconsistent between landscapes, with maximum values in Hueyusca (pre-mountain) followed by the peak (mountain). Thus, as before, maximum RAI and apparent occurrence values were not found in the less disturbed mountain range landscape, nor the less anthropized locations of the slope and Manquemapu (Figures 4 and 5, Table 2).

3.2. *Exploratory Analysis of Secondary Hypothesis*

Deviance analysis of secondary hypotheses showed that none of the variables being considered exhibited relevant marginal effects explaining variability in puma apparent occurrence ($p \geq 0.403$, Table 4), whereas some evidence was found ($p = 0.074$, Table 4) of positive effects from the RAI of fox on that of puma. Model selection procedures also failed to identify an informative model for explaining apparent puma occurrence (Table 5). The most informative model which included positive effects from fox RAI as its only explanatory variable has a probability (AICc-w) of just 0.07 (Table 5) and explained only 8% of the deviance. The other four alternative models, including the null model, all received weak support from the data (ΔAICc ≤ 2, Table 5). Model selection results for puma RAI were more conclusive. The most comprehensive model one that considered the negative effects of INRA and positive effects of livestock apparent abundance exhibited a probability of 0.54 and explained 37% of the observed deviance (Table 6).

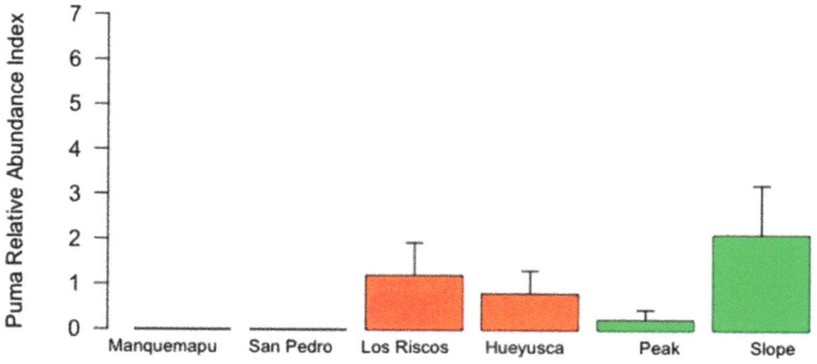

Figure 4. Puma RAI in six locations of the coastal range of southern Chile (whiskers represent 1 standard error). White, red and green bars identify coast, pre-mountain range and mountain range landscapes, characterized by intermediate, high and low degrees of human intervention, respectively.

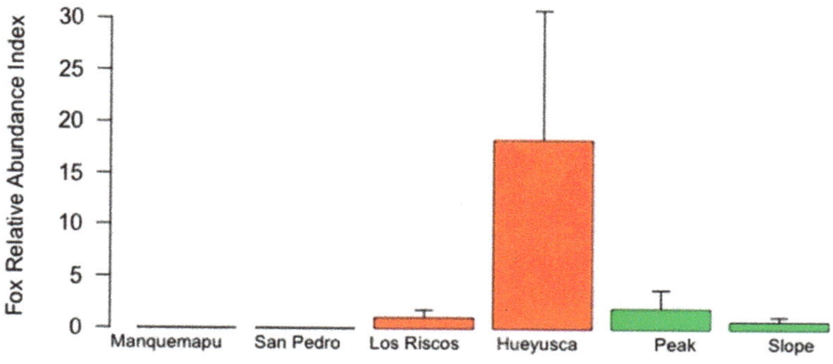

Figure 5. Fox RAI in six locations of the coastal range of southern Chile (whiskers represent 1 standard error). White, red and green bars identify coast, pre-mountain range and mountain range landscapes, characterized by intermediate, high and low degrees of human intervention, respectively.

Table 4. Marginal likelihood ratio tests for secondary hypotheses, which considered the effects of the six quantitative variables considered on the apparent occurrence and RAI of puma (*Puma concolor*) and South American grey fox (*Lycalopex griseus*) in six locations of the Southern Chile coastal range. *p*-values < 0.1 highlighted in bold.

	Degrees of Freedom	Puma				Fox			
		Apparent Occurrence		RAI		Apparent Occurrence		RAI	
		LR X^2	p (>X^2)	LR X^2	p (>X^2)	LR X^2	p (>X^2)	LR X^2	p (>X^2)
Dog	1	0.388	0.533	0.223	0.637	0.03	0.864	−1.126	1.000
Competitor	1	2.793	**0.095**	5.832	0.016	4.527	**0.033**	17.928	**<0.001**
Prey	1	0.602	0.438	0.257	0.612	0.974	0.324	0.134	0.714
Livestock	1	0.701	0.403	0.233	0.629	2.941	**0.086**	0.298	0.585
Human presence Index	1	0.007	0.934	0.114	0.735	0.006	0.939	0.768	0.381
INRA	1	0.136	0.712	0.973	0.324	3.047	**0.081**	10.112	**0.001**

Table 5. Exploratory analysis of potential explanatory variables for observed variability in the apparent occurrence of puma (*Puma concolor*). Only the best five models, as ranked by AICc values, are shown. Positive/negative signs indicate variables and effects considered by each model. H = landscape, L = location, LST = livestock, HPI = human presence index, INRA = relative integrated anthropization index, D2 = explained deviance, K = number of estimated parameters, logLik = loge likelihood, AICc = second order Akaike information criterion, AICc-W = Akaike weight.

Model	Dog	Prey	Fox	LST	HPI	INRA	D^2	K	logLik	AICc	ΔAICc	AICc-W
1			+1.50				8.23%	3	−32.59	71.63	0.00	0.07
2				+1.27			8.09%	3	−32.64	71.72	0.09	0.07
3							3.82%	2	−33.95	72.13	0.50	0.05
4	+095						5.89%	3	−33.32	73.09	1.46	0.03
5			+0.97	+0.77			9.15%	4	−32.30	73.36	1.73	0.03

Table 6. Exploratory analysis of potential explanatory variables for observed variability in in the RAI of puma (*Puma concolor*). Only the best five models, as ranked by AICc values, are shown. Positive/negative signs indicate variables and effects considered by each model. Codes are the same as in Table 5.

Model	Dog	Prey	LST	HPI	INRA	D^2	K	logLik	AICc	ΔAICc	AICc-W
1			+0.17		−0.17	36.58%	6	−65.39	144.42	0.00	0.54
2			+0.17		−0.29	37.04%	7	−64.86	145.97	1.55	0.25
3			+0.15			32.71%	6	−67.14	147.92	3.50	0.09
4		+0.24	+0.24	−0.23	−0.17	45.23%	8	−64.70	148.34	3.91	0.08
5	+0.21		+0.16		−0.26	37.36%	7	−66.78	149.81	5.38	0.04

Deviance analysis applied to fox records showed more significant marginal effects of livestock apparent abundance and INRA on apparent fox occurrence, and puma RAI and INRA on fox RAI. As found for puma, model selection failed to identify a distinct set of explanatory variables accounting for variability in fox apparent occurrence. Five models received similar support from the data (ΔAICc $\leq$ 2): the most informative exhibited probabilities between 0.17 and 0.23 and explained 34–39% of the observed deviance (Table 7). All these models included positive effects of INRA and livestock apparent abundance. Some of them also included positive effects of prey, dog and puma RAI (Table 7). For the RAI of fox, three potentially informative models were identified, with probabilities between 0.22 and 0.33 and explaining 43–46% of the observed deviance (Table 8). While these models included positive effects of INRA and livestock apparent abundance, the top one also included negative effects of apparent dog abundance (Table 8).

Table 7. Exploratory analysis of potential explanatory variables for observed variability in the apparent occurrence of South American grey fox (*Lycalopex griseus*). Only the best five models, as ranked by AICc values, are shown. Positive/negative signs indicate variables and effects considered by each model. Codes are the same as in Table 5.

Model	Dog	Prey	Puma	LST	HPI	INRA	D^2	K	logLik	AICc	ΔAICc	AICc-W
1		+ 2.44		+ 3.18		+2.66	36.72%	5	−22.65	56.46	0.00	0.23
2		+2.36	+1.35	+2.81		+2.98	39.24%	6	−21.48	56.60	0.14	0.21
3	+2.23		+1.53	+2.41		+3.07	39.06%	6	−21.56	56.77	0.31	0.20
4				+3.13		+2.92	33.61%	4	−24.04	54.84	0.38	0.19
5			+1.29	+2.81		+3.22	36.09%	5	−22.94	57.03	0.57	0.17

Table 8. Exploratory analysis of potential explanatory variables for observed variability in the RAI of South American grey fox (*Lycalopex griseus*). Only the best five models, as ranked by AICc values, are shown. Positive/negative signs indicate variables and effects considered by each model. Codes are the same as in Table 5.

Model	Dog	Prey	LST	HPI	INRA	D2	K	logLik	AICc	ΔAICc	AICc-W
1	−0.03	+0.05	+0.05		+0.05	46.01%	8	−79.93	178.79	0.00	0.33
2		+0.03	+0.03		+0.05	42.95%	7	−81.52	179.29	0.50	0.26
3		+0.03	+0.05	−0.02	+0.06	45.23%	8	−80.34	179.62	0.83	0.22
4	−0.03	+0.05	+0.06	−0.01	+0.06	46.44%	9	−79.70	181.14	2.35	0.10
5	−0.03	+0.05	+0.05		+0.05	46.09%	9	−79.88	181.52	2.72	0.09

4. Discussion

The absence of puma and fox records obtained in the two locations from the coast landscape (Manquemapu and San Pedro Bay) was surprising. Nonetheless, it matched results from a parallel study conducted by us, which showed that carnivore feces were scarce in these locations. These sites are isolated, weakly intervened, and with small settlements of fishermen and wood handcrafters (INRA values of 4.167 and 12.500, respectively). When the study was designed, it was assumed that the mountain range would act as a biological corridor [83], but the current situation probably is the opposite, acting as a barrier and limiting dispersal from the coast landscape. In the past, the entire mountain range suffered from several big fires [57]; some people think they came about by natural causes and others that they were man-made to acquire the burned wood from Patagonian cypress, which is protected as a natural monument and can only be exploited when burnt (independently of cause). Currently, the Patagonian cypress forest at the peak is quite open, full of dead trees, a few survivors, and some recruits (F. García-Solís, personal observation). Unfortunately, Patagonian cypress trees take longer to grow, living up to 3600 years [84]. All this renders the peak location of the mountain range a harsh environment, with almost no shelter for herbivores, thus limiting carnivore presence.

Camera trapping of unmarked species can be challenging, as it is difficult to use capture-recapture methods when assessing their relative abundances and could have biased inference estimating abundances [85,86]. In our study, the two carnivore species were unmarked, thus we assumed equal detectability and potential bias, as camera traps cannot record all animal presences in an area [87]. Their camera records were considered as independent events when consecutive images that contained the same species were recognizable as different individuals, a method used in several studies [60,64,68]. The use of lures is a widespread method in camera trapping, but optimizing the detectability of a target species can produce bias in calculating abundances, as the species behavior may be altered, or some species may be attracted whereas others may be repelled [88–91].

4.1. Relative Integrated Anthropization Index (INRA)

The working hypothesis about habitat quality was related to the fact that the locations from the pre-mountain range would have the highest INRA levels, followed by coast and then by mountain range. Our results supported this mostly, except for the Manquemapu and peak locations. The former had lower INRA, affording better habitat quality than the latter. This lower INRA may be accounted for by the operation of the Manquemapu Management Plan, regulated by its Mapuche Huilliche community. This plan considers zoning areas of human use, dead Patagonian cypress recovery harvest, sustainable management, and collection of marine resources [92]. Although the peak has low human intervention, its forest may offer lower habitat quality owing to its past fire history. The current landscape is a very open forest, full of burned trunks which some of which show small brunches with leaves, this harsh environment could be a barrier to the dispersal of carnivores.

4.2. Predator Apparent Occurrence

Since the use of occupancy or co-occurrence models, was not supported by the data, the use of the apparent occurrence was selected. Los Riscos (pre-mountain range) is characterized by the presence of exotic plantations of eucalyptus (*Eucalyptus nitens* and *Eucalytus globulus*), which they are not native forests still afford a habitat for the puma [53], providing shelter from humans in the surroundings, and probably also food, by being populated by hare and pudu. Further, in that particular landscape, there is a vital remnant of native forest [53,93,94], which provides habitat for the puma's prey. In addition, there is the presence of livestock, which pumas may perceive as a potential food resource. The slope location from the mountain range landscape is characterized by scarce human presence and low activity and preserves most of its native vegetation, rendering it relatively unaltered by humans, which may explain the high puma apparent occurrence. Our results from the marginal likelihood ratio test showed that there was not a significant effect of human activity on puma, pumas may tolerate human presence better than expected. In addition, the effect from prey could not show to be influential either, this could be explained for the prey cannot be detected by the cameras since carnivorous attractant was used. In parallel, the models from the exploratory analysis showed some positive relations for RAI of fox and livestock, both being a potential food resource for puma.

Apparent fox occurrence was influenced by RAI of puma, livestock and INRA. These values were higher in the pre-mountain than in the mountain range. The mesopredator release hypothesis [16] may explain the higher fox presence in the former landscape because the higher human activity may interfere with apparent puma occurrence. A complementary explanation is that foxes, being mesopredators with smaller size may tolerate environments with higher human activity [95]. This parallels the positive effect of INRA on fox apparent occurrence. Alternatively, the presence of puma can facilitate the presence of foxes since the puma behavior of burying its prey after eating to store it for later; this buried prey being subsequently scavenged by foxes [96–98], explaining the positive correlation between them. The exploratory analysis supported the positive effects of INRA and puma on fox. Additionally, South American grey foxes are known to visit exotic plantations due to potential prey such as rodents and hares [99].

4.3. Predator Relative Abundance Indexes

Relative Abundance indices are not necessarily the most informative about abundance species and can have some weaknesses such as: be biased due to the different detection among species, especially in elusive ones; species with extensive home range are more detected, increasing RAI values; and bias due to the different responses to the camera setup among the species [100]. Puma RAI was influenced by the variables of location and fox RAI. Los Riscos and slope present higher values, probably due to low levels of human activity and restricted pass policy, in addition to the presence of livestock as potential prey (our data showed a positive but not so strong correlation. The positive and relevant correlation between RAI of fox and puma could be explained by intraguild predation, which is an extreme form of interspecific competition when species that act as competitors also function as predators [101–103]. In this case, the puma is a potential predator of foxes, the latter's RAI may improve an increase in that of puma. The exploratory analysis showed a negative relation with INRA, and one model showed a negative relation also with human activity, which could be explained by the sensitivity of puma to habitat quality.

Despite large differences in fox RAI between landscapes and locations, our analysis suggests that puma RAI and INRA were positively associated with fox RAI. For instance, Hueyusca showed both the highest fox RAI and high levels of human presence and activity (INRA = 29.167), suggesting once more that foxes may flourish in such anthropized situation. The exploratory analysis supported the survivorship of foxes in human-intervened environments and showed a positive relation of foxes with their prey and the presence of livestock.

It was noteworthy that, even though the cameras traps registered numerous dogs, they were not identified as important ecological variables in any of the models depicting RAI and the apparent occurrence of puma and fox. This result was unexpected, as the impact of free-roaming dogs over wildlife by predation, activity alteration (fear-related), hybridization, and spreading of diseases is well known [42,104]. The present data show that dog numbers were larger in the pre-mountain range whereas those of puma were so in the mountain range. Thus, these two carnivores were segregated over the spatial axis so that dogs may not have an important effect over pumas. Nevertheless, dogs and foxes are abundant in the pre-mountain range, but even if they share space, they are segregated over time, foxes being more active during the night and dogs during the day [8,104].

It can be suggested that eucalyptus plantations in the pre-mountain range could offer an adequate habitat for the puma and the fox due to the presence of shelter from humans from the surroundings and prey availability such as hare, pudu, rodents, and potentially livestock. This was not the case of the coastal range, where we obtained almost no animal records, so it is possible that the mountain range could be acting as a biological barrier rather than a biological corridor. Due to the nature of the present data, it was no possible to detect any relevant effect between the two coexisting carnivores, between their respective prey, or the very abundant presence of dogs. Consequently, we recommend further studies in this specific area and habitats, improving the sampling efforts by implementing a considerable number of cameras and for more extended periods to obtain better data and clearer the relations and conclusions.

Author Contributions: Conceptualization, F.G.-S., J.R.R. and E.J.N.; methodology, F.G.-S., J.R.R. and E.J.N.; software, F.G.-S. and E.J.N.; validation, J.R.R. and E.J.N.; formal analysis, E.J.N.; investigation, F.G.-S., J.R.R. and E.J.N.; resources, J.R.R. and E.J.N.; data curation, E.J.N.; writing -original draft preparation, F.G.-S.; writing -review and editing, F.G.-S., J.R.R. and E.J.N.; visualization, J.R.R. and E.J.N.; supervision, J.R.R.; project administration, J.R.R.; funding acquisition, J.R.R. All authors have read and agreed to the published version of the manuscript.

Funding: This research was funded by ANID PIA/BASAL FB0002.

Data Availability Statement: Not applicable.

Acknowledgments: This research was supported by a doctoral fellowship from Universidad de Los Lagos (Chile) to Fernando García-Solís Marchant. The equipment used was provided by LABECOL (Ecology laboratory) from the Department of Biological Sciences and Bio-diversity of Los Lagos University and CAPES (Center of Applied Ecology and Sus-tainability. Special mention goes to the Fabián M. Jaksic who helped in the reviews of this manuscript; and the Municipality of Purranque, specifically to its Environmental Department chief, Carlos Oyarzún, who provided help in the logistic aspects of the study while funding most of the transportation. Finally, thanks to Anchile Lda. (Robinson Hapette), which allowed the entrance and camera installation in their plantation lands.

Conflicts of Interest: The authors declare no conflict of interest.

Appendix A

Table A1. Records of animal species by landscape in the coastal range of southern Chile.

Common Name/Category	Species	Pre-Mountain Range	Mountain Range	Coast	Total
Mammalia, Order Carnivora					
Dog	*Canis familiaris*	753	70	6	829
Domestic cat	*Felis catus*	1	1	0	2
Kodkod	*Leopardus guigna*	9	1	0	10
Molina's Hog-nosed skunk	*Conepatus chinga*	0	4	0	4
Puma	*Puma concolor*	18	26	0	44
South American grey fox	*Lycalopex griseus*	157	12	0	169
Mammalia, Order Cetartiodactyla					

Table A1. *Cont.*

Common Name/Category	Species	Pre-Mountain Range	Mountain Range	Coast	Total
Southern pudu	*Pudu puda*	1	4	0	5
Red deer	*Cervus elaphus*	4	1	0	5
Mammalia, Order Lagomorpha					
European hare	*Lepus europaeus*	97	5	0	102
Mammalia, Order Rodentia					
Unidentified rodent	Muridae?	0	1	0	1
Aves, Order Pelecaniformes					
Buff-necked ibis	*Theristicus caudatus*	1	0	0	1
Aves. Order Cathartiformes					
Black vulture	*Coragyps atratus*	0	2	0	2
Aves, Order Strigiformes					
Owl	Unknown Strigidae	1	0	1	2
Aves, Order Falconiformes					
Southern caracara	*Caracara plancus*	9	2	0	11
Aves, Order Passeriformes					
Austral thrush	*Turdus falcklandii*	5	16	0	21
Aves, Order Columbiformes					
Chilean pigeon	*Patagioenas araucana*	10	1	0	11
Aves, Order Apodiformes					
Unidentified hummingbirds	Unknown Trochilidae	0	5	0	5
Livestock					
Cow	*Bos taurus*	425	217	42	684
Calf	*Bos taurus*	118	48	9	175
Domestic horse	*Equus caballus*	67	6	84	157
Domestic foal	*Equus caballus*	1	0	0	1
Domestic pig	*Sus scrofa domestica*	42	1	0	43
Domestic sheep	*Ovis aries*	16	0	0	16
Indeterminate animals		33	2	0	35

Table A2. Records of human presence by landscape in the coastal range of southern Chile.

Human Presence Sign	Species	Pre-Mountain Range	Mountain Range	Coast	Total
Machinery		18	0	0	18
People	*Homo sapiens*	435	143	244	822
Vehicle		544	47	0	591

References

1. Caruso, N.; Lucherini, M.; Fortin, D.; Casanave, E.B. Species-Specific Responses of Carnivores to Human-Induced Landscape Changes in Central Argentina. *PLoS ONE* **2016**, *11*, e0150488. [CrossRef]
2. Crooks, K.R.; Burdett, C.L.; Theobald, D.M.; Rondinini, C.; Boitani, L. Global patterns of fragmentation and connectivity of mammalian carnivore habitat. *Phil. Trans. R. Soc. B* **2011**, *366*, 2642–2651. [CrossRef] [PubMed]
3. Ordeñana, M.A.; Crooks, K.R.; Boydston, E.E.; Fisher, R.N.; Lyren, L.M.; Siudyla, S.; Haas, C.D.; Harris, S.; Hathaway, S.A.; Turschak, G.M.; et al. Effects of urbanization on carnivore species distribution and richness. *J. Mammal.* **2010**, *91*, 1322–1331. [CrossRef]
4. Ferreira, A.S.; Peres, C.A.; Bogoni, J.A.; Cassano, C.R. Use of agroecosystem matrix habitats by mammalian carnivores (Carnivora): A global-scale analysis. *Mammal. Rev.* **2018**, *48*, 312–327. [CrossRef]
5. Ripple, W.J.; Estes, J.A.; Beschta, R.L.; Wilmers, C.C.; Ritchie, E.G.; Hebblewhite, M.; Berger, J.; Elmhagen, B.; Letnic, M.; Nelson, M.P.; et al. Status and ecological effects of the world's largest carnivores. *Science* **2014**, *343*, 151–162. [CrossRef]
6. van Eeden, L.M.; Eklund, A.; Miller, J.R.B.; López-Bao, J.V.; Chapron, G.; Cejtin, M.R.; Crowther, M.S.; Dickman, C.R.; Frank, J.; Krofel, M.; et al. Carnivore conservation needs evidence-based livestock protection. *PLoS Biol.* **2018**, *16*, e2005577. [CrossRef] [PubMed]

7. Lamichhane, B.R.; Leirs, H.; Persoon, G.A.; Subedi, N.; Dhakal, M.; Oli, B.N.; Reynaert, S.; Sluydts, V.; Pokheral, C.P.; Poudyal, L.P.; et al. Factors associated with co-occurrence of large carnivores in a human-dominated landscape. *Biodivers. Conserv.* **2019**, *28*, 1473–1491. [CrossRef]
8. Iriarte, A.; Jaksic, F. *Los Carnivoros de Chile*, 2nd ed.; Flora & Fauna Chile y CASEB, P.U. Católica de Chile: Santiago, Chile, 2017; p. 260.
9. Dickman, A.J.; Hazzah, L.; Carbone, C.; Durant, S.M. Carnivores, culture and contagious conflict: Multiple factors influence perceived problems with carnivores in Tanzania's Ruaha landscape. *Biol. Conserv.* **2014**, *178*, 19–27. [CrossRef]
10. Treves, A.; Karanth, K.U. Human-Carnivore Conflict and Perspectives on Carnivore Management Worldwide. *Conserv. Biol.* **2003**, *17*, 1491–1499. [CrossRef]
11. Caniglia, R.; Fabbri, E.; Mastrogiuseppe, L.; Randi, E. Who is who? Identification of livestock predators using forensic genetic approaches. *Forensic Sci. Int. Genet.* **2013**, *7*, 397–404. [CrossRef]
12. Gáspero, P.G.; Easdale, M.H.; Pereira, J.A.; Fernández-Arhex, V.; Von Thüngen, J. Human-carnivore interaction in a context of socio-productive crisis: Assessing smallholder strategies for reducing predation in North-west Patagonia, Argentina. *J. Arid. Environ.* **2018**, *150*, 92–98. [CrossRef]
13. Ordiz, A.; Bischof, R.; Swenson, J.E. Saving large carnivores, but losing the apex predator? *Biol. Conserv.* **2013**, *168*, 128–133. [CrossRef]
14. Ripple, W.J.; Beschta, R.L. Large predators limit herbivore densities in northern forest ecosystems. *Eur. J. Wildl. Res.* **2012**, *58*, 733–742. [CrossRef]
15. Prugh, L.R.; Stoner, C.J.; Epps, C.W.; Bean, W.T.; Ripple, W.J.; Laliberte, A.S.; Brashares, J.S. The Rise of the Mesopredator. *BioScience* **2009**, *59*, 779–791. [CrossRef]
16. Crooks, K.R.; Soule, M.E. Mesopredator release and avifaunal extinctions in a fragmented system. *Nature* **1999**, *400*, 563–566. [CrossRef]
17. Barry, J.M.; Elbroch, L.M.; Aiello-Lammens, M.E.; Sarno, R.J.; Seelye, L.; Kusler, A.; Quigley, H.B.; Grigione, M.M. Pumas as ecosystem engineers: Ungulate carcasses support beetle assemblages in the Greater Yellowstone Ecosystem. *Oecologia* **2019**, *189*, 577–586. [CrossRef] [PubMed]
18. Allen, M.L.; Elbroch, L.M.; Wilmers, C.C.; Wittmer, H.U. Trophic facilitation or limitation? Comparative effects of pumas and black bears on the scavenger community. *PLoS ONE* **2014**, *9*, e102257. [CrossRef]
19. Ray, J.C.; Redford, K.H.; Steneck, R.S.; Berger, J. *Large Carnivores and the Conservation of Biodiversity*, 1st ed.; IlandPress: Washington, DC, USA, 2005; p. 544.
20. Bump, J.K.; Peterson, R.O.; Vucetich, J.A. Wolves modulate soil nutrient heterogeneity and foliar nitrogen by configuring the distribution of ungulate carcasses. *Ecology* **2009**, *90*, 3159–3167. [CrossRef]
21. Ripple, W.J.; Beschta, R.L. Linking Wolves and Piants: Aldo Leopold on Trophic Cascades. *BioScience* **2005**, *55*, 613–621. [CrossRef]
22. Brown, J.S.; Laundré, J.W.; Gurung, M. The ecology of fear: Optimal foraging, game theory, and trophic interactions. *J. Mammal.* **1999**, *80*, 385–399. [CrossRef]
23. Creel, S.; Winnie, J.A. Responses of elk herd size to fine-scale spatial and temporal variation in the risk of predation by wolves. *Anim. Behav.* **2005**, *69*, 1181–1189. [CrossRef]
24. Hawlena, D.; Schmitz, O.J. Physiological stress as a fundamental mechanism linking predation to ecosystem functioning. *Am. Nat.* **2010**, *176*, 537–556. [CrossRef] [PubMed]
25. Genovart, M.; Negre, N.; Tavecchia, G.; Bistuer, A.; Parpal, L.; Oro, D. The Young, the Weak and the Sick: Evidence of Natural Selection by Predation. *PLoS ONE* **2010**, *5*, e9774. [CrossRef]
26. Creel, S.; Christianson, D.A.; Winnie, J.A., Jr. A survey of the effects of wolf predation risk on pregnancy rates and calf recruitment in elk. *Ecol. Appl.* **2011**, *21*, 2847–2853. [CrossRef]
27. Terborgh, J.; Estes, J.A.; Paquet, P.; Ralls, K.; Boyd-Herger, D.; Miller, B.J.; Noss, R.F. The Role of Top Carnivores in Regulating Terrestrial Ecosystems. In *Continental Conservation: Scientific Foundations of Regional Reserve Networks*; Islands Press: Washington, DC, USA, 1999; p. 227.
28. Beschta, R.L.; Ripple, W.J. The role of large predators in maintaining riparian plant communities and river morphology. *Geomorphology* **2012**, *157*, 88–98. [CrossRef]
29. Kuijper, D.P.J.; de Kleine, C.; Churski, M.; van Hooft, P.; Bubnicki, J.; Jedrzejewska, B. Landscape of fear in Europe: Wolves affect spatial patterns of ungulate browsing in Bialowieża Primeval Forest, Poland. *Ecography* **2013**, *36*, 1263–1275. [CrossRef]
30. Watson, J.E.M.; Shanahan, D.F.; Di Marco, M.; Allan, J.; Laurance, W.F.; Sanderson, E.W.; Mackey, B.; Venter, O. Catastrophic Declines in Wilderness Areas Undermine Global Environment Targets. *Curr. Biol.* **2016**, *26*, 2929–2934. [CrossRef]
31. George, S.L.; Crooks, K.R. Recreation and large mammal activity in an urban nature reserve. *Biol. Conserv.* **2006**, *133*, 107–117. [CrossRef]
32. Tigas, L.A.; van Vuren, D.H.; Sauvajot, R.M. Behavioral responses of bobcats and coyotes to habitat fragmentation and corridors in an urban environment. *Biol. Conserv.* **2002**, *108*, 299–306. [CrossRef]
33. Zhao, G.; Yang, H.; Xie, B.; Gong, Y.; Ge, J.; Feng, L. Spatio-temporal coexistence of sympatric mesocarnivores with a single apex carnivore in a fine-scale landscape. *Glob. Ecol. Conserv.* **2020**, *21*, e00897. [CrossRef]
34. Gause, G.F. Experimental analysis of Vito Volterra's mathematical theory of the struggle for existence. *Science* **1934**, *79*, 16–17. [CrossRef]

35. Jonathan Davies, T.; Meiri, S.; Barraclough, T.G.; Gittleman, J.L. Species co-existence and character divergence across carnivores. *Ecol. Lett.* **2007**, *10*, 146–152. [CrossRef] [PubMed]
36. Jaksic, F.; Marone, L. *Ecologia de Comunidades*, 2nd ed.; Ediciones de la Universidad Católica: Santiago, Chile, 2007; p. 336.
37. Jiménez, J.; Nuñez-Arjona, J.C.; Mougeot, F.; Ferreras, P.; González, L.M.; García-Domínguez, F.; Muñoz-Igualada, J.; Palacios, M.J.; Pla, S.; Rueda, C.; et al. Restoring apex predators can reduce mesopredator abundances. *Biol. Conserv.* **2019**, *238*, 108234. [CrossRef]
38. O'Brien, T.G.; Kinnaird, M.F.; Wibisono, H.T. Crouching tigers, hidden prey: Sumatran tiger and prey populations in a tropical forest landscape. *Anim. Conserv.* **2003**, *6*, 131–139. [CrossRef]
39. Hughes, J.; Macdonald, D.W. A review of the interactions between free-roaming domestic dogs and wildlife. *Biol. Conserv.* **2013**, *157*, 341–351. [CrossRef]
40. Vanak, A.T.; Gompper, M.E. Dogs *canis familiaris* as carnivores: Their role and function in intraguild competition. *Mammal. Rev.* **2009**, *39*, 265–283. [CrossRef]
41. Vanak, A.T.; Gompper, M.E. Interference competition at the landscape level: The effect of free-ranging dogs on a native mesocarnivore. *J. Appl. Ecol.* **2010**, *47*, 1225–1232. [CrossRef]
42. Young, J.K.; Olson, K.A.; Reading, R.P.; Amgalanbaatar, S.; Berger, J. Is Wildlife Going to the Dogs? Impacts of Feral and Free-roaming Dogs on Wildlife Populations. *BioScience* **2011**, *61*, 125–132. [CrossRef]
43. Montecino-Latorre, D.; San Martín, W. Evidence supporting that human-subsidized free-ranging dogs are the main cause of animal losses in small-scale farms in Chile. *Ambio* **2019**, *48*, 240–250. [CrossRef]
44. Ritchie, E.G.; Dickman, C.R.; Letnic, M.; Vanak, A.T. Dogs as predators and trophic regulators. In *Free-Ranging Dogs and Wildlife Conservation*; Gompper, M.E., Ed.; Oxford University Press: Oxford, UK, 2015. [CrossRef]
45. O'Connell, A.F.; Nichols, J.D.; Karanth, K.U. *Camera Traps in Animal Ecology: Methods and Analyses*; Springer: Berlin/Heidelberg, Germany, 2010; p. 271.
46. Karanth, K.U. Estimating Tiger Panthera tigris Population From Camera-Trap Data Using Capture-Recapture Models. *Biol. Conserv.* **1995**, *71*, 333–338. [CrossRef]
47. Karanth, K.U.; Nichols, J.D. Estimation of tiger densities in India using photographic captures and recaptures. *Ecology* **1998**, *79*, 2852–2862. [CrossRef]
48. Smith-Ramírez, C. The Chilean coastal range: A vanishing center of biodiversity and endemism in South American temperate rainforests. *Biodivers. Conserv.* **2004**, *13*, 373–393. [CrossRef]
49. McAlpin, M. Conservation and community-based development through ecotourism in the temperate rainforest of southern Chile. *Policy Sci.* **2008**, *41*, 51–69. [CrossRef]
50. Köoppen, W.; Volken, E.; Brönnimann, S. The thermal zones of the Earth according to the duration of hot, moderate and cold periods and to the impact of heat on the organic world. *Meteorol. Z.* **2011**, *20*, 351–360. [CrossRef]
51. Universidad Católica de Chile. Cartografiía Interactiva de los Climas de Chile. Available online: http://www7.uc.cl/sw_educ/geografia/cartografiainteractiva/Inicio/Paginas/UntitledFrameset-1.htm (accessed on 27 July 2021).
52. Agrometereologia. Registro Precipitaciones Purranque. Available online: https://agrometeorologia.cl/ (accessed on 27 July 2021).
53. Rodas-Trejo, J.; Rebolledo, G.; Rau, J.R. Uso y selección de hábitat por mamiferos carnívoros y herbívoros en bosque nativo y plantaciones forestales del sur de chile. *Gest. Ambient.* **2010**, *19*, 33–46.
54. Municipalidad de Purranque. *Plan. Regulador Comunal de Purranque: Memoria Explicativa*; Municipalidad de Purranque: Purranque, Chile, 2019; p. 384.
55. Mosyakin, S.L.; Sokolova, I.V.; Tatanov, I.V. A corrected type designation for Lophozonia heterocarpa, with an update on the lectotype of Fagus obliqua (*Nothofagus obliqua, Lophozonia obliqua*) (*Nothofagaceae*). *Phytotaxa* **2019**, *424*, 177–183. [CrossRef]
56. Farías, A.; Wolodarsky, A.; Pliscoff, P.; Tecklin, D. *Análisis de Conservación a Escala de Paisaje en la Cordillera de la Costa Entre el río Toltén y Cucao*; WWF Chile: Santiago, Chile, 2008.
57. Molina Otarola, R.; Correa, M.; Smith-Ramirez, C.; Gainza, A. *Alerceros Huilliches de la Cordillera de la Costa de Osorno*, 1st ed.; Andros: Santiago, Chile, 2006; p. 297.
58. Alcamán, E. Los Mapuches-Huilliche del Futahuillimapu Septentrional. Expansion colonial, Guerras Internas. *Rev. Hist. Indig.* **1997**, *2*, 29–75.
59. Kelly, M.J. Jaguar monitoring in the Chiquibul forest Belice. *Caribb. Geogr.* **2003**, *13*, 19–32.
60. Negrões, N.; Sarmento, P.; Cruz, J.; Eira, C.; Revilla, E.; Fonseca, C.; Sollmann, R.; Tôrres, N.M.; Furtado, M.M.; Jácomo, A.T.A.; et al. Use of Camera-Trapping to Estimate Puma Density and Influencing Factors in Central Brazil. *J. Wildl. Manag.* **2010**, *74*, 1195–1203. [CrossRef]
61. Silver, S.C.; Ostro, L.E.T.; Marsh, L.K.; Maffei, L.; Noss, A.J.; Kelly, M.J.; Wallace, R.B.; Gómez, H.; Ayala, G. The use of camera traps for estimating jaguar Panthera onca abundance and density using capture/recapture analysis. *Oryx* **2004**, *38*, 148–154. [CrossRef]
62. Moreira-Arce, D.; Vergara, P.M.; Boutin, S. Diurnal human activity and introduced species affect occurrence of carnivores in a human-dominated landscape. *PLoS ONE* **2015**, *10*, e0137854. [CrossRef]
63. Brassine, E.; Parker, D. Trapping elusive cats: Using intensive camera trapping to estimate the density of a rare African felid. *PLoS ONE* **2015**, *10*, e0142508. [CrossRef] [PubMed]

64. Kelly, M.J.; Holub, E.L. Camera Trapping of Carnivores: Trap Success Among Camera Types and Across Species, and Habitat Selection by Species, on Salt Pond Mountain, Giles County, Virginia. *Northeast. Nat.* **2008**, *15*, 249–262. [CrossRef]
65. Flora y Fauna Chile Ltda. *Diagnóstico del Estado Poblacional del Puma (Puma concolor) y Evaluación de la Efectividad de Medidas de Prevención de Ataques a Ganado Doméstico en las Provincias de Llanquihue y Osorno, Región de Los Lagos*; Flora & Fauna Chile Ltda: Santiago, Chile, 2013; p. 64.
66. Muñoz Pedreros, A.; Ra u, J.; Valdebenito, M.; Quintana, V.; Martínez, D. Densidad relativa de pumas (*Felis concolor*) en un ecosistema forestal del sur de Chile. *Rev. Chil. Hist. Nat.* **1995**, *68*, 501–507.
67. Rau, J.R.; Muñoz-Pedreros, A. Indices de visitas a estaciones de atraccion olfativa. In *Huellas y Signos de Mamiferos de Chile*; Muñoz-Pedreros, A., Ed.; CEA Ediciones: Valdivia, Chile, 2008; pp. 92–94.
68. Meek, P.D.; Ballard, G.; Claridge, A.; Kays, R.; Moseby, K.; O'Brien, T.; O'Connell, A.; Sanderson, J.; Swann, D.E.; Tobler, M.; et al. Recommended guiding principles for reporting on camera trapping research. *Biodivers. Conserv.* **2014**, *23*, 2321–2343. [CrossRef]
69. Fuentes, E.R.; Jaksic, F.M. Latitudinal Size Variation of Chilean Foxes: Tests of Alternative Hypotheses. *Ecology* **1979**, *60*, 43–47. [CrossRef]
70. Jiménez, J.E.; Lucherini, M.; Novaro, A.J. *Pseudalopex culpaeus. The IUCN Red List of Threatened Species*; International Union for Conservation of Nature: Gland, Switzerland, 2008. [CrossRef]
71. R Core Team. *A Language and Environment for Statistical Computing*; R Foundation for Statistical Computing: Vienna, Austria, 2008.
72. Monterrubio-Rico, T.C.; Charre-Medellín, J.F.; Pérez-Martínez, M.Z.; Mendoza, E. Use of remote cameras to evaluate ocelot (*Leopardus pardalis*) population parameters in seasonal tropical dry forests of central-western Mexico. *Mammalia* **2018**, *82*, 113–123. [CrossRef]
73. McCullagh, P.; Nelder, J.A. *Generalized Linear Models*, 2nd ed.; Chapman and Hall: London, UK, 1989.
74. Fournier, D.A.; Skaug, H.J.; Ancheta, J.; Ianelli, J.; Magnusson, A.; Maunder, M.N.; Nielsen, A.; Sibert, J. AD Model Builder: Using automatic differentiation for statistical inference of highly parameterized complex nonlinear models. *Optim. Methods Softw.* **2012**, *27*, 233–249. [CrossRef]
75. Dunn, P.K.; Smyth, G.K. Randomized Quantile Residuals, Dun. *J. Comput. Graph. Stat.* **1996**, *5*, 236–244.
76. Hartig, F. DHARMa: Residual Diagnostics for Hierarchical (Multi-Level/Mixed) Regression Models. 2021. Available online: https://cran.r-project.org/web/packages/DHARMa/vignettes/DHARMa.html (accessed on 2 August 2021).
77. Wasserstein, R.L.; Lazar, N.A. The ASA's Statement on p-Values: Context, Process, and Purpose. *Am. Stat.* **2016**, *70*, 129–133. [CrossRef]
78. Amrhein, V.; Greenland, S.; Mcshane, B. Scientists rise up against statistical significance. *Nature* **2019**, *567*, 305–307. [CrossRef] [PubMed]
79. Fox, J. *Applied Regression Analysis and Generalized Linear Models*, 2nd ed.; Sage Publications: Thousand Oaks, CA, USA, 2008; Volume 53.
80. Martínez-Dueñas, W. INRA—índice integrado relativo de antropización: Propuesta técnica-conceptual y aplicación. *Intropica Rev. Inst. Investig. Trop.* **2010**, *5*, 37–46. [CrossRef]
81. Steinhardt, U.; Herzog, F.; Lausch, A.; Lehmann, S. Hemeroby index for landscape monitoring and evaluation. In *Environmental Indices—System Analysis Approach*, 1st ed.; Pykh, Y.A., Hyatt, D.E., Lenz, R.J., Eds.; OLSS Publishers Co. Ltd.: Oxford, UK, 1999; pp. 237–254.
82. Burnham, K.P.; Anderson, D.R. *Model. Selection and Multimodel Inference: A Practical Information-Theoretic Approach*; Springer: New York, NY, USA, 2002; p. 655. [CrossRef]
83. Rau, J.R.; Jiménez, J.E. Diet of Puma (*Puma concolor*, Carnivora: Felidae) in Coastal and Andean Ranges of Southern Chile. *Stud. Neotrop. Fauna Environ.* **2002**, *37*, 201–205. [CrossRef]
84. Lara, A.; Villalba, R. A 3620-year temperature record from *Fitzroya cupressoides* tree rings in southern South America. *Science* **1993**, *260*, 1104–1106. [CrossRef] [PubMed]
85. Burton, A.C.; Neilson, E.; Moreira, D.; Ladle, A.; Steenweg, R.; Fisher, J.T.; Bayne, E.; Boutin, S. Wildlife camera trapping: A review and recommendations for linking surveys to ecological processes. *J. Appl. Ecol.* **2015**, *52*, 675–685. [CrossRef]
86. Amburgey, S.M.; Yackel Adams, A.A.; Gardner, B.; Hostetter, N.J.; Siers, S.R.; McClintock, B.T.; Converse, S.J. Evaluation of camera trap-based abundance estimators for unmarked populations. *Ecol. Appl.* **2021**, *31*, 7. [CrossRef]
87. Bischof, R.; Ali, H.; Kabir, M.; Hameed, S.; Nawaz, M.A. Being the underdog: An elusive small carnivore uses space with prey and time without enemies. *J. Zool.* **2014**, *293*, 40–48. [CrossRef]
88. Ferreras, P.; Díaz-Ruiz, F.; Monterroso, P. Improving mesocarnivore detectability with lures in camera-trapping studies. *Wildl. Res.* **2018**, *45*, 505–517. [CrossRef]
89. Mills, D.; Fattebert, J.; Hunter, L.; Slotow, R. Maximising camera trap data: Using attractants to improve detection of elusive species in multi-species surveys. *PLoS ONE* **2019**, *14*, e0216447. [CrossRef]
90. Fidino, M.; Barnas, G.R.; Lehrer, E.W.; Murray, M.H.; Magle, S.B. Effect of Lure on Detecting Mammals with Camera Traps. *Wildl. Soc. Bull.* **2020**, *44*, 543–552. [CrossRef]
91. Holinda, D.; Burgar, J.M.; Burton, A.C. Effects of scent lure on camera trap detections vary across mammalian predator and prey species. *PLoS ONE* **2020**, *15*, e0229055. [CrossRef]
92. Programa de Naciones Unidas para el Desarrollo. *Comunidad Indigena de Manquemapu, Chile*; Serie de Estudios de Casos de la Iniciativa Ecuatorial: New York, NY, USA, 2016.

93. Rebolledo, G.; Rau, J.R. Analisis de la Estructura y Organizacion de un Paisaje Forestal del Sur de Chile. *Gest. Ambient.* **2010**, *19*, 47–66.
94. Pincheira-Ulbrich, J.; Rau, J.; Peña-Cortés, F. Tamaño y forma de fragmentos de bosque y su relación con la riqueza de especies de árboles y arbustos Patch size and shape and their relationship with tree and shrub species richness. *Fyton* **2009**, *78*, 121–128.
95. Roemer, G.W.; Gompper, M.E.; Van Valkenburgh, B. The Ecological Role of the Mammalian Mesocarnivore. *BioScience* **2009**, *59*, 165–173. [CrossRef]
96. Allen, M.L.; Elbroch, L.M.; Wilmers, C.C.; Wittmer, H.U. The comparative effects of large carnivores on the acquisition of carrion by scavengers. *Am. Nat.* **2015**, *185*, 822–833. [CrossRef] [PubMed]
97. Elbroch, L.M.; Wittmer, H.U. Table scraps: Inter-trophic food provisioning by pumas. *Biol. Lett.* **2012**, *8*, 776–779. [CrossRef]
98. Elbroch, L.M.; O'Malley, C.; Peziol, C.M.; Quigley, H.B. Vertebrate diversity benefiting from carrion provided by pumas and other subordinate, apex felids. *Biol. Conserv.* **2017**, *215*, 123–131. [CrossRef]
99. Zúñiga, A.; Muñoz-Pedreros, A.; Fierro, A. Uso de habitat de cuatro carnivoros terrestres en el sur de Chile. *Gayana* **2009**, *73*, 200–210. [CrossRef]
100. Sollmann, R.; Mohamed, A.; Samejima, H.; Wilting, A. Risky business or simple solution—Relative abundance indices from camera-trapping. *Biol. Conserv.* **2013**, *159*, 405–412. [CrossRef]
101. Arim, M.; Marquet, P.A. Intraguild predation: A widespread interaction related to species biology. *Ecol. Lett.* **2004**, *7*, 557–564. [CrossRef]
102. Fedriani, J.M.; Fuller, T.K.; Sauvajot, R.M.; York, E.C. Competition and intraguild predation among three sympatric carnivores. *Oecologia* **2000**, *125*, 258–270. [CrossRef] [PubMed]
103. Polis, G.A.; Myers, C.A.; Holt, R.D. The Ecology and Evolution of Intraguild Predation: Potential Competitors That Eat Each Other. *Annu. Rev. Ecol. Syst.* **1989**, *20*, 297–330. [CrossRef]
104. Silva-Rodríguez, E.A.; Ortega-Solís, G.R.; Jiménez, J.E. Conservation and ecological implications of the use of space by chilla foxes and free-ranging dogs in a human-dominated landscape in southern Chile. *Austral. Ecol.* **2010**, *35*, 765–777. [CrossRef]

land

MDPI

Review

Understanding Species–Habitat Associations: A Case Study with the World's Bears

David Lance Garshelis

IUCN SSC Bear Specialist Group, Cohasset, MN 55721, USA; dgarshelis.bsg@gmail.com

Abstract: Habitat modeling is one of the most common practices in ecology today, aimed at understanding complex associations between species and an array of environmental, bioclimatic, and anthropogenic factors. This review of studies of seven species of terrestrial bears (Ursidae) occupying four continents examines how habitat models have been employed, and the functionality of their predictions for management and conservation. Bear occurrence data have been obtained at the population level, as presence points (e.g., sign surveys or camera trapping), or as locations of individual radio-collared animals. Radio-collars provide greater insights into how bears interact with their environment and variability within populations; they are more commonly used in North America and Europe than in South America and Asia. Salient problematic issues apparent from this review included: biases in presence data; predictor variables being poor surrogates of actual behavioral drivers; predictor variables applied at a biologically inappropriate scale; and over-use of data repositories that tend to detach investigators from the species. In several cases, multiple models in the same area yielded different predictions; new presence data occurred outside the range of predicted suitable habitat; and future range projections, based on where bears presently exist, underestimated their adaptability. Findings here are likely relevant to other taxa.

Keywords: habitat adaptability; habitat model; habitat selection; habitat suitability; use versus availability; Maxent; presence points; GPS radio-collars; anthropogenic variables; species distribution models

Citation: Garshelis, D.L. Understanding Species–Habitat Associations: A Case Study with the World's Bears. *Land* **2022**, *11*, 180. https://doi.org/10.3390/land11020180

Academic Editors: Juan F. Beltrán, Pedro Abellán and John Litvaitis

Received: 21 December 2021
Accepted: 18 January 2022
Published: 23 January 2022

1. Introduction

An intriguing product of evolution is that some species within a given taxonomic family have wide niches, whereas others have much narrower niches, and likewise some species are distributed much more broadly than others. A particularly striking example of this variation in niche and range is the bear family Ursidae [1]. This family consists of only eight species, one of the smallest among the Order Carnivora. Of those eight species, only one, the brown bear (*Ursus arctos*), ranges across multiple continents: North America, Europe, and Asia (Figure 1a). Four species are endemic to Asia (Figure 1b), one endemic to North America (American black bear, *U. americanus*), one endemic to the Andes mountains of South America (Andean bear, *Tremarctos ornatus*) (Figure 1c), and one endemic to the Arctic (polar bear, *U. maritimus*). Among the Asian endemics, the Asiatic black bear (*U. thibetanus*) is wide-ranging across a diverse array of habitats from the Russian Far East to Southeast Asia and westward to Iran. Conversely, sun bears (*Helarctos malayanus*) are restricted to Southeast Asia, and sloth bears (*Melursus ursinus*) are confined to the Indian subcontinent. Giant pandas (*Ailuropoda melanoleuca*) once ranged across a large portion of China but are now limited to a small mountainous area that was once the western edge of their historic range. This paper investigates habitats used by the seven species of terrestrial bears, which excludes the polar bear, or "sea bear" from its Latin name.

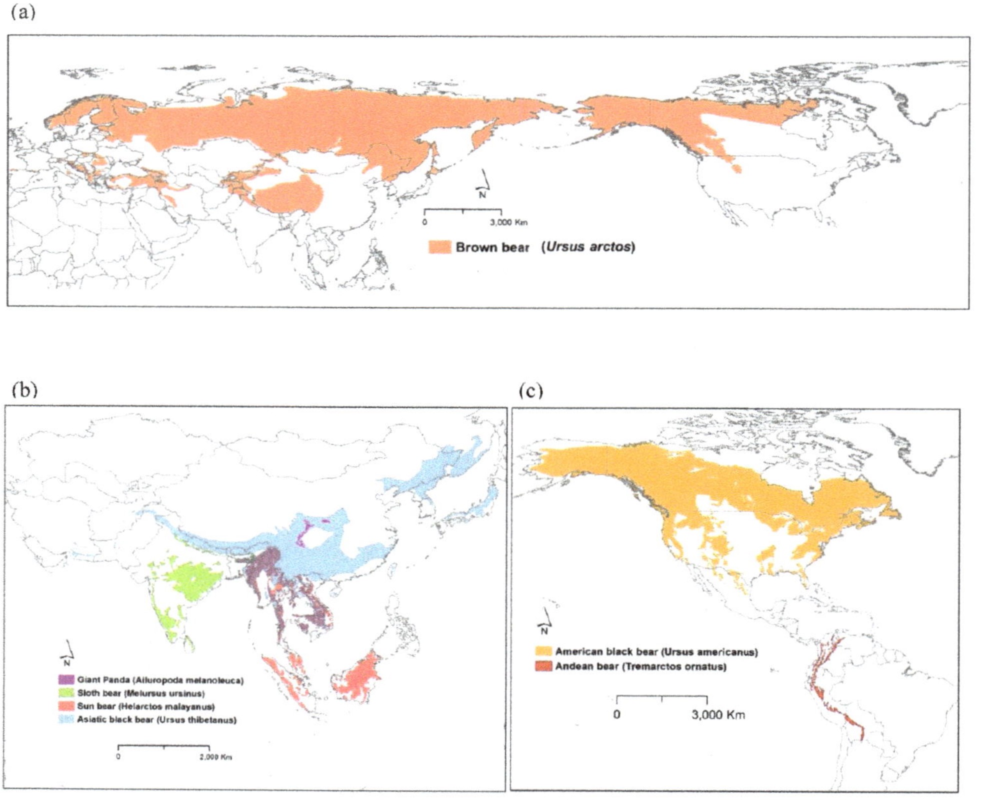

Figure 1. (**a**) The brown bear ranges across three continents (and once ranged into North Africa). (**b**) The Asiatic black bear, sun bear, sloth bear, and giant panda are all endemic to Asia, and their ranges at least partially overlap on a broad scale. (**c**) The American black bear is endemic to North America (3 countries), and the Andean bear is endemic to South America (5 countries). Depicted here is the maximum present range, combining categories for extant and possibly extant range from the IUCN (International Union for Conservation of Nature) Red List.

The aim here is not to define what constitutes suitable habitats for these seven species, but rather to demonstrate the extraordinary diversity, complexity, and adaptability in their use of habitats. Moreover, a co-equal aim is to show that our perceptions of how bears use and select habitats, and thus our perception of what is most suitable, is partially a construct of the methods employed. Habitat is multi-faceted and not static. In its simplest form, bear habitat is the dominant cover type (usually forest). However, to a bear, the value of a habitat lies in the types and quantity of foods produced, safety, and in some places, escape from the heat. Matthiopoulos et al. refer to these three habitat components as resources, risks, and conditions, respectively [2]. Most bears do not have significant non-human predators, aside from other bears, but tigers (*Panthera tigris*) pose a threat to sloth bears, sun bears, Asiatic black bears, and brown bears in parts of their range. Humans, though, are a potential or perceived significant threat in most places. Thus habitat may be judged from a bear's perspective by a combination of the proximity of humans and infrastructure, such as roads and dwellings, plus the availability of concealment. However, humans also may provide bears a source of concentrated food in the form of crops, livestock, beehives, and garbage. All of the bears use these to various degrees (although least for giant pandas, and to a very minor degree for sloth bears). In human-dominated landscapes, it is often difficult to extract the human factor, and typically a goal of habitat studies is to understand how the human factor interplays with habitat selection.

This paper has six main sections. It begins with a definition of habitat, then follows with a general overview of the diets and broad habitats used by the seven terrestrial bears. Next is a review of methods and their corresponding assumptions to examine finer-scale habitat associations. The Section 4 is a summary of papers examining habitat use, selection, or suitability for individual species. The Section 5 draws some conclusions about the adaptability of bears, limitations of the methodologies, and implications for using modeling results in management and conservation. The Section 7 is a list of recommendations for future studies of bear–habitat associations, aimed at reducing biases and misinterpretations, and leading to more reliable real-world applications. The general conclusions and recommendations stemming from this review should apply to other species as well.

2. Definitions

The original meaning of *habitat* is the type of place (meaning the collection of resources and conditions) where each particular species can live. The word is derived from Latin, basically meaning the area that a species inhabits (that which is habitable). Some argue that we should only use this original meaning because that is how it was used by several early ecologists [3,4], even though most modern authors do not use it this way [5]. That being the case, it makes no sense to refer to habitat suitability, as all of a species' habitat is by definition suitable. Likewise, it would make no sense to refer to habitat use; instead, one should refer only to use of specific components of habitat.

Through time, however, scientists began referring to those habitat components as habitat types, sometimes with specific names (e.g., forested habitat). Even the seminal paper by Johnson referred to habitat as a vegetative type [6]. Moreover, looking back at early literature, the purported original meaning is not entirely clear. In 1917, Nichols defined habitat as "any unit area in which the combined influence of climatic, edaphic, and biotic factors is essentially uniform throughout" [7], and in 1922, Yapp commented that most authors at that time were using the term to mean "factors or conditions of the environment," adding that "ecological factors are so numerous and so variable that their possible combinations are bewildering ... [so] nature draws no hard and fast lines" [8].

It is useful to recognize that there are now two distinct definitions of habitat, one being species-specific, and the more-common usage being a way to define environmental space [9]. Gaillard et al. referred to these as the functional and structural meanings, respectively [10]. Here, I use *habitat* in the structural sense: a spatial unit with definable characteristics in terms of resources or conditions (similar to [2,11]). Hence, rather than "bear habitat" (a very ambiguous term, as this paper will demonstrate), I will refer to different kinds of habitats, characterized by environmental features, such as vegetation, topography, human influences, etc. Accordingly, *habitat use* means the extent to which different definable components of the environment are used. *Habitat selection* refers to the process of choosing different habitat types or components. *Habitat preference* refers to the underlying proclivity for choosing a habitat, but can be measured only in controlled situations where animals are provided a range of choices. *Species–habitat association* (as in the title of this paper) is the connection between habitat features and the occurrence or demographics of the target species. *Habitat suitability* is the extent to which the habitat can sustain the species (i.e., population viability). Often it is equated to selection, assuming animals select habitats that are most suitable, but this is a poor assumption in that animals like bears often select habitats that offer abundant resources (food), but where human-caused mortality may reduce their suitability. Habitat suitability should technically be measured via the population growth rate, but this is rarely done, because it requires distinguishing which specific habitat features contributed to the observed growth rate.

3. Broad Differences among Species

3.1. Dietary Patterns

Six of the seven species of terrestrial bears are omnivores, feeding on plant and animal matter, but to different degrees. The giant panda, conversely, is a true herbivore, specializing in bamboo. It is not only the most divergent from the other bears in terms of diet, but also falls within a separate subfamily (Ailuropodinae). Its diet varies mainly in terms of the species (>40) and parts of bamboo consumed ([12,13]).

The Andean bear is the next most vegetarian bear, with over 300 plant species identified in its diet [14]. Like the giant panda, it is in a separate subfamily (Tremarctinae). Members of this nearly extinct subfamily of so-called short-faced bears once ranged widely across North and South America and included species that were far more carnivorous than the sole extant survivor from this line [15,16]. Andean bears also consume some small mammals, and occasionally prey upon or scavenge larger mammals, including some wild ungulates and domestic cattle, where available. However, their main foods are plant meristematic tissues (bromeliads, palms, bamboo, cacti) and fleshy fruits [17,18].

The sloth bear is the most insectivorous bear, with specialized morphological adaptations for myrmecophagy (ant and termite feeding). There are no known places where this species exists without termites or ants as a dietary mainstay, although the extent of ants and termites in the diet can vary significantly by location and season [19]. There are very few places where termites are not a major portion of the diet (i.e., where ants totally replace termites), meaning that the geographic and elevational range of this species within the Indian subcontinent is largely constrained by the distribution of termites [20]. Sloth bears also consume a variety of fruits, but not much vegetative matter [21,22].

The sun bear is second in terms of the extent of insect feeding. This species seems to prefer fruits, and across its range in Southeast Asia, it eats a very wide diversity, depending on what is seasonally available. In one small study site in Borneo, sun bears consumed 115 species of fruits in 30 families [23]. Most of the fruits in Southeast Asia are tree-borne, so these bears spend much time in trees feeding; if they waited for the fruits to fall, many would disappear from competitors or rapid decomposition in the tropical environment. However, in some seasons or years when fruits are lacking, the sun bear's diet is composed largely of insects [24]. They are especially noted for excavating stingless bees from their nests inside hollow trees [25]. While this species is sympatric with Asiatic black bears across most of mainland Southeast Asia, and the two share a diet of the same fruits, sun bears rely more on insects, and this slight difference in diet may allow the two species to coexist [26,27].

Asiatic black bears consume a variety of fruits and nuts, green vegetation, ants and other insects, and prey on some small mammals and small-sized ungulates [28]; their predation on ungulates may be high in some areas [29,30]. Although it is somewhat coincidental that there is another species with the common name "black bear", the American black bear is also quite similar to the Asian counterpart in terms of its general reliance on fruits and nuts, green vegetation, varied dependence on insects, and occasional predation on ungulates, primarily neonates [31]. Unlike the Asiatic black bear, the American black bear may rely heavily on fish in some areas, both coastal and inland [32,33].

The brown bear, having the largest distribution encompassing an enormous range of habitats, exhibits the greatest dietary plasticity, from nearly entirely herbivorous to near-obligate carnivory [34,35]. On the Tibetan steppe, it subsists on marmots (*Marmota himalayana*) and pikas (*Ochotona curzoniae*) because fruits are nearly absent [36,37]. In coastal North America and east Asia, seasonally abundant salmon (*Oncorhynchus* spp.) are an important food, but not the only food. In interior parts of North America, the brown bear (called grizzly bear) preys on burrowing rodents and neonatal ungulates, but also relies heavily on forbs, roots, and fruits. In Eurasia, they also prey on neonatal ungulates and consume vegetable matter and a variety of fruits, but consume more hard mast (nuts) than in North America [38]. Additionally, in Europe, supplemental feeding is common in some countries, providing a reliable, concentrated food source that can constitute an appreciable portion of their diet [39].

3.2. General Patterns of Habitat Use

Whereas bears are normally thought of as forest-dwelling animals, most of the species also occupy treeless environments. Brown bears live in the northern treeless tundra, where their diet is largely vertebrates and green vegetation [38]. One unique population of American black bears occupies treeless tundra of northeastern Canada, where they prey on caribou (*Rangifer tarandus caribou*) [40]; climate change is also enabling northward expansion of American black bears into more treeless environments, where fruits and insects are scarce, so they have adapted by feeding largely on meat [41]. Grizzly bears historically existed in grassland prairies of North America, but were extirpated there, and protected remnant populations are now expanding out of their core forested range into some of these vacant areas [42,43]. A population of Asiatic black bears in Japan lives in a primarily grassland habitat where ants are a major dietary component [44]. Both brown bears and Andean bears inhabit grasslands at high elevations above treeline in their respective ranges; in the Andes, these habitats are referred to either as puna or páramo. In the Himalayas, brown bears have been recorded at >5000 m, and Andean bears recorded above 4000 m. Asiatic black bears match or exceed Andean bears in maximum elevation, but only due to a higher treeline [14,28,35]. However, Asiatic black bears make seasonal forays into treeless alpine areas [45]. Sloth bears use lowland, grasslands with scattered trees in parks in Nepal and India; these areas have a high density of termite mounds, but may flood seasonally pushing the bears to higher elevations [46].

Several bear species also occur in arid regions. A small brown bear population exists in the Gobi desert of Mongolia, where they are dependent on scattered oases [47]. Andean bears occur in a low elevation arid region of Peru, where they depend on waterholes and during some periods of the year must subsist on the woody parts of pasallo trees (*Eriotheca ruizii*) [48]. Asiatic black bears in the western-most part of their geographic range in southern Iran live in a sparsely vegetated area where they are highly dependent on abandoned fruit orchards [49].

All of the terrestrial bear species occupy various kinds of forests. In northern latitudes, brown bears, American black bears, and Asiatic black bears inhabit boreal forests. There they consume fruits when seasonally available, but often just a few species of berries due to low diversity. Otherwise, they consume green vegetation and a large proportion of ants [38,45,50,51]. Brown bears are more predatory on vertebrates than either of the black bear species, although vertebrate consumption varies with ungulate abundance, including domestic species [52].

Temperate forests tend to have a greater diversity and abundance of fruits, especially nuts, than boreal forests, providing each bear species with a richer diet and thus a better buffer from year to year variations in productivity of any single fruiting species [53]. In North America, temperate forests in the eastern half of the continent tend to have richer supplies of fruits and more nuts than in the west, so American black bears tend to be larger and more prolific in the east [54].

Asiatic black bears, sun bears, sloth bears, and Andean bears all occupy tropical and subtropical forests. Andean bears regularly use montane humid forests, elfin, and cloud forests [14]. Asiatic black bears and sun bears co-occur in mixed deciduous and semi-evergreen forests across Southeast Asia, although there is evidence that in montane forests, with rich supplies of oaks, Asiatic black bears exclude sun bears [26]. Sun bears use higher elevations where Asiatic black bears are absent, reaching 2000 m on Borneo [55]. A particularly interesting dichotomy occurs with sun bears in that about half their range falls within seasonal tropical forests north of the Isthmus of Kra on Peninsular Malaysia (10°30′ N), and half in the aseasonal dipterocarp rain forests of the Sundaic region to the south, including the islands of Sumatra and Borneo [56]. The Sundaic region experiences highly synchronized masting events followed by inter-mast periods of 2–9 years when fruits are much less abundant, greatly influencing sun bear diet and health [57]. Sloth bears are the most lowland-dwelling tropical species, generally occurring below 1000 m in Nepal and India (although reaching 2000 m in the Western Ghats) and below 300 m in Sri Lanka [19,58,59]. During the 1990s they were extirpated from Bangladesh, due to the conversion of lowland forests to agriculture, whereas Asiatic black bears and some sun bears continue to persist there [60].

4. Methods of Studying Fine-Grained Habitat Use, Selection, and Suitability

Quantification of habitat use, selection, and suitability occur as a stepwise conceptual process by which ecologists try to understand species–habitat associations. Ecologists may aim to (1) understand why animals (or plants) occur where they do, (2) create maps of their distribution or potential distribution, and (3) make predictions about how the distribution may change in the future [2]. Investigators generally begin by examining environmental and anthropogenic variables (resources, conditions, and risks) around sites that were known to be used, then compare that to random (available) sites in the landscape to gauge selection; scaled selection indices may be used to assess relative habitat suitabilities, which assumes that the animals are making choices that, on whole, maximize their fitness [9].

In a highly influential paper, Johnson proposed four hierarchical orders of habitat selection, where at each level, the animal chooses from within a different window of available resources: the first order determines where the species exists (presumably all suitable habitat); the second order determines the placement of home ranges; the third order pertains to site use within home ranges; and the fourth order regards selection of specific foods or other resources at each site [6]. Often researchers are unclear about which level they are investigating, or employ use at one spatial scale and availability on a different scale, and fail to recognize that animals make choices at multiple scales [61,62]. Furthermore, strict adherence to use–availability comparisons leads to the mistaken assumption that increased availability of preferred habitats should always motivate increased use; i.e., that there is never enough, or even too much of a good thing [9]. This assumption clearly does not fit species like bears, which use many types of foods in different habitats, and where conditions change seasonally, and even week to week [63]. An ironic consequence of measuring selection via a comparison of use to availability is that selection may seem low for frequently used but very common habitats, and high for occasionally used habitats that are rare, just opposite of how it would be perceived from use alone.

Measuring use is the linchpin for all else. Bears researchers have generally used one of two broad methods: (1) collection of presence points, or (2) locations of animals with radio-collars. These two kinds of data are different in many respects, leading to potential differences in interpretation, as described below. Radio-collars are the primary study method in North America and in some parts of Europe, whereas bear studies in Asia and South America have relied primarily on presence points (Figure 2). A new technique is also emerging whereby the estimated local density of the species of interest is compared to the availability of habitats.

4.1. Presence Points

Presence points can be obtained by a variety of means, including camera traps, sign surveys, sightings, surveys of local people who encounter bears or their sign, or records of dead bears. A fundamental assumption is that any presence point is truly a bear, and if more than one bear species occurs in the area, it is actually the species of interest. Local people's accounts of bear presence should be verified [64]. Even photos can be misidentified as to the species, since some bears look alike [65,66]. Some kinds of bear signs can be confused with that of other species, so to avoid misidentification it may be preferable to only rely on types of signs that are definitive (e.g., claw marks on trees; [67]). However, this can lead to a bias if the more reliably identified sign occurs in certain types of habitats (Figure 3).

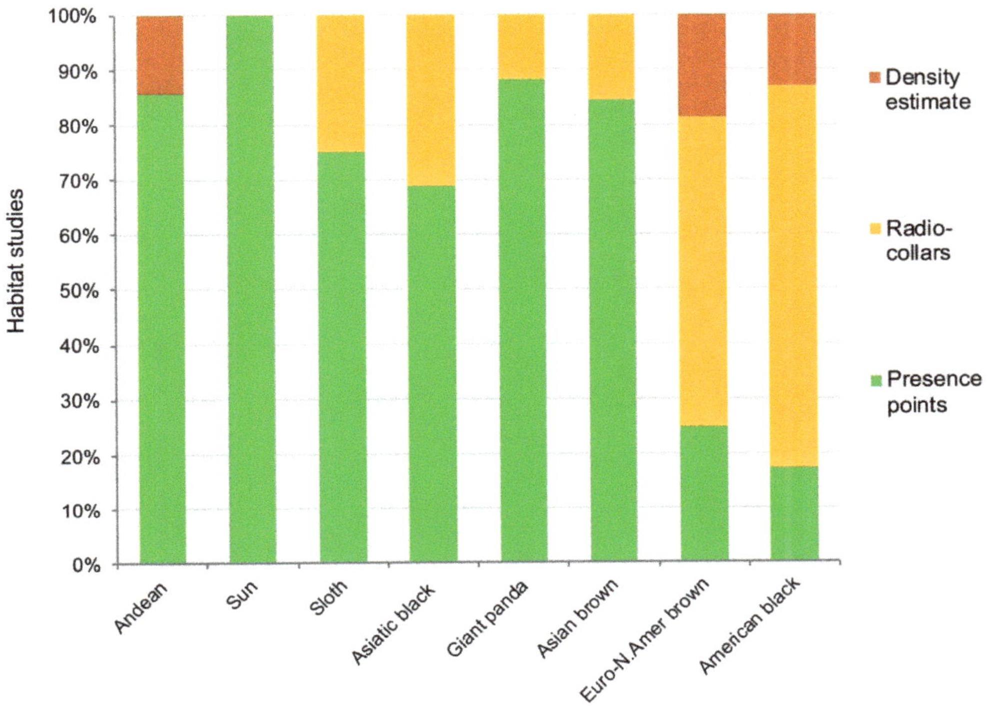

Figure 2. Relative use of presence points, locations of radio-collared bears, and site-specific density estimates in studies of habitat use, selection, and suitability for the seven species of terrestrial bears. Brown bears are split because of continental differences in the use of these methods. Use of density for Andean bears may be exaggerated by small sample size (n = 7 total studies). Data are from published papers, mainly since 2000, cited in text (n = 141), which may not be a perfect representation of all studies.

A complication of presence data is that a lack of detection may signify the absence of use or simply non-detection, and detection can vary by habitat and over time (as sign decays). Lacking true absence data, models are generally fit to the presence-only data to extract covariates that best explain the pattern of use. A fundamental assumption of such models is that the underlying data reflect the true relative use with respect to habitat features. Hence, sampling is a critical component; sampled points that differ in some way from the full inference space would create a potential bias and misinterpretation of habitat selection [68]. It is not the intent here to review the various ecological niche models or species distribution models, or even to comment on general biases and suggested remedies, as there is a vast amount of literature on this [2,69–73]. However, some examples from the bear literature related to these potential issues may be helpful in emphasizing the integrity of the presence points.

Nazeri et al. obtained sun bear presence points (primarily camera trap records) from the Malaysian government, all collected within primary forests and primarily in protected areas in the northern half of Peninsular Malaysia [74]. There is only one bear species in Malaysia, so there is no possibility for species confusion. They used Maxent to extract covariates that best explained where these records occurred, and thereby created a map of suitable habitat. They compared their map to a recent map created via expert opinion. In essence, one map was created by people with little expertise in the species' ecology but having a set of verified presence points and a model, versus another made by a Malaysian field biologist with extensive knowledge of the species ecology but with uncertain presence records and not applied in a quantifiable way. Comparing the two maps showed significant disparities: the Maxent map predicted habitat to be unsuitable in a number of places where the species expert thought they occurred. Unfortunately, there are currently no data indicating which is more correct. However, it is notable that the Maxent map was based on presence points in the best habitats (primary forests in protected areas), so it is no wonder that the model would predict bears not to occur in other situations.

(a)

(b)

Figure 3. Many habitat studies of bears are based on presence points identified from sign. (**a**) A sampling problem arises when uncertain sign is either incorrectly counted as a bear, or always discounted. (**b**) Relying only on definitive sign, such as claw marks on trees prevents identification errors, but may create a bias towards habitats where bears create more of that type of sign (e.g., habitats with many fruit trees that bears climb). Photos: D. Garshelis; R. Steinmetz.

A somewhat tongue-in-cheek niche modeling of Sasquatch sightings makes the point eloquently stated by Baldwin [68]: "just because a model can be built does not mean that it is informative." Sasquatch is one of a number of crypto-zoological large, bipedal mammals that have purportedly been observed but for which no specimens exist. Sasquatch (or Bigfoot) sightings were collected from across the western U.S. states of Washington, Oregon, and California, and through modeling, a predicted range map was created [75]. Using presence points of American black bears from the same three states produced a very similar predictive map, suggesting either that these two "species" occur in the same habitats, or that many purported Sasquatch sightings were, in fact, bears.

Possibly even more bizarre than the niche modeling of Sasquatch is an example with Asiatic black bears [76]. Presence points were obtained from a database encompassing a wide portion of the known distribution from northeastern China, Japan, and South Korea, to Vietnam and Thailand, and westward to India and Pakistan (Figure 1b). Problems, though, included the fact that the sample size outside Korea was small (22 points), and several points were well beyond the known distribution of the species. That is, the coordinates were certainly incorrect (and in one case, possibly a sloth bear instead). This rather shaky dataset was run in a Maxent model, with 19 bioclimatic variables, 16 landcover variables, plus elevation and distance to the road. The model generated predictions of habitat suitability. Remarkably these areas of purported suitable habitat included a large region where this species has never occurred (i.e., there is no historical or even fossil evidence), including a swath from Mongolia through Siberia and Kazakhstan to Tajikistan (occupied by brown bears), central India (occupied by sloth bears), Malaysia and Indonesia (occupied by sun bears), and the Philippines (never occupied by any bears). Once again, this illustrates not just the importance of ensuring the accuracy and geographical balance of the presence data, but also recognizing that well-fitting model results do not mean they accurately depict reality.

Large public databases of presence points may foster novel insights about habitat suitability and distribution of species, while at the same time providing a recipe for people with little direct knowledge of the species to publish misleading findings. One recent high-profile paper based on publicly available data and modeling proposed priority "rewilding" sites for bears in places where they never existed historically, including American black bears in Canadian prairies, Asiatic black bears in Malaysia and Tajikistan, and sloth bears in central Bhutan [77]. Knowledge of the species or consultation with experts would have prevented such obvious mistakes.

4.2. Bears with Radio-Collars

Radio-collars provide point locations of known animals, which can be matched to certain habitat variables. In the 1980–1990s, bear biologists obtained locations from VHF radio-collars, generally during the day, and could classify habitat variables in a very broad sense from visual examination. Since then, GPS collars have become the standard, and a large number of habitat variables can be extracted from GIS layers. However, while this new technology avoids many of the sampling pitfalls often encountered with collecting presence points and enables far greater insights, it also highlights the problem of measuring availability. Viewing habitat from the eye of a radio-collared bear, how far away does it consider its available choices?

Judgments about availability can affect our perception of selection as much as the use side of the equation [6,11,78]. Whereas a bear's home range is technically all "known" to it, and thus available, at each moment in time, it has imperfect knowledge of the exact conditions (e.g., fruit availability) everywhere in its home range, and moreover cannot instantly jump to any other location, but rather must expend time and energy getting there. So, on a short time scale, it makes decisions about habitat use based on what is in proximity. That is, its choices are spatially constrained [79]. Bears are highly mobile, and generally could get to any location within their home range in a day, but that said, even if two different preferred habitats occurred on opposite sides of a bear's home range, it would not be efficient to continually move between them. This problem regarding availability also pertains to use. With VHF collars, locations were typically obtained at time intervals such that an animal could be virtually anywhere in its home range, but sample sizes were thus relatively small. With GPS collars and points obtained hourly (or even more frequently), each location is partly dependent on the previous one, not just in terms of how far the bear could move, but also by its assessment of its immediate surroundings (i.e., patches of food). This so-called auto-correlation problem is actually an opportunity, as the wealth of points are not only an unbiased representation of the animal's use of habitat, but also yield information about fine-scale choices in its movements [80].

One recent solution specifically to deal with and appropriately exploit such fine-scale data is the use of step selection functions. These are now commonly used for data collected from GPS radio-collars [81]. Instead of viewing the points as a sample of independent locations, the method examines the choices the animal made when it moved from each point to the next. These movement steps are compared to other random steps that it could have made, but did not, which may have landed the animal potentially in a different habitat or crossing an unproductive or risky habitat feature (like a road). However, mobile animals like bears may be making decisions at multiple scales: both where it wants to be now, and where it wants to be generally heading, all the while gathering more information and modifying decisions. It uses a combination of memory of specific locations, experiences, and current information to assess what is likely to be available where.

5. Fine-Scale Habitat Associations within Species

Given that fine-scale habitat use tends to be very specific to the situation at each locale, and because most bears are generalist opportunists, it is difficult to extract general trends by reviewing habitat studies of each species. Moreover, researchers have used somewhat different classification systems of habitat variables. Hence, whereas this section shows the wide variation and also draws out some generalities about the habitat suitability of each species, another prime intent is to highlight inconsistent or suspect results. To gain useful and practical information from habitat suitability studies and thereby inform better management and conservation, we need to be wary of potential methodological and interpretational flaws.

The papers examined here are a nearly complete sample for some species, but represent a selected sample for those species where the number of publications would be too large to completely review (Europe and North America). The intent here is not to cite every applicable paper, but rather to provide a multitude of examples of key aspects of species–habitat associations and the methods used to assess them. The cited papers are mainly limited to those published since 2000 so as to gain an appreciation for kinds of studies occurring recently, although with a few important older papers included. This review excludes a fairly large amount of literature on habitat selection of den sites (brown bears and American black bears hibernate in nearly all of their range; Asiatic black bears hibernate in northern and high elevation parts of their range). The bear species are listed roughly from south (tropics) to north and east to west, corresponding to the increased use of more rigorous methods (Figure 2). This increasing northward gradient in scientific rigor is also evident in the total number of publications (not just habitat studies) on each bear species [82].

5.1. Andean Bears

Although this review is mainly limited to papers published in the past 20 years, one cannot discuss the habitat suitability of Andean bears without mentioning the classic field study by Bernard Peyton in 1982 in Machu Pichu, Peru [83,84]. Comparing vegetative and topographic features at sites with bear signs versus random sites without signs, Peyton observed bear sign to be concentrated in certain elevation zones and associated with the availability of bear foods and vegetative hiding cover, and lack of grazing livestock (which trample cover); the only positive association with humans were cornfields in proximity to cover. This study became a model for other sign-based studies of this species.

A field study in Ecuador recorded signs along narrow transects (where detectability was high) and identified montane cloud forests and herbaceous paramo (high altitude grassland) as being the most suitable habitats, the former providing seasonal fruits and the latter year-round terrestrial bromeliads, with surmised seasonal elevational movements by bears between these [85,86]. Oddly, though, a DNA hair-snaring study at the same site captured more individual bears in what was judged as poorer habitats [87].

A sign study in Bolivia also indicated selection of high elevation montane forest, as well as mid-elevation elfin forest, but high elevation paramo was used by bears comparable to its availability [88]. The metric of use in this study was obvious bear trails, so it seems possible that bears used trails more in elfin forests, with high tree density, and were thus more easily detected by investigators than in open grassland where they could walk anywhere and not create a trail.

Velez-Liendo et al. collected bear sign and sighting data (>500 points) across their range in Bolivia, and created two models to predict distribution from suitable habitat: (1) a resource-based model incorporating sources of food, shelter, and water, and (2) a more standard model with assorted variables for topography, forest type associated with major ecoregions, and human features [89]. The two models only agreed on 16% of the predicted range area and 20% of the best-predicted habitat, even though both were based on habitat considerations specific to this species, and both fit the occurrence data well. Differences were in part due to gradients in shelter and key foods (first model) versus abrupt boundaries between distinct forest types (second model). Later, they used the resource-based model to identify 13 key habitat patches, ranging in size from 400 to 5000 km^2 [90].

Meza Mori et al. used interview surveys to obtain presence points of Andean bears in northern Peru, an area with an altitudinal range of 120–4900 m and a wide range of temperatures and rainfall [91]. They used Maxent and 23 environmental variables (mainly bioclimatic), some of which showed complex relationships with presence data, to predict habitat suitability. Model results were driven by three variables (mean temperature, precipitation in driest month, and forest cover). Although the area of occupied range predicted by this model was similar to that of an IUCN (International Union for Conservation of Nature) map, the spatial overlap between the two maps was poor. The IUCN map was based on the model from Bolivia, mentioned above, and also only mapped patches of a certain size and connectivity [89,90]. Another potential source of discrepancy may be a spatial bias in the interview sampling. On the other hand, only about a third of the presence points in this study fell within the IUCN map, indicating that these new data records should help to better define the range of this species.

Another study in Peru, this one in the equatorial dry forest (a narrow low elevational band along the western Andes) used camera traps to individually identify bears by their facial markings, and thereby estimate density; spatial variation in density was compared to four remotely sensed habitat features [92]. Models for two study sites varied in terms of how many of these variables were included and showed differing effects of elevation and forest cover, apparently related to differences in food availability. Using an estimated density of 4 bears/100 km^2 as representing suitable habitat, they found that only 6% of the predicted range was within the IUCN range map; however, the map predicted by the model was mainly informed by a negative association with roads, rather than a positive association with natural habitat features, whereas the IUCN map was based on other features (but fewer presence points).

5.2. Sun Bears

A particularly controversial subject with regard to sun bears is whether they are able to survive in degraded or successional forests, following logging. This species has drawn particular attention in this regard because logging is greatly altering its habitat across much of its range in Southeast Asia. Furthermore, many forests have been converted to agriculture, especially oil palm (*Elaeis guineensis*) in Malaysia and Indonesia, and whereas it is clear that sun bears cannot survive in a monoculture of oil palm, there is mixed information about their use of oil palm plantations.

Scotson et al. accumulated by-catch camera trap photos from across the southern half of the sun bear range and found that a single variable, canopy cover derived from Landsat photos, predicted the rate of photo-capturing sun bears [93]. They did not attempt to include other variables in their model, but considered tree cover a surrogate not only for all other habitat variables, including the availability of food, but also (unexplainably) poaching. An alternate explanation for high rates of photo-capture under a dense canopy could be that bears tended to rest in shady areas to avoid heat stress [94]. Whereas other studies have confirmed that sun bears select for areas with high tree cover, a large number of studies have now demonstrated that sun bears are also widespread in degraded and successional forests in both the mainland and Borneo and Sumatra [67,95–99]. Hwang et al. also pointed to many other camera-trap studies that were not specifically directed at sun bears that have consistently shown high occupancy in forests 6–20 years after logging, and growing evidence that poaching is driving sun bear density more than habitat degradation [67,100].

In an extreme case of habitat alteration, Fredrikkson used sign surveys to monitor sun bear use of a large patch of forest that was severely burned, killing most of the understory (and probably most insects) and 80% of fruit trees that bears had used for food [25]. Sun bears were forced to move to an adjacent unburned forest leaving the burned forest virtually vacant. However, within 7 years after the fire, bears began returning to the burned site; in particular, their sign indicated that use of termites steadily increased in the burned area.

This species' reliance on insects is often underappreciated. For example, Hwang et al. found that sun bears used a previously logged forest more than an adjacent primary forest during a poor fruiting year because the logged forest, with more deadwood, provided a greater abundance of termites and bees [67]. Likewise, at a site in Thailand where sun bears and Asiatic black bears occur sympatrically, Steinmetz et al. found that sun bears may be excluded from a habitat that provides high fruit abundance but can coexist with the larger black bears by concentrating on insects [26].

Only one study has so far attempted to determine habitat suitability for sun bears. Nazeri et al., in peninsular Malaysia, used presence records obtained from other sources and a Maxent model with a large number of variables to try to define suitable habitats [74]. This study, mentioned earlier, relied exclusively on presence points from primary forests. A later set of presence points collected by Abidin et al., namely sites of human–sun bear conflicts, which occurred most often on edges of forest-agricultural areas, showed that large areas deemed unsuitable by Nazeri et al. were occupied by sun bears [101].

Abidin et al.'s study is one of several showing that whereas sun bears require forest, they readily exploit oil palm plantations for food [101]. Indeed, sun bears that utilized oil palm fruits are significantly heavier than bears living in forests with no access to oil palm [56]. Normua et al. first showed that radio-collared bears moved back and forth between a forest and an adjacent oil palm plantation [102]. Cheah demonstrated this more conclusively with GPS-collared bears that used oil palm as their main food source, but spent daytime hours secluded in the adjacent forest [103]. Likewise, at other sites, Tee et al. and Guharajan et al. showed with camera traps and sign that sun bears heavily used a forest edge near an oil palm plantation (Figure 4) [98,104]. Large expanses of agriculture are certainly a barrier to sun bears, as evidenced by genetic segregation, but as Kunde et al. showed in Cambodia, a few individuals still manage to cross [105].

Figure 4. Sun bear habitat has been severely disrupted by oil palm plantations, but a big question is how much they use the fruits in the plantations. Shown here are oil palm fruits that were cut from trees and piled along a road at the boundary of the plantation and forest (before being picked up), easily accessible to sun bears. Bears thus do not have to climb and damage trees to obtain the fruit, and because they visit plantations at night, they are often not detected by workers [106]. Photo: D. Garshelis.

5.3. Sloth Bears

Few studies have been focused specifically on habitat use by sloth bears. In the early 1990s, Joshi et al. monitored radio-collared sloth bears in Chitwan National Park, Nepal and observed seasonal movements mainly by adult male bears between two major habitat types: alluvial tall grasslands in the dry season, and upland sal (*Shorea robusta*) forest in the wet season [46]. Most females and subadult males did not make this shift, and it is not understood why, although social factors were posited. The alluvium became saturated in the wet season, making the excavation of termites more difficult, but those bears that stayed there nevertheless maintained a largely termite-based diet. No other studies of this species reported a seasonal habitat shift.

Twenty years later Ghimire and Thapa conducted a sign survey in this same park, and compared use to availability of four habitat classes, and concluded that bears did not select for either the grasslands or the sal forest, as observed in the study of collared bears, but rather "mixed-forest" [107]. This study, though, highlighted a common problem in relying only on the ratio of use:availability—54% of the sign was found in sal forest, but because this forest was so common (and some was at higher elevations that the bears rarely used), it appeared that this most-used habitat was "selected against", which of course is illogical [9].

Ratnayeke et al. studied the habitat use of radio-collared sloth bears in Sri Lanka, and compared use to availability of five vegetation types [108]. Tall forests, which were least disturbed by humans and had moister ground conditions with less dense understory, were most selected, although some bears also selected secondary forests. Bears in this study had exceedingly small home ranges, indicating that the available habitat provided a rich supply of food (mainly termites, with some seasonal fruits). It is interesting that whereas the earlier Nepal study found the highest use and density [58] in the grasslands, the shorter grass habitat in Sri Lanka was selected least. A Sri Lanka-wide sloth bear survey based on local interviews also indicated that bears were found mainly in forests with tall trees at elevations below 300 m [59]. However, these authors noted that sloth bears also occurred in forests with sparse tree canopy, but only if human disturbance was low, indicating that concealment is likely important.

In contrast to the situation in Sri Lanka, Akhtar et al. found abundant sloth bear signs in highly degraded areas in central India, but mainly in sal forest [109]. A critical resource in this area, identified in a separate study by the same authors but not in their general sign survey, is boulder fields where the bears can find refuge to rest during the heat of the day and when people in the surrounding area are active (Figure 5) [110]. Two studies in central India, both using DNA in scats to identify individual bears, found that some individuals moved substantial distances through narrow forest corridors between protected areas, while others traveled through human-dominated agricultural areas to get from one protected area to another [111,112], raising the question of what prompted bears to go on such long excursions, through unfavorable habitats, and how they managed to know where they were headed.

Three studies of habitat selection of sloth bears have occurred in the Western Ghats of southwestern India. Puri et al. conducted sign surveys, and using occupancy modeling found that forest cover, terrain ruggedness, dryness (less rainfall), and low human use were positively associated with sloth bear presence [113]. Following on these results, Srivastha et al. used occupancy modeling based on camera trapping (with temporal replicates) and sign (with spatial replicates), with forest cover, terrain ruggedness, and NDVI (Normalized Difference Vegetation Index, which is higher in wetter environments) as ecological variables [114]. However, they found that the two different sources of presence data yielded different results, with accordingly different habitat-based distribution maps. They suggested that micro-habitat features that they could not map, such as the presence of fruiting trees and termites (i.e., main food sources), would be better predictors of sloth bear habitat use. Ramesh et al. used camera trapping in one reserve in the Western Ghats, and measured habitat features at each camera trap station, including the fine scale food-producing factors recommended by Srivastha (although Ramesh et al.'s study came before). However, they found that all of their covariates were weak predictors of sloth bear occupancy [115].

Figure 5. Sloth bears in parts of India regularly use crevices in boulder fields to sleep during the day and come out to feed at dusk. This key habitat offers protection from people and from heat, yet sampling bear sign within this habitat would greatly underestimate its importance. Photo: D. Garshelis.

5.4. Asiatic Black Bears

Studies of habitat use have been conducted in at least 8 of 18 countries occupied by Asiatic black bears. Findings varied tremendously not just across but even within these range countries. Japan has a high population of Asiatic black bears on Honshu Island, which has attracted a number of studies aimed at understanding habitat needs. Three studies were conducted in Nagano Prefecture in central Honshu [116–118]. Early studies (1996–2001) with VHF radio-collars accumulated only ~500 total radio-locations over multiple years; one study had to eliminate 1000 of 1500 radio-locations due to uncertainty about the accuracy with respect to habitat patches [116], raising the possibility that the discarded locations may have been in habitats less accessible to the investigators (i.e., potentially creating a bias). A decade later, GPS collars provided far more locations [118,119]. No consensus pattern of habitat use or selection can be derived from the three studies, possibly because they were conducted in different parts of the prefecture. However, the GPS collar-based studies demonstrated that in summer bears selected early successional forests (after cutting, or natural avalanches) or along forest edges or near roads or rivers because berry production was high due to light penetration [118,119]. Another study in the Japanese Alps showed bears moving to alpine areas in summer and feeding on sprouting herbaceous plants [45]. An interesting takeaway from this telemetry study is that seasonal bear home ranges shifted elevationally, and therefore showed dramatic shifts in vegetational composition; the amount of selection at the scale of the telemetry location thus depended on whether the annual or seasonal home range was considered available habitat.

A study in Taiwan using both VHF and satellite-based radio-collars highlighted potential biases caused by VHF collars being more accessible to investigators in certain locations, associated with certain habitats [120]. These bears also showed significant elevational seasonal shifts, although not the same as reported in the Japanese Alps. Overall, bears selected areas below 2000 m, but it is important to note that this study was in a national park, so these lower elevation areas were not impacted by people. Bear use of habitats judged at the level of individual radio-locations tended to match the seasonally shifting availability.

A recent study in China accumulated and modeled presence data collected over 10 years, from various camera trapping studies, combined with absence data reported in some nature reserves and other surveys [121]. Employing 24 candidate variables, they generated a predicted distribution from maps at two different scales, which were based on different sets of predictor variables. The range area in the final map was 78% less than an IUCN map derived from expert opinion, utilizing some of the same data points. To some extent, this difference arose because the habitat-based map consisted of small, sometimes disjunct pixels, which a person creating a range map would necessarily join together into larger polygons. However, another reason for the discrepancy is that the model was informed by records of no confirmed presence (presumed absence) in broad regions classified as extant in the IUCN map.

Nepal and Pakistan have similar habitats for Asiatic black bears, and three studies based on the incidence of sign highlighted selections of pine forests above 1600 m (below that was dominated by agriculture) [122–124]. Awan et al. found that different types of sign were associated with different habitats [124], raising the possibility that if detectability or decay rates of different sign types vary (e.g., feces and footprints vs. claw marks on trees), then that could confound the perception of habitat use (Figure 3). A wider landscape study, involving eight countries in the Hindu Kush mountains, employed presence points from multiple sources and explored the effects of 19 bioclimatic variables and 5 topographic or land use variables in a Maxent model; this study mapped the region of suitable habitat but did not provide specific information about what was considered most suitable, except in terms of elevation (1500–3000 m) [125]. Two Maxent models in a national park in Nepal, using the exact same presence data but employing different variables obtained similar range maps, but one used only variables related to topography and land cover [122], whereas the other used primarily bioclimatic variables and concluded that mean temperature was the strongest predictor of Asiatic black bear presence [126]. Given the wide latitudinal range of Asiatic black bears, from the tropics to the subarctic, it would seem doubtful that temperature per se significantly impacts where they occur, although the temperature was likely coarsely related to habitat features on the ground, especially sources of food.

In tropical Thailand (14° N), Ngoprasert et al. found that fruit abundance was 2.3x higher in the local vicinity of Asiatic black bear sign than where signs wt absent [127]. Other measured habitat variables, including elevation, ground cover, human disturbance, and distance to a park boundary had no influence on occurrence. By contrast, in neighboring Laos, Scotson et al. used locally collected data near bear sign plus remotely-sensed predictors, and did not observe a significant relationship with fruit abundance [128]. Poaching influenced the incidence of bear signs probably more than habitat choices by bears, resulting in more signs in areas less accessible to people. Inexplicably, their model based on bear signs of all ages (some >2 years old) predicted higher habitat suitability over a much wider area than a model based only on recent signs (<1 year old).

At the other end of the habitat spectrum, in the arid region at the westernmost extremity of the range in southern Iran, Almasieh et al. obtained 200 existing presence points from other investigations and then added 70 new points [129]. Points were rarified to exclude those <1 km from each other. These points fell into 31 discrete clusters, which they called habitat cores. They used topographic, elevational, landcover, and climatic variables in Maxent to predict habitat suitability, yielding 45 separate patches and over 50 potential corridors. They reported that canyon bottoms in riparian areas, which provided some food and shade, and steep slopes as security from people, were selected by bears (Figure 6). Farashi and Erfani obtained presence records, all from indirect sources, and created another modeled prediction for this same area, using 32 environmental variables, and a suite of 10 different models, of which Maxent performed the best [130]. Even though they used the same model and apparently many of the same presence points as Almasieh et al., their results were very different, with the predicted distribution being a large continuous region, rather than fragmented small populations. The reason is that this model was most influenced by annual precipitation, which would be more consistent than habitat features across the region. Morovati et al. assembled another group of presence points through field studies; after rarification to reduce autocorrelation, their sample of 95 points was reduced to 53 [131]. Instead of just using presence, they also created pseudo-absence points, and using 15 variables, employed an ensemble of models to assess habitat suitability. This procedure produced a continuous distribution that excluded many of their own data points and which was spatially quite unlike either of the other two model predictions from this region. In this model, presence was positively influenced by increasing altitudes, from 300 to 2500 m, just opposite of Almasieh et al.'s conclusion that these bears seek valley bottoms.

(a) (b)

Figure 6. Asiatic black bears in an arid environment at the western edge of their geographic range in southeastern Iran rely on date palm (*Phoenix dactylifera*) and other abandoned orchards as primary food sources. (**a**) They use valleys with water sources as places to obtain foods and as travel corridors. (**b**) Footprints of a bear in a valley bottom. Photos: A. Ghoddousi; T. Ghadirian.

5.5. Giant Pandas

Giant pandas are a conservation success story. Once an endangered species believed to be on the brink of extinction, they have been provided extraordinary protection through a system of nature reserves in combination with conservation measures outside the reserves, which together have served to provide an increasing area of suitable habitat and negligible human-caused mortality, enabling their population to increase [132]. Their total current range is restricted to a 25,800 km^2 area within six mountain ranges in three Chinese provinces. Within this small area, a large number of habitat assessments have been conducted, some within specific nature reserves, some in a specific mountain range, and a few using data from rangewide population surveys (National Surveys) conducted at ~10 year intervals, covering the entire extent of possible panda range. The intense effort at habitat assessment, by far exceeding that of any other bear species, is in accordance with a strategy of understanding and improving the habitat conditions both inside and outside reserves to increase carrying capacity, and to link available habitat so animals can move more freely.

In 2014, Hull et al. synthesized the findings of 23 giant panda habitat assessments conducted from the mid-1980s to 2009 [133]. Of these, 22 relied on signs (mainly feces) to define presence points, and 1 used radio-collars; 46 different habitat components were investigated. On the whole, these studies indicated that giant pandas selected for a high density of bamboo, gentle to moderate slopes with high solar radiation at mid-elevation and little human disturbance. In synthesizing these studies, however, Hull et al. observed that many of the studies may have introduced a bias in their sampling locations where panda sign was absent, namely being close to trails with easier access; moreover, studies did not account for the variation in sign detection due to habitat (i.e., harder to see in very thick vegetation).

A key study by Zhang et al. in 2011 asked the question: "What happens if [due to issues with sampling or scale] the resulting habitat models are wrong?" [134]. This study, the first of its kind, utilized results of the Third National Panda Survey, conducted during 1999–2003. Field observers hiked transects covering the complete range of the species, located signs (feces, foraging sites, and dens), and collected detailed ecological data at each site plus control sites with no sign. They then constructed and compared models to define those variables that best distinguished a presence point from a control point. They found that the presence of bamboo and forest age (old growth versus secondary growth) were the strongest predictors of panda presence. The novel finding that old-growth was as important a predictor as bamboo was not readily explained. The authors posited that shade may have produced more nutritious bamboo, or tree cavities were needed for den sites. They argued for a revision in previous maps of suitable habitat, giving higher priority to old-growth forests. Ironically, another paper published in 2011, which was unique in examining habitat selection from the perspective of radio-collared giant pandas, found that they only chose areas with relatively low solar radiation (compared to what was available) during the peak of summer, but otherwise sought out sites of high sunlight (i.e., opposite shady old-growth forest), presumably related to ambient temperature [135].

Since the key papers by Hull et al. and Zhang et al., more than a dozen additional panda habitat studies have been conducted. Some of these employed GPS radio-collars, leading to some significant new insights. Prior to that, during 1995–2006, a Chinese government moratorium precluded using radio-collars on giant pandas. Four pandas were tracked for about a year each in 2007–2009 in Foping Nature Reserve and five were tracked for about a year each in 2010–2011 in Wolong Nature Reserve. The Foping data showed extensive individualistic seasonal shifts in elevation [136]. In Wolong, Hull et al. confirmed high use in what previous studies ascertained as suitable habitat: forested, gentle slopes at mid to high elevation with high solar radiation. However, the collared pandas also used non-forested (shrubby) habitats and steep slopes, previously deemed to be unsuitable in sign surveys (possibly because they were undersampled) [137]. Selection of habitats varied among individuals, across seasons, and depended on scale (placement of the home range or choice of habitat within the home range). Bai et al. used these same five collared pandas in Wolong, and compared a host of fine-scale habitat features measured on the ground in core versus secondary (distinguished by the amount of use) parts of the home ranges [138]. One of their new findings was that slope steepness was not a good predictor of core vs. secondary use within home ranges. Previously it was thought that giant pandas avoided steep slopes to save energy. In this study, though, it was found that they selected for areas with animal paths through the thick bamboo, which would help conserve energy.

Other recent papers used presence points from the Third and Fourth (2011–2014) National Surveys. A comparison of the locations of panda sign between these two rangewide surveys revealed that pandas shifted in terms of some key variables [139]. Importantly, the selection for old-growth forest discovered in the Third National Survey disappeared 10 years later in the Fourth Survey because the early successional areas outside the nature reserves had matured somewhat, and surprisingly, pandas were found to be equally likely to use these secondary forests as primary forests; that is, habitat suitability improved significantly outside reserves commensurate with a ban on logging [140]. This positive conservation finding was diminished by a new emerging threat: it was found that elevation of panda use increased, not due to a change in elevation of bamboo, but to increased livestock grazing at some lower elevations.

Livestock grazing deters panda use [141,142], probably owing to their disturbance, and their consumption and trampling of bamboo [143,144]. This problem has been increasing as people switch from farming to livestock [145]. A number of studies have also reported that pandas actively shun human development, including roads [133,146–148], which not only present a hazard but also reduce nearby regeneration of bamboo [149]. It now seems clear that a reduction in human activities, grazing of livestock, and road building would be a top priority for improving panda habitat and thus benefiting its conservation, at least in the short term [150,151]. Longer term challenges remain insofar as effects of climate change, both on the availability of bamboo and also possibly on temperature effects on pandas directly [152], although model predictions differ greatly [153,154]. However, it must be remembered that just a few centuries ago, giant pandas occupied lowland habitats, where temperatures were higher than in the mountains where they live now, and where they consumed different kinds of bamboo and other plants [155]. Although giant pandas are specialist feeders, the many habitat studies that have been conducted have shown that selection can change with differing availability of resources and changing threats, and also that such changes are perceived by pandas at multiple scales [156].

5.6. Brown Bears

Reviewing all of the many habitat studies of brown bears around the world would require a full paper on its own. Here, we take a tour across four countries in Asia, eight in Europe, and two in North America, to get a flavor for the techniques and findings, which sometimes conflict.

In Asia, studies have been based mainly on presence points (Figure 2). The brown bear population in Iran has been the subject of an inordinate number of habitat studies (at least seven), all using species distribution modeling. One Maxent-based study examined habitat in the lush Hyrcanian forests along the southside of the Caspian Sea along the northern slopes of the Alborz Mountains, and found that bears selected for forested habitats away from human settlements and roads [157]. Another study in the same region using the same model concluded that steeper slopes and higher NDVI were the best positive predictors of habitat suitability, with human disturbance a negligible factor [158]. In the northwestern corner of Iran, in the Caucasus region, investigators ran five different models with presence points and found varying results; the most consistently chosen variable was a positive relationship with shelter afforded by topography (e.g., valleys) [159]. Widely spaced patches of suitable habitat suggested low connectivity. Moving south, into the Zagros Mountain range, two studies found that bears selected for higher elevations (>2000 m), away from people and roads. One of these also indicated that bears preferred areas farther from forest [160]. The other study indicated that the habitat requirements were such that within an area of just 35,000 km^2, there were 33 distinct patches of suitable habitat with uncertain connectivity among them [161]. In a semi-arid area in the Zagros of Fars Province, marking the southernmost distribution of brown bears worldwide, the overriding factor in habitat selection was availability of water, but staying away from roads was also important [162]. The most recent study obtained 184 independent presence points across the entire range of brown bears in Iran, some collected opportunistically and some from prior research studies [163]. This model chose different principal predictor variables than previous studies, in part because it was a different model (Random Forest vs. Maxent) at a different scale and covering a wider geographic area, and in part because presence points were obtained from opportunistic observations, which could have been biased. The degree of fragmentation of the predicted populations based on habitat suitability depended on how far these bears were likely to disperse. Contrary to some of the findings from the habitat models, a landscape genetics study across the entire brown bear range in Iran showed high genetic variability and interchange within each mountain range (i.e., no obvious fragmentation), but separation between the Alborz-Caucasus bears and the Zagros bears [164].

A consistent finding of the many studies in Iran is that habitats selected by bears were strongly influenced by human development. However, the perceived impact of this factor depended on exactly what parameters were included in the model. For example, in Turkey, elevation and indices of roughness and ruggedness predicted the occurrence of brown bears, but they may have been surrogates for bears avoiding flat places occupied by people, since the investigators did not include anthropogenic variables in the model [165]. Another investigation in Turkey, employing GPS radio-collars discovered two distinct behavioral types: (1) bears that avoided people and made long-distance seasonal migrations, and (2) more sedentary bears that relied on human sources of food and selected for areas near roads and buildings [166]. This raises the question as to whether investigations that rely solely on presence points would even include observations of bears in and around towns as such areas might not normally be sampled.

Many studies now obtain occurrence data from open access databases, with little understanding as to how they were obtained, and possibly little experience with the species on the ground. For example, a Maxent modeling study covering 11 countries in Central Asia obtained all data from the Global Biodiversity Information Facility (https://www.gbif.org, accessed date (20 December 2021)), then selected points within protected areas, and concluded that annual temperatures and precipitation accounted for >90% of the brown bear's predicted distribution, and that landcover, elevation, slope, and aspect mattered little [167]. Likewise, a Maxent-based study on the Tibetan plateau concluded that climatic variables dictated the bears' distribution, whereas 14 different types of landcover had no apparent effect [168]. Seemingly contrary to this finding, a study by the same investigators in the same study site found that landcover, human population density, and NDVI were the three main factors explaining where bears broke into people's houses, a problem that has become extreme in this area [169]. A particularly thorough study in the Himalayas, where investigators collected 720 presence points through their own field sampling, reported that a mixture of climatic, landcover, and human population variables, optimized at a broad scale (generally 64–128 km radius from the bear), were all important, and all relationships were nonlinear; this region has extreme seasonal differences in temperature, rainfall, snowfall, and topography [170].

Brown bears in Europe face different circumstances than in Asia, as nearly every population is heavily impacted by people, and some are small and isolated, although several are expanding [35,171]. Systematic surveys to look for bears and bear signs have been conducted in both the Cantabrian (Spain) and Pyrenees (shared Spain and France) populations, yielding many thousands of data points (one study obtained >3000 separate video recordings of bears) [172]—in fact, far more than all the presence points combined for brown bears in Asia. Using these data, studies found that bears selected for rugged forests with hard mast, and selected against areas close to roads, homes, and agriculture [173,174]. However, since the Pyrenees population is still expanding, it was not possible to differentiate unsuitable habitat from habitat that simply was not yet occupied. Furthermore, as Mateo Sanchez et al. found, bears respond to some factors close by (e.g., food patches) and others more distantly, so habitat suitability models can be quite sensitive to the scale at which each variable is measured [175].

Studies in central Italy were similar to Spain and France: presence data indicated that bears selected for forests with mast, and against sites near people and agriculture [176–178]. Falcucci et al. overlapped habitat suitability model results (from >2500 presence points) for Apennine brown bears with a risk map based on locations of mortalities (mainly anthropogenic), and found that 43% of the area considered highly suitable habitat was close to roads and was a high mortality risk from anthropogenic causes—that is, one cannot tell from presence points alone whether suitable habitat is also a mortality sink [177]. Indeed, Maiorano et al., modeling a compilation of >5000 independent presence points from radio-collared bears, sightings, tracks, hair, and scats in this population learned that there is substantial overlap in the use of mast-rich forests by bears during fall, and by hunters chasing wild boar (*Sus scrofa*) with dogs, which poses a clear risk to bears [178].

In the Italian Alps, investigators preparing to augment a very small, relict population created a habitat suitability map based on presence data. This predicted that bears would use the forests and stay away from human-developed sites with homes or agriculture. Resource Selection Functions generated from collared bears after the augmentation were compared to the former predictions, and found concordance in some respects, but bears actually used orchards and shrublands more than predicted by the initial model, and also crossed roads more than anticipated [179], again showing that mortality risks may be underestimated in habitat models built from presence points.

In Greece, bears with GPS radio-collars selected for rough terrain away from people, but they came closer to human-related habitat features, such as dwellings, roads, and crops at night [180]. In a number of other European countries, brown bears are attracted to feeding sites with the purpose of either diverting them from using foods in human settlements (diversionary feeding), or congregating bears for hunting or viewing (supplemental feeding). There is now growing evidence that such feeding reduces the size of bear home ranges and likely alters their use of habitats [181,182]. In Slovakia, maize fields have been increasing, and some GPS-collared bears relied on these during fall, whereas other individuals continued to select woodlands and natural foods [183].

In Sweden, bears with GPS radio-collars selected for steep slopes and regenerating forests; those living nearer people selected especially steep slopes, although this selection eased at night when people were not active [184]. The authors explained that regenerating forests had more food (ants, herbaceous plants, berries, and moose [*Alces alces*]). Follow-up studies in the same area showed that bears exhibited consistent individual differences in habitat selection, and moreover that some individual patterns were not apparent at the population level [185,186]. For example, at the population level, bears strongly selected for young forests and against bogs, but on an individual level this dichotomy was not as evident, and may have varied with sex, age, local habitat availability, local population density, learning, and different behavioral responses to the environment.

In North America, a large proportion of habitat selection studies for brown bears/grizzly bears have relied on bears with GPS radio-collars, allowing for investigations of individual, seasonal, and yearly differences. In the Yellowstone Ecosystem, USA, for example, it was well established that grizzly bears were dependent on the seeds of white-barked pine (*Pinus albicaulis*), given that this food factored heavily in their diet and that year-to-year variations in mast crops had significant demographic consequences [187]. However, GPS collar locations showed that about a third of the population made negligible use of white-barked pine, and further, that as availability of this resource declined from trees dying, the bears, unexpectedly, reduced their use of this food, and shifted to alternative foods and habitats [188]. That is, there was a negative trend in the selection for what had been thought to be a preferred habitat.

A big advantage of monitoring individual bears is in being able to decipher different scales, or orders of selection (sensu Johnson [6]), and in understanding mechanisms and motivations for selection and changes in selection. A study in the foothills of Alberta, Canada, found that radio-collared grizzly bears were located in forest clearcuts 23% of the time, and these cuts made up 19% of the landscape, but this broad scale view provided no insights as to their attraction to this habitat: it turned out that they selected it more than expected from availability only during the summer, when feeding largely on ants and ungulate calves [189]. Furthermore, bears selected for certain specific ages and portions of clearcuts (i.e., high light penetration supporting ants on dry warm slopes). Moreover, bears shifted the time of day that they used young clearcuts and more often used older, shadier regenerating cuts, so as to avoid overheating; this result was shown by matching the timing of GPS collar locations in different habitats to temperatures continuously measured there [190]. A follow-up study revealed a high selection for edges between forests of different ages, but this selection was stronger for females than males [191]. Examining data from these collared bears further it was found that their habitat choices were learned over time, and not instinctual [192].

In the same Alberta study site, movements of GPS-collared bears were compared to patches of Canada buffaloberry (*Shepherdia canadensis*), a primary food during late summer and early fall. Bears selected for areas with high contrast in fruiting density between patches of this key food resource [193]. However, their selection for habitat heterogeneity was at a rather small scale, equivalent to their average travel distance within a 5.5 h window (~1900 m).

In neighboring British Columbia, Canada, perceptions of grizzly bear selection varied enormously by the scale of the investigation, suggesting that studies that chose a single scale were prone to misinterpretations. This was demonstrated by several different studies: one used radio-collared bears [194], another used bear visitation at hair snaring sites (visited or not visited during a site check) [195], and yet another used bear densities estimated by DNA from hair-snare sites [196]. To some extent, scale dependency occurs because availability varies by the scale at which it is measured—in other words, the scale issue is a function of the way people examine the selection process that bears make. This causes some habitats to appear to be selected for when viewed at some scales and selected against at other scales. However, additionally, bears may respond differently to habitat variables at different scales. For example, Apps et al. showed that grizzly bears selected for high elevations, steep slopes, and rugged terrain, all to stay away from humans, and at a broad scale, selected for landscapes of higher forest productivity; however, at a finer scale they found a surprising negative association with a "green vegetation index" (from Landsat imagery), because areas of high greenness included both avalanche chutes (which were attractive feeding places) and wetlands (which were avoided) [195].

In another scale-dependency example, a grizzly bear study in a barren-ground environment north of treeline in the Canadian Arctic found that radio-collared bears of both sexes incorporated eskers and tussock/hummock tundra into their home ranges, because these habitats provided the most food in terms of berries, sedges, grasses, and ground squirrels. However, looking at fine scale usage within home ranges revealed that females with young avoided these habitats when males were most likely to be there, indicating that they adjusted their habitat use to avoid infanticide [197].

5.7. American Black Bears

American black bears are routinely studied using radio-collared animals, so it is not surprising that this technique is widely employed in habitat studies (Figure 2). In a particularly unique study, Gould et al. used bear detections at hair-sampling sites in an occupancy model to predict the probability of bear use, then tested and verified the results with radio-collared bears and sites where bears were killed in the sport harvest in New Mexico [198]. However, their model was at a coarse scale, including only a positive association with an index of primary productivity (EVI, enhanced vegetation index) and a negative association with roads. Furthermore, their model over-predicted use in an irrigated agricultural area (with high EVI). Some recent innovative studies examined relationships between estimated bear density and habitat, integrating spatial capture–recapture with locations of GPS-collared animals [199,200]. Welfelt et al. discovered that the greater primary productivity in western Washington than in arid eastern Washington did not translate to a higher bear density, due to higher human densities and hence greater human-caused mortality in places where primary productivity was high [200].

Among telemetry studies, sample sizes in terms of the number of individuals and number of locations vary enormously, but GPS collars are now providing massive amounts of data. A recent exemplary study included >86,000 independent locations (reduced from >277,000 in the full dataset) obtained from 236 adult bears over a period of 36 years across the bear range in Florida [201]. Among various studies, analyses have entailed fewer than 10 to more than 20 habitat variables, including vegetation, water, anthropogenic features, and topography. The inclusion of more variables has often revealed that habitat selection or suitability is not easily defined as a certain forest type, but may be a complex array of factors that changes over the course of a year with changing food availability. Moreover, bears may benefit greatly by visiting very small habitat patches with concentrated foods, not easily detected in studies, and for which availability is difficult to quantify (Figure 7).

(a) (b)

Figure 7. American black bears are attracted to concentrated food sources, such as (a) lush sedge along the shoreline of a coastal bay and (b) a small garden of sunflowers. Most habitat use studies are unlikely to be able to measure either use or availability at such a small scale, and moreover, the short time feeding at such places greatly underestimates the attraction to and importance of these food sources. Photos: D. Garshelis.

Several studies have found that American black bears select for habitats known to produce the fruits and nuts that constitute the main portion of their diet, and may avoid some types of anthropogenic features. The actual forest types producing bear foods vary across regions, with deciduous or mixed forests typically used more due to higher food production than pure coniferous forests [202,203], although with some exceptions for those conifer species that produce fruits important to bears [204]. Only a few studies actually measured fruit production in different habitats to inform a habitat model [205]; others attempted to use "greenness" (NDVI) as a surrogate for food production, with mixed results and difficult interpretations [206]. Many fruits that black bears consume are most productive in forest openings or along forest edges; accordingly, some studies found that forest cutting can increase the diversity and abundance of fruits, thus providing better bear habitat [207,208]. Sollmann et al. found that bear density was highest where high forest cover was broken by openings, but were unsure if the openings enhanced fruit production or provided anthropogenic foods [199].

American black bears also may select for habitats where ungulate prey are seasonally available, as for example during moose calving [209,210]. However, studies in both Alberta and Newfoundland, Canada, found that only some individual GPS-collared bears selected for habitats where caribou calved, whereas most of the population did not [211,212]. Furthermore, the traits of individual bears in terms of whether they exploited caribou remained consistent year to year [213].

In some areas, American black bears selected river valleys as corridors for movement through otherwise developed or agricultural landscapes [205,214,215]. Water availability, used for travelways, forage, drinking, and cooling, may be important in some regions but not others [216–218]. In Louisiana, on the landscape scale, bears selected areas with swamps, which comprised just 2% of the landscape, because of an essential tree species that provided elevated cavities for hibernation; however, on a finer scale and during spring-summer months, bears in one study site selected swamps, whereas bears in another site, just 10 km away, had different resources available and did not use swamps [219].

Poor et al. made the important distinction that habitat suitability is not the same as habitat preference: American black bears exhibit high behavioral plasticity and can often adapt to changing conditions related to human development and forest modifications [201]. As an example, the common finding that most bear species avoid roads may be situation-specific. In Wisconsin, forest roads with little traffic had a positive influence on where radio-collared bears selected their home ranges, although no apparent influence on their habitat use within their home ranges [218], probably indicating that the most roaded portions of the forest were places that happened to offer the best resources for bears, but the roads themselves were not the attraction.

A study of habitat suitability for a group of 18 midwestern U.S. states based on expert opinion highlighted a significant difference of opinion about the effect of roads in particular [220]. Presence points helped validate the model, but 23% of presence locations fell outside the area of most suitable habitat as judged by species experts, suggesting that experts have a narrower sense of what bears need. On the other hand, the expert-based model likely over-estimated the suitability of some portions of as yet unoccupied range due to some fine-scale habitat features that were not included in the broad-scale model.

One purpose for modeling habitat suitability is to predict bear population expansion. This can be particularly difficult because the locations of bears in occupied habitats may not foretell the full suite of habitats that they could potentially occupy; that is, the model may be misinformed by being limited to data from bears inside the current range. Ditmer et al. observed bears at the edge of their range within a sparsely-forested area of Minnesota (considered atypical bear habitat) and used short time windows of habitat use by collared bears to try to predict the minimum amount of forest that they could tolerate; they used this to gauge how far west across a gradient of decreasing forest the population could expand [205]. Gantchoff et al. argued that habitat models could over-estimate the potential range if the sexes are pooled, because females in their expanding population in Missouri were observed to be more constrained to less human-developed areas than males [221]. However, their frame of reference for what females could tolerate was based on where females lived currently. Males are more exploratory and may first venture into novel habitats, but females in more established populations coexist with males in human-developed areas [215].

Population expansion in Mexico offers a particularly intriguing test case for model predictions. Delfín-Alfonso et al. employed a long-term collection of presence points (582 points over 120 years) and examined 19 climatic variables and 4 terrain variables to define a potential distribution, which has been expanding from northern Mexico southward [222]. However, new occurrences of bears have since been reported far south of that model's predicted suitable habitat [223,224]. Another model, restricted to more recent presence points (being more accurately associated with current habitat conditions) and variables related to land cover, elevation and human density, but not climatic variables yielded better predictions relative to these points at the current southern limit of the range (21° N) [225]. On the other hand, the presence of wandering black bears well outside their primary range is not necessarily indicative of a population expansion nor of what constitutes suitable habitat [226]. Only time will tell the difference.

6. Conclusions

This review, covering 141 studies of fine-scale habitat use of the seven species of terrestrial bears, yielded three important conclusions: (1) bears use an enormous array of habitats, and adapt to changing conditions in a variety of ways; (2) biases and weaknesses in some of the techniques used to assess habitat selection and suitability have sometimes produced inconsistent or misleading results; and (3) management or conservation recommendations arising from many of these studies have been very limited. This paper concludes by addressing each of these topics.

6.1. Adaptability of Bears

One commonality among many of the studies is that human disturbance (e.g., distance to settlements or roads) may strongly affect habitat selection, meaning that what is observed in a human-dominated landscape is different than what would be observed in a landscape without people. Habitats judged to be unsuitable might only be avoided because of people.

For example, bears likely do not avoid roads or habitats along roads, but rather avoid the risk of being hit by a moving vehicle [227]. Indeed, when a COVID-induced lockdown reduced human activity, brown bears in an isolated population in the Italian Alps crossed roads more and used a wider array of habitats [228]. Hence, it is not enough to categorize habitat suitability or barriers simply in terms of human infrastructure, but rather to gauge the effects of human presence. Bears apparently recognize the difference.

Bears also weigh the potential threats of people versus the benefits of foods that humans provide. Thus, whereas some bears might be repelled by human disturbances, other segments of the population are attracted to human-related food sources and may profit nutritionally, even if at more risk. Sun bears exploiting oil palm plantations at night are heavier than those consuming only natural foods. Asiatic black bears in an arid habitat in southern Iran with few natural foods rely extensively on abandoned orchards for subsistence. In the Italian Alps, reintroduced brown bears were predicted not to use orchards, but they did. Some rare, recent sightings of brown bears in Syria occurred in orchards [229]. High densities of American black bears occur near anthropogenic food sources, and population expansion may be facilitated by the availability of edible crops.

Bears are very driven by food. However, whereas many studies have investigated habitat selection, few have investigated food selection in the same way, in part because the availability of different foods is much more difficult to measure. The breadth of the diet varies among areas and species, with the simplest case being the giant panda, which focuses completely on one source of food (although different parts, ages, and species of bamboo). Typically we assume that high use of certain foods is equivalent to selection, and that bears adapt their use of habitat according to the availability of preferred foods that occur there. However, as demonstrated by the case in Yellowstone, where brown bears shifted their focus away from a key declining food [188], our predictions about how they will adapt to changing foods may be wrong. Similarly, brown bears relied heavily on salmon (*Oncorhynchus nerka*) in Kodiak, Alaska, and in normal years adjusted the timing and location of foraging commensurate with the availability of spawning salmon in different streams [230,231]. However, in years when red elderberry (*Sambucus racemosa*) fruited unusually early due to higher than normal spring temperatures, then coinciding with the main salmon run, bears unexpectedly departed the salmon streams to feed on these berries in the hillsides [232]. Likewise, sun bears, which clearly benefit from a bounty of diverse fruits during masting events in primary forests, nevertheless selected for a neighboring secondary forest that produced more insects when fruits were less available [67]. The same seems true for shelter: whereas sun bears often choose resting sites in the canopy of tall trees in primary forests, they use other types of trees to rest in secondary forests [97], and there is no evidence that their health or survival is reduced by doing so. This makes the point that there is a difference between what bears select when offered certain options, and what they need. Most habitat studies (not limited to bears) do not investigate what is truly needed [9]. This may lead to an underappreciation for what bears are capable of adapting to in a changing landscape. That is not to say that we should not be alarmed by trends in declining forest cover in Asia, or by climate change-related projections, but rather to inject some caution on conclusions from habitat suitability modeling.

This review has shown that in a number of cases, bears have been observed in places that models have classified as unsuitable habitat—Andean bears in Peru, sun bears in Malaysia, sloth bears in non-forested areas in India, American black bears in central Mexico. In some cases there may be a distinction between a few wandering individuals and a viable population, but on the other hand, population expansion often begins with the exploratory movements of a few individuals into habitats that differ from where the core population resides [233].

6.2. Methodological Shortcomings

The basic method for assessing the suitability of habitats involves measuring use and availability, and comparing the two. Each of these is difficult to measure, and errors or biases can lead to significant misperceptions. The least biased method for measuring use involves GPS radio-collars. Unlike VHF collars, where locational data were often limited by investigator access and certain times of day, GPS collars provide data wherever the bears go. For 11 years investigators were forbidden from using collars on giant pandas, but when this moratorium was lifted, the new GPS collar data showed that pandas frequently used habitats previously thought to be unsuitable. Radio-collars also provide a view into sex, age, and individual differences. Furthermore, by tracing the routes of bears, one can examine the speed of travel and surmise decisions that bears make with respect to habitats (e.g., clustering of points vs. quick travel through). Moreover, since investigators can obtain locations in near real time via satellite transmission, it is possible to visit points where the bear had just been and examine micro-habitat features that explain why it was there [234].

Sign surveys also can provide information about micro-habitats, but not connected to an individual bear in real time. Furthermore, it is more difficult to ensure unbiased sampling, since the sign is only visible where the investigator looks, and investigators may spend more effort searching in places where they think the bears are likely to be. Further, and importantly, detectability varies by type of sign and by habitat, and different types of signs are apt to be more prevalent in different habitats. Asiatic black bears and sun bears climb more trees and leave easily recognizable claw marks in a forest with many fruiting trees, but in forests where they rely more on insects, signs are harder to detect or to reliably distinguish as being from a bear (Figure 3). Some studies of sloth bears and Asian brown bears have relied on scats, which are easy to distinguish, but scats are more detectable in open habitats.

It is now becoming increasingly common for investigators to obtain data that they did not collect, including camera trapping records from various sources (e.g., so-called by-catch data from studies not focused on bears) and open-access data repositories. These sources provide even less control against sampling biases, and also force the investigator to rely on habitat features obtained from remote sensing rather than on the ground. This limits the scale and nature of the data and also disengages the investigator from the species and conditions on the ground.

Availability is even harder to measure because it involves interjecting a human perception into what the bear perceives as choices. However, the investigator's measure of availability often drives perceived selection. Presence-only models, like Maxent, rely on sampling the background, where the investigator sets the frame of that sampling. Often that sampling is done on a computer, but sometimes investigators have gone to the field to examine "pseudo-absence" points, which are intended to be a random selection of points with no confirmed presence. Even in the best of efforts, though, as shown with giant pandas, investigators may introduce biases by choosing sampling points that are less difficult to get to. Studies have come a long way in recognizing that availability is scale-dependent, and new sophisticated techniques allow for varying the areal window around presence points to try to find the scale at which habitat covariates are most influential.

Availability, and hence selection, changes over time. In the short term, food conditions change within habitats, sometimes week to week. Temperatures within habitats can also change. Hence, certain habitats or portions of habitats are selected only at certain seasons and times of day. If that time frame is short, then on a coarse level it may appear that the habitat is unimportant. Changes also occur over longer time frames. A common example is forest regeneration. When logged forests were newly regenerating outside panda reserves, pandas selected for old-growth forests. However, the perception of old-growth being a key characteristic of habitat suitability turned out to be wrong, as shown by pandas' increasing use of later successional forests that were protected from logging.

Threats also can change over time and vary spatially. Whereas panda habitat improved outside reserves, increased livestock grazing inside some reserves affected bamboo and pandas directly. The most obvious threat is direct killing by people, which for many bear populations (except pandas) is a main population driver. Human-caused mortality reduces density and may thereby alter bear distribution relative to habitat. It can thus appear that habitats subject to the highest levels of human-caused mortality are actively selected against by bears. Conversely, bears may choose habitats that unknown to them have a high mortality risk, and in that sense, these "attractive sinks" may really be viewed as less suitable, even if they provide abundant resources [235]. Ciarniello et al. found that grizzly bears in British Columbia, Canada, existed at one-quarter the density in an area with better food resources than a more mountainous area where foods were poorer because in the food-rich area they foraged close to roads where human-caused mortality was high, whereas in the mountainous area they foraged farther from people [236]. Likewise, in another site in British Columbia, Lamb et al. estimated demographic parameters from DNA mark–recapture information and found that in a local area rich in production of some key berries, grizzly bear survival was low and population growth negative, due to high human-caused mortality, but bears actively moved into the area due to the attractive foods [237].

6.3. Implications for Habitat Management

The number of studies employing niche models, habitat suitability models, or species distribution models has been exploding, with over 1000 new papers published yearly on a variety of species [238–240]. In this Section 7, we explore some issues that may limit the implications of models for the management and conservation of bears worldwide.

A basic issue is whether model results are likely to be reliable and useful. For many papers, it is hard to assess the quality of the data, and why certain predictor variables were chosen, due to poor or unclear protocols, and general lack of standardization and documentation [240]. Presence points are often obtained with sampling methods that are not fully described, or retrieved from public databases where multiple sorts of errors may occur [241]. The number of points used in the models reviewed here ranged from several thousand to less than 100. Predictor variables are likewise either collected in the field (micro-habitats near points), or from easily-accessible databases. The number of predictor variables in the models reviewed here ranged from more than 20 to just 1.

Investigators can create models with easy-to-use programs (e.g., Maxent) and high computing power. This has the advantage of allowing investigation of a wide range of potential influences at various scales. It is now increasingly evident that scale optimization is both species and habitat dependent, and essential for creating models that better mimic nature [62,242]. However, this review has also highlighted some associations that are hard to explain. For example, is bioclimatic variability the actual factor driving the distribution of brown bears in Central Asia, or does temperature and precipitation affect local food availability (e.g., [243]), or are they related to other habitat characteristics that were not measured [244]? Models commonly use environmental or bioclimatic surrogates, but biologists should ideally consider the foods that bears rely on in each habitat plus risks they are likely to face. The profusion of available data in the hands of investigators who have limited experience with the species has the danger of creating what Nielsen et al. [245] called a "technology trap", leading to misinterpretations and misguidance for practitioners. Studies on the ground are still needed to either verify or correct the deductions stemming from models [67].

Some studies, mainly in North America, have linked habitat associations with density, which is ultimately what managers and conservationists are interested in [196,199,200,246–249]. On the other hand, density can be as much a function of human-caused mortality and social factors within bear populations as other components of the landscape [200,236,246,249]. Furthermore, it may be difficult to relate density, which inherently pertains to some larger area, to fine-scale habitat features. In a dry forest in Peru, cliffside travel corridors, waterholes, and patches of fruit were observed to be important for individually identifiable Andean bears, but density was too blunt an instrument to pick these out as crucial habitat features [48,92]. Better than density would be linking habitat to fitness [10]. In a unique such study in the Canadian Rockies of Alberta, Boulanger et al. found that radio-collared grizzly bears gained weight more quickly if they had access to forests of multiple seral stages, but as a consequence of living in this area suffered higher rates of mortality [250].

Models built from presence points yield a snapshot of how animals are likely to be distributed on the landscape but may not be good at predicting how populations are apt to respond to landscape changes. The concept of a gradient of suitable habitats may lead practitioners to assume that increasing the availability of the most suitable habitats should increase population growth, but for generalist species, this may not hold true. If bears have enough of the preferred habitat, then providing more of it will not necessarily increase use, or population growth [9]. Generalist omnivores are likely to benefit from variety, even if some habitats provide more food or better shelter. Additionally, a number of studies are beginning to show that individual or sex-age related variability plays a part in selection, so what seems best on a population level, might not be best for all. Moreover, it is worth being cautious of results of any single study given the variability and discrepancies among studies.

Discrepancies may arise for a number of reasons. Some datasets may be biased. Equally important, we might expect models supplied with different arrays of predictor variables to gauge habitat suitability differently. These factors may account for the inconsistent results obtained among studies of sloth bears in the Western Ghats of India, Asiatic black bears in central Japan and southern Iran, and brown bears in the Zagros Mountains of Iran. As this review has demonstrated, combinations of variables at different scales can help explain where bears occur on the landscape, but these should be viewed as imperfect surrogates for how bears actually discriminate habitat suitability. Population-level views may inaccurately represent the variation in behavior that exists within populations. Further, being highly adaptable, bears are likely to find ways of coping with site-specific circumstances in different ways. As such, it is not surprising that a model generated from observations in one area may not readily transfer to another. A good example is the rangewide map of Andean bears generated from a Bolivia-based model which poorly matched results of site-based models in Peru.

Scharf and Fernández compiled data from habitat studies of brown bears across Europe, and tested predictions from a composite model against previous individual site models [251]. They found transferability of results to be high among sites close together, but weaker for distant sites, where conditions are more different. In particular, individual models used different surrogates for human impacts; furthermore, the way humans interact with bears varies geographically. The composite model for European brown bears predicted that only 56% of the area actually occupied by bears is suitable habitat, suggesting that the model greatly under-estimated what bears view as suitable (i.e., what they can adapt to). By contrast, a study that compared habitat suitability models for snow leopards (*Panthera uncia*), a more specialist species than brown bears, concluded that despite some differences in sampling and predictor variables, models were in basic agreement about that species' habitat needs [252]. This is also generally true for giant pandas, the most specialist bear; nevertheless, some significant differences exist between panda habitat studies, some of which may be due to ecological differences among the six mountain ranges that they inhabit, some due to changing environmental conditions, and some due to different sampling of points, different variables, and different scales.

A recent review paper by Lee-Yaw et al. investigated 201 studies, spanning a diverse array of animals and plants, that tested the predictions of species distribution models against independent assessments of occurrence, abundance, population performance, and genetic diversity [253]. Of these, predictions of occurrence were by far the simplest, yet only half the model predictions were accurate. The authors emphasized that predictions from these models should be treated as hypotheses in need of testing with other data, rather than informing management and conservation directly.

Many of the bear studies reviewed here made predictions about the future—often dire consequences of habitat alterations from logging or climate change (e.g., [93,167]). What is generally lacking, though, are real tests of these predictions (but for other taxa see [254,255]). In the particularly unique case of two rangewide giant panda surveys conducted 10 years apart, the first showed a strong reliance on old-growth forest, while the next survey (essentially testing the first survey's prediction) indicated that once early successional forests had a chance to mature, pandas used them to the same extent as old-growth. Even tests of present predicted distributions are uncommon. This review revealed some cases where genetic data indicated that habitats were not as fragmented as models predicted (brown bears in Iran), or where bears were living outside the predicted range (e.g., sun bears in Malaysia, American black bears in Mexico); but often such contrary information is not published. It would be harder yet to demonstrate the absence of bears in places where models predicted that they should occur.

Rarer still are cases where conservation actions stemmed directly from results of habitat suitability studies, even though most modeling papers justify this as the main reason for the study. There tends to be a disconnect between modelers and decision-makers [238], which is evident in general between ecological theory and conservation practice—what Knight et al. called the "research–implementation gap" [256]. Bridging this gap requires that the scientific process yielding the actionable recommendations be understood and considered credible by practitioners. This means that models should be transparent and repeatable (not sensitive to data inputs and methodological decisions), and that they employ knowledge of the biology of the target species and fully consider the inherent complexity of the situation on the ground; further, there needs to be a more active exchange of information between modelers, species experts, conservationists, and decision-makers [257–260]. That is not to say that this has not occurred in a few exemplary cases. For example, some studies predicted and then created habitat corridors to connect bear populations, which have been instrumental in a successful conservation effort [261,262]. However, more work needs to be done to check and verify habitat suitability studies and to ensure that credible results are implemented in management and conservation programs.

7. Recommendations

This paper closes with 10 concrete recommendations for future investigations of bear–habitat associations apparent from this review.

1. Ensure that sampling is representative of the available resources and conditions. Generally, this will be the case for data from GPS collars, but biases can arise with point sampling, especially using sign (which is created and decays at different rates in different habitats), and potentially to a large extent in data repositories where the investigator has no control over the data collection. The veracity of the data should be examined, not simply assumed.
2. Choose candidate covariates that have probable biological connections to the species, not just ones that fit a model. The British statistician George Box is famously quoted as saying "all models are wrong, but some are useful." It is hard to imagine a useful model in which the predictors explained much of the variation in the model, but did not actually relate biologically to the target species.
3. Test habitat availability at multiple scales. Human investigators cannot know how a bear perceives its world—what specific resources it seeks, what threats it is trying to avoid, what it knows or remembers as to where resources and threats are located, and how it weighs these various factors. The best we can do is test various spatial windows.
4. Employ both ground-based and remote-sensing-based variables. Ground-based variables bring the investigator closer to what the bear perceives in its environment, especially the foods. Investigators should have a connection with their target species. One way of doing that is to examine sites used by GPS-collared bears. Remote-sensing variables enable investigators to have a wider view of environmental variables and measure things that cannot be measured at ground level.
5. Search for variables that meaningfully measure risks, and distinguish selection from suitability. Human-related factors are commonly included in models, but they may

be difficult to interpret. Investigators commonly measure distances to roads or settlements, and assume that negative associations imply bears' perceptions of risk. However, in many cases, bears are attracted to human foods, or to habitats where roads are built, so their selection may be maladaptive. It is important to recognize that selection does not equate to suitability.

6. Be aware that population-level associations may hide important individual-level differences. This review has pointed to a number of cases where individual bears or bears of different sex-age groups in the same area responded to resources differently, including both natural and human-related foods.
7. Compare results of multiple models, explain discrepancies, and build composite models. This review highlighted a number of cases where multiple studies in the same region achieved contradictory results, but authors typically ignored these. Models are one approach for deciphering complex data, but that very complexity means that models do not mimic nature precisely. Increasing knowledge requires not just constructing more models, but understanding why results differ among models. Efforts to systematically compare individual models and build composite models are likely to increase the reliability of outcomes.
8. Test model predictions. Model predictions can be compared to each other, but better yet, compared to actual bear occurrence or demography. This has been accomplished in very few cases, as this review revealed, and often key aspects of model predictions have not been upheld. A concerted effort to test predictions of published models would be highly worthwhile.
9. Increase transparency to enable practitioners to utilize results. Habitat modeling papers are often written with the professed goal of benefitting bear management or conservation, but connections between research papers and actions on the ground are scarce. This disconnect may arise from the practitioners' view that models are unreliable, not understandable, not realistic, or not clear insofar as to what actions should be taken.
10. Look for associations between habitat and demography, not just relative use of different habitats. Habitat use is a potentially misleading parameter by which to judge habitat suitability. Cases were mentioned here where bears were attracted to habitats where survival is poor, or conversely, where habitats used for just a short period of time provided a crucial resource.

A final key point is to appreciate not just the complexity and variability of the environment, but also the adaptability of bears. This makes characterizing the suitability of their habitats very difficult. It also underscores why it is so difficult to explain what "bear habitat" is (at least for most of the species), and why the older, functional definition of habitat as a species-specific collection of resources is not very useful for this taxa. Further, it also means that we should be wary of habitat-based model projections about the future.

Funding: This research received no external funding.

Institutional Review Board Statement: Not applicable.

Acknowledgments: I thank Michael Proctor for creating the range maps from the shape files available from the IUCN Red List of Threatened Species (https://www.iucnredlist.org/; accessed 31 December 2021). I am grateful to the members of the IUCN SSC Bear Specialist Group for continual insights about bear habitats, providing the motivation for this paper, and Juan Beltrán for the invitation to write it. I appreciate the constructive comments and suggestions of three anonymous reviewers that improved the paper.

Conflicts of Interest: The author declares no conflict of interest.

References

1. Ma, Y.; Wang, M.; Wei, F.; Nie, Y. Geographic distributions shape the functional traits in a large mammalian family. *Ecol. Evol.* **2021**, *11*, 13175–13185. [CrossRef]
2. Matthiopoulos, J.; Fieberg, J.; Aarts, G. *Species-Habitat Associations: Spatial Data, Predictive Models, and Ecological Insights*; University of Minnesota Libraries Publishing: Minneapolis, MN, USA, 2020; ISBN 978-1-946135-68-1.
3. Hall, L.S.; Krausman, P.R.; Morrison, M.L. The habitat concept and a plea for standard terminology. *Wildl. Soc. Bull.* **1997**, *25*, 173–182.
4. Krausman, P.R.; Morrison, M.L. Another plea for standard terminology. *J. Wild. Mgmt.* **2016**, *80*, 1143–1144. [CrossRef]
5. Darracq, A.K.; Tandy, J. Misuse of habitat terminology by wildlife educators, scientists, and organizations. *J. Wild. Mgmt.* **2019**, *83*, 782–789. [CrossRef]
6. Johnson, D.H. The comparison of usage and availability measurements for evaluating resource preference. *Ecology* **1980**, *61*, 65–71. [CrossRef]
7. Nichols, G.E. The interpretation and application of certain terms and concepts in the ecological classification of plant communities. *Plant World* **1917**, *20*, 305–319. [CrossRef]
8. Yapp, R.H. The concept of habitat. *J. Ecol.* **1922**, *10*, 1–17. [CrossRef]
9. Garshelis, D.L. Delusions in habitat evaluation: Measuring use, selection, and importance. In *Research Techniques in Animal Ecology: Controversies and Consequences*; Boitani, L., Fuller, T.K., Eds.; Columbia University Press: New York, NY, USA, 2000; pp. 111–164, ISBN 978-0-231-11341-0.
10. Gaillard, J.-M.; Hebblewhite, M.; Loison, A.; Fuller, M.; Powell, R.; Basille, M.; Van Moorter, B. Habitat—Performance relationships: Finding the right metric at a given spatial scale. *Phil. Trans. R. Soc. B* **2010**, *365*, 2255–2265. [CrossRef]
11. Beyer, H.L.; Haydon, D.T.; Morales, J.M.; Frair, J.L.; Hebblewhite, M.; Mitchell, M.; Matthiopoulos, J. The interpretation of habitat preference metrics under use–availability designs. *Phil. Trans. R. Soc. B* **2010**, *365*, 2245–2254. [CrossRef]
12. Wang, H.; Zhong, H.; Hou, R.; Ayala, J.; Liu, G.; Yuan, S.; Yan, Z.; Zhang, W.; Liu, Y.; Cai, K.; et al. A diet diverse in bamboo parts is important for giant panda (*Ailuropoda melanoleuca*) metabolism and health. *Sci. Rep.* **2017**, *7*, 3377. [CrossRef]
13. Swaisgood, R.R.; McShea, W.M.; Wildt, D.; Hull, V.; Zhang, J.; Owen, M.A.; Zhang, Z.; Dvornicky-Raymond, Z.; Valitutto, M.; Li, D.; et al. Giant panda (*Ailuropoda melanoleuca*). In *Bears of the World: Ecology, Conservation and Management*; Penteriani, V., Melletti, M., Eds.; Cambridge University Press: Cambridge, UK, 2021; pp. 63–77, ISBN 978-1-108-48352-0.
14. Vela-Vargas, I.M.; Jorgenson, J.P.; González-Maya, J.F.; Koprowski, J.L. *Tremarctos ornatus* (Carnivora: Ursidae). *Mamm. Species* **2021**, *53*, 78–94. [CrossRef]
15. Soibelzon, L.H.; Grinspan, G.A.; Bocherens, H.; Acosta, W.G.; Jones, W.; Blanco, E.R.; Prevosti, F. South American giant short-faced bear (*Arctotherium angustidens*) diet: Evidence from pathology, morphology, stable isotopes, and biomechanics. *J. Paleontol.* **2014**, *88*, 1240–1250. [CrossRef]
16. Bocherens, H. Isotopic tracking of large carnivore palaeoecology in the mammoth steppe. *Quat. Sci. Rev.* **2015**, *117*, 42–71. [CrossRef]
17. García-Rangel, S. Andean bear *Tremarctos ornatus* natural history and conservation. *Mammal Rev.* **2012**, *42*, 85–119. [CrossRef]
18. Velez-Liendo, X.; Jackson, D.; Ruiz-García, M.; Castellanos, A.; Espinosa, S.; Laguna, A. Andean bear (*Tremarctos ornatus*). In *Bears of the World: Ecology, Conservation and Management*; Penteriani, V., Melletti, M., Eds.; Cambridge University Press: Cambridge, UK, 2021; pp. 78–87, ISBN 978-1-108-48352-0.
19. Seidensticker, J.; Yoganand, K.; Johnsingh, A.J.T. Sloth bears living in seasonally dry tropical and moist broadleaf forests and their conservation. In *The Ecology and Conservation of Seasonally Dry Forests in Asia*; McShea, W.J., Davies, S., Bhumpakphan, N., Eds.; Smithsonian Institution Scholarly Press: Washington, DC, USA, 2011; pp. 217–236.
20. Steinmetz, R.; Garshelis, D.L.; Choudhury, A. Adaptations and competitive interactions of tropical Asian bear species define their biogeography: Past, present, and future. In *Bears of the World: Ecology, Conservation and Management*; Penteriani, V., Melletti, M., Eds.; Cambridge University Press: Cambridge, UK, 2021; pp. 45–52, ISBN 978-1-108-48352-0.
21. Ramesh, T.; Sankar, K.; Qureshi, Q. Additional notes on the diet of sloth bear *Melursus ursinus* in Mudumalai Tiger Reserve as shown by scat analysis. *J. Bombay Nat. Hist. Soc.* **2009**, *106*, 204–206.
22. Baskaran, N.; Desai, A.A. Does indigestible food remains in the scats of sloth bear *Melursus ursinus* (Carnivora: Ursidae) represent actual contribution of various diet items? *J. Threat. Taxa* **2010**, *2*, 1387–1389. [CrossRef]
23. Fredriksson, G.M.; Wich, S.A.; Trisno. Frugivory in sun bears (*Helarctos malayanus*) is linked to El Niño-related fluctuations in fruiting phenology, East Kalimantan, Indonesia. *Biol. J. Linn. Soc.* **2006**, *89*, 489–508. [CrossRef]
24. Wong, S.T.; Servheen, C.; Ambu, L. Food habits of Malayan sun bears in lowland tropical forests of Borneo. *Ursus* **2002**, *13*, 127–136.
25. Fredriksson, G.M. Effects of El Niño and Large-Scale Forest Fires on the Ecology and Conservation of Malayan Sun Bears (*Helarctos malayanus*) in East Kalimantan, Indonesian Borneo. Ph.D. Thesis, University of Amsterdam, Amsterdam, The Netherlands, 2012.
26. Steinmetz, R.; Garshelis, D.L.; Chutipong, W.; Seuaturien, N. The shared preference niche of sympatric Asiatic black bears and sun bears in a tropical forest mosaic. *PLoS ONE* **2011**, *6*, e14509. [CrossRef]
27. Steinmetz, R.; Garshelis, D.L.; Chutipong, W.; Seuaturien, N. Foraging ecology and coexistence of Asiatic black bears and sun bears in a seasonal tropical forest in southeast Asia. *J. Mammal.* **2013**, *94*, 1–18. [CrossRef]

28. Kozakai, C.; Seryodkin, I.; Pigeon, K.E.; Yamazaki, K.; Wangchuk, S.; Koike, S.; Tsubota, T.; Jamtsho, Y. Asiatic black bear (*Ursus thibetanus*). In *Bears of the World: Ecology, Conservation and Management*; Penteriani, V., Melletti, M., Eds.; Cambridge University Press: Cambridge, UK, 2021; pp. 110–121, ISBN 978-1-108-48352-0.
29. Hwang, M.-H.; Garshelis, D.L.; Wang, Y. Diets of Asiatic black bears in Taiwan, with methodological and geographical comparisons. *Ursus* **2002**, *13*, 111–125.
30. Narita, R.; Sugimoto, A.; Takayanagi, A. Animal components in the diet of Japanese black bears *Ursus thibetanus japonicus* in the Kyoto Area, Japan. *Wildl. Biol.* **2006**, *12*, 375–384. [CrossRef]
31. Clark, J.D.; Beckmann, J.P.; Boyce, M.S.; Leopold, B.D.; Loosen, A.E.; Pelton, M.R. American black bear (*Ursus americanus*). In *Bears of the World: Ecology, Conservation and Management*; Penteriani, V., Melletti, M., Eds.; Cambridge University Press: Cambridge, UK, 2021; pp. 122–138, ISBN 978-1-108-48352-0.
32. Fortin, J.K.; Farley, S.D.; Rode, K.D.; Robbins, C.T. Dietary and spatial overlap between sympatric ursids relative to salmon use. *Ursus* **2007**, *18*, 19–29. [CrossRef]
33. Fortin, J.K.; Schwartz, C.C.; Gunther, K.A.; Teisberg, J.E.; Haroldson, M.A.; Evans, M.A.; Robbins, C.T. Dietary adjustability of grizzly bears and American black bears in Yellowstone National Park. *J. Wildl. Manag.* **2013**, *77*, 270–281. [CrossRef]
34. Haroldson, M.A.; Clapham, M.; Costello, C.C.; Gunther, K.A.; Kendall, K.C.; Miller, S.D.; Pigeon, K.E.; Proctor, M.F.; Rode, K.D.; Servheen, C.; et al. Brown bear (*Ursus arctos*; North America). In *Bears of the World: Ecology, Conservation and Management*; Penteriani, V., Melletti, M., Eds.; Cambridge University Press: Cambridge, UK, 2021; pp. 162–195, ISBN 978-1-108-48352-0.
35. Swenson, J.E.; Ambarlı, H.; Arnemo, J.M.; Baskin, L.; Ciucci, P.; Danilov, P.I.; Delibes, M.; Elfström, M.; Evans, A.L.; Groff, C.; et al. Brown bear (*Ursus arctos*; Eurasia). In *Bears of the World: Ecology, Conservation and Management*; Penteriani, V., Melletti, M., Eds.; Cambridge University Press: Cambridge, UK, 2021; pp. 139–161, ISBN 978-1-108-48352-0.
36. Xu, A.; Jiang, Z.; Li, C.; Guo, J.; Wu, G.; Cai, P. Summer food habits of brown bears in Kekexili Nature Reserve, Qinghai–Tibetan Plateau, China. *Ursus* **2006**, *17*, 132–137. [CrossRef]
37. Aryal, A.; Hopkins, J.B.; Raubenheimer, D.; Ji, W.; Brunton, D. Distribution and diet of brown bears in the Upper Mustang Region, Nepal. *Ursus* **2012**, *23*, 231–236. [CrossRef]
38. Bojarska, K.; Selva, N. Spatial patterns in brown bear *Ursus arctos* diet: The role of geographical and environmental factors: Biogeographical variation in brown bear diet. *Mammal Rev.* **2012**, *42*, 120–143. [CrossRef]
39. Kavčič, I.; Adamič, M.; Kaczensky, P.; Krofel, M.; Kobal, M.; Jerina, K. Fast food bears: Brown bear diet in a human-dominated landscape with intensive supplemental feeding. *Wildl. Biol.* **2015**, *21*, 1–8. [CrossRef]
40. Chaulk, K.; Bondrup-Nielsen, S.; Harrington, F. Black bear, *Ursus americanus*, ecology on the northeast coast of Labrador. *Can. Field Nat.* **2005**, *119*, 164. [CrossRef]
41. Bonin, M.; Dussault, C.; Côté, S.D. Increased trophic position of black bear (*Ursus americanus*) at the northern fringe of its distribution range. *Can. J. Zool.* **2020**, *98*, 127–133. [CrossRef]
42. Environment Canada. *Canada Environment Canada Recovery Strategy for the Grizzly Bear (Ursus arctos), Prairie Population, in Canada*; Environment Canada: Ottawa, ON, Canada, 2009.
43. Bjornlie, D.D.; Haroldson, M.A. Grizzly bear occupied range in the Greater Yellowstone Ecosystem, 1990–2020. In *Yellowstone Grizzly Bear Investigations: Annual Report of the Interagency Grizzly Bear Study Team, 2020*; van Manen, F.T., Haroldson, M.A., Karabensh, B.E., Eds.; U.S. Geological Survey: Bozeman, MT, USA, 2021; pp. 24–27.
44. Yamazaki, K.; Kozakai, C.; Koike, S.; Morimoto, H.; Goto, Y.; Furubayashi, K. Myrmecophagy of Japanese black bears in the grasslands of the Ashio area, Nikko National Park, Japan. *Ursus* **2012**, *23*, 52–64. [CrossRef]
45. Izumiyama, S.; Shiraishi, T. Seasonal changes in elevation and habitat use of the Asiatic black bear (*Ursus thibetanus*) in the Northern Japan Alps. *Mammal Study* **2004**, *29*, 1–8. [CrossRef]
46. Joshi, A.R.; Garshelis, D.L.; Smith, J.L.D. Home ranges of sloth bears in Nepal: Implications for conservation. *J. Wildl. Manag.* **1995**, *59*, 204. [CrossRef]
47. Tumendemberel, O.; Proctor, M.; Reynolds, H.; Boulanger, J.; Luvsamjamba, A.; Tserenbataa, T.; Batmunkh, M.; Craighead, D.; Yanjin, N.; Paetkau, D. Gobi bear abundance and inter-oases movements, Gobi Desert, Mongolia. *Ursus* **2015**, *26*, 129–142. [CrossRef]
48. Appleton, R.D.; Van Horn, R.C.; Noyce, K.V.; Spady, T.J.; Swaisgood, R.R.; Arcese, P. Phenotypic plasticity in the timing of reproduction in Andean bears. *J. Zool.* **2018**, *305*, 196–202. [CrossRef]
49. Ghadirian, T.; Qashqaei, A.T.; Soofi, M.; Abolghasemi, H.; Ghoddousi, A. Diet of Asiatic black bear in its westernmost distribution range, Southern Iran. *Ursus* **2017**, *28*, 15–19. [CrossRef]
50. Mosnier, A.; Ouellet, J.-P.; Courtois, R. Black bear adaptation to low productivity in the boreal forest. *Écoscience* **2008**, *15*, 485–497. [CrossRef]
51. Romain, D.A.; Obbard, M.E.; Atkinson, J.L. Temporal variation in food habits of the american black bear (*Ursus americanus*) in the boreal forest of Northern Ontario. *Can. Field Nat.* **2013**, *127*, 118. [CrossRef]
52. Stenset, N.E.; Lutnæs, P.N.; Bjarnadóttir, V.; Dahle, B.; Fossum, K.H.; Jigsved, P.; Johansen, T.; Neumann, W.; Opseth, O.; Rønning, O.; et al. Seasonal and annual variation in the diet of brown bears *Ursus arctos* in the boreal forest of Southcentral Sweden. *Wildl. Biol.* **2016**, *22*, 107–116. [CrossRef]

53. Garshelis, D.; Noyce, K. Seeing the world through the nose of a bear—Diversity of foods fosters behavioral and demographic stability. In *Frontiers in Wildlife Science: Linking Ecological Theory and Management Applications*; Fulbright, T., Hewitt, D., Eds.; CRC Press: Boca Raton, FL, USA, 2008; pp. 139–163, ISBN 978-0-8493-7487-6.
54. Beston, J.A. Variation in life history and demography of the American black bear: Life history of black bears. *J. Wildl. Manag.* **2011**, *75*, 1588–1596. [CrossRef]
55. Tumbelaka, L.; Fredriksson, G. The status of sun bears in Indonesia. In *Understanding Bears to Save Their Future*; Japan Bear Network: Ibaraki, Japan, 2006; pp. 73–78.
56. Crudge, B.; Lees, C.; Hunt, M.; Steinmetz, R.; Fredriksson, G. Sun Bears: Global Status Review & Conservation Action Plan, 2019–2028; IUCN SSC Bear Specialist Group/IUCN SSC Conservation Planning Specialist Group/Free the Bears/TRAFFIC: 2019. Available online: http://www.cbsg.org/content/sun-bears-global-status-review-conservation-action-plan-2019-2028 (accessed on 20 December 2021).
57. Wong, S.T.; Servheen, C.; Ambu, L.; Norhayati, A. Impacts of fruit production cycles on Malayan sun bears and bearded pigs in lowland tropical forest of Sabah, Malaysian Borneo. *J. Trop. Ecol.* **2005**, *21*, 627–639. [CrossRef]
58. Garshelis, D.L.; Joshi, A.R.; Smith, J.L.D. Estimating density and relative abundance of sloth bears. *Ursus* **1999**, *11*, 87–98.
59. Ratnayeke, S.; van Manen, F.T.; Pieris, R.; Pragash, V.S.J. Landscape characteristics of sloth bear range in Sri Lanka. *Ursus* **2007**, *18*, 189–202. [CrossRef]
60. Islam, M.A.; Uddin, M.; Aziz, M.A.; Muzaffar, S.B.; Chakma, S.; Chowdhury, S.U.; Chowdhury, G.W.; Rashid, M.A.; Mohsanin, S.; Jahan, I.; et al. Status of bears in Bangladesh: Going, going, gone? *Ursus* **2013**, *24*, 83–90. [CrossRef]
61. Mayor, S.J.; Schneider, D.C.; Schaefer, J.A.; Mahoney, S.P. Habitat selection at multiple scales. *Écoscience* **2009**, *16*, 238–247. [CrossRef]
62. McGarigal, K.; Wan, H.Y.; Zeller, K.A.; Timm, B.C.; Cushman, S.A. Multi-scale habitat selection modeling: A review and outlook. *Landsc. Ecol.* **2016**, *31*, 1161–1175. [CrossRef]
63. Davis, H.; Weir, R.D.; Hamilton, A.N.; Deal, J.A. Influence of phenology on site selection by female American black bears in Coastal British Columbia. *Ursus* **2006**, *17*, 41–51. [CrossRef]
64. Liu, F.; McShea, W.; Garshelis, D.; Zhu, X.; Wang, D.; Gong, J.; Chen, Y. Spatial distribution as a measure of conservation needs: An example with Asiatic black bears in South-Western China. *Divers. Distrib.* **2009**, *15*, 649–659. [CrossRef]
65. Ngoprasert, D.; Steinmetz, R. Differentiating Asiatic black bears and sun bears from camera-trap photographs. *Int. Bear News* **2012**, *21*, 18–19.
66. Sharp, T.; Dharaiya, N.A.; Garshelis, D.L. Differentiating sloth bears and Asiatic black bears in camera-trap photos. *Int. Bear News* **2016**, *25*, 10–12.
67. Hwang, M.; Ditmer, M.A.; Teo, S.; Wong, S.T.; Garshelis, D.L. Sun bears use 14-year-old previously logged forest more than primary forest in Sabah, Malaysia. *Ecosphere* **2021**, *12*, e03769. [CrossRef]
68. Baldwin, R. Use of maximum entropy modeling in wildlife research. *Entropy* **2009**, *11*, 854–866. [CrossRef]
69. Elith, J.; Leathwick, J.R. Species distribution models: Ecological explanation and prediction across space and time. *Annu. Rev. Ecol. Evol. Syst.* **2009**, *40*, 677–697. [CrossRef]
70. Elith, J.; Phillips, S.J.; Hastie, T.; Dudík, M.; Chee, Y.E.; Yates, C.J. A statistical explanation of MaxEnt for ecologists: Statistical explanation of MaxEnt. *Divers. Distrib.* **2011**, *17*, 43–57. [CrossRef]
71. Guisan, A.; Thuiller, W.; Zimmermann, N.E. *Habitat Suitability and Distribution Models: With Applications in R*; Cambridge University Press: New York, NY, USA, 2017; ISBN 978-1-108-50849-0.
72. Fieberg, J.R.; Forester, J.D.; Street, G.M.; Johnson, D.H.; ArchMiller, A.A.; Matthiopoulos, J. Used-habitat calibration plots: A new procedure for validating species distribution, resource selection, and step-selection models. *Ecography* **2018**, *41*, 737–752. [CrossRef]
73. Hao, T.; Elith, J.; Guillera-Arroita, G.; Lahoz-Monfort, J.J. A review of evidence about use and performance of species distribution modelling ensembles like BIOMOD. *Divers. Distrib.* **2019**, *25*, 839–852. [CrossRef]
74. Nazeri, M.; Jusoff, K.; Madani, N.; Mahmud, A.R.; Bahman, A.R.; Kumar, L. Predictive modeling and mapping of Malayan sun bear (*Helarctos malayanus*) distribution using maximum entropy. *PLoS ONE* **2012**, *7*, e48104. [CrossRef]
75. Lozier, J.D.; Aniello, P.; Hickerson, M.J. Predicting the distribution of Sasquatch in Western North America: Anything goes with ecological niche modelling. *J. Biogeogr.* **2009**, *36*, 1623–1627. [CrossRef]
76. Kim, T.-G.; Yang, D.; Cho, Y.; Song, K.-H.; Oh, J.-G. Habitat distribution change prediction of Asiatic black bears (*Ursus thibetanus*) using Maxent modeling approach. *Korean J. Ecol. Environ.* **2016**, *49*, 197–207. [CrossRef]
77. Wolf, C.; Ripple, W.J. Rewilding the world's large carnivores. *R. Soc. Open Sci.* **2018**, *5*, 172235. [CrossRef]
78. Northrup, J.M.; Hooten, M.B.; Anderson, C.R.; Wittemyer, G. Practical guidance on characterizing availability in resource selection functions under a use—availability design. *Ecology* **2013**, *94*, 1456–1463. [CrossRef]
79. Martin, J.; Calenge, C.; Quenette, P.-Y.; Allainé, D. Importance of movement constraints in habitat selection studies. *Ecol. Model.* **2008**, *213*, 257–262. [CrossRef]
80. Fieberg, J.; Matthiopoulos, J.; Hebblewhite, M.; Boyce, M.S.; Frair, J.L. Correlation and studies of habitat selection: Problem, red herring or opportunity? *Phil. Trans. R. Soc. B* **2010**, *365*, 2233–2244. [CrossRef]
81. Thurfjell, H.; Ciuti, S.; Boyce, M.S. Applications of step-selection functions in ecology and conservation. *Mov. Ecol.* **2014**, *2*, 4. [CrossRef]

82. Garshelis, D.; Steinmtez, R. Publication gradient among bear species tied to conservation needs. *Int. Bear News* **2015**, *24*, 7–9.
83. Peyton, B. Criteria for assessing habitat quality of the spectacled bear in Machu Picchu, Peru. *Bears Biol. Manag.* **1987**, *7*, 135. [CrossRef]
84. Peyton, B. Habitat components of the spectacled bear in Machu Picchu, Peru. *Bears Biol. Manag.* **1987**, *7*, 127. [CrossRef]
85. Cuesta, F.; Peralvo, M.F.; van Manen, F.T. Andean bear habitat use in the Oyacachi River Basin. *Ecuador* **2003**, *14*, 198–209.
86. Peralvo, M.F.; Cuesta, F.; van Manen, F. Delineating priority habitat areas for the conservation of Andean bears in Northern Ecuador. *Ursus* **2005**, *16*, 222–233. [CrossRef]
87. Garshelis, D.L. Andean bear density and abundance estimates—How reliable and useful are they? *Ursus* **2011**, *22*, 47–64. [CrossRef]
88. Ríos-Uzeda, B.; Gómez, H.; Wallace, R.B. Habitat preferences of the Andean bear (*Tremarctos ornatus*) in the Bolivian Andes. *J. Zool.* **2006**, *268*, 271–278. [CrossRef]
89. Velez–Liendo, X.; Strubbe, D.; Matthysen, E. Effects of variable selection on modelling habitat and potential distribution of the Andean bear in Bolivia. *Ursus* **2013**, *24*, 127–138. [CrossRef]
90. Velez–Liendo, X.; Adriaensen, F.; Matthysen, E. Landscape assessment of habitat suitability and connectivity for Andean bears in the Bolivian Tropical Andes. *Ursus* **2014**, *25*, 172–187. [CrossRef]
91. Meza Mori, G.; Barboza Castillo, E.; Torres Guzmán, C.; Cotrina Sánchez, D.A.; Guzman Valqui, B.K.; Oliva, M.; Bandopadhyay, S.; Salas López, R.; Rojas Briceño, N.B. Predictive modelling of current and future potential distribution of the spectacled bear (*Tremarctos ornatus*) in Amazonas, Northeast Peru. *Animals* **2020**, *10*, 1816. [CrossRef]
92. Morrell, N.; Appleton, R.D.; Arcese, P. Roads, forest cover, and topography as factors affecting the occurrence of large carnivores: The case of the Andean bear (*Tremarctos ornatus*). *Glob. Ecol. Conserv.* **2021**, *26*, e01473. [CrossRef]
93. Scotson, L.; Fredriksson, G.; Ngoprasert, D.; Wong, W.-M.; Fieberg, J. Projecting range-wide sun bear population trends using tree cover and camera-trap bycatch data. *PLoS ONE* **2017**, *12*, e0185336. [CrossRef]
94. Schneider, M.; Ziegler, T.; Kolter, L. Thermoregulation in Malayan sun bears (*Helarctos malayanus*) and its consequences for in situ conservation. *J. Therm. Biol.* **2020**, *91*, 102646. [CrossRef]
95. Lindsell, J.A.; Lee, D.C.; Powell, V.J.; Gemita, E. Availability of large seed-dispersers for restoration of degraded tropical forest. *Trop. Conserv. Sci.* **2015**, *8*, 17–27. [CrossRef]
96. Sethy, J.; Chauhan, N.P.S. Assessing habitat use by sun bears in Namdapha Tiger Reserve, Arunachal Pradesh, India. *Appl. Ecol. Env. Res.* **2016**, *14*, 215–236. [CrossRef]
97. Lee, D.C.; Powell, V.J.; Lindsell, J.A. Understanding landscape and plot-scale habitat utilisation by Malayan sun bear (*Helarctos malayanus*) in degraded lowland forest. *Acta Oecologica* **2019**, *96*, 1–9. [CrossRef]
98. Guharajan, R.; Arnold, T.W.; Bolongon, G.; Dibden, G.H.; Abram, N.K.; Teoh, S.W.; Magguna, M.A.; Goossens, B.; Wong, S.T.; Nathan, S.K.S.S.; et al. Survival strategies of a frugivore, the sun bear, in a forest-oil palm landscape. *Biodivers Conserv.* **2018**, *27*, 3657–3677. [CrossRef]
99. Guharajan, R.; Mohamed, A.; Wong, S.T.; Niedballa, J.; Petrus, A.; Jubili, J.; Lietz, R.; Clements, G.R.; Wong, W.-M.; Kissing, J.; et al. Sustainable forest management is vital for the persistence of sun bear *Helarctos malayanus* populations in Sabah, Malaysian Borneo. *For. Ecol. Manag.* **2021**, *493*, 119270. [CrossRef]
100. Tilker, A.; Abrams, J.F.; Mohamed, A.; Nguyen, A.; Wong, S.T.; Sollmann, R.; Niedballa, J.; Bhagwat, T.; Gray, T.N.E.; Rawson, B.M.; et al. Habitat degradation and indiscriminate hunting differentially impact faunal communities in the Southeast Asian tropical biodiversity hotspot. *Commun. Biol.* **2019**, *2*, 396. [CrossRef] [PubMed]
101. Abidin, K.Z.; Lihan, T.; Taher, T.M.; Nazri, N.; Zaini, I.-H.A.; Mansor, M.S.; Topani, R.; Nor, S.M. Predicting potential conflict areas of the Malayan sun bear (*Helarctos malayanus*) in Peninsular Malaysia using maximum entropy model. *Mammal Study* **2019**, *44*, 193. [CrossRef]
102. Normua, F.; Higashi, S.; Ambu, L.; Mohamed, M. Notes on oil palm plantation use and seasonal spatial relationships of sun bears in Sabah, Malaysia. *Ursus* **2004**, *15*, 227–231. [CrossRef]
103. Cheah, C.P.I. The Ecology of Malayan Sun Bears (*Helarctos malayanus*) at the Krau Wildlife Reserve, Pahang, Malaysia and Adjacent Plantations. Ph.D. Thesis, University Putra Malaysia, Selangor, Malaysia, 2013.
104. Tee, T.L.; van Manen, F.T.; Kretzschmar, P.; Sharp, S.P.; Wong, S.T.; Gadas, S.; Ratnayeke, S. Anthropogenic edge effects in habitat selection by sun bears in a protected area. *Wildl. Biol.* **2021**, *2021*, wlb.00776. [CrossRef]
105. Kunde, M.N.; Martins, R.F.; Premier, J.; Fickel, J.; Förster, D.W. Population and landscape genetic analysis of the Malayan sun bear *Helarctos malayanus*. *Conserv. Genet.* **2020**, *21*, 123–135. [CrossRef]
106. Guharajan, R.; Abram, N.K.; Magguna, M.A.; Goossens, B.; Wong, S.T.; Nathan, S.K.S.S.; Garshelis, D.L. Does the vulnerable sun bear *Helarctos malayanus* damage crops and threaten people in oil palm plantations? *Oryx* **2019**, *53*, 611–619. [CrossRef]
107. Ghimire, D.; Thapa, T.B. Distribution and habitat preference of sloth bear in Chitwan National Park, Nepal. *J. Nat. Hist. Mus.* **2015**, *28*, 9–17. [CrossRef]
108. Ratnayeke, S.; van Manen, F.T.; Padmalal, U.K.G.K. Home ranges and habitat use of sloth bears *Melursus ursinus inornatus* in Wasgomuwa National Park, Sri Lanka. *Wildl. Biol.* **2007**, *13*, 272–284. [CrossRef]
109. Akhtar, N.; Singh Bargali, H.; Chauhan, N.P.S. Sloth bear habitat use in disturbed and unprotected areas of Madhya Pradesh, India. *Ursus* **2004**, *15*, 203–211. [CrossRef]

110. Akhtar, N.; Bargali, H.S.; Chauhan, N.P.S. Characteristics of sloth bear day dens and use in disturbed and unprotected habitat of North Bilaspur Forest Division, Chhattisgarh, central India. *Ursus* **2007**, *18*, 203–208. [CrossRef]
111. Dutta, T.; Sharma, S.; Maldonado, J.E.; Panwar, H.S.; Seidensticker, J. Genetic variation, structure, and gene flow in a sloth bear (*Melursus ursinus*) meta-population in the Satpura-Maikal landscape of central India. *PLoS ONE* **2015**, *10*, e0123384. [CrossRef] [PubMed]
112. Thatte, P.; Chandramouli, A.; Tyagi, A.; Patel, K.; Baro, P.; Chhattani, H.; Ramakrishnan, U. Human footprint differentially impacts genetic connectivity of four wide-ranging mammals in a fragmented landscape. *Divers. Distrib.* **2020**, *26*, 299–314. [CrossRef]
113. Puri, M.; Srivathsa, A.; Karanth, K.K.; Kumar, N.S.; Karanth, K.U. Multiscale distribution models for conserving widespread species: The case of sloth bear *Melursus ursinus* in India. *Divers. Distrib.* **2015**, *21*, 1087–1100. [CrossRef]
114. Srivathsa, A.; Puri, M.; Kumar, N.S.; Jathanna, D.; Karanth, K.U. Substituting space for time: Empirical evaluation of spatial replication as a surrogate for temporal replication in occupancy modelling. *J Appl Ecol* **2018**, *55*, 754–765. [CrossRef]
115. Ramesh, T.; Kalle, R.; Sankar, K.; Qureshi, Q. Factors affecting habitat patch use by sloth bears in Mudumalai Tiger Reserve, Western Ghats, India. *Ursus* **2012**, *23*, 78–85. [CrossRef]
116. Carr, M.M.; Yoshizaki, J.; van Manen, F.T.; Pelton, M.R.; Huygens, O.C.; Hayashi, H.; Maekawa, M. A multi-scale assessment of habitat use by Asiatic black bears in central Japan. *Ursus* **2002**, *13*, 1–9.
117. Doko, T.; Fukui, H.; Kooiman, A.; Toxopeus, A.G.; Ichinose, T.; Chen, W.; Skidmore, A.K. Identifying habitat patches and potential ecological corridors for remnant Asiatic black bear (*Ursus thibetanus japonicus*) populations in Japan. *Ecol. Model.* **2011**, *222*, 748–761. [CrossRef]
118. Takahata, C.; Nishino, S.; Kido, K.; Izumiyama, S. An evaluation of habitat selection of Asiatic black bears in a season of prevalent conflicts. *Ursus* **2013**, *24*, 16–26. [CrossRef]
119. Takahata, C.; Nielsen, S.E.; Takii, A.; Izumiyama, S. Habitat selection of a large carnivore along human-wildlife boundaries in a highly modified landscape. *PLoS ONE* **2014**, *9*, e86181. [CrossRef] [PubMed]
120. Hwang, M.-H.; Garshelis, D.L.; Wu, Y.-H.; Wang, Y. Home ranges of Asiatic black bears in the Central Mountains of Taiwan: Gauging whether a reserve is big enough. *Ursus* **2010**, *21*, 81–96. [CrossRef]
121. Shen, Y.; Liu, M.; Wang, D.; Shen, X.; Li, S. Using an integrative mapping approach to identify the distribution range and conservation needs of a large threatened mammal, the Asiatic black bear, in China. *Glob. Ecol. Conserv.* **2021**, *31*, e01831. [CrossRef]
122. Bista, M.; Panthi, S.; Weiskopf, S.R. Habitat overlap between Asiatic black bear *Ursus thibetanus* and red panda *Ailurus fulgens* in Himalaya. *PLoS ONE* **2018**, *13*, e0203697. [CrossRef] [PubMed]
123. Goursi, U.H.; Anwar, M.; Bosso, L.; Nawaz, M.A.; Kabir, M. Spatial distribution of the threatened Asiatic black bear in northern Pakistan. *Ursus* **2021**, *2021*, 1–5. [CrossRef]
124. Awan, M.N.; Awan, M.S.; Nawaz, M.A.; Hameed, S.; Kabir, M.; Lee, D.C. Landscape associations of Asiatic black bears in Kashmir Himalaya, Pakistan. *Ursus* **2021**, *2021*, 1–10. [CrossRef]
125. Zahoor, B.; Liu, X.; Kumar, L.; Dai, Y.; Tripathy, B.R.; Songer, M. Projected shifts in the distribution range of Asiatic black bear (*Ursus thibetanus*) in the Hindu Kush Himalaya due to climate change. *Ecol. Inform.* **2021**, *63*, 101312. [CrossRef]
126. Su, H.; Bista, M.; Li, M. Mapping habitat suitability for Asiatic black bear and red panda in Makalu Barun National Park of Nepal from Maxent and GARP models. *Sci. Rep.* **2021**, *11*, 14135. [CrossRef]
127. Ngoprasert, D.; Steinmetz, R.; Reed, D.H.; Savini, T.; Gale, G.A. Influence of fruit on habitat selection of Asian bears in a Tropical Forest. *J. Wildl. Manag.* **2011**, *75*, 588–595. [CrossRef]
128. Scotson, L.; Ross, S.; Arnold, T.W. Monitoring sun bears and Asiatic black bears with remotely sensed predictors to inform conservation management. *Oryx* **2021**, *55*, 131–138. [CrossRef]
129. Almasieh, K.; Kaboli, M.; Beier, P. Identifying habitat cores and corridors for the Iranian black bear in Iran. *Ursus* **2016**, *27*, 18–30. [CrossRef]
130. Farashi, A.; Erfani, M. Modeling of habitat suitability of Asiatic black bear (*Ursus thibetanus gedrosianus*) in Iran in future. *Acta Ecol. Sin.* **2018**, *38*, 9–14. [CrossRef]
131. Morovati, M.; Karami, P.; Bahadori Amjas, F. Accessing habitat suitability and connectivity for the westernmost population of Asian black bear (*Ursus thibetanus gedrosianus*, Blanford, 1877) based on climate changes scenarios in Iran. *PLoS ONE* **2020**, *15*, e0242432. [CrossRef]
132. Swaisgood, R.R.; Wang, D.; Wei, F. Panda downlisted but not out of the woods. *Cons. Lett.* **2017**, *11*, e12355. [CrossRef]
133. Hull, V.; Roloff, G.; Zhang, J.; Liu, W.; Zhou, S.; Huang, J.; Xu, W.; Ouyang, Z.; Zhang, H.; Liu, J. A synthesis of giant panda habitat selection. *Ursus* **2014**, *25*, 148–162. [CrossRef]
134. Zhang, Z.; Swaisgood, R.R.; Zhang, S.; Nordstrom, L.A.; Wang, H.; Gu, X.; Hu, J.; Wei, F. Old-growth forest is what giant pandas really need. *Biol. Lett.* **2011**, *7*, 403–406. [CrossRef]
135. Liu, X.; Cheng, X.; Skidmore, A.K. Potential solar radiation pattern in relation to the monthly distribution of giant pandas in Foping Nature Reserve, China. *Ecol. Model.* **2011**, *222*, 645–652. [CrossRef]
136. Zhang, Z.; Sheppard, J.K.; Swaisgood, R.R.; Wang, G.; Nie, Y.; Wei, W.; Zhao, N.; Wei, F. Ecological scale and seasonal heterogeneity in the spatial behaviors of giant pandas. *Integr. Zool.* **2014**, *9*, 46–60. [CrossRef]
137. Hull, V.; Zhang, J.; Huang, J.; Zhou, S.; Viña, A.; Shortridge, A.; Li, R.; Liu, D.; Xu, W.; Ouyang, Z.; et al. Habitat use and selection by giant pandas. *PLoS ONE* **2016**, *11*, e0162266. [CrossRef]

138. Bai, W.; Huang, Q.; Zhang, J.; Stabach, J.; Huang, J.; Yang, H.; Songer, M.; Connor, T.; Liu, J.; Zhou, S.; et al. Microhabitat selection by giant pandas. *Biol. Conserv.* **2020**, *247*, 108615. [CrossRef]
139. Wei, W.; Swaisgood, R.R.; Dai, Q.; Yang, Z.; Yuan, S.; Owen, M.A.; Pilfold, N.W.; Yang, X.; Gu, X.; Zhou, H.; et al. Giant panda distributional and habitat-use shifts in a changing landscape. *Conserv. Lett.* **2018**, *11*, e12575. [CrossRef]
140. Yang, H.; Viña, A.; Tang, Y.; Zhang, J.; Wang, F.; Zhao, Z.; Liu, J. Range-wide evaluation of wildlife habitat change: A demonstration using giant pandas. *Biol. Conserv.* **2017**, *213*, 203–209. [CrossRef]
141. Li, C.; Connor, T.; Bai, W.; Yang, H.; Zhang, J.; Qi, D.; Zhou, C. Dynamics of the giant panda habitat suitability in response to changing anthropogenic disturbance in the Liangshan mountains. *Biol. Conserv.* **2019**, *237*, 445–455. [CrossRef]
142. Liu, Z.-X.; Dayananda, B.; Jeffree, R.A.; Tian, C.; Zhang, Y.-Y.; Yu, B.; Zheng, Y.; Jing, Y.; Si, P.-Y.; Li, J.-Q. Giant panda distribution and habitat preference: The influence of sympatric large mammals. *Glob. Ecol. Conserv.* **2020**, *24*, e01221. [CrossRef]
143. Hull, V.; Zhang, J.; Zhou, S.; Huang, J.; Viña, A.; Liu, W.; Tuanmu, M.-N.; Li, R.; Liu, D.; Xu, W.; et al. Impact of livestock on giant pandas and their habitat. *J. Nat. Conserv.* **2014**, *22*, 256–264. [CrossRef]
144. Wang, X.; Huang, J.; Connor, T.A.; Bai, W.; Zhang, J.; Wei, W.; Zhang, Z.; Liu, D.; Zhou, C. Impact of livestock grazing on biodiversity and giant panda habitat. *J. Wildl. Manag.* **2019**, *83*, 1592–1597. [CrossRef]
145. Li, B.V.; Kim, M.J.; Xu, W.; Jiang, S.; Yu, L. Increasing livestock grazing, the unintended consequence of community development funding on giant panda habitat. *Biol. Conserv.* **2021**, *257*, 109074. [CrossRef]
146. Qi, D.; Xu, C.; Hou, R.; Chen, P.; Owens, J.R.; Zhang, Z.; Gu, X.; Yang, Z.; Chen, L. Using habitat models to evaluate protected area designing for giant pandas. *Folia Zool.* **2015**, *64*, 56–64. [CrossRef]
147. Zhao, C.; Yue, B.; Ran, J.; Moermond, T.; Hou, N.; Yang, X.; Gu, X. Relationship between human disturbance and endangered giant panda *Ailuropoda melanoleuca* habitat use in the Daxiangling Mountains. *Oryx* **2017**, *51*, 146–152. [CrossRef]
148. Ruan, T.; Han, H.; Wei, W.; Qiu, L.; Hong, M.; Tang, J.; Zhou, H.; Zhang, Z. Habitat suitability evaluation for giant panda in Liziping National Nature Reserve, Sichuan Province. *Glob. Ecol. Conserv.* **2021**, *30*, e01780. [CrossRef]
149. Kang, D.; Zhao, Z.; Chen, X.; Lin, Y.; Wang, X.; Li, J. Evaluating the effects of roads on giant panda habitat at two scales in a typical nature reserve. *Sci. Total Environ.* **2020**, *710*, 136351. [CrossRef]
150. Kang, D. A review of the impacts of four identified major human disturbances on the habitat and habitat use of wild giant pandas from 2015 to 2020. *Sci. Total Environ.* **2021**, *763*, 142975. [CrossRef]
151. Qiu, L.; Han, H.; Zhou, H.; Hong, M.; Zhang, Z.; Yang, X.; Gu, X.; Zhang, W.; Wei, W.; Dai, Q. Disturbance control can effectively restore the habitat of the giant panda (*Ailuropoda melanoleuca*). *Biol. Conserv.* **2019**, *238*, 108233. [CrossRef]
152. Zhang, Y.; Mathewson, P.D.; Zhang, Q.; Porter, W.P.; Ran, J. An ecophysiological perspective on likely giant panda habitat responses to climate change. *Glob. Change Biol.* **2018**, *24*, 1804–1816. [CrossRef]
153. Huang, Q.; Lothspeich, A.; Hernández-Yáñez, H.; Mertes, K.; Liu, X.; Songer, M. What drove giant panda *Ailuropoda melanoleuca* expansion in the Qinling mountains? An analysis comparing the influence of climate, bamboo, and various landscape variables in the past decade. *Environ. Res. Lett.* **2020**, *15*, 084036. [CrossRef]
154. Wang, F.; Zhao, Q.; McShea, W.J.; Songer, M.; Huang, Q.; Zhang, X.; Zhou, L. Incorporating biotic interactions reveals potential climate tolerance of giant pandas. *Conserv. Lett.* **2018**, *11*, e12592. [CrossRef]
155. Han, H.; Wei, W.; Hu, Y.; Nie, Y.; Ji, X.; Yan, L.; Zhang, Z.; Shi, X.; Zhu, L.; Luo, Y.; et al. Diet evolution and habitat contraction of giant pandas via stable isotope analysis. *Curr. Biol.* **2019**, *29*, 664–669.e2. [CrossRef]
156. Sun, X.; Long, Z.; Jia, J. A multi-scale maxent approach to model habitat suitability for the giant pandas in the Qionglai Mountain, China. *Glob. Ecol. Conserv.* **2021**, *30*, e01766. [CrossRef]
157. Farashi, A. Identifying key habitats to conserve the threatened brown bear in northern Iran. *Russ. J. Ecol.* **2018**, *49*, 449–455. [CrossRef]
158. Kouchali, F.; Nezami, B.; Goshtasb, H.; Rayegani, B.; Ramezani, J. Brown bear (*Ursus arctos*) habitat suitability modelling in the Alborz Mountains. *IJESB* **2019**, *12*, 45–54. [CrossRef]
159. Habibzadeh, N.; Ashrafzadeh, M.R. Habitat suitability and connectivity for an endangered brown bear population in the Iranian Caucasus. *Wildl. Res.* **2018**, *45*, 602. [CrossRef]
160. Ahmadipari, M.; Yavari, A.; Ghobadi, M. Ecological monitoring and assessment of habitat suitability for brown bear species in the Oshtorankooh Protected Area, Iran. *Ecol. Indic.* **2021**, *126*, 107606. [CrossRef]
161. Almasieh, K.; Rouhi, H.; Kaboodvandpour, S. Habitat suitability and connectivity for the brown bear (*Ursus arctos*) along the Iran-Iraq border. *Eur. J. Wildl. Res.* **2019**, *65*, 57. [CrossRef]
162. Ansari, H.M.; Ghoddousi, A. Water availability limits brown bear distribution at the southern edge of its global range. *Ursus* **2018**, *29*, 13–24. [CrossRef]
163. Mohammadi, A.; Almasieh, K.; Nayeri, D.; Ataei, F.; Khani, A.; López-Bao, J.V.; Penteriani, V.; Cushman, S.A. Identifying priority core habitats and corridors for effective conservation of brown bears in Iran. *Sci. Rep.* **2021**, *11*, 1044. [CrossRef]
164. Ashrafzadeh, M.-R.; Khosravi, R.; Ahmadi, M.; Kaboli, M. Landscape heterogeneity and ecological niche isolation shape the distribution of spatial genetic variation in Iranian brown bears, *Ursus arctos* (Carnivora: Ursidae). *Mamm. Biol.* **2018**, *93*, 64–75. [CrossRef]
165. Suel, H. Brown bear (*Ursus arctos*) habitat suitability modelling and mapping. *Appl. Ecol. Env. Res.* **2019**, *17*, 4245–4255. [CrossRef]

166. Cozzi, G.; Chynoweth, M.; Kusak, J.; Çoban, E.; Çoban, A.; Ozgul, A.; Şekercioğlu, Ç.H. Anthropogenic Food resources foster the coexistence of distinct life history strategies: Year-round sedentary and migratory brown bears. *J. Zool.* **2016**, *300*, 142–150. [CrossRef]
167. Su, J.; Aryal, A.; Hegab, I.M.; Shrestha, U.B.; Coogan, S.C.P.; Sathyakumar, S.; Dalannast, M.; Dou, Z.; Suo, Y.; Dabu, X.; et al. Decreasing brown bear (*Ursus arctos*) habitat due to climate change in central Asia and the Asian highlands. *Ecol. Evol.* **2018**, *8*, 11887–11899. [CrossRef]
168. Dai, Y.; Hacker, C.E.; Zhang, Y.; Li, W.; Zhang, Y.; Liu, H.; Zhang, J.; Ji, Y.; Xue, Y.; Li, D. Identifying climate refugia and its potential impact on Tibetan brown bear (*Ursus arctos pruinosus*) in Sanjiangyuan National Park, China. *Ecol. Evol.* **2019**, *9*, 13278–13293. [CrossRef]
169. Dai, Y.; Hacker, C.E.; Zhang, Y.; Li, W.; Li, J.; Zhang, Y.; Bona, G.; Liu, H.; Li, Y.; Xue, Y.; et al. Identifying the risk regions of house break-ins caused by Tibetan brown bears (*Ursus arctos pruinosus*) in the Sanjiangyuan region, China. *Ecol. Evol.* **2019**, *9*, 13979–13990. [CrossRef]
170. Dar, S.A.; Singh, S.K.; Wan, H.Y.; Kumar, V.; Cushman, S.A.; Sathyakumar, S. Projected climate change threatens Himalayan brown bear habitat more than human land use. *Anim. Conserv.* **2021**, *24*, 659–676. [CrossRef]
171. Chapron, G.; Kaczensky, P.; Linnell, J.D.C.; von Arx, M.; Huber, D.; Andrén, H.; López-Bao, J.V.; Adamec, M.; Álvares, F.; Anders, O.; et al. Recovery of large carnivores in Europe's modern human-dominated landscapes. *Science* **2014**, *346*, 1517–1519. [CrossRef] [PubMed]
172. Zarzo-Arias, A.; Delgado, M.d.M.; Ordiz, A.; García Díaz, J.; Cañedo, D.; González, M.A.; Romo, C.; Vázquez García, P.; Bombieri, G.; Bettega, C.; et al. Brown bear behaviour in human-modified landscapes: The case of the endangered Cantabrian population, NW Spain. *Glob. Ecol. Conserv.* **2018**, *16*, e00499. [CrossRef]
173. Martin, J.; Revilla, E.; Quenette, P.-Y.; Naves, J.; Allainé, D.; Swenson, J.E. Brown bear habitat suitability in the Pyrenees: Transferability across sites and linking scales to make the most of scarce data. *J. Appl. Ecol.* **2012**, *49*, 621–631. [CrossRef]
174. Piédallu, B.; Quenette, P.-Y.; Bombillon, N.; Gastineau, A.; Miquel, C.; Gimenez, O. Determinants and patterns of habitat use by the brown bear *Ursus arctos* in the French Pyrenees revealed by occupancy modelling. *Oryx* **2019**, *53*, 334–343. [CrossRef]
175. Mateo Sánchez, M.C.; Cushman, S.A.; Saura, S. Scale dependence in habitat selection: The case of the endangered brown bear (*Ursus arctos*) in the Cantabrian range (NW Spain). *Int. J. Geogr. Inf. Sci.* **2014**, *28*, 1531–1546. [CrossRef]
176. Posillico, M.; Meriggi, A.; Pagnin, E.; Lovari, S.; Russo, L. A Habitat model for brown bear conservation and land use planning in the central Apennines. *Biol. Conserv.* **2004**, *118*, 141–150. [CrossRef]
177. Falcucci, A.; Ciucci, P.; Maiorano, L.; Gentile, L.; Boitani, L. Assessing habitat quality for conservation using an integrated occurrence-mortality model. *J. Appl. Ecol.* **2009**, *46*, 600–609. [CrossRef]
178. Maiorano, L.; Boitani, L.; Monaco, A.; Tosoni, E.; Ciucci, P. Modeling the distribution of Apennine brown bears during hyperphagia to reduce the impact of wild boar hunting. *Eur. J. Wildl. Res.* **2015**, *61*, 241–253. [CrossRef]
179. Peters, W.; Hebblewhite, M.; Cavedon, M.; Pedrotti, L.; Mustoni, A.; Zibordi, F.; Groff, C.; Zanin, M.; Cagnacci, F. Resource selection and connectivity reveal conservation challenges for reintroduced brown bears in the Italian Alps. *Biol. Conserv.* **2015**, *186*, 123–133. [CrossRef]
180. de Gabriel Hernando, M.; Karamanlidis, A.; Grivas, K.; Krambokoukis, L.; Papakostas, G.; Beecham, J. Habitat use and selection patterns inform habitat conservation priorities of an endangered large carnivore in southern Europe. *Endang. Species. Res.* **2021**, *44*, 203–215. [CrossRef]
181. De Angelis, D.; Huber, D.; Reljic, S.; Ciucci, P.; Kusak, J. Factors affecting the home range of dinaric-pindos brown bears. *J. Mammal.* **2021**, *102*, 481–493. [CrossRef]
182. Penteriani, V.; Lamamy, C.; Kojola, I.; Heikkinen, S.; Bombieri, G.; del Mar Delgado, M. Does artificial feeding affect large carnivore behaviours? The case study of brown bears in a hunted and tourist exploited subpopulation. *Biol. Conserv.* **2021**, *254*, 108949. [CrossRef]
183. Skuban, M.; Find'o, S.; Kajba, M. Human impacts on bear feeding habits and habitat selection in the Pol'ana Mountains, Slovakia. *Eur. J. Wildl. Res.* **2016**, *62*, 353–364. [CrossRef]
184. Martin, J.; Basille, M.; Van Moorter, B.; Kindberg, J.; Allainé, D.; Swenson, J.E. Coping with human disturbance: Spatial and temporal tactics of the brown bear (*Ursus arctos*). *Can. J. Zool.* **2010**, *88*, 875–883. [CrossRef]
185. Leclerc, M.; Vander Wal, E.; Zedrosser, A.; Swenson, J.E.; Kindberg, J.; Pelletier, F. Quantifying consistent individual differences in habitat selection. *Oecologia* **2016**, *180*, 697–705. [CrossRef]
186. Hertel, A.G.; Leclerc, M.; Warren, D.; Pelletier, F.; Zedrosser, A.; Mueller, T. Don't poke the bear: Using tracking data to quantify behavioural syndromes in elusive wildlife. *Anim. Behav.* **2019**, *147*, 91–104. [CrossRef]
187. Schwartz, C.C.; Haroldson, M.A.; White, G.C.; Harris, R.B.; Cherry, S.; Keating, K.A.; Moody, D.; Servheen, C. Temporal, spatial, and environmental influences on the demographics of grizzly bears in the Greater Yellowstone Ecosystem. *Wildl. Monogr.* **2006**, *161*, 1–68. [CrossRef]
188. Costello, C.M.; Manen, F.T.; Haroldson, M.A.; Ebinger, M.R.; Cain, S.L.; Gunther, K.A.; Bjornlie, D.D. Influence of whitebark pine decline on fall habitat use and movements of grizzly bears in the Greater Yellowstone Ecosystem. *Ecol. Evol.* **2014**, *4*, 2004–2018. [CrossRef]
189. Nielsen, S.E.; Boyce, M.S.; Stenhouse, G.B. Grizzly bears and forestry. *For. Ecol. Manag.* **2004**, *199*, 51–65. [CrossRef]

190. Pigeon, K.E.; Cardinal, E.; Stenhouse, G.B.; Côté, S.D. Staying cool in a changing landscape: The influence of maximum daily ambient temperature on grizzly bear habitat selection. *Oecologia* **2016**, *181*, 1101–1116. [CrossRef] [PubMed]
191. Stewart, B.P.; Nelson, T.A.; Laberee, K.; Nielsen, S.E.; Wulder, M.A.; Stenhouse, G. Quantifying grizzly bear selection of natural and anthropogenic edges: Grizzly bear edge selection. *J. Wild. Mgmt.* **2013**, *77*, 957–964. [CrossRef]
192. Nielsen, S.E.; Shafer, A.B.A.; Boyce, M.S.; Stenhouse, G.B. Does learning or instinct shape habitat selection? *PLoS ONE* **2013**, *8*, e53721. [CrossRef] [PubMed]
193. Denny, C.K.; Stenhouse, G.B.; Nielsen, S.E. Scales of selection and perception: Landscape heterogeneity of an important food resource influences habitat use by a large omnivore. *Wildl. Biol.* **2018**, *2018*, wlb.00409. [CrossRef]
194. Ciarniello, L.M.; Boyce, M.S.; Seip, D.R.; Heard, D.C. Grizzly bear habitat selection is scale dependent. *Ecol. Appl.* **2007**, *17*, 1424–1440. [CrossRef]
195. Apps, C.D.; McLellan, B.N.; Woods, J.G.; Proctor, M.F. Estimating grizzly bear distribution and abundance relative to habitat and human influence. *J. Wildl. Manag.* **2004**, *68*, 138–152. [CrossRef]
196. Nams, V.O.; Mowat, G.; Panian, M.A. Determining the spatial scale for conservation purposes—An example with grizzly bears. *Biol. Conserv.* **2006**, *128*, 109–119. [CrossRef]
197. McLoughlin, P.D.; Case, R.L.; Gau, R.J.; Cluff, D.H.; Mulders, R.; Messier, F. Hierarchical habitat selection by barren-ground grizzly bears in the central Canadian arctic. *Oecologia* **2002**, *132*, 102–108. [CrossRef]
198. Gould, M.J.; Gould, W.R.; Cain, J.W.; Roemer, G.W. Validating the performance of occupancy models for estimating habitat use and predicting the distribution of highly-mobile species: A case study using the American black bear. *Biol. Conserv.* **2019**, *234*, 28–36. [CrossRef]
199. Sollmann, R.; Gardner, B.; Belant, J.L.; Wilton, C.M.; Beringer, J. Habitat associations in a recolonizing, low-density black bear population. *Ecosphere* **2016**, *7*, e01406. [CrossRef]
200. Welfelt, L.S.; Beausoleil, R.A.; Wielgus, R.B. Factors associated with black bear density and implications for management. *J. Wild. Mgmt.* **2019**, *83*, 1527–1539. [CrossRef]
201. Poor, E.E.; Scheick, B.K.; Mullinax, J.M. Multiscale consensus habitat modeling for landscape level conservation prioritization. *Sci. Rep.* **2020**, *10*, 17783. [CrossRef]
202. Obbard, M.E.; Coady, M.B.; Pond, B.A.; Schaefer, J.A.; Burrows, F.G. A distance-based analysis of habitat selection by American black bears (*Ursus americanus*) on the Bruce Peninsula, Ontario, Canada. *Can. J. Zool.* **2010**, *88*, 1063–1076. [CrossRef]
203. Karelus, D.L.; McCown, J.W.; Scheick, B.K.; Oli, M.K. Microhabitat features influencing habitat use by Florida black bears. *Glob. Ecol. Conserv.* **2018**, *13*, e00367. [CrossRef]
204. Onorato, D.P.; Hellgren, E.C.; Mitchell, F.S.; Skiles, J.R. Home range and habitat use of American black bears on a desert montane island in Texas. *Ursus* **2003**, *14*, 120–129.
205. Ditmer, M.A.; Noyce, K.V.; Fieberg, J.R.; Garshelis, D.L. Delineating the ecological and geographic edge of an opportunist: The American black bear exploiting an agricultural landscape. *Ecol. Model.* **2018**, *387*, 205–219. [CrossRef]
206. Duquette, J.F.; Belant, J.L.; Wilton, C.M.; Fowler, N.; Waller, B.W.; Beyer, D.E.; Svoboda, N.J.; Simek, S.L.; Beringer, J. Black bear (*Ursus americanus*) functional resource selection relative to intraspecific competition and human risk. *Can. J. Zool.* **2017**, *95*, 203–212. [CrossRef]
207. Brodeur, V.; Ouellet, J.-P.; Courtois, R.; Fortin, D. Habitat selection by black bears in an intensively logged boreal forest. *Can. J. Zool.* **2008**, *86*, 1307–1316. [CrossRef]
208. Rettler, S.J.; Tri, A.N.; St-Louis, V.; Forester, J.D.; Garshelis, D.L. Three decades of declining natural foods alters bottom-up pressures on American black bears. *For. Ecol. Manag.* **2021**, *493*, 119267. [CrossRef]
209. Garneau, D.E.; Boudreau, T.; Keech, M.; Post, E. Habitat use by black bears in relation to conspecifics and competitors. *Mamm. Biol.* **2008**, *73*, 48–57. [CrossRef]
210. Garneau, D.E.; Boudreau, T.; Keech, M.; Post, E. Black bear movements and habitat use during a critical period for moose calves. *Mamm. Biol.* **2008**, *73*, 85–92. [CrossRef]
211. Latham, A.D.M.; Latham, M.C.; Boyce, M.S. Habitat selection and spatial relationships of black bears (*Ursus americanus*) with woodland caribou (*Rangifer tarandus caribou*) in northeastern Alberta. *Can. J. Zool.* **2011**, *89*, 267–277. [CrossRef]
212. Bastille-Rousseau, G.; Rayl, N.D.; Ellington, E.H.; Schaefer, J.A.; Peers, M.J.L.; Mumma, M.A.; Mahoney, S.P.; Murray, D.L. Temporal variation in habitat use, co-occurrence, and risk among generalist predators and a shared prey. *Can. J. Zool.* **2016**, *94*, 191–198. [CrossRef]
213. Rayl, N.D.; Fuller, T.K.; Organ, J.F.; Mcdonald, J.E.; Otto, R.D.; Bastille-Rousseau, G.; Soulliere, C.E.; Mahoney, S.P. Spatiotemporal variation in the distribution of potential predators of a resource pulse: Black bears and caribou calves in Newfoundland. *J. Wild. Mgmt.* **2015**, *79*, 1041–1050. [CrossRef]
214. Karelus, D.L.; McCown, J.W.; Scheick, B.K.; Kerk, M.; van de Oli, M.K. Home ranges and habitat selection by black bears in a newly colonized population in Florida. *Southeast. Nat.* **2016**, *15*, 346. [CrossRef]
215. Tri, A.N.; Edwards, J.W.; Strager, M.P.; Petty, J.T.; Ryan, C.W.; Carpenter, C.P.; Ternent, M.A.; Carr, P.C. Habitat use by American black bears in the urban–wildland interface of the Mid-Atlantic, USA. *Ursus* **2016**, *27*, 45–56. [CrossRef]
216. Fecske, D.M.; Barry, R.E.; Precht, F.L.; Quigley, H.B.; Bittner, S.L.; Webster, T. Habitat use by female black bears in western Maryland. *Southeast. Nat.* **2002**, *1*, 77–92. [CrossRef]

217. Carter, N.H.; Brown, D.G.; Etter, D.R.; Visser, L.G. American black bear habitat selection in northern Lower Peninsula, Michigan, USA, using discrete-choice modeling. *Ursus* **2010**, *21*, 57–71. [CrossRef]
218. Sadeghpour, M.H.; Ginnett, T.F. Habitat selection by female American black bears in northern Wisconsin. *Ursus* **2011**, *22*, 159–166. [CrossRef]
219. Benson, J.F.; Chamberlain, M.J. Space use and habitat selection by female Louisiana black bears in the Tensas River Basin of Louisiana. *J. Wildl. Manag.* **2007**, *71*, 117–126. [CrossRef]
220. Smith, J.B.; Nielsen, C.K.; Hellgren, E.C. Suitable habitat for recolonizing large carnivores in the midwestern USA. *Oryx* **2016**, *50*, 555–564. [CrossRef]
221. Gantchoff, M.; Conlee, L.; Belant, J. Conservation implications of sex-specific landscape suitability for a large generalist carnivore. *Divers. Distrib.* **2019**, *25*, 1488–1496. [CrossRef]
222. Delfín-Alfonso, C.A.; López-González, C.A.; Equihua, M. Potential distribution of American black bears in northwest Mexico and implications for their conservation. *Ursus* **2012**, *23*, 65–77. [CrossRef]
223. Rojas-Martínez, A.E.; Juárez-Casillas, L.A. First record of American black bear (*Ursus americanus*) from Hidalgo, Mexico. *Rev. Mex. De Biodivers.* **2013**, *84*, 1018–1021. [CrossRef]
224. Aguilar-López, M.; Monter-Vargas, J.L.; Cornejo-Latorre, C.; Hernández-Saintmartin, A. First photo evidence of the American black bear (*Ursus americanus*) in the southwestern limit of its distribution. *West. North Am. Nat.* **2019**, *79*, 124–129. [CrossRef]
225. Monroy-Vilchis, O.; Castillo-Huitrón, N.M.; Zarco-González, M.M.; Rodríguez-Soto, C. Potential distribution of *Ursus americanus* in Mexico and its persistence: Implications for conservation. *J. Nat. Conserv.* **2016**, *29*, 62–68. [CrossRef]
226. Ditmer, M.A.; Iannarilli, F.; Tri, A.N.; Garshelis, D.L.; Carter, N.H. Artificial night light helps account for observer bias in citizen science monitoring of an expanding large mammal population. *J. Anim. Ecol.* **2021**, *90*, 330–342. [CrossRef]
227. Ditmer, M.A.; Rettler, S.J.; Fieberg, J.R.; Iaizzo, P.A.; Laske, T.G.; Noyce, K.V.; Garshelis, D.L. American black bears perceive the risks of crossing roads. *Behav. Ecol.* **2018**, *29*, 667–675. [CrossRef]
228. Corradini, A.; Peters, W.; Pedrotti, L.; Hebblewhite, M.; Bragalanti, N.; Tattoni, C.; Ciolli, M.; Cagnacci, F. Animal movements occurring during COVID-19 lockdown were predicted by connectivity models. *Glob. Ecol. Conserv.* **2021**, *32*, e01895. [CrossRef]
229. Issam, H. The syrian bear still lives in Syria. *Int. Bear News* **2011**, *20*, 7–11.
230. Deacy, W.; Leacock, W.; Armstrong, J.B.; Stanford, J.A. Kodiak brown bears surf the salmon red wave: Direct evidence from GPS collared individuals. *Ecology* **2016**, *97*, 1091–1098. [CrossRef]
231. Deacy, W.W.; Leacock, W.B.; Stanford, J.A.; Armstrong, J.B. Variation in spawning phenology within salmon populations influences landscape-level patterns of brown bear activity. *Ecosphere* **2019**, *10*, e02575. [CrossRef]
232. Deacy, W.W.; Armstrong, J.B.; Leacock, W.B.; Robbins, C.T.; Gustine, D.D.; Ward, E.J.; Erlenbach, J.A.; Stanford, J.A. Phenological synchronization disrupts trophic interactions between Kodiak brown bears and salmon. *Proc. Natl. Acad. Sci. USA* **2017**, *114*, 10432–10437. [CrossRef]
233. Nellemann, C.; Støen, O.-G.; Kindberg, J.; Swenson, J.E.; Vistnes, I.; Ericsson, G.; Katajisto, J.; Kaltenborn, B.P.; Martin, J.; Ordiz, A. Terrain use by an expanding brown bear population in relation to age, recreational resorts and human settlements. *Biol. Conserv.* **2007**, *138*, 157–165. [CrossRef]
234. Hertel, A.G.; Steyaert, S.M.J.G.; Zedrosser, A.; Mysterud, A.; Lodberg-Holm, H.K.; Gelink, H.W.; Kindberg, J.; Swenson, J.E. Bears and berries: Species-specific selective foraging on a patchily distributed food resource in a human-altered landscape. *Behav. Ecol. Sociobiol.* **2016**, *70*, 831–842. [CrossRef]
235. Nielsen, S.E.; Stenhouse, G.B.; Boyce, M.S. A habitat-based framework for grizzly bear conservation in Alberta. *Biol. Conserv.* **2006**, *130*, 217–229. [CrossRef]
236. Ciarniello, L.M.; Boyce, M.S.; Heard, D.C.; Seip, D.R. Components of grizzly bear habitat selection: Density, habitats, roads, and mortality risk. *J. Wildl. Manag.* **2007**, *71*, 1446–1457. [CrossRef]
237. Lamb, C.T.; Mowat, G.; McLellan, B.N.; Nielsen, S.E.; Boutin, S. Forbidden fruit: Human settlement and abundant fruit create an ecological trap for an apex omnivore. *J. Anim. Ecol.* **2017**, *86*, 55–65. [CrossRef]
238. Guisan, A.; Tingley, R.; Baumgartner, J.B.; Naujokaitis-Lewis, I.; Sutcliffe, P.R.; Tulloch, A.I.T.; Regan, T.J.; Brotons, L.; McDonald-Madden, E.; Mantyka-Pringle, C.; et al. Predicting species distributions for conservation decisions. *Ecol. Lett.* **2013**, *16*, 1424–1435. [CrossRef]
239. Vaz, U.L.; Cunha, H.F.; Nabout, J.C. Trends and biases in global scientific literature about ecological niche models. *Braz. J. Biol.* **2015**, *75*, 17–24. [CrossRef]
240. Zurell, D.; Franklin, J.; König, C.; Bouchet, P.J.; Dormann, C.F.; Elith, J.; Fandos, G.; Feng, X.; Guillera-Arroita, G.; Guisan, A.; et al. A standard protocol for reporting species distribution models. *Ecography* **2020**, *43*, 1261–1277. [CrossRef]
241. Aubry, K.B.; Raley, C.M.; McKelvey, K.S. The importance of data quality for generating reliable distribution models for rare, elusive, and cryptic species. *PLoS ONE* **2017**, *12*, e0179152. [CrossRef]
242. Ashrafzadeh, M.R.; Khosravi, R.; Adibi, M.A.; Taktehrani, A.; Wan, H.Y.; Cushman, S.A. A multi-scale, multi-species approach for assessing effectiveness of habitat and connectivity conservation for endangered felids. *Biol. Conserv.* **2020**, *245*, 108523. [CrossRef]
243. Qin, A.; Jin, K.; Batsaikhan, M.-E.; Nyamjav, J.; Li, G.; Li, J.; Xue, Y.; Sun, G.; Wu, L.; Indree, T.; et al. Predicting the current and future suitable habitats of the main dietary plants of the Gobi bear using Maxent modeling. *Glob. Ecol. Conserv.* **2020**, *22*, e01032. [CrossRef]

244. Bucklin, D.N.; Basille, M.; Benscoter, A.M.; Brandt, L.A.; Mazzotti, F.J.; Romañach, S.S.; Speroterra, C.; Watling, J.I. Comparing species distribution models constructed with different subsets of environmental predictors. *Divers. Distrib.* **2015**, *21*, 23–35. [CrossRef]
245. Nielsen, S.E.; McDermid, G.; Stenhouse, G.B.; Boyce, M.S. Dynamic wildlife habitat models: Seasonal foods and mortality risk predict occupancy-abundance and habitat selection in grizzly bears. *Biol. Conserv.* **2010**, *143*, 1623–1634. [CrossRef]
246. Boulanger, J.; Nielsen, S.E.; Stenhouse, G.B. Using spatial mark-recapture for conservation monitoring of grizzly bear populations in Alberta. *Sci. Rep.* **2018**, *8*, 5204. [CrossRef]
247. Lamb, C.T.; Mowat, G.; Reid, A.; Smit, L.; Proctor, M.; McLellan, B.N.; Nielsen, S.E.; Boutin, S. Effects of habitat quality and access management on the density of a recovering grizzly bear population. *J. Appl. Ecol.* **2018**, *55*, 1406–1417. [CrossRef]
248. Mowat, G.; Heard, D.C.; Schwarz, C.J. Predicting grizzly bear density in Western North America. *PLoS ONE* **2013**, *8*, e82757. [CrossRef]
249. Stetz, J.B.; Mitchell, M.S.; Kendall, K.C. Using spatially-explicit capture–recapture models to explain variation in seasonal density patterns of sympatric ursids. *Ecography* **2019**, *42*, 237–248. [CrossRef]
250. Boulanger, J.; Cattet, M.; Nielsen, S.E.; Stenhouse, G.; Cranston, J. Use of multi-state models to explore relationships between changes in body condition, habitat and survival of grizzly bears *Ursus arctos horribilis*. *Wildl. Biol.* **2013**, *19*, 274–288. [CrossRef]
251. Scharf, A.K.; Fernández, N. Up-scaling local-habitat models for large-scale conservation: Assessing suitable areas for the brown bear comeback in Europe. *Divers. Distrib.* **2018**, *24*, 1573–1582. [CrossRef]
252. Atzeni, L.; Cushman, S.A.; Bai, D.; Wang, J.; Chen, P.; Shi, K.; Riordan, P. Meta-replication, sampling bias, and multi-scale model selection: A case study on snow leopard (*Panthera uncia*) in western China. *Ecol. Evol.* **2020**, *10*, 7686–7712. [CrossRef]
253. Lee-Yaw, J.A.; McCune, J.L.; Pironon, S.; Sheth, S.N. Species distribution models rarely predict the biology of real populations. *Ecography*, 2021; *Early View*. [CrossRef]
254. Hirzel, A.H.; Le Lay, G.; Helfer, V.; Randin, C.; Guisan, A. Evaluating the ability of habitat suitability models to predict species presences. *Ecol. Model.* **2006**, *199*, 142–152. [CrossRef]
255. Cianfrani, C.; Le Lay, G.; Hirzel, A.H.; Loy, A. Do habitat suitability models reliably predict the recovery areas of threatened species? *J. Appl. Ecol.* **2010**, *47*, 421–430. [CrossRef]
256. Knight, A.T.; Cowling, R.M.; Rouget, M.; Balmford, A.; Lombard, A.T.; Campbell, B.M. Knowing but not doing: Selecting priority conservation areas and the research–implementation gap. *Conserv. Biol.* **2008**, *22*, 610–617. [CrossRef]
257. Cook, C.N.; Mascia, M.B.; Schwartz, M.W.; Possingham, H.P.; Fuller, R.A. Achieving conservation science that bridges the knowledge–action boundary. *Conserv. Biol.* **2013**, *27*, 669–678. [CrossRef]
258. Villero, D.; Pla, M.; Camps, D.; Ruiz-Olmo, J.; Brotons, L. Integrating species distribution modelling into decision-making to inform conservation actions. *Biodivers Conserv.* **2017**, *26*, 251–271. [CrossRef]
259. Sofaer, H.R.; Jarnevich, C.S.; Pearse, I.S.; Smyth, R.L.; Auer, S.; Cook, G.L.; Edwards, T.C.; Guala, G.F.; Howard, T.G.; Morisette, J.T.; et al. Development and delivery of species distribution models to inform decision-making. *BioScience* **2019**, *69*, 544–557. [CrossRef]
260. Ferraz, K.M.P.M.d.B.; Morato, R.G.; Bovo, A.A.A.; Costa, C.O.R.; Ribeiro, Y.G.G.; Paula, R.C.; Desbiez, A.L.J.; Angelieri, C.S.C.; Traylor-Holzer, K. Bridging the gap between researchers, conservation planners, and decision makers to improve species conservation decision-making. *Conserv. Sci. Pr.* **2021**, *3*, e330. [CrossRef]
261. Proctor, M.F.; Nielsen, S.E.; Kasworm, W.F.; Servheen, C.; Radandt, T.G.; Machutchon, A.G.; Boyce, M.S. Grizzly bear connectivity mapping in the Canada–United States trans-border region. *J. Wild. Mgmt.* **2015**, *79*, 544–558. [CrossRef]
262. Proctor, M.F.; Kasworm, W.F.; Annis, K.M.; MacHutchon, A.G.; Teisberg, J.E.; Radant, T.G.; Servheen, C. Conservation of threatened Canada–USA trans-border grizzly bears linked to comprehensive conflict reduction. *Hum.–Wildl. Interact.* **2018**, *12*, 348–372.

land

Article

European Ground Squirrels at the Edge: Current Distribution Status and Anticipated Impact of Climate on Europe's Southernmost Population

Dimitra-Lida Rammou [1,†], Christos Astaras [2,†], Despina Migli [1], George Boutsis [1], Antonia Galanaki [1], Theodoros Kominos [1] and Dionisios Youlatos [1,*]

1 School of Biology, Aristotle University of Thessaloniki, GR-54124 Thessaloniki, Greece; rdimitra@bio.auth.gr (D.-L.R.); despmigk@bio.auth.gr (D.M.); gboutsis@bio.auth.gr (G.B.); agalanaki@bio.auth.gr (A.G.); tkominos@bio.auth.gr (T.K.)
2 Forest Research Institute, ELGO-DIMITRA, GR-57006 Vasilika, Greece; christos.astaras@fri.gr
* Correspondence: dyoul@bio.auth.gr
† These authors contributed equally to this work.

Abstract: The European ground squirrel (*Spermophilus citellus*) is an endangered semifossorial small mammal of grassland/agricultural ecosystems. In the last few decades, the species' population has declined throughout its range in Europe. The Greek populations represent the southernmost limit of the species' range and are notably small, scattered, and located mainly in human-modified areas. The goal of the present research is to understand the environmental and anthropogenic variables associated with its distribution in the Mediterranean habitats, assess possible drivers of observed local extinctions, and propose conservation and land-use management actions in light of near-future climate change scenarios. We used presence records since 2000 across all known populations (107 colonies) and maximum entropy conditional probability models (MaxEnt) to calculate both the habitat suitability (bioclimatic variables) and habitat availability (anthropogenic/land-use variables) within the European ground squirrel's historical range in northern Greece. We report a projected 39% to 94.3% decrease in habitat suitability by 2040–2060 due to climate change. Based on our findings, we provide guidance by proposing nascent conservation actions to protect the few existing colonies in Greece via improved land management practices and identify in situ climate refugia that could be prioritized as sites for future reintroductions.

Keywords: *Spermophilus citellus*; maximum entropy modeling; species distribution modeling; climate change refugia

Citation: Rammou, D.-L.; Astaras, C.; Migli, D.; Boutsis, G.; Galanaki, A.; Kominos, T.; Youlatos, D. European Ground Squirrels at the Edge: Current Distribution Status and Anticipated Impact of Climate on Europe's Southernmost Population. *Land* **2022**, *11*, 301. https://doi.org/10.3390/land11020301

Academic Editors: Juan F. Beltrán, Pedro Abellán and John Litvaitis

Received: 30 January 2022
Accepted: 14 February 2022
Published: 16 February 2022

Publisher's Note: MDPI stays neutral with regard to jurisdictional claims in published maps and institutional affiliations.

1. Introduction

Agriculture in Europe underwent significant transformations since the middle of the 20th century. Intensification of farming practices due to increased mechanization, together with rapid industrial, urban, and transportation network development, has led to a homogenization of farmlands and the fragmentation of natural and semi-natural habitats, especially in lowland areas [1–6]. At the same time, the shift of human activities near cities was followed by a decline in traditional pastoral activities, with extensive grazing reducing or replaced by intensive grazing [7]. Furthermore, the abandoned grasslands became gradually encroached by shrubland or forest, especially at mountainous landscapes [8,9].

While some wildlife species have benefited from these land-use changes [10–13], the overall biodiversity of European agroecosystems has decreased, threatening a range of grassland species due to habitat loss or degradation and a reduction in habitat connectivity [3,6,14]. Moreover, human-induced global warming is likely to cause shifts in the distribution of many species in the near future [15]. Species with narrow niche breadth could be more susceptible to such changes due to their limited geographic range and low dispersal capacity [15,16]. Identifying currently available and future suitable habitats for these species is key to developing effective conservation priorities for them.

The European ground squirrel (*Spermophilus citellus*) is one such narrow niche breadth species that inhabits natural and anthropogenic grasslands of central and southeastern Europe. Its populations have been declining across its range due mostly to the conversion of suitable habitat to intensively cultivated fields and urban areas and the abandonment of grazing areas [17–21]. A group-living, semifossorial, and mostly herbivorous rodent, the European ground squirrel hibernates in individual burrows from early fall to early spring, depending on the altitude and latitude [22,23]. It is considered an ecosystem engineer and keystone species of grasslands, as its burrowing activity aerates the soil, increases plant composition, and creates microhabitats for other species, while being an important prey for raptors [19,24–27]. The species is listed as endangered in the IUCN Red List Data Book [28] and is protected by the European framework (Bern Convention—Appendix II, Directive 92/43/EEC—Annexes II and IV).

The Greek populations of the European ground squirrel constitute the species' southern distributional limit, having adapted to the Mediterranean climate [23]. The populations occur in three discrete sub-regions at the north of the country, with most colonies being concentrated in central Macedonia and fewer in western Macedonia and Thrace [29,30]. Surveys over the last decade have documented significant reductions, both in the overall range (62.4%) and the number of active colonies (74.6%), compared to records at the end of the previous century [30]. Moreover, the remaining colonies are isolated and significantly smaller in size compared to older records [30]. While some previously unknown populations were discovered during the latest surveys, large areas of the species' historical range remain unoccupied. Climate change is an especially pertinent threat for the Greek populations, given their presence at the southern edge of the species' range. Predictions for anticipated climate changes by 2100 in the Mediterranean region include a 2–5 °C temperature rise, depending on the season, and an overall decrease in precipitation [31,32]. Rising ambient temperatures could affect the circannual rhythms of the species during hibernation, resulting in a loss of body mass that, in turn, will likely negatively affect individual fitness [33,34]. Improving our understanding of current and future European ground squirrel habitat suitability would be key for prioritizing conservation actions and areas that are likely to yield long-term benefits to the species in the face of anticipated climate changes.

Ecological niche modeling (ENM) is a well-established approach for predicting both the distribution of a species across a geographical area, where presence information is limited or imperfect [35], and for assessing the relation of environmental (biotic, abiotic, bioclimatic) and anthropogenic parameters and the species' habitat suitability [36–39]. The maximum entropy algorithm (MaxEnt; [40–42]) is a popular ecological niche modeling method for endangered species, including ground squirrels [43–50], as it has high prediction accuracy, even with relatively small datasets, and it requires only presence data [37,51–54]. These characteristics make it an ideal tool for the European ground squirrel dataset that we have developed.

In this study, we first used all available data on the historical and current distribution of the European ground squirrel in Greece to infer the species' current habitat availability (i.e., suitable land-use/habitat) and habitat suitability (i.e., bioclimatically suitable areas) using MaxEnt ecological niche models, and to assess which parameters limit its distribution. Second, we examined possible anthropogenic and environmental causes for the observed local extinctions of European ground squirrel colonies. Third, we used the current habitat suitability model values to predict the effect of climate change on the species' potential distribution in the near future (2040–2060) under a variety of possible scenarios, with the aim of identifying priority areas for conservation within its historical range (i.e., in situ climate change refugia). Lastly, we examined the results from the above analyses to produce a list of targeted conservation actions for the European ground squirrel in Greece, which we hope will help reverse the very real prospect of the species' country-wide extinction.

2. Materials and Methods

2.1. Study Area

The study area extended across the administrative regions of Western Macedonia, Central Macedonia, and Eastern Macedonia and Thrace in northern Greece (41.44° N, 26.12° E to 39.95° N, 21.41° E; Figure 1), which enclose the historical range of the European ground squirrel in the country. The geomorphology of the area is complex, including mountain ranges, alpine plateaus, and lowland valleys. The mean elevation is 485 m (range 0–2918 m). Main land-uses in the lowlands are monoculture agriculture, settlements, and industrial developments, while mid-to-higher elevation areas are characterized by semi-natural and natural environments. The climate is typical Mediterranean with hot semi-arid to cold semi-arid summers in lowlands and humid subtropical and continental in mountainous areas [55]. The mean annual precipitation is 561.3 ± 108.7 mm (range 411−1071 mm), occurring mainly during the winter months. The mean monthly temperature of the coldest quarter is 3.5 ± 1.9 °C (range −6.1 to 8.8 °C), while in the warmest quarter it is 21.7 ± 2.6 °C (9−25.3 °C) [56]. The European ground squirrel shares its habitat with common small mammal species of the Talpidae, Soricidae, Cricetidae, and Muridae families, as well as the European hedgehog (*Erinaceus europaeus*) and the European hare (*Lepus europaeus*). Potential natural mammalian and avian predators include the least weasel (*Mustela nivalis*), stone marten (*Martes foina*), red fox (*Vulpes vulpes*), European wild cat (*Felis silvestris*), Golden eagle (*Aquila chrysaetos*), and buzzards (*Buteo* spp.) [27]. Domestic dogs and cats are also present across the range, especially in lowland agricultural areas near settlements.

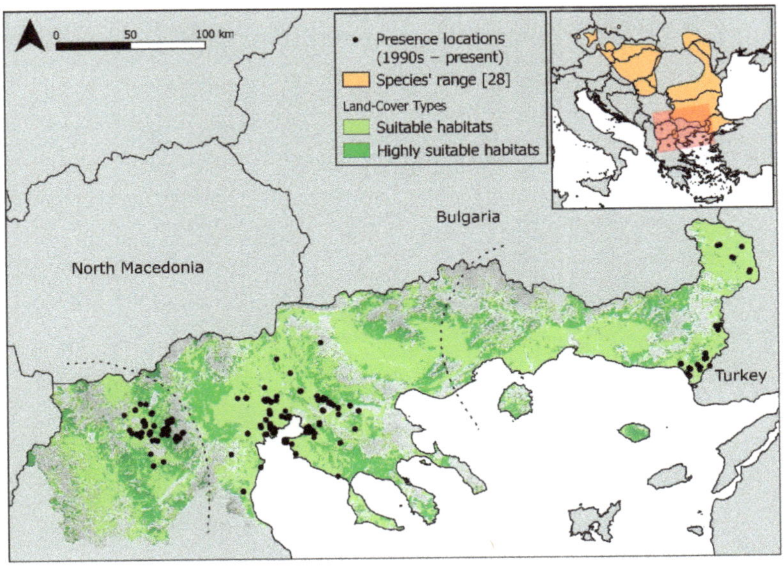

Figure 1. The distribution of the presence records (1990−2021) of *Spermophilus citellus* in Northern Greece. The suitable habitats of the species in the study area are schematically marked in green (light green for suitable and dark green for highly suitable habitats, please see Table 1 for more details), and the unsuitable habitats are in grey. Habitat suitability was assessed based on ecological criteria from [17,19]. Dashed lines separate the three sub-populations of Western Macedonia (**left**), Central Macedonia (**center**), and Thrace (**right**).

Table 1. Bioclimatic, environmental, and anthropogenic variables considered in either or both of the *Spermophilus citellus* ecological niche models (model 1: habitat availability 100 m resolution; model 2: habitat suitability 4500 m resolution). Note: asterisks mark models in which a given variable was used in the final model.

Model	Category	Variable	Type	Source	Initial Resolution
1 *	Environmental—Abiotic	Elevation	Continuous	https://land.copernicus.eu (EU-DEM v1.1)	25 m
1 *,2 *	Environmental—Abiotic	Slope	Continuous	Developed using the EU-DEM v1.1 layer and the QGIS Slope function	25 m
1 *	Environmental—Abiotic	Aspect	Continuous	Developed using the EU-DEM v1.1 layer and the QGIS Aspect function	25 m
1 *	Environmental—Abiotic	Tree cover density	Continuous	https://land.copernicus.eu (Tree Cover Density 2018)	10 m
1 *	Environmental—Abiotic	EGS suitable land-cover	Categorical	https://land.copernicus.eu (Corine Land Cover 2018) (Reclassified/suitable: 2–4, 9, 12–14, 19–22, 35, 37, 38; highly suitable: 6, 10, 11, 15–18, 26, 28, 32)	100 m
1 *	Anthropogenic—Abiotic	Soil imperviousness (soil sealing)	Continuous	https://land.copernicus.eu	10 m
1 *	Anthropogenic—Abiotic	Road density	Continuous	https://geodata.gov.gr; converted tarred roads to 50 m interval point layer, and applied heat map/kernel density estimation function in QGIS (200 m radius/quartic kernel shape)	Vector
2	Environmental—Biotic	Normalized difference vegetation index (NDVI)—20-year mean (1999–2019)	Continuous	https://land.copernicus.eu	1000 m
2	Bioclimatic—Abiotic	(Bio1) Annual mean temperature	Continuous	https://www.worldclim.org/	4500 m
2	Bioclimatic—Abiotic	(Bio2) Annual mean diurnal range	Continuous	https://www.worldclim.org/	4500 m
2 *	Bioclimatic—Abiotic	(Bio4) Temperature seasonality SD	Continuous	https://www.worldclim.org/	4500 m
2	Bioclimatic—Abiotic	(Bio7) Annual temp range	Continuous	https://www.worldclim.org/	4500 m
2 *	Bioclimatic—Abiotic	(Bio8) Mean temp of wettest quarter	Continuous	https://www.worldclim.org/	4500 m
2	Bioclimatic—Abiotic	(Bio9) Mean temp of driest quarter	Continuous	https://www.worldclim.org/	4500 m
2	Bioclimatic—Abiotic	(Bio10) Mean temp of warmest quarter	Continuous	https://www.worldclim.org/	4500 m
2	Bioclimatic—Abiotic	(Bio11) Mean temp of coldest quarter	Continuous	https://www.worldclim.org/	4500 m
2	Bioclimatic—Abiotic	(Bio12) Annual precipitation	Continuous	https://www.worldclim.org/	4500 m
2 *	Bioclimatic—Abiotic	(Bio15) Precipitation seasonality (CV)	Continuous	https://www.worldclim.org/	4500 m
2	Bioclimatic—Abiotic	(Bio16) Precipitation of wettest quarter	Continuous	https://www.worldclim.org/	4500 m
2	Bioclimatic—Abiotic	(Bio17) Precipitation of driest quarter	Continuous	https://www.worldclim.org/	4500 m
2 *	Bioclimatic—Abiotic	(Bio18) Precipitation of warmest quarter	Continuous	https://www.worldclim.org/	4500 m
2	Bioclimatic—Abiotic	(Bio19) Precipitation of coldest quarter	Continuous	https://www.worldclim.org/	4500 m
2	Anthropogenic—Biotic	Population density	Continuous	https://ec.europa.eu/eurostat	1000 m
2 *	Environmental—Abiotic	Soil bulk density	Continuous	https://esdac.jrc.ec.europa.eu (LUCAS Database)	500 m
2 *	Environmental—Abiotic	Soil texture (USDA classification)	Categorical	https://esdac.jrc.ec.europa.eu (LUCAS Database)	500 m

2.2. Species Data

We compiled presence records of the European ground squirrel across the species' historical range in Greece, as indicated in Youlatos [29], from technical reports, interviews, photographs, and field surveys. The oldest records were from the mid-90s, whereas most were from the last decade (2011–2021). The 2427 records obtained (Figure 1) refer to active burrow entrances or observations of individual animals. All were validated for this project [30] and have exact coordinates. In addition, we compiled 403 records of species absence, which were used to correct for sampling bias during ecological niche modeling (MaxEnt). All absence data correspond to true absences validated in the field. The presence records were spatially filtered by randomly retaining a single record per 100 m and 4500 m grid cells, resulting in 425 and 85 presence records for inferring habitat availability (i.e., suitable land-use/habitat) and habitat suitability (i.e., suitable bioclimatic areas) models, respectively (as defined by Gür [44]).

2.3. Ecological Niche Modeling Variables

The variables for inferring the European ground squirrel's habitat availability were used at a resolution of 100 m, while those used for modeling habitat suitability were at a 4.5 km resolution. We resampled variables available at a higher resolution to one of a model using the nearest-neighbor-joining method. Variables were masked to the extent of the study area (three administrative regions of northern Greece) and converted to ASCII format, as required by the MaxEnt software. To reduce the risk of model overfitting due to variable collinearity, which can affect the model transferability spatially or temporally [57], we kept only one of the highly correlated variables (Pearson correlation coefficient $r > 0.7$ or $r < -0.7$). All file conversions and data processing were performed using Quantum GIS v.3.16.14 [58].

In total, seven variables were considered for the habitat availability (100 m resolution) models: two anthropogenic (road density, soil imperviousness) and five environmental (elevation, slope, aspect, tree cover density, European ground squirrel suitable land cover). The land-cover types were categorized into suitable and highly suitable, representing artificial/arable land and land-uses with special management (i.e., airports, pastures)/permanent crops/semi-natural and natural land-cover types respectively, based on what is known about the species' ecology [17,19] and our field observations. Nineteen variables were considered for the habitat suitability (4500 m) models: 14 bioclimatic variables (WorldClim database version 2.1; [56]), 1 anthropogenic (population density), and 4 environmental (20-year mean normalized difference vegetation index (NDVI), soil bulk density, soil texture, and slope). We excluded 5 of the 19 available WorldClim bioclimatic variables from the analysis based on preliminary tests of collinearity, consideration of recent variables included in ecological niche models for the European ground squirrel and its congener, namely, the Anatolian ground squirrel (*Spermophilus xanthoprymnus*) [43,44], and an emphasis on seasonal mean or range rather than min–max values. Details of the source, initial resolution, and model for which a variable was considered are provided in Table 1.

We also downloaded future bioclimatic data for the study area for the period 2041−2060 (near future) from the Coupled Model Intercomparison Project Phase 6 (CMIP6) based on three global climate models (BCC-CSM2-MR2, CNRM-CM6-1, CanESM5; representing long-term average Earth surface temperature rises, resulting from a doubling of atmospheric CO_2, of 3 °C, 4.3 °C, and 5.6 °C respectively) and three combinations of shared socio-economic pathways (SSP) and representative concentration pathways (RCP) by 2100 (SSP2/RCP4.5, SSP3/RCP7.0, SSP5/RCP8.5). The SSPs 2, 3, and 5 represent different narratives of global and regional efforts to combat climate change (SSP2—medium challenges to mitigation and adaptation, SSP3—high challenges to mitigation and adaptation, SSP5—high challenges to mitigation, low challenges to adaptation) (see [59]). The RCPs 4.5, 7.0, and 8.5 refer to slowly declining, slowly rising, and rising CO_2 emissions (for more information refer to https://www.carbonbrief.org/, accessed on 10 January 2022). Using different global models and SSP/RCP scenarios, we captured the uncertainty of the anticipated climate change pathways in the future European ground squirrel ecological niche models.

2.4. Model Implementation and Processing

We used the software Maxent 3.4.4 [42], available from https://biodiversityinformatics.amnh.org/open_source/maxent/, accessed on 15 December 2021), to model the habitat suitability and availability of the European ground squirrel throughout the historical range of the species in Greece and to examine the species' response in relation to the environmental, bioclimatic, and anthropogenic parameters considered. We opted for the use of MaxEnt because it shows good predictive performance, even with small presence datasets, and it does not require real absence data [60]. It is, therefore, suitable for our dataset. Moreover, MaxEnt was shown to produce similar results to more complicated "black box ensemble models" [61].

Since there have been significant concerns raised against using MaxEnt software's default feature classes and regularization parameter options (e.g., [62]), we used the ENMVeval R package [63] to run a combination of model settings (i.e., "tuning"). We tested 36 candidate models for both the habitat availability (100 m resolution) and the habitat suitability (4500 m resolution) modeling process by combining five feature classes (linear; linear and quadratic; hinge; linear, quadratic, and hinge; linear, quadratic, hinge, and product) and nine regularization multiplier values (1 to 5 in 0.5 increments). We used a fixed set of non-correlated environmental, anthropogenic, and/or bioclimatic variables. To address possible survey biases of our dataset for the habitat availability models, we defined the background extent (within which 10,000 background points would be randomly selected) as a polygon enclosing a 10 km buffer around our presence and absence points [64]. This was not done for the habitat suitability models, as the grid cell resolution of 4500 m limited the number of available background points. Therefore, in this case, we used all of the study area as the background extent. Model evaluation statistics were calculated by using the random k-fold methods, which partitioned the data into "bins" (k = 5) for training and testing for cross-validation. To identify the model with the optimal model settings, we used the model with the lowest Akaike information criterion [65] corrected for small sample sizes (AICc) value, which penalizes for model overfitting. In order to evaluate the model, we chose two metrics: the average of the area under the receiver operator curve (AUC) [51] and the continuous Boyce index (CBI) [66,67]. Higher AUC values denote models that discriminate better between conditions at occurrence locations withheld for testing and those at background points [63]. Values close to 0.5 are as informative as random models. We considered models with AUC > 0.9 as excellent, 0.8–0.9 as good, 0.7–0.8 as fair, and <0.7 as poor [68]. However, the usefulness of using only AUC for accuracy measurement has been criticized when true absence data are not available [69]. The CBI is considered more appropriate for the evaluation of presence-only models, as in our case, as it only requires presences [67]. The CBI values range from −1 to +1, with positive values indicative of the model output being positively correlated with the true probability of presence, values near zero the output being not different from a random model, and negative values the output being negatively correlated with the true probability of presence, i.e., counter predictions [70].

Once we had determined the optimal set of model settings, we ran the selected model in the MaxEnt GUI using ten cross-validated replications with no threshold values and the same bias file (in the case of habitat availability); the remaining settings were left at default values. We selected jackknife testing to assess each variables' contribution to the model and selected for response curves to be produced to assess how each variable affected the European ground squirrel's ecological niche model. We used MaxEnt to map habitat suitability/availability using a Cloglog output, with values of 0.0 to 1.0 indicating low to high suitability/availability, respectively. In the case of the habitat suitability model, we also projected the model results onto future conditions, in addition to the current ones across the study area, by providing the MaxEnt software with the predicted variable layers for the 2040–2060 period. In total, nine future habitat suitability maps were projected for the European ground squirrel; one for each of the three global climate models and three SSP/RCP scenario combinations. We averaged the three model results for each SSP/RCP scenario, as per Gür [44], using QGIS. Finally, we categorized the habitat suitability/availability maps into five classes (also as per Gür [44]): very low suitability (<0.2), low suitability (0.2−0.4), moderate suitability (0.4−0.6), high suitability (0.6−0.8), and very high suitability (>0.8). This was done to facilitate interpretation and reporting. We considered areas with values $\geq$0.6 as suitable (or available) for the European ground squirrel. We identified areas suitable for the species both at present and under all future climate scenarios (2041−2060 period) as in situ climate change refugia [71].

2.5. Drivers of Colony Extinction

To examine possible drivers of colony extinction, we assigned colonies as either active or inactive based on the presence or not of the species during the 2019−2021 field visits. We excluded from the analysis colonies that were not visited. A colony was defined as the total number of the burrow entrances, which were loosely distributed in a location, creating aggregations of several individuals that live and interact in the same area [72,73]. We used binomial regression models (link—logit) in R (v.4.1.2; [74]) with active and inactive colonies (0 and 1) being the response variable. We considered all the ecological, anthropogenic, and bioclimatic variables as potential predictor variables considered in the MaxEnt ecological niche models; these were calculated as their mean value within a 1 km buffer from the centroid of each colony, considering the maximum dispersal of the species [75]. In addition, we considered the percent cover by areas modeled as having habitat availability of 60% and above, and the percent change in the number of free-ranging small ruminants over the past 20 years (available at the prefecture level). The latter variable was included to examine whether a reduction in grazing livestock and the ensuing abandonment of traditional grazing areas could explain the observed decline in the number of colonies at the landscape level. We first ran univariate models and compared them against the null (intercept only) model in order to select an optimal set of informative variables while also managing the model complexity. We used the Akaike information criterion (AIC) [65] for the model selection. Among the correlated variables, we kept the one with the lowest univariate model AIC for further consideration. Once a final set of fixed variables was selected, we ran all possible multivariate combinations, again using the AIC for the model selection.

3. Results

3.1. Ecological Niche Modeling

We considered an equal number (36) of candidate models of varying model settings when assessing the European ground squirrel's current habitat availability (model 1, 100 m resolution). The final habitat availability model used aspect, elevation, slope, road density, soil imperviousness, tree cover, and suitable land-cover (categorical) for the European ground squirrel as input variables. The model was developed using linear, quadratic, and hinge feature classes, and a regularization multiplier of 1.0 (Supplementary Table S1). No other model had ΔAICc $\leq$ 2. The average test AUC value for the ten replicates was 0.82 $\pm$ 0.02 and the CBI value was 0.94 $\pm$ 0.02.

The univariate response curves of most variables were either bell-curved or linear, with no truncations or significant differences in shape with the marginal curves (Supplementary Figure S1). The available habitat for the species is typically in south-facing, lowland, slightly sloped (<3°), and treeless areas within human-modified landscapes (i.e., moderate-to-high road density, low-to-moderate soil imperviousness). An exception was natural grasslands at a higher elevation. All variables, except soil imperviousness and aspect, contributed significantly to the final model (i.e., >10% contribution and/or permutation importance; see Table 2). While, according to the model habitat, availability extends over 11.3% of northern Greece, only 43.8% of those areas (5%, 2119 km^2) were within the suitable habitat areas, which accounted for approximately one-fifth of the suitable areas. A large part of the available vs. suitable area disparity certainly stemmed from the difference in the resolution of these models, but it was apparent that current human landscape modifications (buildings, road infrastructure, mining) and land-use practices had reduced the available habitat to a fraction of the European ground squirrel's historical potential. The largest clusters of available habitats are found at the Axios River valley and agricultural areas east of the city of Thessaloniki (Central Macedonia), and the alpine meadows of Mount Vermio (Western Macedonia) (areas 2 and 1 in Figure 2, respectively). Mount Vermio is home to the sole European ground squirrel mountainous population in Greece. While the model predicts habitat availability at several additional mountains and roadless [76] plateaus (e.g., Grammos, Kaimaktsalan, Krystallopigi, Menikio, Sfika), none were within suitable habitat according to the bioclimatic and soil model. The available habitat at the Evros Delta (Thrace) is highly fragmented.

Table 2. Percentage contribution (Pc) and permutation importance (Pi) values of variables used to predict the distribution of available (model 1) and suitable habitats (model 2) of *Spermophilus citellus*.

Variables	Model 1: Habitat Availability		Model 2: Habitat Suitability	
	Pc (%)	Pi (%)	Pc (%)	Pi (%)
Road density	30.9	26.7	–	–
Elevation	29.1	25.7	–	–
EGS suitable land-cover	14	16	–	–
Tree cover density	12.7	9.5	–	–
Slope	8.8	18.4	52.5	29.8
Aspect	2.4	1.5	–	–
Soil imperviousness	2.1	2.3	–	–
Precipitation seasonality (Bio15)	–	–	22.4	34.7
Soil texture	–	–	11.1	6.1
Soil bulk density	–	–	7.2	3.6
Precipitation of warmest quarter (Bio18)	–	–	4.4	18.7
Temperature seasonality SD (Bio4)	–	–	1.5	3.3
Mean temp of wettest quarter (Bio8)	–	–	0.9	3.7

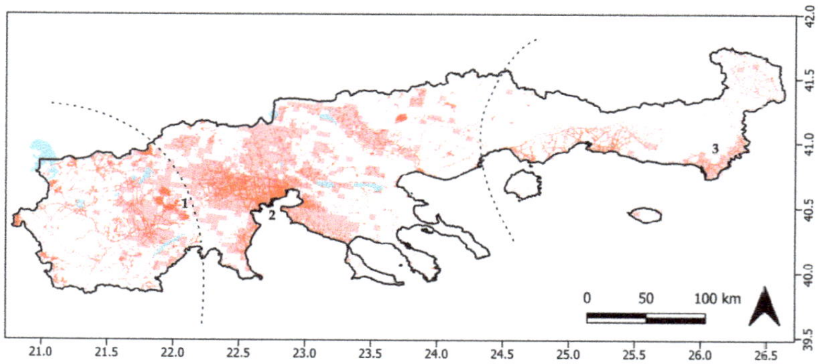

Figure 2. Map of *Spermophilus citellus* current habitat availability (dark red, resolution 100 m) and habitat suitability (light red, resolution 4.5 km) across the species' historical range. The dashed lines separate the three sub-populations (Western Macedonia, Central Macedonia, and Thrace), while the numbers indicate the main available habitats of the species in (**1**) Mount Vermio, (**2**) Axios River valley and eastern Thessaloniki, and (**3**) Delta Evros River.

We considered 36 candidate models of varying model settings to assess the European ground squirrel's current habitat suitability (model 2, 4.5 km resolution) within the species' historical range. The final habitat availability model used slope, soil bulk density, soil texture (categorical), temperature seasonality (Bio4), mean temperature of the wettest quarter (Bio8), precipitation seasonality (Bio15), and precipitation of warmest quarter (Bio18) as input variables. The model was developed using linear and quadratic feature classes and a regularization multiplier of 1.0 (Supplementary Table S1). No other model had ΔAICc ≤ 2. The average test AUC value for the ten replicates was 0.76 ± 0.05 and the CBI value was 0.73 ± 0.15.

The univariate response curves of most variables were also either bell-curved or linear without truncations and differed in overall shape from the marginal curves only for temperature seasonality (Bio4) (Supplementary Figure S1). The model showed that areas most suitable for the European ground squirrel were flat (or with slope <5°) with silt clay-loam or clay-loam soil of high bulk density, with cold winters (i.e., wettest quarter) and dry summers (i.e., warmest quarter), and low variation in seasonal precipitation. Slope, precipitation seasonality, summer precipitation, and soil texture were the variables that contributed the most to the final model (see percent contribution and/or permutation importance in Table 2).

Based on the MaxEnt model, the currently suitable areas (predicted suitability ≥ 0.6; ~20 km^2 grid area at the equator) for the European ground squirrel extend over 25.3% of northern Greece (10,327 km^2) (Figure 2). The distribution of suitable habitat coincided broadly with the three known sub-populations in Greece: Western Macedonia, Central Macedonia, and Thrace, with a clear and extended (100–300 km) discontinuity in habitat suitability between the populations of Central Macedonia and Thrace. There were also no suitable areas at the western edge of the study area, with the city of Kozani and surrounding areas being the westernmost limit, which matched the known historical distribution of the species.

Under the future climate change scenarios, the European ground squirrel's suitable habitat will significantly contract (range 39% to 94.3%) by 2041–2060 (Figure 3), affecting all three sub-populations. There are no predicted areas of habitat suitability expansion. Conservatively, i.e., under the most pessimistic scenario, the climate refugia for the European ground squirrel within its historical range in Greece will be limited to along the Axios River in Central Macedonia and the delta of the Evros River in Thrace. In Western Macedonia, suitable habitats will be fragmented near Mount Vermio and semi-mountain areas near the city of Ptolemaida.

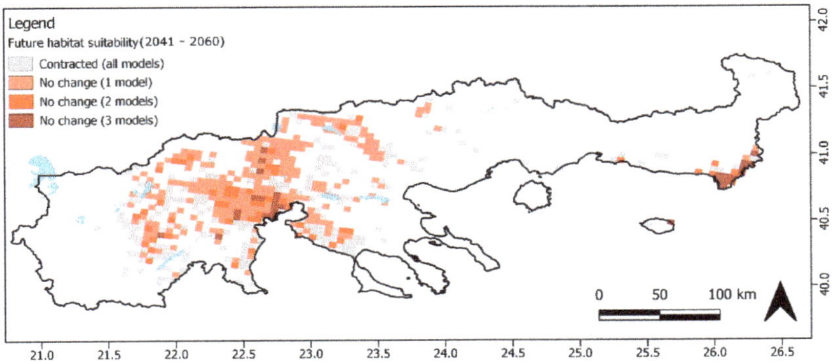

Figure 3. Near-future (2041–2060) *Spermophilus citellus* predicted habitat suitability contraction (grey) and persistence (light to dark red) under three future scenarios (SSP/RCP 2–4.5, 3–7.0, 5–8.5; mean suitability of BCC-CSM2-MR2, CNRM-CM6-1, CanESM5 global climate models ≥0.6). Dark red grid cells denote areas predicted to remain suitable for the species under all future scenarios and can therefore be considered most likely to be climate refugia. Map resolution: 4.5 km.

3.2. Drivers of Colony Extinction

For this study, we analyzed data from 107 colonies (Figure 1 and Supplementary Table S2). Most (68.2%) were within agricultural landscapes consisting of a mosaic of arable land, permanent crops, and pastures. One in five colonies (19.63%) was located in artificial areas (e.g., discontinuous urban fabric, industrial or commercial units, airports, sports leisure facilities, and construction sites). The remaining colonies were at wetlands (8.4%), where the species lives in elevated, well-drained areas (e.g., canal banks, flood zone dikes), and semi-natural grasslands (3.7%). While most of the colonies were on public lands (57%), less than one-third of the total (29%) were within the Natura 2000 network of protected areas. During the 2019–2021 surveys, only 37 (34.6%) of the colonies were still active.

The model that best explained the characteristics of colonies that went extinct over the past two decades consisted of two environmental variables: soil imperviousness and percent cover by high suitability habitat (see Table 3). Both variables were significant ($p < 0.001$) and had a negative relation to a colony's probability of extinction. Nevertheless, there was significant unexplained variance ($\chi^2 = 13.621$, df = 8, $p = 0.09$), which suggests that we were unable to effectively explain the drivers of local extinction.

Table 3. Model estimates and significance of environmental variables for the extinction of *Spermophilus citellus* colonies.

Variables	Estimate	SE	z-Values	Pr (>∣z∣)
Intercept (β)	1.555	0.337	4.613	<0.0001
Soil imperviousness	−0.0813	0.029	−2.754	<0.001
High suitability habitats	−3.639	1.260	−2.887	<0.001

4. Discussion

Our results provide the most complete assessment to date of the conservation status of the European ground squirrel population at its southernmost range, combining data from multiple sources to incorporate all known colonies since the mid-1990s. The reported colony extinction rate over the past two decades, combined with the low habitat-availability-to-habitat-suitability ratio and significant forecasted habitat suitability contraction by 2041–2060, build a bleak picture of the species' prospect for survival. Nevertheless, the study also provided the information required for prioritizing actions, areas, and land-uses for the urgently needed conservation efforts to save the species from extinction.

The large unexplained variance in causes of colony extinction suggests that additional, not tested, variables may be responsible, and/or that the drivers of the observed rapid decline may not be universal or detectable at the spatial scale examined. While the area occupied by a colony was not an important predictor of extinction, in many cases, the colonies were not monitored frequently enough (or at all) to be able to document population trends leading to extinction. Considering that many of these colonies had very small populations (<20 adult animals) at the time of last count [30], stochasticity (e.g., due to weather, predation, disease) alone could explain their eventual demise [77,78]. In fact, outright habitat loss or overall land-use change was rarely observed to be the case of a colony's extinction during field visits. Given the European ground squirrel's low vagility and fragmented distribution of colonies, even at the last strongholds of the species in Central and West Macedonia, the possibilities of recolonizing these areas are very low [79]. The limited connectivity and natural emigration between colonies (and even more between sub-populations) could have already led to genetic isolation and inbreeding depression, which, in turn, affected the population fitness [80]. An important first step to protecting the few remaining colonies is the adoption of a regular, statistically robust, monitoring protocol that is not limited to population counts, but that extends also to measuring genetic variation and parasitic charge. This will allow for the early identification and reduction of threats. Such small-scale monitoring projects have already been launched in 2020 within two national parks in Central Macedonia (Axios Delta National Park and Koroneia and Volvi National Park), but it is important that they secure long-term funding and that they are expanded to all colonies.

Perhaps counterintuitively, most of the colonies assessed for the study (87.5%) were within human-modified landscapes, even adjacent to human settlements or industrial areas, while colonies in highly suitable (natural) habitats, such as grasslands and sclerophyllous vegetation with sparse trees, were few and more likely to go extinct. Field surveys during

2019–2021 showed one in ten collapsed colonies having dense tallgrass vegetation with no signs of grazing or mowing. The importance of grazing in maintaining open abandoned fields is known [81,82], and Greece has experienced significant declines in the number and size of extensive grazing herds [7,9]. Abandonment of rural land has been linked to a decline in farmland species and biodiversity [83–86], and this may be the cause for at least some of the observed extinctions within European ground squirrel natural habitats. The systematic management of natural or semi-natural grasslands is needed, ensuring that areas with European ground squirrel colonies are either mowed or grazed frequently enough to maintain a suitable food vegetation structure [87,88]. For such a measure to be sustainable, areas with the European ground squirrel's presence should be recognized as high-value farmland (HNVF) to increase the viability of extensive livestock farming [89,90]. Moreover, the national grazing management plans currently under development should explicitly take into account the presence of grazing-dependent species, such as the European ground squirrel.

The prevalence of the remaining colonies in areas near human settlements indicated the current dependence of the species on human activities [30,50]. Human-managed areas provide short grass, steppe-like habitats that could be an important factor for the survival of colonies and their connectivity [18,50,91–93]. Therefore, in addition to actions aimed at grasslands, European-ground-squirrel-friendly practices should be adopted in agricultural areas. Agri-environmental management schemes promoted by the Common Agricultural Policy (CAP) [94,95] aim to substantially enhance or restore farmland species' habitats. Such efforts need to be long-term and adapted to local conservation needs [96]. Measures that could be beneficial to the European ground squirrel are, among others, low or no pesticide and fertilizer use, no plowing of fallows fields until October if the fields are to be cultivated next year, no burning of fallow vegetation, vegetation cut at least once annually (preferably before June), maintenance of unploughed strips at the edge of fields, intercropping, and the selection of crops that do not require dressed seeds [95,97]. Studies of another endangered, fossorial, small mammal, namely, the European hamster (*Cricetus cricetus*), also indicated that increasing the crop variety and farmland habitat mosaic improved the density and fitness of hamster populations [98,99]. The European ground squirrel could serve as a "flagship species" for farmland biodiversity (as they are charismatic and attractive to the public [100] and a keystone species), contributing toward sustainable agricultural landscapes.

Our results also showed an important role of roads in European ground squirrel habitat availability. Lowland areas (e.g., within the Axios River and Evros River deltas, and areas east of Thessaloniki) are, on the one hand, significant clusters of available habitats, but on the other, fragmented by considerable road networks. Despite the negative effects of roads on wildlife [101,102], the grass strips along roads could act as corridors for the expansion and connectivity of nearby colonies. It is important that such dispersal corridors are identified to take measures for reducing roadkill risk (e.g., speed bumps, signposts, fencing) and maintaining suitable vegetation along them (e.g., via mowing, planting appropriate grass/forbs, and ensuring sufficient soil drainage). Similar measures have been suggested for providing "stepping-stone" habitats along rivers [50]. Another measure for lowland habitats could be to improve the quality of abandoned, underutilized, or undeveloped plots around settlements and industrial infrastructure, where several colonies persist, albeit with a small number of individuals, in order to provide more suitable microhabitats for nesting and foraging of the populations [103]. In these areas, the invasive plant silverleaf nightshade (*Solanum elaeagnifolium*) abounds. This North American toxic invasive species is frequently encountered within European ground squirrel colonies in Central Macedonia and, to a lesser extent, in the colonies of Western Macedonia and Thrace [104]. It is considered a pest that outcompetes native species of the Mediterranean [104,105]. Based on our field observations, European ground squirrels feed on shoots, leaves, flowers, and seeds of the silverleaf nightshade, potentially affecting (negatively or positively) its spread. We do not fully understand the effect of the plant's toxins on the European ground squirrel's physiology or the vegetation composition, which are concerns that have been documented for other invasive plant species [106]. More studies are needed to assess the need, or not, to control this and other invasive plants within European ground squirrel colonies [107].

Contrary to the high level of habitat fragmentation reported for lowland areas, semi-mountainous and mountainous areas contain large tracts of roadless [76], available habitat for the European ground squirrel. Unfortunately, most of these areas do not appear to be suitable habitats according to the bioclimatic model. Nevertheless, surveys should be undertaken there in the near future. In Western Macedonia, the incongruence between available and suitable habitats for semi-mountainous areas near the cities of Kozani and Ptolemaida is due to the large-scale open-pit mining fueling the soon-to-be-closed coal energy production plants. Although there is no information on whether the species was present in these areas, we propose that the planned restoration activities for the mines should explicitly take into consideration the potential for these areas' natural or assisted recolonization by the European ground squirrel, as it could help increase the connectivity of the Western Macedonian sub-population.

Another land-use development of concern for the population of Western Macedonia is the proposed construction of wind farms in the Mount Vermio area that contains in its entirety the sole mountainous population of the country. While the impact of the construction and operation of wind farms on European ground squirrels is not yet well understood [47,108–110], we consider that measures to protect this high-altitude-adapted population and its large, fragmented, natural alpine environment (without invasive plant species) should be a conservation priority for the species. A solution would be to either move the wind farms to areas that exclude the land of current colonies and their potential connection corridors outside the Natura 2000 (GR1210001) [111] or to require the construction and operation companies to adhere to specific operation protocols that will mitigate all impact on this unique population and safeguard its long-term conservation.

According to our future habitat suitability projections for the European ground squirrel, there will be contractions in the broader Mount Vermio region, which is one more reason that the population there should be protected. The high-altitude adaptation of this population could be key in captive breeding programs aimed at establishing additional high-altitude colonies within the historical range of the species [112]. The most conservative future climate change scenarios predict the larger-in-size climate change refugia for the European ground squirrel to be in the lowland areas of the Axios River and Evros River delta regions. While the former area still supports some large colonies, the Evros population is in critical condition, with just a handful of small colonies persisting. It is imperative that they are urgently protected on the ground, with measures aimed at buffering them from stochastic events (e.g., flooding, food scarcity, possibly predators, and accidental eradication due to land-use changes). Similar future range contraction to lowland areas due to climate change was also reported for other ground squirrels, such as the Anatolian ground squirrel (*Spermophilus xanthoprymnus*) [44]. However, our study's findings disagree with the predicted suitable habitat expansion of the European ground squirrel in Greece reported by Demirtaş [43]. That study examined the past, present, and future distribution of the European ground squirrel across Europe, including the southern lineage containing the Greek populations. Only six locations from the country were used. Since our study was based on a much larger dataset, we believe that our predictions are likely more realistic, while acknowledging the considerable uncertainty that climate change predictions inherently contain.

Overall, our results identified three areas that incorporate (a) the genetic variation of the Greek sub-population [Rammou et al., in preparation], (b) both lowland and mountain population adaptations, and (c) most remaining individuals [30]. These are the Axios River valley, the Evros River delta, and the Mount Vermio alpine meadows. Therefore, we propose these three areas to be the focal areas where the core breeding of populations will sustain and probably expand the species' current distribution, while they will constitute future climate refugia as well. These focal areas could furthermore receive via translocation individuals from very small colonies (<10 animals; 21 of 37 known colonies) occurring in less favorable habitats and faced with a high probability of extinction. At this point, Bulgarian, Czech, Slovakian, Polish, and Hungarian specialists have accumulated over 30 years of expertise on population reinforcement conservation activities [75,112–117]. However, before any translocation or population reinforcement takes place in Greece, a national legal framework (Species Action Plan) is required in order to align all conservation activities with national and EU directives, such as the CAP strategic plan and the Habitat Directive (92/43/EEC). This is especially important, as many European ground squirrel populations are outside the protected area network, with many in public lands managed by different public bodies (e.g., municipalities, airports, archaeological sites, military facilities), for which a national management plan is needed in order to expedite and facilitate cross-agency collaborations.

5. Conclusions

This study reiterated the importance of developing species-specific conservation approaches, especially for populations at the edge of a species' range, which are most likely to be affected by anticipated climate changes. Our analysis shows that the status of the European ground squirrel's southernmost population is deteriorating, with most known colonies having been lost over the past decade. While the species persists in all three of the previously reported sub-populations in West Macedonia, Central Macedonia, and Thrace, our ecological niche model predicts habitat suitability contraction in the next twenty to forty years across all of these regions. The species' forecast climate refugia are in need of different conservation interventions. The already scattered lowland colonies in Central Macedonia (Axios River valley) and Thrace (Evros River delta) face increased isolation, and therefore future conservation efforts should emphasize maintaining or establishing sub-population connectivity. On the other hand, the sole remaining mountainous colony on Mount Vermio (West Macedonia) occurs in good quality natural habitat, though it requires protection of its habitat from forest encroachment and proposed large-scale energy production developments. Halting the observed population decline of all colonies is a universal priority, however, which will involve maintaining, and eventually expanding, habitat availability and identifying colony-specific drivers of extinction. For each planned activity, the trade-offs of the prioritization process should be considered (see conservation triage [118]). Such coordinated and well-planned actions require a currently lacking national action plan for the species.

Supplementary Materials: The following supporting information can be downloaded at: https://www.mdpi.com/article/10.3390/land11020301/s1, Table S1: Model evaluation statistics of model 1: habitat availability (100 m resolution) and model 2: habitat suitability (4.5 km resolution) with delta AICc values of $\leq$2. The variables are referred to as FC, feature classes (L—linear, Q—quadratic, H—hinge; RM, regularization multiplier; AUC_{DIFF}, the difference between training and testing AUC; validation AUC, the validation set to estimate prediction error for model selection; OR_{10}, 10% training omission rate; AICc, the Akaike information criterion corrected for small sample sizes; delta AICc, the difference between the lowest AICc and each AICc; N.coef, the number of coefficients. Figure S1: Marginal (above group of diagrams) and univariate (below group of diagrams) response curves of the variables that were used in the habitat availability ecological niche model 1 (left) and habitat suitability ecological niche model 2 (right). The numbers of categorical variables indicate suitable land-cover for EGS: 0, the unsuitable habitats; 1, the suitable habitats; and 2, the highly suitable habitats, according to our classification, and for soil texture: 3, silt clay-loam; 5, sandy clay-loam; 6, clay-loam; 9, loam; and 12, sandy loam, according to USDA classification (for more details, please see Section 2.3 of Materials and Methods). Table S2: Area characteristics (property, Corine land-cover, and protection status) of the present and absent colonies of *Spermophilus citellus* in three sub-populations in Greece that were used in the analysis.

Author Contributions: D.-L.R. and C.A. contributed equally to this project (†). Conceptualization, D.-L.R., C.A. and D.Y.; methodology, D.-L.R. and C.A.; software, C.A.; validation, C.A. and D.Y.; formal analysis, C.A.; investigation, D.-L.R., D.M., G.B. and D.Y.; resources, D.-L.R., C.A., D.M., G.B.

and D.Y.; data curation, D.-L.R. and C.A.; writing—original draft preparation, D.-L.R. and C.A.; writing—review and editing, D.-L.R., C.A., D.M., G.B., A.G., T.K. and D.Y.; visualization, D.-L.R. and C.A.; supervision, D.Y.; project administration, D.Y.; funding acquisition, D.-L.R., C.A. and D.Y. All authors have read and agreed to the published version of the manuscript.

Funding: This research is co-financed by Greece and the European Union (European Social Fund-ESF) through the Operational Programme "Human Resources Development, Education and Lifelong Learning" in the context of the project "Strengthening Human Resources Research Potential via Doctorate Research" (MIS-5000432), implemented by the State Scholarships Foundation (IKY).

Data Availability Statement: Data may be available upon request to the authors.

Acknowledgments: We would like to thank Juan F. Beltrán for inviting us to participate in the special issue "Wildlife Protection and Habitat Management: Practice and Perspectives". Field research complied with the laws of the Ministry of Environment and Energy (research permits 178107/64 and 8817/242). Many thanks go to C.G. Georgiadis, M. Kachamakova, and M. Psaralexi for their great support during fieldwork. We express our gratitude to Juan F. Beltrán and the three anonymous reviewers whose comments substantially improved the manuscript.

Conflicts of Interest: The authors declare no conflict of interest.

References

1. Antrop, M. Landscape change and the urbanization process in Europe. *Landsc. Urban Plan.* **2004**, *67*, 9–26. [CrossRef]
2. Butchart, S.H.M.; Walpole, M.; Collen, B.; Van Strien, A.; Scharlemann, J.P.W.; Almond, R.E.A.; Baillie, J.E.M.; Bomhard, B.; Brown, C.; Bruno, J.; et al. Global biodiversity: Indicators of recent declines. *Science* **2010**, *328*, 1164–1168. [CrossRef] [PubMed]
3. Foley, J.A.; DeFries, R.; Asner, G.P.; Barford, C.; Bonan, G.; Carpenter, S.R.; Chapin, F.S.; Coe, M.T.; Daily, G.C.; Gibbs, H.K.; et al. Global consequences of land use. *Science* **2005**, *309*, 570–574. [CrossRef]
4. Geri, F.; Amici, V.; Rocchini, D. Human activity impact on the heterogeneity of a mediterranean landscape. *Appl. Geogr.* **2010**, *30*, 370–379. [CrossRef]
5. Jongman, R.H.G. Homogenisation and fragmentation of the european landscape: Ecological consequences and solutions. *Landsc. Urban Plan.* **2002**, *58*, 211–221. [CrossRef]
6. Leu, M.; Hanser, S.E.; Knick, S.T. The human footprint in the West: A large-scale analysus of anthropogenic impacts. *Ecol. Appl.* **2008**, *18*, 1119–1139. [CrossRef]
7. Hadjigeorgiou, I. Past, present and future of pastoralism in Greece. *Pastor. Res. Policy Pract.* **2011**, *1*, 24. [CrossRef]
8. Malandra, F.; Vitali, A.; Urbinati, C.; Garbarino, M. 70 Years of land use/land cover changes in the apennines (Italy): A meta-analysis. *Forests* **2018**, *9*, 551. [CrossRef]
9. Sidiropoulou, A.; Karatassiou, M.; Galidaki, G.; Sklavou, P. Landscape pattern changes in response to transhumance abandonment on mountain vermio (North Greece). *Sustainability* **2015**, *7*, 15652–15673. [CrossRef]
10. Farina, A. Recent changes of the mosaic patterns in a montane landscape (North Italy) and consequences on vertebrate fauna. *Options Méditerr. Série Sémin.* **1991**, *15*, 121–134.
11. Francis, R.A.; Chadwick, M.A. What makes a species synurbic? *Appl. Geogr.* **2012**, *32*, 514–521. [CrossRef]
12. Robledano, F.; Esteve, M.A.; Farinós, P.; Carreño, M.F.; Martínez-Fernández, J. Terrestrial birds as indicators of agricultural-induced changes and associated loss in conservation value of mediterranean wetlands. *Ecol. Indic.* **2010**, *10*, 274–286. [CrossRef]
13. Russo, D.; Ancillotto, L. Sensitivity of bats to urbanization: A Review. *Mamm. Biol.* **2015**, *80*, 205–212. [CrossRef] [PubMed]
14. Fischer, J.; Lindenmayer, D.B. Landscape modification and habitat fragmentation: A synthesis. *Glob. Ecol. Biogeogr.* **2007**, *16*, 265–280. [CrossRef]
15. Schloss, C.A.; Nuñez, T.A.; Lawler, J.J. Dispersal will limit ability of mammals to track climate change in the western hemisphere. *Proc. Natl. Acad. Sci. USA* **2012**, *109*, 8606–8611. [CrossRef]
16. Broennimann, O.; Thuiller, W.; Hughes, G.; Midgley, G.F.; Alkemade, J.M.R.; Guisan, A. Do geographic distribution, niche property and life form explain plants' vulnerability to global change? *Glob. Chang. Biol.* **2006**, *12*, 1079–1093. [CrossRef]
17. Coroiu, C.; Kryštufek, B.; Vohralík, V.; Zagorodnyuk, I. Spermophilus Citellus. In *IUCN 2012, IUCN Red List of Threatened Species*; Version 2012.1; IUCN: Gland, Switzerland, 2008; Available online: https://www.iucnredlist.org/species/20472/9204055 (accessed on 10 January 2022).
18. Hoffmann, I.E.; Millesi, E.; Pieta, K.; Dittami, J.P. anthropogenic effects on the population ecology of european ground squirrels (Spermophilus citellus) at the periphery of their geographic range. *Mamm. Biol.* **2003**, *68*, 205–213. [CrossRef]
19. Janák, M.; Marhoul, P.; Matějů, J. *Action Plan for the Conservation of the European Ground Squirrel Spermophilus Citellus in the European Union*; European Commission: Brussels, Belgium, 2013; pp. 5–8.
20. Kryštufek, B.; Vohralík, V. *Mammals of Turkey and Cyprus. Rodentia I: Sciuridae, Dipodidae, Gliridae, Arvicolinae*; Zgodovinsko društvo za južno Primorsko: Ljubljana, Slovenia, 2005; pp. 42–43.
21. Kryštufek, B.; Glasnović, P.; Petkovski, S. The status of a rare phylogeographic lineage of the vulnerable european souslik spermophilus citellus, endemic to central macedonia. *Oryx* **2012**, *46*, 442–445. [CrossRef]

22. Grulich, I. Sysel obecný citellus citellus, L. v ČSSR. Práce Brněn. *Základny ČSAV* **1960**, *32*, 473–563.
23. Youlatos, D.; Boutsis, Y.; Pantis, J.D.; Hadjicharalambous, H. Activity patterns of european ground squirrels (Spermophilus citellus) in a cultivated field in Northern Greece. *Mammalia* **2007**, *71*, 183–186. [CrossRef]
24. Davidson, A.D.; Lightfoot, D.C. Interactive effects of keystone rodents on the structure of desert grassland arthropod communities. *Ecography* **2007**, *30*, 515–525. [CrossRef]
25. Lindtner, P.; Gömöryová, E.; Gömöry, D.; Stašiov, S.; Kubovčík, V. Development of physico-chemical and biological soil properties on the european ground squirrel mounds. *Geoderma* **2019**, *339*, 85–93. [CrossRef]
26. Lindtner, P.; Svitok, M.; Ujházy, K.; Kubovčík, V. Disturbances by the European ground squirrel enhance diversity and spatial heterogeneity of plant communities in temperate grassland. *Biodivers. Conserv.* **2020**, *29*, 853–867. [CrossRef]
27. Ramos-Lara, N.; Koprowski, J.L.; Kryštufek, B.; Hoffmann, I.E. Spermophilus citellus (Rodentia: Sciuridae). *Mamm. Species* **2014**, *46*, 71–87. [CrossRef]
28. Hegyeli, Z. *Spermophilus Citellusi*, IUCN Red List Threatment Species. 2020. Available online: https://www.iucnredlist.org/species/20472/91282380 (accessed on 10 January 2022).
29. Youlatos, D. *Spermophilus Citellus* (Linnaeus, 1766). In *The Red Book of Endangered Animals of Greece*; Legakis, A., Maragou, P., Eds.; Hellenic Zoological Society: Athens, Greece, 2009; p. 528.
30. Rammou, D.-L.; Kavroudakis, D.; Youlatos, D. Distribution, population size, and habitat characteristics of the endangered european ground squirrel (Spermophilus citellus, Rodentia, Mammalia) in its southernmost range. *Sustainability* **2021**, *13*, 8411. [CrossRef]
31. Giannakopoulos, C.; Le Sager, P.; Bindi, M.; Moriondo, M.; Kostopoulou, E.; Goodess, C.M. Climatic changes and associated impacts in the mediterranean resulting from a 2 °C global warming. *Glob. Planet. Chang.* **2009**, *68*, 209–224. [CrossRef]
32. Giorgi, F.; Lionello, P. Climate change projections for the Mediterranean region. *Glob. Planet. Chang.* **2008**, *63*, 90–104. [CrossRef]
33. Goldberg, A.R.; Conway, C.J. Hibernation behavior of a federally threatened ground squirrel: Climate change and habitat selection implications. *J. Mammal.* **2021**, *102*, 574–587. [CrossRef]
34. Németh, I.; Nyitrai, V.; Altbäcker, V. Ambient temperature and annual timing affect torpor bouts and euthermic phases of hibernating european ground squirrels (Spermophilus citellus). *Can. J. Zool.* **2009**, *87*, 204–210. [CrossRef]
35. Guisan, A.; Tingley, R.; Baumgartner, J.B.; Naujokaitis-Lewis, I.; Sutcliffe, P.R.; Tulloch, A.I.T.; Regan, T.J.; Brotons, L.; McDonald-Madden, E.; Mantyka-Pringle, C.; et al. Predicting species distributions for conservation decisions. *Ecol. Lett.* **2013**, *16*, 1424–1435. [CrossRef]
36. Austin, M.P. Spatial prediction of species distribution: An interface between ecological theory and statistical modelling. *Ecol. Model.* **2002**, *157*, 101–118. [CrossRef]
37. Cianfrani, C.; Broennimann, O.; Loy, A.; Guisan, A. More than range exposure: Global otter vulnerability to climate change. *Biol. Conserv.* **2018**, *221*, 103–113. [CrossRef]
38. Guisan, A.; Zimmermann, N.E. Predictive habitat distribution models in ecology. *Ecol. Model.* **2000**, *135*, 147–186. [CrossRef]
39. Guisan, A.; Thuiller, W. Predicting species distribution: Offering more than simple habitat models. *Ecol. Lett.* **2005**, *8*, 993–1009. [CrossRef]
40. Phillips, S.J.; Dudík, M.; Schapire, R.E. A maximum entropy approach to species distribution modeling. In Proceedings of the Twenty-First International Conference On Machine Learning, Banff, AB, Canada, 4–8 July 2004; pp. 655–662.
41. Phillips, S.J.; Anderson, R.P.; Schapire, R.E. Maximum entropy modeling of species geographic distributions. *Ecol. Model.* **2006**, *190*, 231–259. [CrossRef]
42. Phillips, S.J.; Anderson, R.P.; Dudík, M.; Schapire, R.E.; Blair, M.E. Opening the black box: An open-source release of maxent. *Ecography* **2017**, *40*, 887–893. [CrossRef]
43. Demirtaş, S. Estimation of the climate preference between two lineages of europe-an ground squirrel using maximum entropy modeling. *J. Adv. Res. Nat. Appl. Sci.* **2020**, *6*, 328–341. [CrossRef]
44. Gür, H. The future impact of climate and land-use changes on anatolian ground squirrels under different scenarios. *bioRxiv* **2021**, 460244. [CrossRef]
45. Dilts, T.E.; Weisberg, P.J.; Leitner, P.; Matocq, M.D.; Inman, R.D.; Nussear, K.E.; Esque, T.C. Multiscale connectivity and graph theory highlight critical areas for conservation under climate change. *Ecol. Appl.* **2016**, *26*, 1223–1237. [CrossRef]
46. Holt, A.C.; Salkeld, D.J.; Fritz, C.L.; Tucker, J.R.; Gong, P. Spatial analysis of plague in California: Niche modeling predictions of the current distribution and potential response to climate change. *Int. J. Health Geogr.* **2009**, *8*, 38. [CrossRef]
47. Inman, R.D.; Esque, T.C.; Nussear, K.E.; Leitner, P.; Matocq, M.D.; Weisberg, P.J.; Dilts, T.E.; Vandergast, A.G. Is there room for all of us? Renewable energy and *Xerospermophilus Mohavensis*. *Endanger. Species Res.* **2013**, *20*, 1–18. [CrossRef]
48. Kryštufek, B.; Stanciu, C.; Ivajnšič, D.; Cherkaoui, S.I.; Janžekovič, F. Facts and misconceptions on the palaearctic existence of the striped ground squirrel. *Mammalia* **2018**, *82*, 248–255. [CrossRef]
49. Tian, L. Relationship between environmental factors and the spatial distribution of *Spermophilus* dauricus during 2000–2015 in China. *Int. J. Biometeorol.* **2018**, *62*, 1781–1789. [CrossRef] [PubMed]
50. Tzvetkov, J.; Koshev, Y. GIS habitat model of potential distribution of european ground squirrel (*Spermophilus citellus*) in Bulgaria. In Proceedings of the 6th European Ground Squirrel Meeting, Belgrade, Serbia, 4–6 November 2016.

51. Elith, J.; Graham, C.H.; Anderson, R.P.; Dudík, M.; Ferrier, S.; Guisan, A.; Hijmans, R.J.; Huettmann, F.; Leathwick, J.R.; Lehmann, A.; et al. Novel methods improve prediction of species' distributions from occurrence data. *Ecography* **2006**, *29*, 129–151. [CrossRef]
52. Pearson, R.G.; Thuiller, W.; Araújo, M.B.; Martinez-Meyer, E.; Brotons, L.; McClean, C.; Miles, L.; Segurado, P.; Dawson, T.P.; Lees, D.C. Model-based uncertainty in species range prediction. *J. Biogeogr.* **2006**, *33*, 1704–1711. [CrossRef]
53. Rebelo, H.; Jones, G. Ground validation of presence-only modelling with rare species: A case study on Barbastelles Barbastella Barbastellus (Chiroptera: Vespertilionidae). *J. Appl. Ecol.* **2010**, *47*, 410–420. [CrossRef]
54. Wisz, M.S.; Hijmans, R.J.; Li, J.; Peterson, A.T.; Graham, C.H.; Guisan, A.; NCEAS Predicting Species Distributions Working Group. Effects of sample size on the performance of species distribution models. *Divers. Distrib.* **2008**, *14*, 763–773. [CrossRef]
55. Beck, H.E.; Zimmermann, N.E.; McVicar, T.R.; Vergopolan, N.; Berg, A.; Wood, E.F. Present and future köppen-geiger climate classification maps at 1-km resolution. *Sci. Data* **2018**, *5*, 180214. [CrossRef]
56. Fick, S.E.; Hijmans, R.J. WorldClim 2: New 1-km spatial resolution climate surfaces for global land areas. *Int. J. Climatol.* **2017**, *37*, 4302–4315. [CrossRef]
57. Feng, X.; Park, D.S.; Liang, Y.; Pandey, R.; Papeş, M. Collinearity in ecological niche modeling: Confusions and challenges. *Ecol. Evol.* **2019**, *9*, 10365–10376. [CrossRef]
58. QGIS Development Team QGIS Geographic Information System. 2021. Available online: http://www.qgis.org (accessed on 10 January 2022).
59. Riahi, K.; Van Vuuren, D.P.; Kriegler, E.; Edmonds, J.; O'Neill, B.C.; Fujimori, S.; Bauer, N.; Calvin, K.; Dellink, R.; Fricko, O.; et al. The shared socioeconomic pathways and their energy, land use, and greenhouse gas emissions implications: An overview. *Glob. Environ. Chang.* **2017**, *42*, 153–168. [CrossRef]
60. Phillips, S.J.; Dudík, M. Modeling of species distributions with maxent: New extensions and a comprehensive evaluation. *Ecography* **2008**, *31*, 161–175. [CrossRef]
61. Kaky, E.; Nolan, V.; Alatawi, A.; Gilbert, F. A comparison between ensemble and MaxEnt species distribution modelling approaches for conservation: A case study with egyptian medicinal plants. *Ecol. Inform.* **2020**, *60*, 101150. [CrossRef]
62. Morales, N.S.; Fernández, I.C.; Baca-González, V. MaxEnt's parameter configuration and small samples: Are we paying attention to recommendations? A systematic review. *PeerJ* **2017**, *5*, e3093. [CrossRef]
63. Muscarella, R.; Galante, P.J.; Soley-Guardia, M.; Boria, R.A.; Kass, J.M.; Uriarte, M.; Anderson, R.P. ENMeval: An R package for conducting spatially independent evaluations and estimating optimal model complexity for maxent ecological niche models. *Methods Ecol. Evol.* **2014**, *5*, 1198–1205. [CrossRef]
64. Boria, R.A.; Olson, L.E.; Goodman, S.M.; Anderson, R.P. Spatial filtering to reduce sampling bias can improve the performance of ecological niche models. *Ecol. Model.* **2014**, *275*, 73–77. [CrossRef]
65. Burnham, K.P.; Anderson, D.R. Model Selection and Multimodel Inference. In *A Practical Information-Theoretic Approach*, 2nd ed.; Springer: New York, NY, USA, 2002.
66. Boyce, M.S.; Vernier, P.R.; Nielsen, S.E.; Schmiegelow, F.K.A. Evaluating resource selection functions. *Ecol. Model.* **2002**, *157*, 281–300. [CrossRef]
67. Hirzel, A.H.; Le Lay, G.; Helfer, V.; Randin, C.; Guisan, A. Evaluating the ability of habitat suitability models to predict species presences. *Predict. Species Distrib.* **2006**, *199*, 142–152. [CrossRef]
68. Araújo, M.B.; Pearson, R.G.; Thuiller, W.; Erhard, M. Validation of species–climate impact models under climate change. *Glob. Chang. Biol.* **2005**, *11*, 1504–1513. [CrossRef]
69. Lobo, J.M.; Jiménez-Valverde, A.; Real, R. AUC: A misleading measure of the performance of predictive distribution models. *Glob. Ecol. Biogeogr.* **2008**, *17*, 145–151. [CrossRef]
70. Sun, X.; Long, Z.; Jia, J. A multi-scale maxent approach to model habitat suitability for the giant pandas in the Qionglai mountain, China. *Glob. Ecol. Conserv.* **2021**, *30*, e01766. [CrossRef]
71. Ashcroft, M.B. Identifying refugia from climate change. *J. Biogeogr.* **2010**, *37*, 1407–1413. [CrossRef]
72. Armitage, K.B. Sociality as a life-history tactic of ground squirrels. *Oecologia* **1981**, *48*, 36–49. [CrossRef]
73. Koshev, Y.S. Distribution and status of the european ground squirrel (Spermophilus citellus) in Bulgaria. *Lynx* **2008**, *39*, 251–261.
74. R Core Team. *R: A Language and Environment for Statistical Computing*; R Foudation for Statistical Computing: Vienna, Austria, 2021.
75. Matějů, J.; Hulová, Š.; Nová, P.; Cepáková, E.; Marhoul, P.; Uhlíková, J. *Action Plan for the European Ground Squirrel (Spermophilus citellus) in the Czech Republic*; Charles University and Agency for Nature and Landscape Protection of the Czech Republic: Prague, Czech Republic, 2010; p. 80.
76. Kati, V.; Kassara, C.; Psaralexi, M.; Tzortzakaki, O.; Petridou, M.; Galani, A.; Hoffmann, M.T. Conservation policy under a roadless perspective: Minimizing fragmentation in Greece. *Biol. Conserv.* **2020**, *252*, 108828. [CrossRef]
77. Brown, J.H.; Kodricbrown, A. Turnover rates in insular biogeography—Effect of immigration on extinction. *Ecology* **1977**, *58*, 445–449. [CrossRef]
78. Pulliam, H.R. On the relationship between niche and distribution. *Ecol. Lett.* **2000**, *3*, 349–361. [CrossRef]
79. Hubbell, S.P. *The Unified Neutral Theory of Biodiversity and Biogeography*; Princeton University Press: Princeton, NJ, USA, 2001; pp. 152–200.

80. Říčanová, Š.; Bryja, J.; Cosson, J.-F.; Gedeon, C.; Choleva, L.; Ambros, M.; Sedláček, F. Depleted genetic variation of the european ground squirrel in Central Europe in both microsatellites and the major histocompatibility complex gene: Implications for conservation. *Conserv. Genet.* **2011**, *12*, 1115–1129. [CrossRef]
81. Davidson, A.D.; Ponce, E.; Lightfoot, D.C.; Fredrickson, E.L.; Brown, J.H.; Cruzado, J.; Brantley, S.L.; Sierra-Corona, R.; List, R.; Toledo, D.; et al. Rapid response of a grassland ecosystem to an experimental manipulation of a keystone rodent and domestic livestock. *Ecology* **2010**, *91*, 3189–3200. [CrossRef]
82. Luoto, M.; Pykälä, J.; Kuussaari, M. Decline of landscape-scale habitat and species diversity after the end of cattle grazing. *J. Nat. Conserv.* **2003**, *11*, 171–178. [CrossRef]
83. Tzanopoulos, J.; Mitchley, J.; Pantis, J.D. Vegetation dynamics in abandoned crop fields on a Mediterranean island: Development of succession model and estimation of disturbance thresholds. *Agric. Ecosyst. Environ.* **2007**, *120*, 370–376. [CrossRef]
84. Zomeni, M.; Tzanopoulos, J.; Pantis, J.D. Historical analysis of landscape change using remote sensing techniques: An explanatory tool for agricultural transformation in Greek rural areas. *Landsc. Urban Plan.* **2008**, *86*, 38–46. [CrossRef]
85. Zakkak, S.; Kakalis, E.; Radović, A.; Halley, J.M.; Kati, V. The impact of forest encroachment after agricultural land abandonment on passerine bird communities: The case of Greece. *J. Nat. Conserv.* **2014**, *22*, 157–165. [CrossRef]
86. Zakkak, S.; Halley, J.M.; Akriotis, T.; Kati, V. Lizards along an agricultural land abandonment gradient in pindos mountains, Greece. *Amphib. Reptil.* **2015**, *36*, 253–264. [CrossRef]
87. Blüthgen, N.; Dormann, C.F.; Prati, D.; Klaus, V.H.; Kleinebecker, T.; Hölzel, N.; Alt, F.; Boch, S.; Gockel, S.; Hemp, A.; et al. A quantitative index of land-use intensity in grasslands: Integrating mowing, grazing and fertilization. *Basic Appl. Ecol.* **2012**, *13*, 207–220. [CrossRef]
88. Petluš, P.; Petlušová, V.; Baláž, I.; Ševčík, M.; Lešová, A.; Hapl, E. Impact of management measures on the european ground squirrel population development. *Folia Oecol.* **2021**, *48*, 169–179. [CrossRef]
89. European Environment Agency. *High Nature Value Farmland: Characteristics, Trends, and Policy Challenges*; Institute European Environmental Policy: Copenhagen, Denmark, 2004.
90. Herzon, I.; Birge, T.; Allen, B.; Povellato, A.; Vanni, F.; Hart, K.; Radley, G.; Tucker, G.; Keenleyside, C.; Oppermann, R.; et al. Time to look for evidence: Results-based approach to biodiversity conservation on farmland in Europe. *Land Use Policy* **2018**, *71*, 347–354. [CrossRef]
91. Kis, J.; Váczi, O.; Katona, K.; Altbäcker, V. A növényzet magasságának hatása a cinegési ürgék élőhelyválasztására. The effect of vegetation height to habitat selection of ground squirrels in cinegés. *Termvéd. Közlemények* **1998**, *7*, 117–123.
92. Matějů, J.; Nová, P.; Uhlíkovà, J.; Hulová, Š.; Cepáková, E. Distribution of the European ground squirrel (Spermophilus citellus) in the Czech Republic in 2002–2008. *Lynx* **2008**, *39*, 277–294.
93. Ricankova, V.; Fric, Z.; Chlachula, J.; Stastna, P.; Faltynkova, A.; Zemek, F. Habitat requirements of the long-tailed ground squirrel (*Spermophilus undulatus*) in the Southern Altai. *J. Zool.* **2006**, *270*, 1–8. [CrossRef]
94. European Commission. *Agri-Environment Measures—Overview on General Principles, Types of Measures, and Application*; European Commission: Brussels, Belgium, 2005.
95. Kleijn, D.; Baquero, R.A.; Clough, Y.; Díaz, M.; De Esteban, J.; Fernández, F.; Gabriel, D.; Herzog, F.; Holzschuh, A.; Jöhl, R.; et al. Mixed biodiversity benefits of agri-environment schemes in five european countries. *Ecol. Lett.* **2006**, *9*, 243–254. [CrossRef]
96. Walker, L.K.; Morris, A.J.; Cristinacce, A.; Dadam, D.; Grice, P.V.; Peach, W.J. Effects of higher-tier agri-environment scheme on the abundance of priority farmland birds. *Anim. Conserv.* **2018**, *21*, 183–192. [CrossRef]
97. Sokos, C.K.; Mamolos, A.P.; Kalburtji, K.L.; Birtsas, P.K. Farming and wildlife in mediterranean agroecosystems. *J. Nat. Conserv.* **2013**, *21*, 81–92. [CrossRef]
98. Bald, V.; Boetzl, F.A.; Krauss, J. Where do hamsters go after cereal harvest? A case study. *Basic Appl. Ecol.* **2021**, *54*, 98–107. [CrossRef]
99. Tissier, M.L.; Kletty, F.; Robin, J.-P.; Habold, C. Sustainable agriculture: Nutritional benefits of wheat–soybean and maize–sunflower associations for hibernation and reproduction of endangered common hamsters. *Sustainability* **2021**, *13*, 1352. [CrossRef]
100. Liordos, V.; Kontsiotis, V.J.; Anastasiadou, M.; Karavasias, E. Effects of attitudes and demography on public support for endangered species conservation. *Sci. Total Environ.* **2017**, *595*, 25–34. [CrossRef]
101. Clevenger, A.P.; Chruszcz, B.; Gunson, K.E. Spatial patterns and factors influencing small vertebrate fauna road-kill aggregations. *Biol. Conserv.* **2002**, *109*, 15–26. [CrossRef]
102. D'Amico, M.; Román, J.; De los Reyes, L.; Revilla, E. Vertebrate road-kill patterns in mediterranean habitats: Who, when and where. *Biol. Conserv.* **2015**, *191*, 234–242. [CrossRef]
103. Kenyeres, Z.; Bauer, N.; Nagy, L.; Szabó, S. Enhancement of a declining european ground squirrel (Spermophilus citellus) population with habitat restoration. *J. Nat. Conserv.* **2018**, *45*, 98–106. [CrossRef]
104. Krigas, N.; Tsiafouli, M.A.; Katsoulis, G.; Votsi, N.-E.; Van Kleunen, M. Investigating the invasion pattern of the alien plant solanum elaeagnifolium cav. (Silverleaf Nightshade): Environmental and human-induced drivers. *Plants* **2021**, *10*, 805. [CrossRef]
105. Krigas, N.; Kokkini, S. A survey of the alien vascular flora of the urban and suburban area of thessaloniki, N Greece. *Willdenowia* **2004**, *34*, 81–99. [CrossRef]
106. Kajzer-Bonk, J.; Szpiłyk, D.; Woyciechowski, M. Invasive goldenrods affect abundance and diversity of grassland ant communities (Hymenoptera: Formicidae). *J. Insect Conserv.* **2016**, *20*, 99–105. [CrossRef]

107. Dueñas, M.-A.; Hemming, D.J.; Roberts, A.; Diaz-Soltero, H. The threat of invasive species to IUCN-listed critically endangered species: A systematic review. *Glob. Ecol. Conserv.* **2021**, *26*, e01476. [CrossRef]
108. Inman, R.D.; Esque, T.C.; Nussear, K.E.; Leitner, P.; Matocq, M.D.; Weisberg, P.J.; Dilts, T.E. Impacts of climate change and renewable energy development on habitat of an endemic squirrel, Xerospermophilus mohavensis, in the Mojave Desert, USA. *Biol. Conserv.* **2016**, *200*, 112–121. [CrossRef]
109. Łopucki, R.; Łopucki, R.; Klich, D.; Ścibior, A.; Gołębiowska, D.; Perzanowski, K. Living in habitats affected by wind turbines may result in an increase in corticosterone levels in ground dwelling animals. *Ecol. Indic.* **2018**, *84*, 165. [CrossRef]
110. Lovich, J.E.; Ennen, J.R. Assessing the state of knowledge of utility-scale wind energy development and operation on non-volant terrestrial and marine wildlife. *Appl. Eng.* **2013**, *103*, 52–60. [CrossRef]
111. Kati, V.; Kassara, C.; Vrontisi, Z.; Moustakas, A. The biodiversity-wind energy-land use nexus in a global biodiversity hotspot. *Sci. Total Environ.* **2021**, *768*, 144474. [CrossRef]
112. Koshev, Y.; Kachamakova, M.; Arangelov, S.; Ragyov, D. Translocations of european ground squirrel (Spermophilus citellus) along altitudinal gradient in Bulgaria—An Overview. *Nat. Conserv.* **2019**, *35*, 63–95. [CrossRef]
113. Balaz, I.; Jancova, A.; Ambros, M. Restitution of the european ground squirrel (Spermophilus citellus) in Slovakia. *Lynx* **2008**, *39*, 235–240.
114. Gedeon, C.I.; Váczi, O.; Koósz, B.; Altbäcker, V. Morning release into artificial burrows with retention caps facilitates success of European ground squirrel (Spermophilus citellus) translocations. *Eur. J. Wildl. Res.* **2011**, *57*, 1101–1105. [CrossRef]
115. Kachamakova, M.; Koshev, Y. Post-release settlement and survival of endangered European ground squirrel after conservation reinforcement. *J. Nat. Conserv.* **2021**, *63*, 126048. [CrossRef]
116. Löbbová, D.; Hapl, E. Conservation of European ground squirrel (Mammalia: Rodentia) in Slovakia: Results of current reintroduction programme. *Slovak Raptor J.* **2014**, *8*, 105. [CrossRef]
117. Matějů, J.; Říčanová, Š.; Poláková, S.; Ambros, M.; Kala, B.; Matějů, K.; Kratochvíl, L. Method of releasing and number of animals are determinants for the success of European ground squirrel (Spermophilus citellus) reintroductions. *Eur. J. Wildl. Res.* **2012**, *58*, 473–482. [CrossRef]
118. Bottrill, M.C.; Joseph, L.N.; Carwardine, J.; Bode, M.; Cook, C.; Game, E.T.; Grantham, H.; Kark, S.; Linke, S.; McDonald-Madden, E.; et al. Is conservation triage just smart decision making? *Trends Ecol. Evol.* **2008**, *23*, 649–654. [CrossRef] [PubMed]

land

MDPI

Project Report

Habitat Management of the Endemic and Critical Endangered Montseny Brook Newt (*Calotriton arnoldi*)

Daniel Guinart [1], Sònia Solórzano [1], Fèlix Amat [2], Jordina Grau [1], Daniel Fernández-Guiberteau [3] and Albert Montori [3,*]

[1] Oficina Tècnica de Parcs Naturals, Diputació de Barcelona, The Montseny Natural Park and Biosphere Reserve, Masia Mariona, Ctra 5119, Km 2,5 (Mosqueroles), Fogars de Montclús, 08479 Barcelona, Spain; guinartsd@diba.cat (D.G.); solorzanohs@diba.cat (S.S.); grausj@diba.cat (J.G.)

[2] Àrea d'Herpetologia, BiBIO, Museu de Granollers–Ciències Naturals, Palaudàries 102, Granollers, 08402 Barcelona, Spain; famat@mcng.cat

[3] CREAC, Centre de Recerca i Educació Ambiental de Calafell, GRENP/Aj, Calafell, 43882 Tarragona, Spain; info@guibernatura.com

* Correspondence: amontori@grenp.org

Abstract: *Calotriton arnoldi* is an endemic amphibian inhabiting Montseny Natural Park and Biosphere Reserve (PNRBM), listed as "critically endangered (CR)" by IUCN. At the end of 2016, the Life Tritó del Montseny (LIFETM) project (LIFE15 NAT/ES/000757) was launched. The aim of the project was to promote around fifty actions to ensure the conservation of *C. arnoldi* and its natural habitat, and this entailed five strategic lines: (1) Increasing the scientific and technical knowledge with regard to *C. arnoldi*'s conservation status and its habitat management. (2) Expanding its geographic distribution. (3) Involving and engaging stakeholders in the conservation of the Montseny brook newt. (4) Eliminating or minimizing threats that exist in the riparian habitat. (5) Establishing proper legal coverage and defining long-term strategic planning. The successes and failures experienced throughout the process provide us with essential information that will enable us to develop an adaptive management of the habitat. In order to eliminate or minimize threats to the newt's habitat, some of the actions that are currently being carried out are: (a) Land acquisitions and land exchanges with private properties. (b) Land stewardship procedures, with two custody agreements being signed. (c) Reduction of water withdrawal with nine water catchments and distribution being remodeled. (d) Improvement of water treatments and storage by installing ecological wastewater treatment facilities. (e) Ensuring ecological connectivity and riparian forest restoration. Here, we present an evaluation of the actions carried out to improve the habitat of this species, including the necessary considerations for them to be implemented correctly and to be successful in a natural area, which is under public-private management.

Keywords: urodela; salamandridae; caudata; biosphere reserve; habitat restoration; species management; life project

Citation: Guinart, D.; Solórzano, S.; Amat, F.; Grau, J.; Fernández-Guiberteau, D.; Montori, A. Habitat Management of the Endemic and Critical Endangered Montseny Brook Newt (*Calotriton arnoldi*). *Land* **2022**, *11*, 449. https://doi.org/10.3390/land11030449

Academic Editors: Juan F. Beltrán, Pedro Abellán, John Litvaitis and Francesco M. Angelici

Received: 25 January 2022
Accepted: 17 March 2022
Published: 21 March 2022

1. Introduction

One of the major environmental effects of anthropic activity is the alteration of the hydrological cycle, which directly affects the quantity and quality of freshwater supplies [1,2]. Water resources, usually scarce and unevenly distributed throughout the Mediterranean region, are highly susceptible to Global Change. In fact, a decrease in streamflow discharge in several Spanish rivers over the period 1921–1996 has already been reported [3].

Calotriton arnoldi is an aquatic newt described in 2005 [4] as endemic in the Montseny Natural Park and Biosphere Reserve, (PNRBM from now) and which has a fragmented distribution in two areas located on the Tordera river's eastern and western slopes (Figure 1). Both subpopulations are separated genetically and morphologically [5,6]. After this description, the conservation status changed due to the low number of populations and individuals, and thus, it is now listed as critically endangered "CR" [7]. *C. arnoldi* is only found at altitudes above 600 m asl in clean, cold and well-oxygenated streams surrounded by well-structured beech and oak forests. It is a small newt (maximum total length is 103 mm), the dorsum is dark, and the head is heavily flattened [4] (Figure 2).

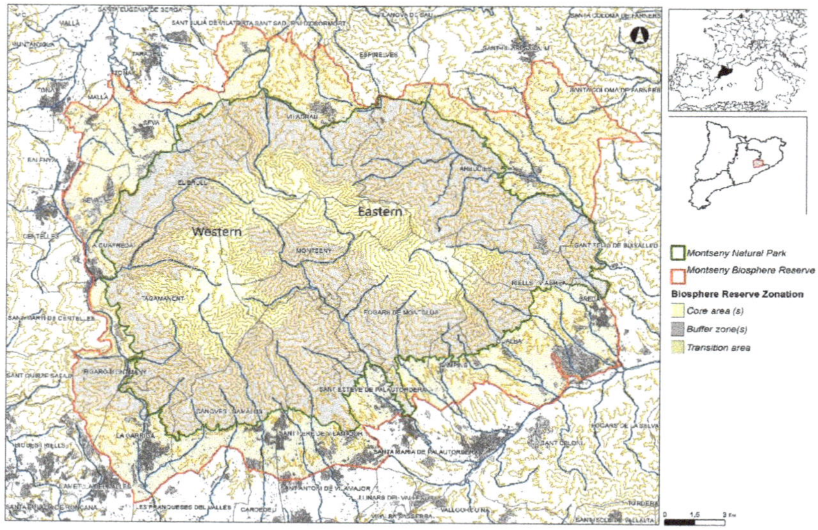

Figure 1. Location of the PNMRB and the two isolated subpopulations (Western and Eastern) of *C. arnoldi* in La Tordera river basin.

Figure 2. (**Right**) Male of eastern subpopulation. (**Left**) Male of western subpopulation.

With the elaboration of the PNRBM's Conservation Plan in 2011 [8], managers were able to strengthen the monitoring program for the Montseny brook newt and its habitat, perform population health surveys, increase and improve the breeding program's functionality and start the release of captive newts in uninhabited streams.

In 2016, the LIFE Tritó del Montseny project (LIFETM henceforth) (LIFE15 NAT/ES/000757) began with Diputació de Barcelona (DiBa), Diputació de Girona (DiGi), Generalitat de Catalunya (GC), Zoo de Barcelona (BCNZoo) and Forestal Catalana S.A. (FC) as partners in the project. The LIFETM provides the necessary tools for a suitable management of the habitat and the species so as to obtain the necessary data, and thus, plan for the future management of this species' populations and habitats (Supplementary Materials Table S1).

Surface water diversion [9] is one of the most dramatic and immediate threats to this species since large amounts of water are being extracted from PNRBM for human consumption and livestock along all the river secondary basins. It is likely that water overexploitation is a severe threat because of the species' ecological requirements [10,11].

In fact, the current environmental legislation is not being enforced efficiently with regard to the maintenance of ecological flows and their monitoring and this constitutes a significant difficulty when dealing with this threat [12].

Wood plantations [9] and, more specifically, those that consist of fast-growing allochthonous conifers, may be behind the waterflow reduction in the La Tordera basin. They require large amounts of water and take up 11.40 ha (12%) within the natural range of *C. arnoldi* [13]. Global warming and other severe climatic or weather events outside the natural range of variation [9] are other threats currently affecting the species. For instance, the beech (*Fagus sylvatica*) forest, an excellent habitat for *C. arnoldi*, has shifted upwards by 70 m at the highest altitudes (1600–1700 m asl) since 1945, and it is being replaced by a holm oak (*Quercus ilex*) forest at lower altitudes (800–1400 m) [14].

The recent appearance of an isolated focus of *B. salamandrivorans* very close to the *C. arnoldi* populations [15], forces us to be very strict with regard to preventive biosecurity measures.

The goals proposed by the LIFETM are: (1) Increasing the scientific and technical knowledge of *C. arnoldi*. (2) Expanding its geographic distribution. (3) Involving and engaging stakeholders in the conservation of the Montseny brook newt. (4) Eliminating or minimizing threats to the riparian habitat. (5) Establishing proper legal coverage and defining long-term strategic planning. This publication aims to be an operational paper for scientists and conservation managers. Fulfilling our objectives (Figure 3) will depend on how we think and act as project managers while being aware of what it means to solve problems through action-oriented approaches (Figure 3) [16].

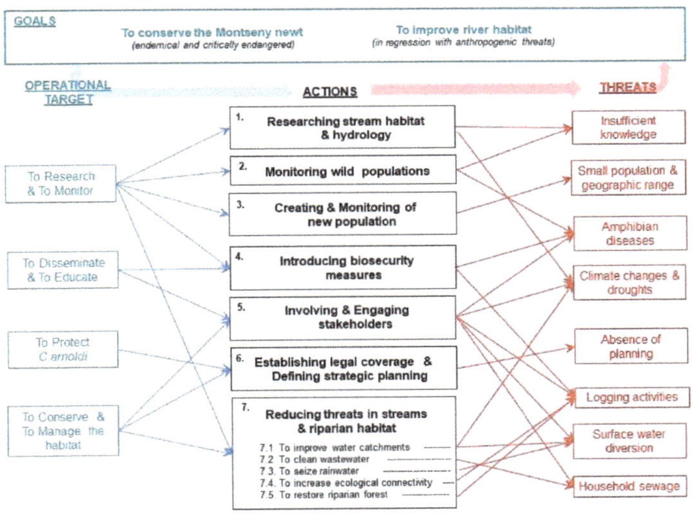

Figure 3. Goals, targets, actions and threats. Two big goals have been established. Of these, four operational targets are specified. To achieve these targets, seven groups of actions have been executed for the different threats to be addressed.

2. Materials and Methodology

Each stream's toponymic name and coordinates have been omitted and population codes are used in all sections in order to protect the species.

2.1. Monitoring Methodology

2.1.1. Researching the Stream's Habitat and Hydrology

A surface water hydrological monitoring network (http://www.lifetrivers.eu (accessed on 14 March 2022)) has been implemented to provide a continuous time series of the brooks. This network currently includes: 5 meteorological stations, 7 (U20-001-04 Data Logger Hobo®) water level sensors, 15 (UA-002-64 Data Logger. Hobo®) light and temperature sensors, 1 (Be-U-4 Hobo® optical base) to download data Logger data and Software (BHW-PRO-DLD Software HoboWare Pro® Windows®/Mac) for data analysis (Figure 4).

Figure 4. Network of monitoring abiotic variables. (**A**) Technicians downloading hydrological data. (**B**) Meteorological station located in eastern subpopulation area.

Stretches of 30 m were established in each brook to perform the hydrological description [17]. A Generalized Linear Model (GLM) with logistic function for each variable has been applied while the registered variables are as follows: Altitude, Slope, LBOM (leaf), FBOM, Bed structure (rocks, stones, pebbles, gravel, sand), Flow, Depth, Wet width, Maximal water speed, Stream structure (runs, falls, pools, dry stretches) and hydraulic status (dry, hyporheic, arheic, oligorheic with some subterranean flow stretches, oligorheic, eurehic, hiperheic).

To analyze the water's chemical components, a total of 59 samples were obtained (40 in streams without newts), and 12 chemical parameters were analyzed: pH, alkalinity, Na^+, K^+, Ca^{2+}, Mg^{2+}, NH^{+4}, $NO_3{}^-$, $SO_4{}^{-2}$, Cl^-, Cu and conductivity [17,18]. A geomorphologic, geotectonic and hydromorphologic study on the streams where *C. arnoldi* is located compared to where it is not present was conducted to outline the parameters that define the fluvial habitat of the newt. Sixteen streams were analyzed and in each river course, and there were 1, 2 or 3 control stations, resulting in a total of 29 sections of ten meters in length [19].

2.1.2. Monitoring of Wild Populations

In order to avoid the spread of emerging infectious diseases, biosecurity protocols were implemented [20,21] throughout the sampling. The materials on the field were disinfected by using Virkon before and after the visits for each population. Three kinds of surveys were designed to gather ecological data of all known populations.

Intensive Survey

In the chosen A2 stream, a 150 m stretch was selected and divided into sections of 10 m each. The surveys were performed from 2018 to 2020. The stream was regularly surveyed on a monthly basis and newts were actively searched for by moving upwards. Newts were marked by injecting a visible implant elastomer (VIE, Northwest Marine Technology, Inc., Anacortes, WA, USA) and using a visual code based on ten injection points: four in the abdomen, four in the legs and two in caudal region [22] (Figure 5). The captured individuals were either georeferenced, measured and sexed [23], or assigned to one of the four immature age-classes that were taken into consideration (larvae, metamorphic, juvenile and subadult). The sex ratio was calculated as the proportion of mature males in relation to the total number of adults [24]. However, immature individuals were not included in estimated population size models. To estimate population size, we assumed that populations were closed. This assumption is based on the fact that *Calotriton* newts are not particularly mobile organisms [25], and the selected stretch includes the upper limit for the presence of water and the lower limit for the presence of the species in this stream [22].

Figure 5. Marked individual with elastomer in caudal area.

The POPAN model estimator (adaptation of the Jolly-Seber model) was used to estimate population parameters using the MARK^{TM} software [26]. The selected model with lower Akaike value was φ (.) p (t) pent (t) N (Supplementary Materials Table S2), where apparent survival (φ) was considered to be constant over time, while the probability of capture (p) variable over time, and the probability of entry into the population by chance (pent) varied during the period in which they were studied.

Extensive Survey

C. arnoldi is currently known to be inhabiting eight streams, three in the eastern area and five in the western one. Field surveys were performed at night by a team that consisted of the same researcher and two rangers of the GC who are trained in the detection of the species. Samplings were performed either in spring and autumn (55.5%) or only in spring due to the severe droughts experienced in the autumn season. The amount of time spent was proportional to the length of the stream ($\bar{x}$: 973 m ± 95%CI: 140 m, Range: 150–1558 m) to maintain a constant survey effort. No active searching was performed and only specimens which were detected visually were taken into consideration. All observed newts were georeferenced and sexed where possible [23], and three individual classes were considered: adults, immature and larvae. We summarized the gathered information by computing two variables: the percentage of stream length inhabited by newts and the relative abundance of adults (number of adults/stream length) in relation to the survey numbers.

2.1.3. Creating and Monitoring New Population

Due to the species' critical situation shortly after their description, in 2007 a captive stock of newts was started by placing 20 newts from the two subpopulations in the facilities of the Wildlife Center Recovery of Torreferrusa (GC) [27].

The species' potential distribution area was modeled by employing the Maxent maximum entropy method [28]. The lack of biological and environmental representativeness in the available data was also taken into account when calibrating the models. Comparative multidisciplinary studies were performed between the streams which the newt inhabits and those which it does not. The plant structure, hydrology, trophic availability, geomorphology and the presence of predators, among others, were analyzed [17–19,29–32]. A New Populations Analyses Commission was created and worked in parallel with the experts' commission to decide, after several field surveys in optimal candidate streams, on how, where and when to release newts as well as the number and which age-classes should be released. A cost evaluation was also important to determine the project's effort capacity and the viability of its objectives.

New population surveys employ a similar methodology to that of an intensive survey. All of the newts released were taken from breeding centers. Prior to this, the newts were sexed, measured, weighted and marked using Trovan Ltd. microchips. Release points were marked with iron flags, while GPS coordinates were also registered for each point. Once the specimens were released, two active surveys per year were carried out.

2.2. Management Methodology

2.2.1. Introducing Biosecurity Measures

To control and monitor emerging diseases affecting amphibians and the Montseny brook newt, a Biosecurity Commission was created with three main purposes: (i) establishing the protocols to be followed in all the activities carried out in the PNRBM (educational, economic, sports, leisure, etc.) [20,21]; (ii) setting up space and time monitoring processes to determine the presence or absence of pathogens in the species' habitat as well as a security perimeter; and (iii) training of PNRBM workers and raising users' awareness of biosecurity protocols (Figure 6).

Figure 6. Biosecurity procedures to prevent the entry of pathogens in the *C. arnoldi* distribution area. The disinfection of all material and machinery with Virkon S is mandatory before and after all activities.

2.2.2. Involving and Engaging Stakeholders

Only 20% of the surface area of PNRBM is under public ownership. Meetings were arranged with forest owners to reach agreements between both parties. A commission was created to disseminate the LIFETM project among the locals and the general public. In the digital sphere, a website (https://lifetritomontseny.eu (accessed on 14 March 2022)), Youtube channel and Twitter were developed. In terms of educational actions, a travelling exhibition was created: "My name is *Calotriton* and I only live in Montseny" and the educational program "El Montseny a l'Escola" (http://www.elmontsenyalescola.cat (accessed on 14 March 2022)) includes supporting material for the educational community. BCNZoo also offers an educational program on *C. arnoldi*. The involvement of landowners was achieved through plenary meetings, and the mechanisms to involve landowners resulted in the signing of land stewardship and the purchase or exchange of land contracts and agreements.

2.2.3. Establishing Proper Legal Coverage

The Montseny brook newt was included in Annex IV of the Habitat Directive as a *Euproctus asper*. Letters were sent to the Spanish Government in order to pursue the explicit recognition of *C. arnoldi* in the Habitat Directive (92/43/CEE), ensure the conservation of its habitat and develop actions that improve its conservation status. I order to improve the river habitat, a protection zone was drawn, land was obtained, and a recovery plan is awaiting approval.

2.2.4. Eliminating or Minimizing Threats in Streams and Riparian Habitat

Prior to the launch of the LIFETM, the PNRBM Conservation Plan [8] analyzed the main threats, set by the Unified Classifications of Threats [9], and these are: (1) water management/use; (2) household sewage; (3) climate change and droughts; (4) logging roads; and (5) wood plantations. Specific actions were specified in order to reduce the negative impact on the species mainly caused by habitat anthropogenic changes. We grouped habitat restoration actions by using five typologies.

Reducing and Improving Water Catchments

We are promoting the removal of shallow water catchments, the remodeling of existing water catchments, the change from sprinkler irrigation to trickle irrigation and the utilization of rainwater. LIFETM provided the relevant legal information regarding water catchments to the landowners and sent a report to the Catalan Water Agency for their dismantling based on their negative impact.

Cleaning Wastewater

LIFETM improved wastewater that had been released into the streams which the species inhabits by employing tertiary treatment of wastewater from isolated houses and public amenities. After establishing primary and secondary treatment processes, we built artificial marshlands which consist of shallow ponds with gravel and aquatic vegetation (70% *Phragmites australis* and 30% *Iris pseudacorus*) which means that water emerging from the artificial marshland is discharged into the forest, where it is filtered below ground.

Seizing Rainwater

In order to diminish water over-exploitation, we are currently implementing two actions: rainwater harvesting and changing current irrigation systems. To collect water rain, we employed the catchment and conduction of run-off water, filtering it and storing it in tanks so that it can pumped and used for gardening purposes, to fill swimming pools and for livestock.

Increasing Ecological Connectivity

To increase the number of stream sections that have optimal connectivity for all aquatic organisms, LIFETM has removed forest tracks and rebuilt bridges of different types according to their level of use, while recovering the natural streambed to allow the movement of aquatic fauna along it.

Restoring the Riparian Forest

The restoration of the riparian forest was performed by removing allochthonous conifer plantations, carrying out slope stabilization and promoting the replanting of autochthonous species. Exotic conifers require huge amounts of water and, for this reason, the LIFETM cut out exotic conifers (*Pseudotsuga menziesii*, *Pinus ponderosa*, *P. sylvestris* and *Cedrus* sp). The open spaces created by this logging will be managed to naturally regrow an autochthonous forest with a high diversity of plant species (*Sambucus nigra*, *Alnus glutinosa*, *Fraxinus excelsior* and *Corylus avellana*).

3. Results

3.1. Monitoring Results

3.1.1. Stream Habitat and Hydrology

The long-term monitoring network of hydrological variables has been successful by means of 264 visits to 13 streams. This has provided information on temperature and flow discharge, along with the aquatic status for three years. Currently, this network is fully operational, and the first results have been published [32]. Results show that the brooks with newt populations have not completely dried out despite there being summer hydric stress (Figure 7). The pools and a little surface flow (less than 0.5 L/s) remain, even in the streams with the lowest discharge. Large variations in discharge have been recorded, from 0.5 L/s to 10 L/s, following major rainfalls, suggesting a possible direct relationship between waterflow and precipitation (r = 0.892; $p < 0.001$; $n = 27$), with catchment groundwater having a residual role on the brook flow. Data shows the basal flow rate for most of the streams is lower than 4 L/s, with values below 1 L/s in one of them [32]. There is a certain resilience in the flow at times when rainfall is scarce. Streams do not dry out completely because of the underground circulation of water through the fissured rocks is maintained.

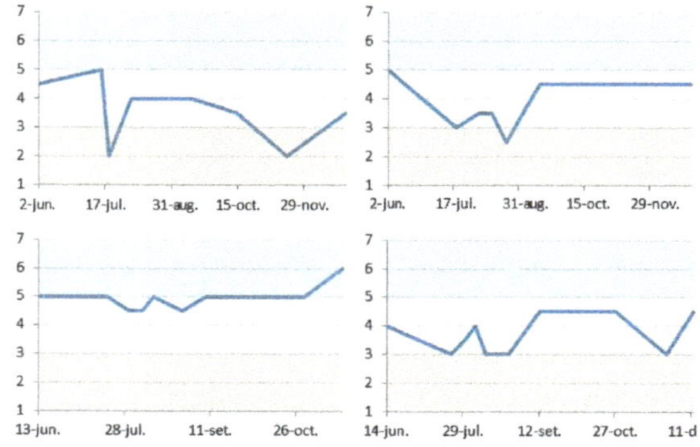

Figure 7. Hydrological temporal variation of four representative streams inhabited by *C. arnoldi*. Values of Y-axis: 1: dried; 2: Hiporheic; 3: arheic; 4; oligorheic whith streaches whith only subterranean running water; 5: oligorheic; 6: eurheic; 7: hiperrheic. Modified from [32].

The water temperature shows the expected annual cycle values and different patterns between the streams [17,32]. The recorded temperatures (Table 1) are within the preferred temperature values (Tp) described for the species (Males, $\bar{x}$: 14.98 °C, Range: 11.7–19.48; Females, $\bar{x}$: 17.51 °C, Range: 15.22 to 20.88) [33].

Table 1. Water surface temperature of some streams which *C. arnoldi* inhabits or where it is absent. Max: Maximal year temperature. Min: Minimal year temperature. Ampl: Average of daily thermic amplitude. Values are expressed in degrees centigrade. Data resumed from [17].

Inhabited	Max	Min	Ampl	Absent	Max	Min	Ampl
B5	17.5	2.52	3.41	Stream 1	17.92	3.26	2.99
B1	15.2	7.38	2.07	Stream 2	17.35	4.10	2.61
B2	20.3	1.76	5.06	Stream 3	16.76	3.05	3.48
A1	14.6	2.09	6.28	Stream 4	17.46	1.33	7.24
B4	16.8	1.76	3.43	Stream 5	16.37	0.35	9.11
A2	17.1	3.58	4.76				
A3	16.7	2.73	6.90				

No significant differences in the habitat's hydromorphology have been found between brooks which are inhabited by the newt and those where the newt is not present (Figure 8). The application of a Generalized Linear Model (GLM) with logistic function for each variable does not show a significant model (chi-square test $p > 0.05$) for all cases [17]. The Montseny's headwater streams are characterized by low mineralization (Table 2). Ammonium has been undetectable in many cases; therefore, it will not be taken into consideration in the statistical analyses [18]. The two groups of streams do not show any significant differences (Figure 8).

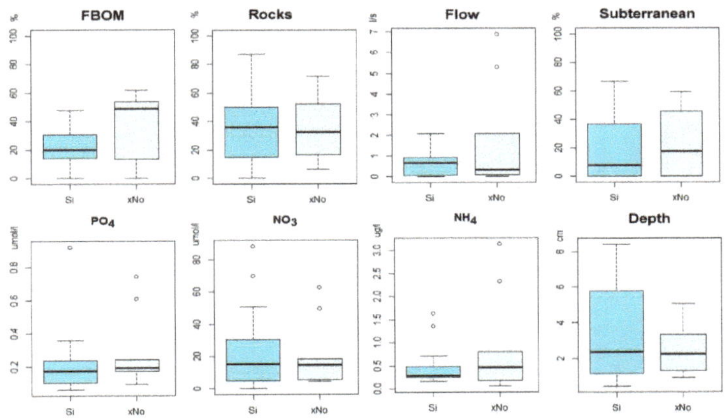

Figure 8. Some hydrolomorphological variables analyzed. FBOM: fine organic matter (expressed as percentage of the riverbed where this matter is found). Si: Streams with presence of newts. xNo: Streams with absence of newts. Depth: Average of Depth in the center of streambed. All comparisons are nonsignificant.

Table 2. Chemical composition of the water in the streams inhabited by *C. arnoldi* and streams were *C. arnoldi* is absent. (Cond): Conductivity in μS/cm^{-1}; (Alc.): Alkalinity. Na^+, K^+, Ca^{2+}, Mg^{2+}, $NH_4{}^+$, $NO_3{}^-$, $SO_4{}^{-2}$ and Cl^- in μeq/L^{-1}. Data from [32].

	n	Cond x̄	Cond Std	pH x̄	pH Std	Alc. x̄	Alc. Std	Na^+ x̄	Na^+ Std	K^+ x̄	K^+ Std	Ca^{2+} x̄	Ca^{2+} Std
Absent	40	78.5	41.4	7.4	0.4	520	334	237.1	82.2	10.1	5.8	410.3	295
Inhab.	19	63.5	11.8	7.3	0.1	342	97.1	227.9	44.1	8.2	3.1	264.6	58.3

	n	Mg^{2+} x̄	Mg^{2+} Std	$NH_4{}^+$ x̄	$NH_4{}^+$ Std	$NO_3{}^-$ x̄	$NO_3{}^-$ Std	$SO_4{}^{-2}$ x̄	$SO_4{}^{-2}$ Std	Cl^- x̄	Cl^- Std
Absent	40	193.1	104.7	0.2	0.5	17.4	25.7	121.7	72.2	118.8	85.9
Inhab.	19	175.5	34.8	0.1	0	17.8	23.6	136.8	57.7	108.3	14.3

The dense surrounding vegetation mainly consists of deciduous trees, which lead to a significant input of fallen leaves in the riverbed. This is the main organic input in the stream system, as the autochthonous primary production is really low (Chlorophyll a concentration under 60 μg/cm^2), due to low sunlight penetration. Relatively high concentrations of large organic matter (LBOM) and fine organic matter (FBOM) were recorded. Vegetation cover plays an important role in water temperature, preventing an increase where there is a greater degree of cover.

The flooding waters have extremely low conductivity (Table 2) and are poor in inorganic nutrients. The geomorphological data of the torrents [19] indicate that they have exclusive geological and hydrological characteristics. The presence of large blocks and colluvium that form morphological gaps and fissural and cavernous porosity, leading to subway water circulation and the presence of porosity between the blocks, determine the presence of newt. There is a preference for areas with a large accumulation of blocks, with at least 2 levels of stratification, and with large blocks without imbrication as well as a high occupation of the bed. Areas where the bed with rocky substrate is occupied by blocks that form cavities of decimetric order, together with the presence of fracture fissural cavities, may determine the presence of newts [19].

3.1.2. Monitoring of Wild Populations

Intensive Surveys

In the surveyed stream (A2), the largest number of specimens were caught in spring, mainly May–June (Figure 9A). Stream surface groundwater activity decreases during the cold period from January to March and quickly during the summer (July and August), coinciding with the torrent's dry period. These data are in accordance with those provided by different studies [10,18,34]. There is also a significant increase in activity in autumn and early winter [22]. The species has the tendency of inhabiting underground aquifers, between the cracked rocks in the stream, and this influences the timing of surveys because during dry periods, surface activity is almost non-existent.

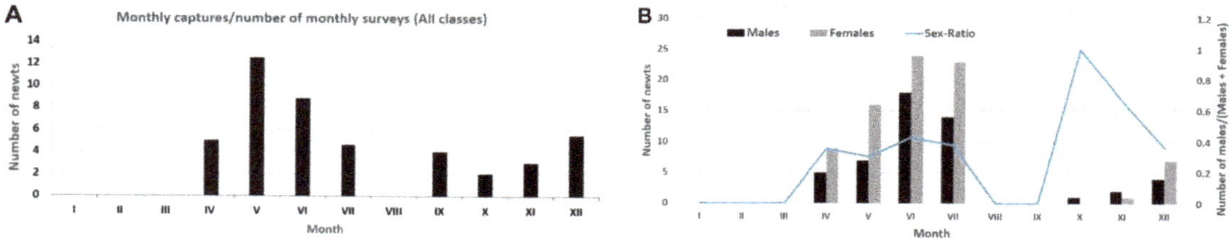

Figure 9. (**A**): Number of individuals from 2018 to 2020 in the A2 stream, captured by surveys in relation to number of surveys for each month. (**B**): Monthly Sex-Ratio in the A2 stream and absolute number of males and females captured each month.

The sex-ratio average obtained (♂♂/(♂♂+ ♀♀)) is favorable to females (0.42), even throughout the year (Figure 9B). These observations therefore match [4,11], which found that sex-ratio is clearly favorable to females in the eastern and western nuclei. After two years, the average displacement is 7.05 m. This mostly consists of males, whose preference is to go upstream.

In the A2 stream, 70 newts were captured, marked and released from 2019 to 2020, and the recapture rate reached 34.3%. Their estimated survival is very high (φ = 0.9984 CI95%: 0.9965–0.9994) and the average density of newts was 0.63 newts/m (CI95%: 0.54–0.79).

Extensive Surveys

All eight populations were monitored by means of extensive surveying (Table 3). The number of observed newts per number of nights spent in stream sections for the period 2010 to 2020 and the samplings during this period showed a decline after there had been increases for two years (Figure 10A). This trend was more marked in eastern populations, which exhibited larger numbers of newts than in western ones.

Table 3. Length of the streams where wild populations of Montseny brook newt were located. Potential: potential inhabited total length of stream (in m). Inhabited: Estimated inhabited lengths (in meters). I/P: percentage of occupied length (indicator of relative stream occupancy). A codes: Eastern populations. B codes: Western populations.

Population Code	Potential	Inhabited	% I/P
B1a	821	83	10.11
B1b	1117	587	52.55
B2	895	475	53.07
B3	932	107	11.48
B4	2220	782	35.23
B5	804	654	81.34
A1	3150	697	22.13
A2	2250	178	7.91
A3	1180	34	2.81
Western Range	6789	2688	39.59
Eastern Range	6580	909	13.82

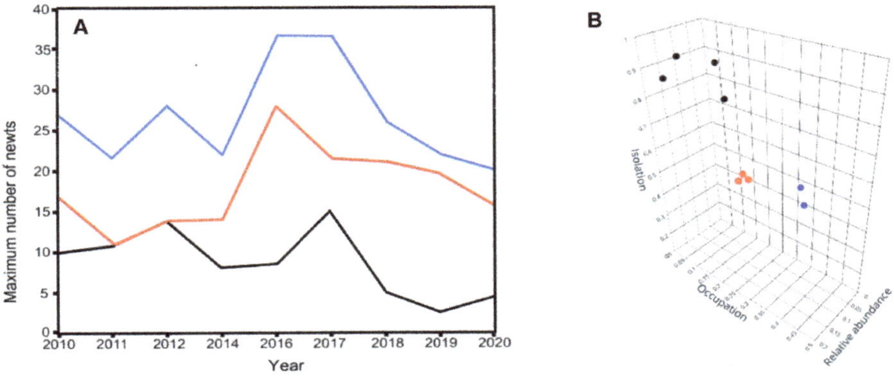

Figure 10. (**A**): Maximum number of newts observed at night for every year. Lack of data during 2013 and 2015 is due to the severe drought affecting the populations that prevented us from gathering enough data (black: eastern populations; red: western populations; blue: total). (**B**): Tridimensional plot of the values for each Montseny brook newt population on three indicators: relative abundance (number of newts/m). occupation (percentage of stream occupied by the population) and isolation (dendritic index between pairs of geographically close populations). Black dots show the tridimensional position of populations with high risk of extinction given their high levels of isolation and low percentage of stream occupation. The best combination of favorable indicators, which are high connectivity, stream occupation and relative abundance, are found only in two populations marked with blue dots, whereas three populations (red dots) showed intermediate risk of extinction.

The tridimensional space generated by the three indicators provides a clear picture when assessing the vulnerability of *C. arnoldi* populations (Figure 10B). Four populations are very isolated from the neighboring ones and in total, two of them occupy a very small proportion of the stream and yet, quite remarkably, relative abundances are high in all cases, which indicates a strong concentration of newts in small sections. The other four populations seem to be in a better situation from a conservation standpoint and show better connectivity and extensive occupation of the stream. However, and perhaps quite strikingly, only two of them additionally have a large relative abundance. Therefore, this ideal combination is found in 25% of the populations and both belong to the western area.

3.1.3. Creating and Monitoring New Populations

The field studies [17–19,29–32] confirm that streams where *C. arnoldi* is absent have similar biotic and abiotic characteristics when compared to streams where the newt is present. Table 4 shows the new populations created prior and during to the LIFETM.

Table 4. New populations established in the *C. arnoldi* potential area of distribution. Only population codes are written because of conservation policies for this CR species. Range: slope location in La Tordera river basin. First release: year when first release was made. Last released: last booster. Last recapture: year when *C. arnoldi* was captured in the stream. Property: Land property where the new population and stream stretch are located.

Code	Range	Number of Released Newts	First Released	Last Released	Last Recapture	Property
A4	Eastern	166	2011	2014	2016	Private
A5	Eastern	63	2020	2020	2020	Public
A6	Eastern	127	2014	2020	2021	Public
B6	Western	436	2010	2020	2021	Private
B7	Western	106	2014	2015	2019	Private
B8	Western	261	2019	2021	2021	Public
B9	Western	267	2021	2021	—	Public

The average displacement of newts throughout 2019 in B8 stream was 2.44 m downstream. However, displaced recaptured newts move in average 14.75 m downstream and 7.44 m upstream. Despite wild *C. arnoldi* specimens showing very little mobility and dispersal capacity in natural habitats [35,36], it is expected that newts that were born in captivity and released into a stream will disperse and explore more while seeking suitable microhabitats. However, the results do not confirm this (Figure 10B). The dispersion observed in the first recapture in the B8 stream compared to the estimate for the wild population A2 within two consecutive captures is not significant (*t*-test: −1.21; df: 56; $p = 0.229$). Preliminary results indicate that the population is well established. The distribution of recaptures in the new population B8 is very similar to the distribution of recaptured individuals in the A2 population (Figure 11A). Figure 11B shows the dispersion of the recaptured specimens relative to their release point.

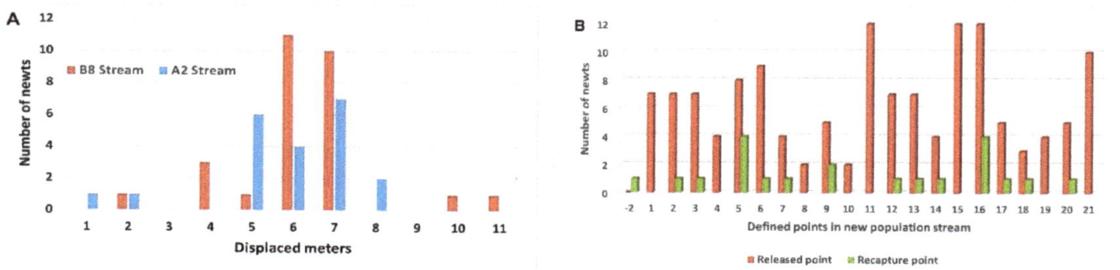

Figure 11. (**A**) Displacements observed for the newly released individuals in a newly created population (A8) in the first year and in the wild population (A2). Horizontal axis represents the number of displaced meters from release to recapture. The positive values indicate upstream displacements and negative values indicate downstream displacements. (**B**) Distribution of newly released newts in the stream B8 and the recaptures during the first year after release. Red bars indicate the release point and green bars the recapture point.

3.2. *Management Results*

3.2.1. Biosecurity Measures

Throughout 2018, several theoretical-practical training sessions were held for workers and users of the DiBa. At the same time, two biosecurity protocols were developed to be applied in the activities. An extensive one that would cover all LIFETM activities and those that would be developed in the PNRBM (i.e., fishing, surveys, works, etc.), while another protocol focused on educational activities to be carried out in the aquatic environment, aimed primarily at nature schools and educational centers that carry out activities in the PNRBM [20,21]. In 2021, information posters were placed to increase awareness of PNRBM's ban on bathing in rivers and torrents.

3.2.2. Involving and Engaging Stakeholders

The negative effect of visiting the riverside habitat and the dangers associated with the unintentional introduction of infectious diseases by amphibians (Bd, Bsal, Ranavirus, among others) have been highlighted. People who want to see the Montseny brook newt are advised to visit the BCNZoo and CRPS breeding centers. With the aim of disseminating the project, a variety of informative material has been published (Table 5). The monthly number of visits to the website increases with time and are correlated with the number of published news each month ($R = 0.670$; $DF = 40$; $p < 0.001$).

Table 5. Dissemination actions and informative material made during LIFETM. (En. Ct. Cs): English, Catalan and Castilian languages. (*): Data from 01/01/2017 to 12/31/2021. (**): Data from 1 May 2018 to 31 December 2021.

	Number of Items	Number of Media or Sites	Number of Receptors/Visitors
Permanent Exhibitions			
Informative panels	8	8	Visitors of PNRBM, BCNZoo, CRFPS and CRFTF.
Temporal Exhibitions			
My name is *Calotriton* and I only live in Montseny	8 panels	23	20,712 schoolchildren/individuals **
Photography (Iñaki Relanzón)	1	1	584 (Web)
Press			
National Press (Spain)	182	18	(55 Written press and 127 digital press) *
National Press (Catalonia)	182	39	(55 Written press and 127 digital press) *
Newsletters	11	3 (En. Ct. Cs)	234 registered *
Information leaflet		3 (En. Ct. Cs)	16,500 unities
Web			
http://www.lifetritomontseny.eu (accessed on 14 March 2022)	125	1 (En. Ct. Cs)	>74,000 visits (9066 users) *
Video capsules	8	1 (En. Ct. Cs)	3163
YouTube Chanel	166	1	12,705 (including video capsules)
Education			
El Montseny a l'escola	3	18	1372 pupils (2021/22)
Workshops	1	1	224 children
BCNZoo	1	1	13,800 schoolchildren/individuals *

The exhibition "My name is *Calotriton* and I only live in Montseny" has travelled to the 18 municipalities that make up the PNMRB. From September 2021, the exhibition will continue to travel around the Library Network of the Diputació de Barcelona. The educational program "El Montseny a l'escola" (http://www.elmontsenyalescola.cat (accessed on 14 March 2022)) has allowed 2450 pupils to participate in the didactic material and 224 children in the workshop to make a plaster newt (Table 5). The new education and breeding center at BCNZoo has become the benchmark facility for the Montseny brook newt and has received almost 500,000 visitors since June 2018. A total of 13,800 people (mostly school groups) have carried out activities on the Montseny brook newt, guided by the BCNZoo educational staff.

Seventy meetings have been held with forest owners. This has led to the involvement of nine estates which the newt inhabits in the Montseny area. Three land stewardship contracts have been signed with local landowners who have *C. arnoldi* in their properties while contracts have been exchanged with two more. The project has been disseminated in 46 conferences, presented at local, national and European technical conferences and congresses (4945 attendees in total). Three hydrology conferences have been held with the participation of 20 scientists from 14 different institutions [37,38]. With regard to foresters and forest managers, a Manual of Good Environmental Practices has been developed [39].

3.2.3. Establishing Proper Legal Coverage

Habitat Directive technicians, from the EU and Spain Government, established that since *C. arnoldi* is a split of the *Euproctus asper*, included in the Annex IV of DH 92/43/CEE, E. asper would be divided into two species, *C. arnoldi* (Code 6920) and C. asper (Code 6944), to prepare the sexennial reports (Art. 17 of DH). A technical document on the recovery plan has been developed, and it is expected to be formally approved before the end of LIFETM in 2022. To improve the protection of the streams where *C. arnoldi* lives, work is being done on two strict protection measures within its distribution area. (i) Supervised management zone has been established on private land and this involves a land stewardship contract to manage habitat conservation targets. These zones have a width of between 50 m and 100 m on each side of the streams. (ii) The nature reserve zone is a legal figure where exploitation of natural resources is banned. As part of the drafting of the PNRBM's protection plan and within the framework of new legislation for the Montseny Natural Park (D.127/2021), it is expected that most brook newt populations will be included inside this zone. Due to the fact that the majority of the newt populations are on private land, 87 ha with brook newt populations have been purchased.

3.2.4. Eliminating or Minimizing Threats in Streams and Riparian Habitat

Reducing and Improving Water Catchments

During the last four years we have removed six water catchments directly affecting populations of *C. arnoldi* and these were entirely financed by the landowners to avoid future fines. From June 2020, the owners of legalized water catchments must ensure sustainable water use and the maintenance of ecological water flow (based on legislation 2000/60/CE. DL. 3/2003. D. 380/2006 and D. 1/2017). The LIFETM advised and promoted the modification and improvement of four legalized catchments. In order to do so, mechanisms avoiding water extraction when water tanks are full were installed. In addition, distribution boxes were installed to restrict the amount of water available for exploitation and which is legally allowed (Figure 12).

Figure 12. Installed mechanisms which regulate water extraction (**A**) and distribution boxes (**B**) restricting the amount of legally allowed water for exploitation.

Cleaning Wastewater

One tertiary treatment has been installed (Figure 13) and when the water is discharged it is in good condition: pH (6.7) and optimal DQO (74 mgO_2/L).

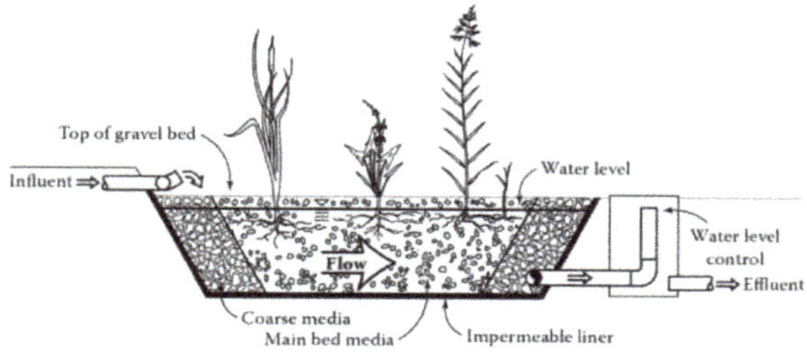

Figure 13. (**Left**) Artificial marshlands formed by shallow ponds with gravel and aquatic vegetation (70% *Phragmites australis* and 30% *Iris pseudacorus*). (**Right**) Schematic draft with permission from [40]. Copyright 2006 by the Water Environment Research Fundation.

Seizing Water Rain

The LIFETM has removed drip irrigation in an area of 6500 m^2 which produces 60,000 plants with a mean water consumption of 20,000 m^3/year [41]. Currently, this area is irrigated by employing 65,000 droppers therefore leading to, at least theoretically, a reduction in water use of 11,700 m^3/year. The LIFETM is promoting rainwater harvesting by reconditioning roofs and roads that cover 5830 m^2 in total. This adaptation may retrieve 4110 m^3/year to tanks, which have been installed in four properties at lower altitudes while owners also use electric or solar-powered pumps to move water to other tanks placed at higher altitudes. Overall, we have installed 14 tanks which has led to a storage of 895 m^3 in total, which will result in economizing 4000 m^3/year of water from the streams inhabited by Montseny brook newts. As an example, for a plant nursery where rainwater is collected and stored in four tanks (total capacity 314 m^3), the estimate is that there will be complete hydric autonomy for 6–8 months per year [41]. Another practical implementation was carried out in a camping area, where we estimate that the system may meet all of the camping area's water demands [42].

Increasing Ecological Connectivity

The LIFETM removed eight forest tracks and built fifteen bridges (Figure 14) of different types according to their level of use, while recovering the natural streambed to allow the movement of aquatic fauna along it. The most relevant work was done in two of the eastern streams, which had sections with under 50 lineal meters of connectivity and this was improved to more than 6 km without any barrier.

Figure 14. Modified bridges to recover the natural stream bed and to allow the displacement of aquatic fauna along them. Left images: before recovering structure. Right images: the same locality after recovering structure.

Restoring the Riparian Forest

About 700 m^2 of riparian forest has been recovered and the growth of autochthonous species has been encouraged (*Sambucus nigra, Alnus glutinosa, Fraxinus excelsior, Corylus avellana*) [39]. More than a thousand meters of barriers were installed to reduce the discharge of erosion sediments by using bioengineering techniques. The LIFETM has already cut down 3.7 ha of exotic conifers, thereby promoting the natural regrowth of an autochthonous forest.

4. Discussion

4.1. Stream Habitat and Hydrology

In PNRBM the holm oak forest appears to be expanding upwards at the expense of beech forests and heathlands [14], which may contribute further to a reduction of the streamflow. Thus, both climate and vegetation cover changes may threaten *C. arnoldi* populations. [43] conclude that in the Montseny Massif, the frequency and length of low streamflow events will increase dramatically. Montseny brook's hydrology is highly dependent on precipitation, and therefore its evolution is linked to climate change and the predictions have been ominous thus far [12,44,45]. The presence of blocks and fractures in the stream bed result in a continuous underground runoff flow that ensures the persistence of newts during dry seasons in which surface water flow disappears [17,19]. The data obtained indicate that the populations of *C. arnoldi* are resilient in a hypogean habitat during dry summer periods, but the current trend is towards population decline [7,11,34,46].

Over the previous 50 years, a decline in river discharges has been observed in Spain and it has been shown that this decline was due to an expansion of shrubs and forests in formerly cultivated areas [43,47–49]. A simplistic conclusion would be that it is beneficial to remove all forests, but this statement is completely misguided because vegetation recovery is generally thought to provide important environmental benefits, such as increases in carbon sequestration [50], and is thought to be the best alternative for nature conservation and biodiversity purposes [51]. Moreover, forests noticeably reduce runoff coefficients and slightly decrease annual flooding, a fact which can be explained by the effects of rainfall interception and forest water consumption that reduce soil water content and limit catchment hydrological responses [47,51]. Changing water availability and demand will require the application of innovative land- and water-management strategies, as well as forest management within the watershed scale, and water use and management [12]. A reduction of between 50% to 44% in relation to the current catchment has been achieved [40], and in some cases, the water automation will possibly last the 12 months of the year thanks to rainwater harvesting [41].

As previously mentioned, within the species' potential distribution area in the upper La Tordera river basin, there are no differences between the torrents where the species lives and where it is not currently found. Therefore, the most plausible hypothesis is that land use in the last 150 years may have caused the rarefaction and disappearance of the species in many torrents (Mining, carbonization, wood extraction, the transformation of forest into pastures with the consequent increase in the insolation of torrents and the contribution of sediments and chemical pollutants).

4.2. Monitoring of Wild Populations

Despite the species' very small geographic range, its distinct populations experience environmental spatial heterogeneity [25,52], which may be behind the differences we found in their demographic parameters [25,46]. The extensive survey highlights that the estimated survival rate is lower in the eastern population than in the western population, whereas population size follows the opposite pattern. One potential explanation is that larger population density increases competition for resources which leads to higher mortality [53]. Firstly, newts spend most of their time within the interstitial hypogeum environment of the stream bed. Secondly, we avoided removing rocks to locate newts to limit the impact of the study, but this is likely to have decreased capture and recapture probabilities.

The intensive monitoring of one eastern population through active search [22,35] resulted in higher population density values when compared to an extensive survey [46] due to the higher number of captures and recaptures. This could suggest that an active search by lifting stones is a more suitable methodology for demographic studies. While this may be the case, the effort required to carry it out for all populations makes it unaffordable for the LIFETM and therefore, it is better to use the suggested methodology for extensive monitoring to estimate global trends. However, the demographic data obtained with intensive monitoring are essential to estimate the ecological parameters needed for modeling new populations and to estimate the probability of survival and the carrying capacity of the streams selected to create it [13].

Prey availability [29] in streams running over beech forests may be greater, as it was also recorded for the sister species *Calotriton asper* [54,55]. The eastern stream also has greater availability of interstitial microhabitats as refuges, and these differences may altogether result in larger population densities in the eastern population [34,35,46]. These results match those provided by [46], which highlight that the differences in *C. arnoldi*'s habitat would explain the differences in demographic parameters between the two analyzed Montseny brook newt populations. [56,57] concludes that the Pyrenean brook newt's presence in streams seems to depend both on the morphology of the stream and on introduced or translocated fish density.

4.3. Creating and Monitoring New Populations

One of the LIFETM's main successes has been to work with multidisciplinary teams and the creation of committees based on expertise, working together to evaluate the hydrological dynamics and actions to create new populations [16,58].

In 2021, the presence of newts has been confirmed in 3 of the 6 new established populations (Table 4) despite the intense drought suffered throughout the year. The adjusted demographic data obtained in the field studies with regard to the wild populations [22,34,35] have allowed us to obtain very precise theoretical models [31]. However, it is still too early to say that the models developed for the creation of new populations are actually fulfilling their function and that the newly created populations are actually establishing themselves.

The new population established in B8 shows similar dispersal characteristics to the wild population in A2. Thus, it cannot be confirmed whether previous life in captivity increases dispersal, searching for optimal habitats in a new environment. It could indicate a good choice of a torrent for the creation of a new population [36]. On the other hand, the weight of the recaptured individuals after one year in relation to the time of release gives an average weight gain of 1.36% [35].

4.4. Biosecurity Measures

Biosecurity measures and protocols have clearly proven their effectiveness. Despite the few cases detected of Bd, in one pool it was caused by the illegal release of a non-native amphibian species of the PNRBM (far from *C. arnoldi* distribution area). None of the three diseases surveyed (Bd, Bsal or Ranavirus) has been detected in any *C. arnoldi* wild or newly created populations. This indicates that our methodology with regard to biosecurity protocols is the appropriate one to prevent the entry of pathogens into the species' wild habitat. The location of Bsal in a nearby area outside the PNRBM [15] requires us to be extremely strict when implementing biosafety protocols in the basin inhabited by *C. arnoldi*. On the other hand, there are no data on the effects that Ranavirus has on *C. arnoldi*, and its recent expansion in the north of the Iberian Peninsula means that prevention is of the utmost importance [59].

4.5. Involving and Engaging Stakeholders

The actions proposed in LIFETM have in many cases exceeded the capacity of the project itself, mainly due to problems that have arisen in the management of *C. arnoldi*'s habitat, which is mainly located on private land. The species' recovery plans carried out, as well as the reference guides on species recovery plans, highlight the importance of public property for the management of the habitat and the species [58]. However, land purchase or exchange and stewardship agreements as well as the actions related to the efficient use of water resources, although incomplete, can be regarded as another success and as a good result for the work carried out as part of the project.

4.6. Establishing Proper Legal Coverage

Genetic results indicate that there is an ancient isolation between the eastern and western subpopulations. For this reason, the LIFETM has treated both subpopulations as two distinct units both in terms of population management, breeding centers and the creation of new populations. However, if Global Change endangers the species in its natural area in the future, there would be the possibility of creating hybrid populations because what would really be endangered is the species itself. This is an objective not raised in the LIFETM, but should be discussed in an experts committee, in order to reach a consensus and it would require the GC's approval. It is the administration responsible for the species conservation, as the Recovery and Captive Breeding Plan prepared by them raises the treatment of the two subpopulations as differentiated evolutionary units [27].

4.7. Eliminating or Minimizing Threats in Streams and Riparian Habitat

New population hydrological and expert commissions are a very effective way to make a threat assessment like TRA (Threat Reduction Assessment) [60]. Furthermore, the long-term monitoring of this species is mandatory in order to provide managers with useful information to repeatedly evaluate the state of conservation and threats of *C. arnoldi* populations. However, the detectability of *C. arnoldi* greatly depends on the occurrence of environmental conditions such as temperature and hydroperiod. For these reasons, only the persistent absence of observation year after year during a long period of sampling can be understood as evidence of population extinction.

The restoration of the riparian habitat is a job that does not yield immediate results. To determine how and where the interventions are working, threat reduction assessments should be performed to measure the conservation project's success [60]. However, some actions such as creating barriers against erosion and the removal of barriers that cross streams have been effective. For example, the intensive study of the A2 wild population [35] revealed that the bridge that crossed the population exerted a barrier effect for the connectivity between both sides of the torrent and produces the connectivity of both populational subgroups.

Changes in temperature or precipitation volumes have the potential to produce shifts in phenology and the timing of reproduction [61,62]. The models of climatic change [12,45] for the Mediterranean area forecast more irregularity in precipitation patterns, specifically more rainfall in autumn than in spring, and concentrated over a few days in some years while in others, there will be heavy droughts. Air temperature is rising [63] and the Montseny massif in particular experienced an increase of around 0.3 °C per decade in the second half of the 20th century [12]. A temperature increase can lead to enhanced evapotranspiration [63], which eventually will result in reduced streamflow. Besides climate change, landscape cover type has an influence on water resources as well. Several catchment studies have shown that an expansion of forest cover leads to a decrease in the water yield [64,65]. A reduction in tree canopy cover shows an increase in annual water yield, higher for conifers than for deciduous hardwood forests [3]. Trees intercept large quantities of rainfall, from 18% (scrub) to 32% (forest) vegetation, which has great repercussions on vegetation cover and the availability of water resources [66,67].

The maintenance of riparian habitat has been described as one of most important action to maintain *C. arnoldi* populations. However, data published on the effects of global warming predicts an increase in slow flows and drought periods [43]. Similar results highlighted the importance of flow intermittence on *Calotriton asper* distribution, with a decrease in habitat suitability when the frequency of zero-flow events increases [57].

The SWOT (strengths, weaknesses, opportunities and threats) of LIFETM is showed in Supplementary Materials Table S3.

4.8. Management Proposals

- Research and monitoring should be promoted in order to manage the natural heritage and resources with technical and objective criteria. However, for effective conservation to be achieved, collaboration from all the participants is needed (researchers, managers, stakeholders).
- Changing water availability and demand and mitigation strategies to improve the management of water resources will need to continue being implemented. This requires the application of innovative land- and water-management strategies on the all-watershed scale.
- Open spaces such as pastures and grasslands should be recovered in the river basin, promoting a landscape with a diversity of habitats that respect the riparian and autochthonous forests on erodible and sloping soils.
- It is highly recommended to promote the rewilding of habitats in order to have an old and mature forest, between pastures and grassland, mainly because tree transpiration is sometimes much higher in younger than in older trees.
- Coniferous plantations, especially of fast-growing species (such as *Pseudotsuga menziesii, Pinus radiate, P. ponderosa, P. nigra* or *P. uncinate*), must be removed in the protected areas, in order to increase the streamflow where *C. arnoldi* is present, because the largest water yield increases appear after removal of conifer forest.
- To conclude, the intensive human exploitation of natural resources in the Montseny is a threat to the survival of the *C. arnoldi*. Therefore, we should reduce water abstraction from the streams, and we should restore the diversity of natural habitats, especially by promoting meadows, respecting the natural evolution of the forest and encouraging the development of the autochthonous riparian forest.

Supplementary Materials: The following supporting information can be downloaded at: https://www.mdpi.com/article/10.3390/land11030449/s1. Table S1: Actions and objectives of LIFETM. Table S2: POPAN estimators and additional results. Table S3: SWOT (strengths. weaknesses. opportunities and threats).

Author Contributions: Conceptualization: D.G. and A.M.; methodology: D.G., D.F.-G., F.A. J.G., S.S. and A.M.; validation: D.G., D.F.-G., F.A. J.G., S.S. and A.M.; investigation: all authors; resources: D.G., D.F.-G., F.A. J.G., S.S. and A.M.; writing—original draft preparation: all authors; writing—review and editing: D.F.-G., F.A. J.G., S.S. and A.M.; supervision. D.F.-G., F.A. and A.M.; project administration: D.F.-G., J.G. and S.S. All authors have read and agreed to the published version of the manuscript.

Funding: This research was funded by Diputació de Barcelona. Life projects of the European Commission (LIFE15 NAT/ES/000757), the Amphibian Ark and the Bin Zayed Species Foundation project number 14258012.

Institutional Review Board Statement: The study did not require ethical approval. The animal study protocol was approved by the Institutional Review Board of Life Programme (LIFE15 NAT/ES/000757 in October 1 of 2016).

Informed Consent Statement: Not applicable.

Acknowledgments: A project of this magnitude involves the participation of many people without whom it could not have been carried out. In this section, we wish to express our gratitude, for their important role in the project, to the professional groups of PNRBM's rangers, maintenance brigade, technical and administrative; the Staff of Natural Parks of Diputació de Barcelona and Diputació de Girona, of Zoo de Barcelona, of Wildlife Service and Agents Rurals of Generalitat de Catalunya and of Forestal Catalana, and the private owners of PNRBM, GRENP, Agència Catalana de l'Aigua, Chester Zoo and CREAF. We also want to thank all the specialists who participated in the numerous workshops on hydrology, introductions or new populations and biosecurity. The scientific capture permits were granted by the Wildlife Service of the Generalitat de Catalunya and the Area of Natural

Parks of the Diputació de Barcelona. We would like to give special thanks to the people who have been involved the most, namely JMª. Alonso, A. Álvarez, J.C. Àngel, J.A.S. Carranza, R. Casanovas, X. Comas, C. Comajuana, J. Comellas, M. Comellas, R. Fradera, P. Fruitós, F. Gallart, G. Garcia, M. Giol, J. Gomà, C. Guillot, J. Latron, L. López, E. Martí, J. Melero, A. Oltra, G. Pascual, A. Pérez, I. Planas, M. Pomarol, D. Riera, J. Santaella, Ó. Soler, A. Tarragó, E. Valbuena, P. Varona, N. Vicens and D. Villero. The scientific input provided by the anonymous reviewers is gratefully acknowledged.

Conflicts of Interest: The authors declare no conflict of interest. The funders had no role in the design of the paper, in the writing of the manuscript or in the decision to publish the results.

References

1. IPCC 2021. Code Red' for Human Driven Global Heating, Warns UN Chief. IPCC Report. Available online: https://news.un.org/en/story/2021/08/1097362 (accessed on 21 December 2021).
2. Foley, J.A.; DeFries, R.; Asner, G.P.; Barford, C.; Bonan, G.; Carpenter, S.R.; Chapin, F.S.; Coe, M.T.; Daily, G.C.; Gibbs, H.K.; et al. Global consequences of land use. *Science* **2005**, *309*, 570–574. [CrossRef] [PubMed]
3. Gallart, F.; Llorens, P. Observations on land cover changes and water resources in the headwaters of the Ebro catchment, Iberian Peninsula. *Phys. Chem. Earth* **2004**, *29*, 769–773. [CrossRef]
4. Carranza, S.; Amat, F. Taxonomy, biogeography and evolution of *Euproctus* (Amphibia: Salamandridae), with the resurrection of the genus *Calotriton* and the description of a new endemic species from the Iberian Peninsula. *Zool. J. Linn. Soc.* **2005**, *145*, 555–582. [CrossRef]
5. Valbuena-Ureña, E.; Amat, F.; Carranza, S. Integrative phylogeography of *Calotriton* newts (Amphibia, Salamandridae), with special remarks on the conservation of the endangered Montseny brook newt (*Calotriton arnoldi*). *PLoS ONE* **2013**, *8*, e62542. [CrossRef] [PubMed]
6. Valbuena-Ureña, E.; Soler-Membrives, A.; Steinfartz, S.; Orozco-TerWengel, P.; Carranza, S. No signs of inbreeding despite long-term isolation and habitat fragmentation in the critically endangered Montseny brook newt (*Calotriton arnoldi*). *Heredity* **2017**, *118*, 424435. [CrossRef] [PubMed]
7. Carranza, S.; Martínez-Solano, I. *Calotriton arnoldi*. The IUCN Red List of Threatened Species 2009 e.T136131A4246722. Available online: https://www.iucnredlist.org/species/136131/4246722 (accessed on 3 October 2021).
8. Guinart, D.; Solórzano, S.; Vicens, N.; Anton, M. *Pla de Conservació del Parc Natural del Montseny. Reserva de la Biosfera*; Diputació de Barcelona: Barcelona, Spain, 2014; 561p.
9. Salafsky, N.; Salzer, D.; Stattersfield, A.J.; Hilton-Taylor, C.; Neugarten, R.; Butchart, S.H.M.; Collen, B.; Cox, N.; Master, L.L.; O'Connor, S.; et al. A standard lexicon for biodiversity conservation: Unified classifications of threats and actions. *Conserv. Biol.* **2008**, *22*, 897–911. [CrossRef] [PubMed]
10. Montori, A.; Campeny, R. Situación actual de las poblaciones de tritón pirenaico (*Euproctus asper*) en el macizo del Montseny. *Boletín De La Asoc. Herpetol. Española* **1991**, *2*, 10–12.
11. Amat, F. Estat de conservació del tritó pirinenc (*Euproctus asper*) a la Reserva de la Biosfera i Parc Natural del Montseny. In *VI Trobada d'Estudiosos del Montseny*; Diputació de Barcelona: Barcelona, Spain, 2005; pp. 119–121.
12. Peñuelas, J.; Germain, J.; Álvarez, E.; Aparicio, E.; Arús, P.; Basnou, C.; Blanché, C.; Bonada, N.; Canals, P.; Capodiferro, M.; et al. Impacts of Use and Abuse of Nature in Catalonia with Proposals for Sustainable Management. *Land* **2021**, *10*, 144. [CrossRef]
13. Villero, D.; Anton, M.; Garcia, E.; Munné, A.; Solà, C.; Bonada, N.; Fortuño, P.; Rotchés, R.; Guinart, D.; Solórzano, S.; et al. Aigües continentals. In *Estat de la Natura a Catalunya 2020*; Lluís Brotons, Núria Pou, Pau Sainz de la Maza & Sara Pont (Coord.); Departament de Territori i Sostenibilitat, Generalitat de Catalunya: Barcelona, Spain, 2020; pp. 35–56.
14. Peñuelas, J.; Boada, M. A global change-induced biome shift in the Montseny mountains (NE Spain). *Glob. Change Biol.* **2003**, *9*, 131–140. [CrossRef]
15. Martel, A.; Vila-Escarré, M.; Fernández-Giberteau, D.; Martinez-Silvestre, A.; Canessa, S.; Van Praet, S.; Pannon, P.; Chiers, K.; Ferran, A.; Kelly, M.; et al. Integral chain management of wildlife diseases. *Conserv. Lett.* **2020**, *13*, e12707. [CrossRef]
16. Battisti, C.; Amori, G.; Luiselli, L. Toward a new generation of effective problem solvers and project-oriented applied ecologists. *Web Ecol.* **2020**, *20*, 11–17. [CrossRef]
17. Martí, E.; Ribot, M. *Caracterització Hidromorfològica de les Rieres de la Conca Alta de la Tordera per Entendre L'hàbitat Potencial del Tritó del Montseny (Calotriton Arnoldi)*; Unpublished Report; CEAB-CSIC-Diputació de Barcelona: Barcelona, Spain, 2018; 22p.
18. Ávila, A. *Estudi de la Dinàmica Hidro-Química de la Conca del Torrent de la Mina Com a Zona Potencial de Reintroducció del Tritó i Caracterització de la Composició Química de Torrents del Montseny Amb Presència i Absència del Tritó*; Unpublished Report; Centre de Recerca Ecològica i Aplicacions Forestals (CREAF)-DIBA: Barcelona, Spain, 2018; 37p.
19. Mata, R.; Puiguriguer, M. *Caracterització Geològica de L'hàbitat del Tritó del Parc Natural del Montseny*; Unpublished Report; Axial Geologia i Medi Ambient S.L.-Diputació de Barcelona: Barcelona, Spain, 2019; 158p.
20. Fernández-Guiberteau, D.; Montori, A. *Les Malalties Infeccioses en Amfibis Manual de Bones Pràctiques en Les Activitats Educatives de Descoberta*; Gemma Pascual, Narcís Vicens i Daniel Guinart (Coord.); Diputació de Girona, Diputació de Barcelona, Life Tritó Montseny, Eds.; LIFETM Docs: Barcelona, Spain, 2020; 20p. Available online: https://lifetritomontseny.eu/altres-documents (accessed on 24 January 2022).

21. Fernández-Guiberteau, D.; Montori, A.; Pérez-Sorribes, L.; Carranza, S. *Protocols Sanitaris per a les Activitats que Impliquin la Interacció Directa o Indirecta amb les Poblacions D'amfibis al Medi Natural*; LIFETM Docs: Barcelona, Spain, 2020; 15p. Available online: https://lifetritomontseny.eu/-/manual-de-bones-pr%C3%A0ctiques (accessed on 24 January 2022).
22. Fernández-Guiberteau, D.; Montori, A.; Amat, F. *Estudi Demogràfic i Seguiment de la Població de Calotriton Arnoldi al Torrent A2. 2018*; Unpublished Report; LIFETM-GRENP: Barcelona, Spain, 2019; 29p.
23. Amat, F.; Oromí, N.; Sanuy, D.; Carranza, S. Sexual dimorphism and age structure of the Montseny newt (*Calotriton arnoldi*). *Amphibia-Reptilia* **2015**, *36*, 245–252. [CrossRef]
24. Wilson, K.; Hardy, I.C.W. Statistical analysis of sex ratios: An introduction. In *Sex Ratios—Concepts and Research Methods*; Hardy, I.C.W., Ed.; Cambridge University Press: Cambridge, UK, 2002; pp. 48–92. [CrossRef]
25. Montori, A.; Llorente, G.A.; Richter-Boix, A. Habitat features affecting the small-scale distribution and longitudinal migration patterns of *Calotriton asper* in a pre-Pyrenean population. *Amphibia-Reptilia* **2008**, *29*, 371–381. [CrossRef]
26. Schwarz, C.J.; Arnason, A.N. A General Methodology for the Analysis of Capture-Recapture Experiments in Open Populations. *Biometrics* **1996**, *52*, 860–873. [CrossRef]
27. Carbonell-Buira, F.; Aresté, M.; Fradera, R. *Final Report Captive Breeding Program. Life Tritó Montseny*; Unpublished Report; Centre de Fauna de Torreferrussa (Forestal Catalana), Zoològic de Barcelona, Centre de Fauna de Pont de Suert: Barcelona, Spain, 2021.
28. Phillips, S.; Anderson, R.; Schapire, R. Maximum entropy modeling of species geographic distributions. *Ecol. Modell.* **2006**, *190*, 231–259. [CrossRef]
29. Gomà, J.; Sabater, F. *Estudi dels Macroinvertebrats i Meiofauna dels Torrents Potencialment Habitables pel Tritó del Montseny (Calotriton arnoldi)*; Unpublished Report; Universitat de Barcelona-FEMH-Diputació de Barcelona: Barcelona, Spain, 2018; 17p.
30. Villero, D.; Hermoso, V.; Solorzano, S.; Amat, F.; Guinart, D.; Brotons, L. *Caracterització de L'hàbitat del Tritó del Montseny*; Unpublished Report; Diputació de Barcelona-CTFC: Barcelona, Spain, 2018.
31. Villero, D.; Amat, F.; Canesa, S.; Guinart, D.; Hermoso, V.; Salgado-Rojas, J.; Solórzano, S.; Brototons, L. *Avaluació de L'estratègia per Ampliar L'àrea de Distribució del Tritó del Montseny. Planificació de la Creació de Noves Poblacions*; Unpublished Report; CREAF-Diputació de Barcelona: Barcelona, Spain, 2019; 19p.
32. Gomà, P.; Avila, A.; Comas, X.; Guinart, D.; Grau, J.; Solorzano, S. *Qualitative and Quantitative Hydrological Monitoring. Annual Report (LIFE15 NAT/ES/000757)*, Unpublished Report; Barcelona, Spain, 2020; 148p.
33. Contreras, J. *Maximum Critical Temperature, Cold Tolerance and Thermopreferendum of the Montseny Newt (Calotriton Arnoldi)*; Tesis del Màster de Biodiversitat, Universitat de Barcelona: Barcelona, Spain, 2019; 28p.
34. Amat, F.; Carranza, S. *Estudi Demogràfic del Tritó del Montseny (Calotriton Arnoldi) al Parc Natural i Reserva de la Biosfera del Montseny*; Unpublished Report; Diputació de Barcelona: Barcelona, Spain, 2005.
35. Fernández-Guiberteau, D.; Montori, A.; Amat, F.; Carbonell, F.; López-López, L.M.; Comas, X. *Seguiment Demogràfic de Calotriton Arnoldi de la Població Natural A2 i de les de Nova Creació B8, A5 i A6. 2019*; Unpublished Report; GRENP-DiBa: Barcelona, Spain, 2020; 68p.
36. Carbonell-Buira, F.; Fernández-Guiberteau, D.; Montori, A. Establishment of a new western population of Montseny Brook Newt in 2019. *AArk Newsl.* **2020**, *49*, 25–26.
37. Broekman, A.; Sánchez, A. *Resum Executiu del Cicle de Trobades del Comitè Assessor d' Hidròlegs pel PNRB Montseny*; Unpublished Report; PNRBM-CREAF: Barcelona, Spain, 2019.
38. Cristobal-Roselló, J. *Document Marc per a la Integració de la Recerca Hidrològica en la Gestió del Parc Natural del Montseny*; Reserva de la Biosfera (PNM-RB). Unpublished Report; PNRBM-LIFETM: Barcelona, Spain, 2020.
39. Vayreda, J.; Comas, L.; Solorzano, S.; Guinart, D. *Buenas Prácticas Forestales en los Bosques de Ribera del Montseny*; Unpublished Report; PNRBM-LIFETM: Barcelona, Spain, 2020.
40. Wallace, S.; Knight, R.L. *Small Scale Constructed Wetland Treatment Systems: Feasibility, Design Criteria and O&m Requirements*, 1st ed.; WERF: London, UK, 2006; ISBN 1843397285.
41. Miralpeix, M. *Estudi a la Finca de Can Rubi, de la Transformació del Sistema de Reg del CULTIU de plantes en Contenidor Existent i de L'aprofitament de les Aigües Pluvials i Escorrentia dels Camps*; Unpublished Report. PNRBM-LIFETM 2018. Servei assistència tècnica reg, s.l. Núm. de valoració: 17-v/09242. Núm. d' informe: I-18/04072; Dputació de Barcelona: Barcelona, Spain, 2018; 37p.
42. Sorolla, A.; Mota, E.; Rueda, I.; Sorolla, G.; Unzeta, C.; Ribera, S. *Avantprojecte per la Millora Ecològica de la Gestió L'aigua al Xalet i el Càmping de Fontmartina*; Unpublished Report; Naturalea-Informe LIFE Trito Montseny: Barcelona, Spain, 2018.
43. Ledesma, J.L.J.; Montori, A.; Altava-Ortiz, V.; Barrera-Escoda, A.; Cunillera, J.; Àvila, A. Future hydrological constraints of the Montseny brook newt (*Calotriton arnoldi*) under changing climate and vegetation cover. *Ecol. Evol.* **2019**, *9*, 9736–9747. [CrossRef] [PubMed]
44. Minuartia. *Estudi Dels Efectes del Canvi Climàtic en el Montseny: Diagnosi, Impactes i Vulnerabilitats*; LIFE Clinomics. Acció A1. Unpublished Report; Diputació de Barcelona: Barcelona, Spain, 2016; 146p.
45. González-Hidalgo, J.C.; López-Bustins, J.A.; Štepánek, P.; Martín-Vide, J.; de Luisa, M. Monthly precipitation trends on the Mediterranean fringe of the Iberian Peninsula during the second half of the twentieth century (1951–2000). *Int. J. Climatol.* **2009**, *29*, 1415–1429. [CrossRef]
46. Amat, F.; Fernández-Guiberteau, D.; Montori, A.; Oro, D. Spatial heterogeneity in the demography of the critically endangered Montseny brook newt (*Calotriton arnoldi*). *Salamandra* **2021**, *57*, 309–316.

47. Gallart, F.; Llorens, P.; Latron, J.; Regüés, D. Hydrological processes and their seasonal controls in a small Mediterranean mountain catchment in the Pyrenees. *Hydrol. Earth Syst. Sci.* **2002**, *6*, 527–537. [CrossRef]
48. Beguería, S.; López-Moreno, J.I.; Lorente, A.; Seeger, M.; García-Ruiz, J.M. Assessing the effect of climate change and land-use changes on streamflow in the central Spanish Pyrenees. *Ambio* **2003**, *32*, 283–286. [CrossRef] [PubMed]
49. López-Moreno, J.I.; Vicente-Serrano, S.M.; Moran-Tejeda, E.; Zabalza, J.; Lorenzo-Lacruz, J.; García-Ruiz, J.M. Impact of climate evolution and land use changes on water yield in the Ebro basin. *Hydrol. Earth Syst. Sci.* **2011**, *15*, 311–322. [CrossRef]
50. Nabuurs, G.J.; Schelhaas, M.J.; Mohren, G.M.J.; Field, C.B. Temporal evolution of the European Forest sector carbon sink 1950-1999. *Glob. Change Biol.* **2003**, *9*, 152–160. [CrossRef]
51. Llorens, P.; Gallart, F.; Cayuela, C.; Roig-Planasdemunt, M.; Casellas, E.; Molina, A.J.; Moreno-de-las-Heras, M.; Bertran, G.; Sánchez-Costa, E.; Latron, J. What have we learnt about Mediterranean catchment hydrology? 30 years observing hydrological processes in the vallcebre research Catchments. *Geogr. Res. Lett.* **2018**, *4*, 475–502. [CrossRef]
52. Amat, F.; Carranza, S.; Valbuena-Ureña, E.; Carbonell, F. Saving the Montseny brook newt (*Calotriton arnoldi*) from extinction: An assessment of eight years of research and conservation. *Froglog* **2014**, *22*, 55–57.
53. Harper, E.B.; Semlitsch, R.D. Density dependence in the terrestrial life history stage of two anurans. *Oecologia* **2007**, *153*, 879–889. [CrossRef] [PubMed]
54. Montori, A. Alimentación de los adultos de *Euproctus asper* (Dugès, 1852) en la montaña media del Prepirineo catalán (España). *Rev. Esp. Herpetol.* **1991**, *5*, 23–36.
55. Sánchez-Hernández, J.; Montori, A.; Llorente, G.A. Ontogenetic dietary shifts and food resource partitioning in a stream-dwelling urodela community: Mechanisms to allow coexistence across seasons. *Russ. J. Herpetol.* **2019**, *26*, 135–149. [CrossRef]
56. Dalibard, M.; Buisson, L.; Riberon, A.; Laffaille, P. Identifying threats to Pyrenean brook newt (Calotriton asper) to improve decision making in conservation management: A literature review complemented by expert-driven knowledge. *J. Nat. Conserv.* **2020**, *54*, 125801. [CrossRef]
57. Dalibard, M.; Laffaille, P.; Sánchez-Pérez, J.M.; Sauvage, S.; Buisson, L. Accounting for flow intermittence in freshwater species distribution modelling. *Ecohydrology* **2021**, *14*, 8,e2346. [CrossRef]
58. Machado, A. *Guidelines for Action Plans for Animal Species: Planning Recovery. Nature and Environment No. 92*; Council of Europe Publishing: Strasbourg, France, 1997; 82p, ISBN 978-92-871-3472-1.
59. Bosch, J.; Mora-Cabello, A.; Marquínez, S.; Price, S.J.; Thumsova, B.; Bielby, J. Long-term monitoring of amphibian populations of a National Park in Northern Spain reveals negative persisting effects of Ranavirus, but not Batrachochytrium dendrobatidis. *Front. Vet. Sci.* **2021**, *8*, 645491. [CrossRef] [PubMed]
60. Salafsky, N.; Margoluis, R. Threat reduction assessment: A practical and cost-effective approach to evaluating conservation and development projects. *Conserv. Biol.* **1999**, *13*, 830–841. [CrossRef]
61. Beebee, T.J.C. Amphibian breeding and climate. *Nature* **1995**, *374*, 219–220. [CrossRef]
62. Blaustein, A.R.; Walls, S.C.; Bancroft, B.A.; Lawler, J.J.; Searle, C.L.; Gervasi, S.S. Direct and Indirect Effects of Climate Change on Amphibian populations. *Diversity* **2010**, *2*, 281–313. [CrossRef]
63. Vicente-Serrano, S.M.; Lopez-Moreno, J.I.; Beguería, S.; Lorenzo-Lacruz, J.; Sanchez-Lorenzo, A.; García-Ruiz, J.M.; Espejo, F. Evidence of increasing drought severity caused by temperature rise in southern Europe. *Environ. Res. Lett.* **2014**, *9*, 9. [CrossRef]
64. Wang, K.C.; Dickinson, R.E.; Liang, S.L. Global atmospheric evaporative demand over and from 1973 to 2008. *J. Clim.* **2012**, *25*, 8353–8361. [CrossRef]
65. Bosch, J.M.; Hewlett, J.D. A review of catchment experiments to determine the effect of vegetation changes on water yield and evapotranspiration. *J. Hydrol.* **1982**, *55*, 3–23. [CrossRef]
66. Sahin, V.; Hall, M.J. The effects of afforestation and deforestation on water yields. *J. Hydrol.* **1996**, *178*, 293–309. [CrossRef]
67. Llorens, P.; Domingo, F. Rainfall partitioning by vegetation under Mediterranean conditions. A review of studies in Europe. *J. Hydrol.* **2007**, *335*, 37–54. [CrossRef]

land

MDPI

Article

Effects of Population Declines on Habitat Segregation and Activity Patterns of Rabbits and Hares in Doñana National Park, Spain

Juan F. Beltrán [1,*], Jaime R. Rau [2], Ramón C. Soriguer [3], Maura B. Kufner [4], Miguel Delibes [5] and Francisco Carro [3]

1 Departamento de Zoología, Universidad de Sevilla, 41012 Seville, Spain
2 Laboratorio de Ecología, Departamento de Ciencias Biológicas y Biodiversidad, Universidad Los Lagos, Campus Osorno, Osorno 5290000, Chile; jrau@ulagos.cl
3 Departamento de Etología y Conservación de la Biodiversidad, Estación Biológica de Doñana, CSIC, 41092 Seville, Spain; soriguer@ebd.csic.es (R.C.S.); pcarro@ebd.csic.es (F.C.)
4 Center Dr. Ricardo Luti of Ecology and Renewable Natural Resources Research (CERNAR), National University of Córdoba, Córdoba X5000GYA, Argentina; beti0611@gmail.com
5 Departamento de Biología de la Conservación de la Biodiversidad, Estación Biológica de Doñana, CSIC, 41092 Seville, Spain; mdelibes@ebd.csic.es
* Correspondence: beltran@us.es

Abstract: Competition, predation, and diseases are key factors shaping animal communities. In recent decades, lagomorphs in Europe have been impacted by virus-borne diseases that have caused substantial declines in their populations and, subsequently, in many of their predators. We examined activity and habitat-use patterns of sympatric European rabbits (*Oryctolagus cuniculus* L.) and Iberian hares (*Lepus granatensis* R.) in Doñana National Park, Spain, (DNP) during two periods of disease outbreak. In the first period (1984–1985), fecal pellet counts and roadside counts indicated that lagomorph species were segregated, with rabbits occurring in scrublands and hares in marshlands. Both species also occupied rush and fern belt ecotones. Roadside counts at sunrise, midday, sunset, and midnight revealed that rabbits and hares had the same activity patterns (crepuscular and nocturnal) in the zone of sympatry. During the second period (2005–2016), roadside counts showed that rabbits and hares were mainly nocturnal in scrublands and border marshlands. Hares occupied scrublands; a habitat previously occupied only by rabbits. These results are interpreted in light of the competition theory and predation pressure. The disease-caused decline of rabbits has likely favored hares that moved into scrublands, a vegetation type previously occupied exclusively by rabbits. The decline of rabbits in DNP has also caused the almost disappearance of this area of the Iberian lynx (*Lynx pardinus*), a rabbit specialist, thus enabling generalist predators to increase. Generalist predators have subsequently increased predation pressure on both rabbits and hares, causing them to switch to nocturnal activity.

Keywords: activity patterns; *Lepus granatensis*; population decline; niche; *Oryctolagus cuniculus*; roadside census; predator–prey relationships; spatio-temporal behavior

Citation: Beltrán, J.F.; Rau, J.R.; Soriguer, R.C.; Kufner, M.B.; Delibes, M.; Carro, F. Effects of Population Declines on Habitat Segregation and Activity Patterns of Rabbits and Hares in Doñana National Park, Spain. *Land* **2022**, *11*, 461. https://doi.org/10.3390/land11040461

Academic Editors: Christine Fürst and Ian G. McLean

Received: 21 December 2021
Accepted: 22 March 2022
Published: 24 March 2022

1. Introduction

Competition for resources is considered one of the main factors shaping the coexistence of species in natural communities [1]. Among lagomorphs, competition is expected to be greater among closely related species, either phylogenetically or ecologically [2,3]. Predation can also play a role in how species co-exist by increasing mortality or modifying spatio-temporal activity patterns of lagomorphs [4–6]. Additionally, diseases may affect community structure, especially if they impact a keystone species [7–9]. In Europe, populations of rabbits and hares have been affected by imported diseases that have substantially reduced their populations, and subsequently caused a reduction of predators dependent on them [9,10].

After the emergence of myxomatosis in 1950s, rabbit (*Oryctolagus cuniculus*) populations recovered, only to crash again in late 1980s from rabbit hemorrhagic disease (RHD) [7,11]. In several countries, the first epizootic occurrence of myxomatosis was accompanied by an increase in European hare (*Lepus europaeus*) populations, followed by a concomitant decline of European rabbits [12–18]. This pattern was interpreted as indirect evidence of competition between both species [19–22]. Overall, three possible mechanisms of competition between rabbits and hares have been documented: (a) diseases that harm hares, but not rabbits (e.g., the stomach worm (*Graphidium strigosum*) [17]); (b) despite the known aggressiveness of hares [23], behavioral observations that suggested rabbits are the "winners" when directly fighting hares [21], or rabbits driving hares away from their burrows (but see Broekhuizen [16]); and (c) competition for food [24].

European rabbits were impacted again in 2010 [25,26] with the arrival of a novel genotype of the calicivirus RHDV (RHDV2 or RHDVb or *Lagovirus europaeus*), ref. [27] that reached Spain in 2011 [9], and Doñana National Park (DNP) more specifically in 2013. A long-term monitoring program at DNP detected a decline in rabbit numbers during 2013. In Coto del Rey (northern DNP), there was a decline of >80% of rabbits during 2012–2013 [28]. Similar declines were detected in all populations surveyed within DNP (F. Carro, unp. obs.). On the other hand, the Iberian hare population in the DNP underwent a moderate decline in the period 1996–2012 that was attributed to varying flooding cycles of the marshlands, changes in vegetation cover, and predation pressure [29]. Since 2003, hare numbers have decreased by 88% in DNP [29], and they were also affected by the first outbreaks of the novel ha-MYXV in 2018 [30].

Interspecific competition can be indirectly assessed from both allopatric and contiguous spatial distributions [31–33]. To minimize competition, potential competitors segregate along predictable niche parameters of diet, habitat use, and activity periods [31]. Therefore, the effects of myxomatosis and RHD among sympatric populations of hares and rabbits were expected to not only go beyond changes in their respective abundance patterns, but also to affect habitat use and activity periods of both lagomorphs. Long-term studies may shed light on vertebrate population dynamics [34]. In particular, changes in abundance, especially when those changes affect species differentially, may be seen as a "natural experiment" [35].

In this paper, we use a natural experiment to examine niche relationships between rabbits and hares in southwestern Spain. Our objective was to evaluate the effects of rabbit population crashes (mid 1980s and 2011–2013) on hare populations, specifically if habitat use and activity patterns varied in the context of the competitive exclusion hypothesis. We speculated that the collapse of rabbit populations would not only reduce competitive interactions between rabbits and hares, but also increase predator pressure on remaining lagomorphs. Specifically, we expected hares to expand into habitats previously occupied only by rabbits [3]. We also expected changes in activity, with rabbits becoming nocturnal in response to increased predation pressure [4].

2. Materials and Methods

2.1. Study Area

Our study was conducted in Doñana Biological Reserve (DBR, Figure 1), a restricted area of the Doñana National Park (DNP), situated on the right bank of the mouth of Guadalquivir River (approximately 37° N, 6°30′ W), spanning ≈ 1220 km^2, including the peripheral zone of protection. The marshland or "marisma" is usually flooded from October–November to May–June, and spans ≈ 55% of the DNP area. Around 30% of DNP is Mediterranean scrubland, and 15% of DNP is sand dunes with scattered pine forest (*Pinus pinea*). A detailed description of the DNP area can be found in Valverde (1958) [36], Aguilar-Amat et al. (1979) [37], and Green et al. (2018) [38]. The climate is Mediterranean with a slight Atlantic influence. Summers are warm and dry, and winters mild and wet. The average annual rainfall is 500–600 mm, 87% of which falls from October to April.

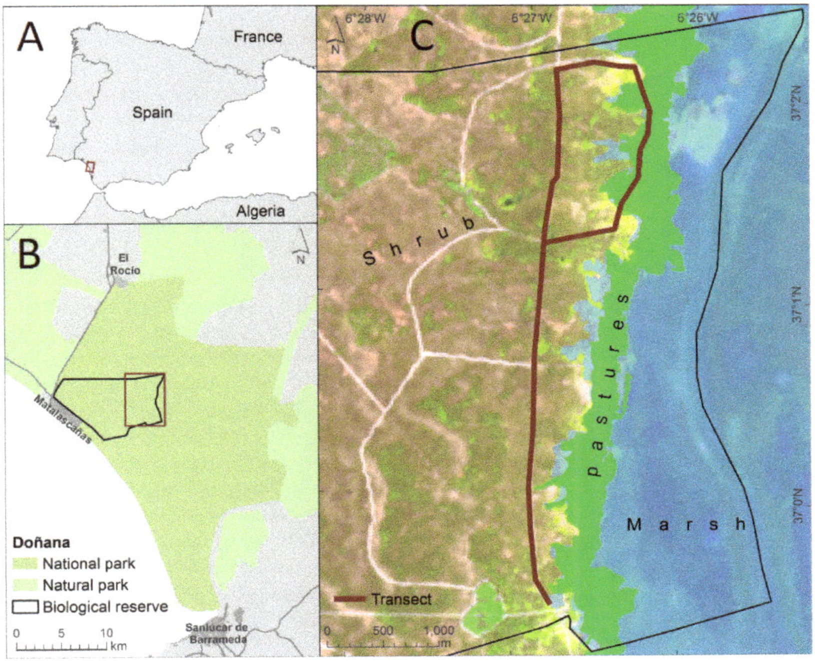

Figure 1. Location of the study area (Biological Reserve of Doñana) in Doñana National Park ((**A**,**B**), southwest Spain). The transect ((**C**), ~8 km long) used for roadside counts of lagomorphs.

Vegetation types present include pure scrublands, border scrublands, a fern belt, a rush belt, border marshlands, and pure marshlands (Figure 1) [39,40]. Each one of the four transitional bands in the ecotone (known locally as "La Vera") ranges from nearly 15 m to more than 120 m wide.

2.2. *First Period of the Study: 1983–1985*

The first period of the study was characterized by a pre-epizootic decline of rabbits and hares. Beginning in early November 1983 (just before the flooding of the *marisma* that year), 10 plots of 1 m^2 spaced 1 m apart were randomly placed in each of the six vegetation bands. To assess spatial distributions of rabbits and hares, all lagomorph fecal pellets were counted and removed from each plot (see [41] for a review). In addition, in August 1984 (mid-summer), 10 similar plots were established in a pure marsh area (Leo Biaggi area) approximately 10 km east of the ecotone (Figure 1). Hare pellets are usually larger than those of rabbits [42,43]. To differentiate between them, we collected samples of fresh pellets in rabbit warrens (January 1984) and hare bedding sites (October 1983), and recorded three measurements (thickness = minimum diameter, width, and length = maximum diameter) using a caliper. To assess the frequency of both species within each sampling plot, we used a filter approach based on the three measurements to differentiate pellets of rabbits from those of hares.

Roadside counts of both lagomorphs [44,45] were conducted along a 5.2 km transect in scrublands and a 2.8 km ecotone transect (hereafter called border marshland transect; Figure 1) using a 4 $\times$ 4 vehicle that traveled under 20 km/h. Counts were made at sunset, approximately four times a month from November 1983 to November 1984. The abundance of hares and rabbits was indexed by individuals tallied/10 km driven (kilometric abundance index, KAI). To examine circadian activity, roadside counts of both species were performed four times a day at sunrise, midday, sunset, and midnight for three consecutive days during the winter of 1985, along the previously described transects. To express lagomorph abundance on a comparative biomass basis, average autumn weights of both species in the study area (both sexes pooled, rabbits, $n = 20$, mean $\pm$ standard error = 0.91 $\pm$ 0.04 kg; hares, $n = 18$, m $\pm$ SE = 1.98 $\pm$ 0.06 kg) were obtained from the records of the Doñana Biological Station scientific collections.

Statistical analyses followed the procedures of Zar [46]. We generally considered equal sample sizes with two-tailed hypotheses, therefore the t-tests (both paired-sample t-test and two-sample t-test) were used. However, when equal sample size or other requirements could not be met, nonparametric equivalents were used.

2.3. Second Period of Study: 2005–2016

The second period of the study occurred from 2005–2007 and 2014–2016 when populations of both lagomorphs had suffered dramatic declines. Hares and rabbits counts (individuals/10 km, KAI) were obtained from the same transects used in 1983–1985. Surveys were also conducted via a 4 $\times$ 4 vehicle, both at sunset (from 1.5 h before sunset to sunset) and at night (1.5 h after sunset) with the aid of a handheld 100-watt spotlight after dusk [47,48]. Surveys were conducted in spring, summer, and early autumn during 2005 and 2007, and in spring and autumn during 2014 and 2016 by at least two people [49–51]. Observers were seated on the roof of the vehicle $\approx$ 3 m above ground level [52]. Hares and rabbits were identified using binoculars. We also measured the perpendicular distance of animals from the transect line using a laser telemeter. The maximum width of the contact strip was 200 m, and depended on the height of the vegetation at the time. The majority of contacts occurred within 100 m of the observer. Pellet counts were performed only during the first period of study, and were used to examine habitat segregation. Roadside counts were used complementarily, both in the first and the second period, as an efficient method to assess not only habitat segregation, but also estimate population trends [7,44,53].

3. Results

3.1. First Period: 1983–1985

3.1.1. Pellet Characteristics

Pellet-size distributions resembled normality, enabling us to calculate 95% confidence intervals (Figure 2). Hare pellets were longer ($t = 3.9$, $p < 0.01$), wider ($t = 11.1$, $p < 0.01$), and thicker ($t = 12.2$, $p < 0.01$) than rabbit pellets, and our measurements were consistent with those reported elsewhere [42,43]. However, overlap did occur in thickness (24%), therefore we used 7 mm to distinguish rabbit ($\leq$7 mm) from hare (>7 mm) pellets, but the overlap of both species at 5–7.5 mm was evident (Figure 2). Hence, we distinguished "probable rabbit pellets" from "probable hare pellets".

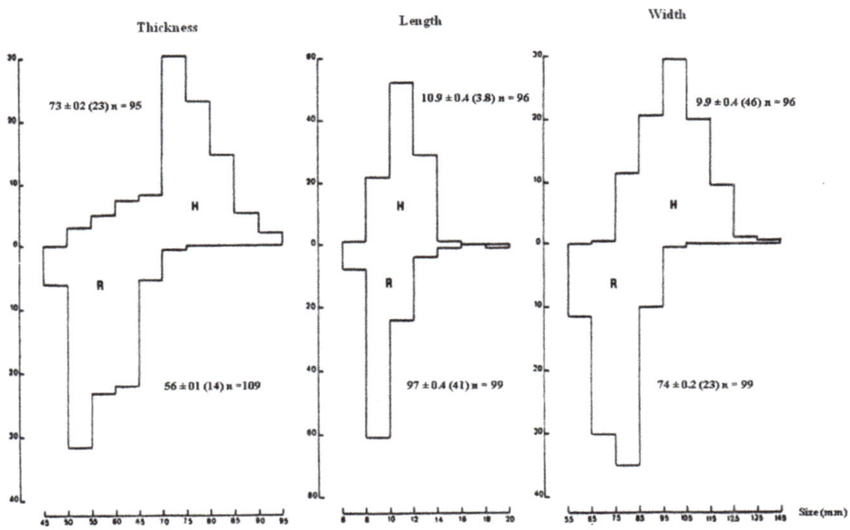

Figure 2. Frequency histograms of Iberian hare (above, H) and European rabbit (below, R) pellet dimensions from samples collected in rabbit warrens (January 1984 and hare bedding sites (October 1983). The mean ± SE, 95%, confidence intervals, and sample sizes are indicated.

3.1.2. Habitat Use

Based on species-specific traits of fecal pellets (Figure 2), habitat use differed by species (Figure 3). Hares mainly used marshlands, and rabbits scrublands. Both species used transitional zones in the ecotone in the same proportion (i.e., there were no significant differences). This habitat segregation was also confirmed when an independence analysis was applied to the marshland and scrubland pellet abundances (2 × 2 contingency test corrected for continuity; $p < 0.001$). Additional pellet counts made at the marshland habitat (see Section 2.1 and Section 2.2) did not detect rabbit pellets, while 3.0 + 0.98 hare pellets/m_2 were found.

Roadside counts indicated that rabbit abundance was four times higher in the scrubland than in the ecotone (Table 1: the means differ statistically; paired sample *t*-test; t = 3.3, $p = 0.02$). Hares did not occupy scrublands during the first period of study, and both species were observed in the ecotone (Table 1: paired-sample *t*-test; t = 3.3, $p = 0.02$). In this zone, peaks in hare abundance appeared to be related to the start of the marsh flooding (e.g., November 1983), and probably linked to the beginning of the breeding season (April 1984). By considering only the six months in which both species occur together each year in the study area (Table 1), no statistical differences were found between the abundances of rabbits and hares in the ecotone (paired-sample *t*-test; t = 0.41, 220 $p = 0.69$), but when biomass was considered (Table 1), hare biomass was 2.6 times greater than rabbit biomass.

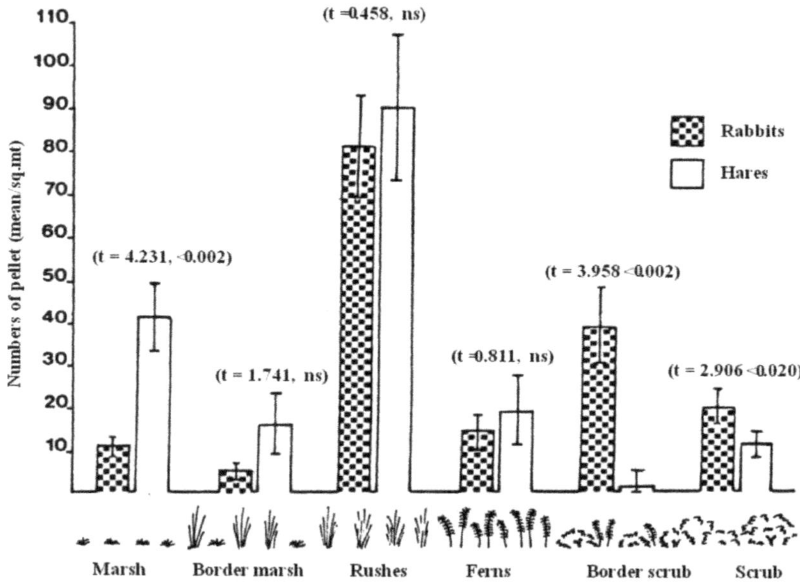

Figure 3. Habitat use by European rabbits and Iberian hares in six vegetation zones at Doñana National Park, SW Spain. Statistical *t*-values are indicated for each vegetation type and are provided at the top of the bars. Note that transitional zones were used by both species, whereas hares were more abundant in the marshland and rabbits in the scrubland.

Table 1. Relative abundances based on roadside counts (individuals/10 km) of European rabbits and Iberian hares in two vegetation types in Doñana National Park, 1983 and 1984. Means ± SE and sample size are indicated for each month.

		Scrublands	Border Marshlands	
Month	n	Rabbits	Rabbits	Hares
November 1983	3	10.7 ± 8.7	8.3 ± 3.3	45.0 ± 35.5
December 1983	2	33.0 ± 19.1	0.0	2.5 ± 2.5
April 1984	3	32.7 ± 10.5	11.7 ± 9.3	18.3 ± 8.3
May 1984	4	66.5 ± 18.0	2.5 ± 2.5	5 ± 3.5
July 1984	4	67.0 ± 15.7	27.5 ± 7.8	1.2 ± 1.2
November 1984	3	18.7 ± 4.4	3.3 ± 3.3	1.7 ± 1.7
Grand mean	19	41.4 ± 7.4	10.0 ± 3.1	11.8 ± 6.1
Average biomass(kg/10 km)		37.7	9.1	23.4

3.1.3. Circadian Activity

In the border marshland where both species occurred during winter, rabbits and hares were both most active at dusk and night (Figure 4). However, in the scrublands, rabbits appeared to be most active from midday to sunset, at least in winter; few individuals were seen active at night.

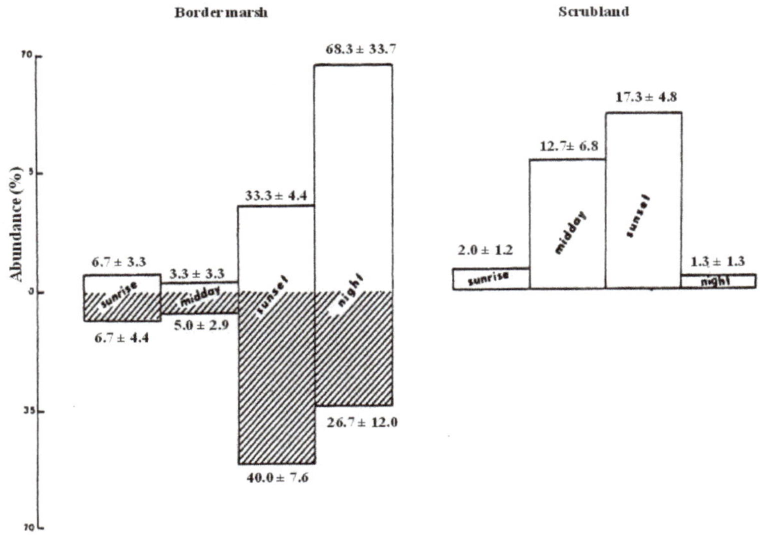

Figure 4. Circadian activity of European rabbits (white bars) and Iberian hares (stripped bars) at the border marshland and scrubland zones, estimated based on roadside counts in four time intervals over three consecutive days in winter 1985. Means ± SE of individuals/10 km are provided, and pairwise comparisons by species within a time period were not significant.

3.2. Second Period: 2005–2016

After the sharp decline of rabbits, changes in habitat use were detected. In scrublands, the peak number of rabbits was observed at sunset in 1985 (Figure 4), but this changed to night in 2005–2006 (Tables A1 and A2, Figure 5). However, small sample sizes limited these comparisons. In border marshlands, where rabbits showed a mainly nocturnal pattern in 1985, they seemed to maintain this pattern (t = 1.77, p = 0.10).

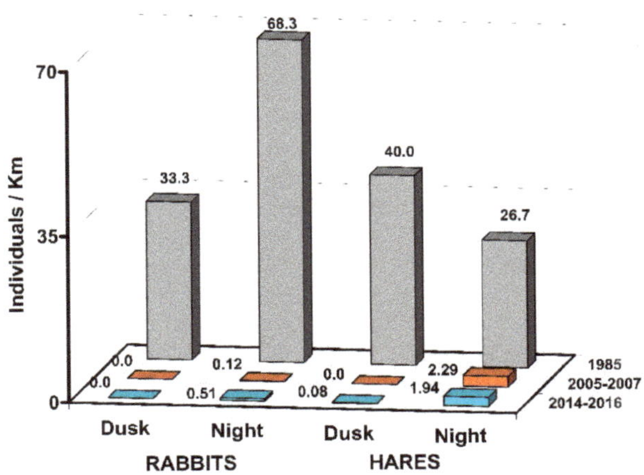

Figure 5. Abundances of European rabbits and Iberian hares in the border marshland during sunset and night, as estimated by roadside counts in the three intervals of study (data from 1985 are the same as in Figure 4; the rest of the data are from Tables A1 and A2). Note that, in spite of the two-order of magnitude drop in population numbers since 2005, rabbits have maintained the same pattern of activity (being mostly nocturnal in this habitat).

For hares, observed changes included both habitat use and activity patterns. In the first study period (1984–1985), no hares were observed in the scrublands, and their activity peaked in the border marshlands at sunset (Figure 4). From 2005 to 2006, hares tended toward a nocturnal activity (Figure 5), not only in marshlands (although differences there were not significant, t = 0.14, p = 0.69), but also in the scrublands, where they had been absent during the roadside counts in 1984–1985.

4. Discussion

In the initial study period, rabbits and hares were spatially segregated. This could indicate that exploitative competition occurs between both species [54]. The distinct associations of rabbits with scrublands and hares with marshlands may be one of the factors enabling the species to coexist [55]. In a previous study, Rogers and Myers [56] also did not detect rabbits in open marshes. Based on resting time (daytime) habitat selection, Vidus-Rosin et al. [56] also reported segregation between European hares (*Lepus europaeus*), and introduced eastern cottontails (*Sylvilagus floridanus*) as a possible strategy to reduce interspecific competition. However, two criticisms can be made regarding the results on differential habitat use based on fecal-pellet counts. First, our data correspond to "probable" rabbit and hare pellets, since no definite distinction of pellet size was found. Second, Rowland et al. [57] suggested that fecal pellet distributions may not be directly associated with habitat use by herbivores due to the patchy distribution of foraging areas.

Using sand-track records in summer, Alvarez et al. [58] reported that the circadian activity of rabbits in Doñana in the scrublands was mainly crepuscular and nocturnal. Similarly, Villafuerte et al. [59], using roadside counts in scrublands, also showed that the majority of rabbit activity occurred during twilight periods and night. High levels of rabbit activity at sunset were constant throughout the year. When occupying the same vegetation type (border marshlands), our observations indicated that rabbits and hares foraged at the same time (sunset-night), thus increasing the probability for interference competition [60]. Within border marshlands, grasses and forbs account for ~70% of the diet of rabbits in Doñana [61]. Since this foraging area is spatially (see Figure 1) and temporally limited (especially when the nutritional value of grasses is taken into account [62]), interspecific competition can result in a species–resource specialization [63], where rabbits became scrubland specialists and hares became marshland specialists.

The observation that rabbits and hares use different but neighboring habitats (i.e., contiguous allopatry) when one species is dominant to the other would be an example of "type 1 coexistence" (*sensu* [27]), and enables long-term coexistence. Several other factors, including disease outbreaks or differential predation, could modify the competitive superiority of rabbits [17,21]. An infected rabbit would be easier for predators to capture than a healthy hare. Predators may then modify patterns of coexistence of rabbits and hares by exerting greater predation pressure on rabbits [64]. Specifically, the coexistence of rabbits and hares may have been mediated by specialist predators upon rabbits, such as the Iberian lynx (*Lynx pardinus* T.) and the Iberian imperial eagle (*Aquila adalberti* B.) [65,66]. Further, the abundance of hares in DNP may have been controlled by such factors as seasonal habitat reduction (i.e., the flooding of the marshland [67]), direct interference by rabbits (i.e., agonistic interactions [21]), and infestation by stomach parasites (17].

The situation and interpretation described above was unexpectedly tested with the decline of the population numbers of both rabbits (due to the arrival of two successive epizootics of RHDV) and hares (mainly due to the varying flooding cycles of the marshlands and increased predation pressure [53]). This "natural experiment" gave us an opportunity to observe responses at temporal and spatial scales, and evaluate previous interpretations.

Our data show that rabbits have become mainly nocturnal, a behavior previously restricted to a more open border scrubland habitat. This was likely a response to increased predation pressure among remaining rabbits [4,68,69]. Nocturnal activity is also currently observed at a much larger spatial scale (i.e., DNP, Carro et al., unp. obs.), and contrasts to previous studies that showed rabbits as mostly crepuscular (sunset and dawn) with some activity at night (mostly during February–March) [59].

Among hares, the population response observed was two-fold. First, on a spatial basis, recent (2005–2016) roadside counts showed the presence of hares in scrublands. Hares occupied scrublands year-round, rather than as a seasonal response to flooding of the marshlands in autumn (Tables A1 and A2). Second, circadian activity by hares also changed from crepuscular to nocturnal (Figure 4). Such patterns have been reported via the use of radio-collared hares [70]. Contrary to reported by Katona et al. [24], we did not observe an increase in the hare abundance in the study area.

The implications of the dramatic declines of rabbits and hares in the Doñana ecosystem are likely complex. Perhaps most striking are implications to endangered species in the area. The Iberian lynx (*Lynx pardinus*) is a rabbit specialist, with rabbits representing 85–95% of their diet [66]. As a result of the reduced rabbit abundance, the lynx population in DNP has nearly been extirpated. A similar fate is threatening the Iberian imperial eagle (*Aquila adalberti*) [71]. This pattern of severe decline contrasts with generalist carnivores in the area, including red foxes (*Vulpes vulpes*), badgers (*Meles meles*), and mongooses (*Herpestes ichneumon*). These predators are able to switch among available food sources, making them resilient to the decline of rabbits. Field observations suggest that the fox abundance in DNP has increased in recent years (F. Carro, *unp. obs.*). This increase may have been a response to the reduction in control of foxes by lynxes or mesopredators [72].

5. Conclusions

We conclude that, after the population decline of both species, their ecological overlap has increased substantially. Their convergence in nocturnal behavior is most likely a response to the increase in predation pressure [73,74]. Most recent numbers (2005–2016) of roadside counted rabbits are two orders of magnitude lower than in 1985, and hare numbers are only slightly higher (Appendix A, Tables A1 and A2). The remaining individuals of both species have selected the night as the circadian period that presents the minimum risk of predation. Bakker et al. [75] have shown that European rabbits are sensitive to perceived predation risk [76]. Most of the terrestrial predators of rabbits and hares in DNP have activity peaks at sunset and dawn [4,68,77].

The observed move by hares into vegetation types exclusively inhabited by rabbits supports the implications of relaxing competitive exclusion. Research focusing on both species at sympatry (e.g., the Doñana ecotone), allopatry (rabbits and hares feeding only in their respective preferred habitats), and removal experiments [31] will be valuable. Our study shows the importance of maintaining long-term monitoring of wildlife populations using standard procedures. This endeavor is enormously facilitated in protected areas, and needs the commitment and effort of their people.

Author Contributions: Conceptualization, methodology, and field sampling, J.R.R., J.F.B., F.C., R.C.S., M.D. and M.B.K.; resources, J.R.R., J.F.B., F.C., R.C.S., M.B.K. and M.D.; writing—original draft preparation, J.R.R. and J.F.B.; writing—review and editing, J.F.B., J.R.R., F.C., M.D. and R.C.S. All authors have read and agreed to the published version of the manuscript.

Funding: During the study, J.F.B. enjoyed a predoctoral fellowship from the Consejo Superior de Investigaciones Científicas (CSIC), and J.R.R. and M.B.K enjoyed a grant from the Instituto de Cooperación Iberoamericana. Financial assistance for the trip of M.D. to Canada was received from the Consejería de Educación de la Junta de Andalucía. The first period of study (1983–1985) was included in the project 944 of CSIC-CAICYT. During the second study interval (2005–2016), this research was partially supported by the Dirección General de Espacios Naturales y Participación Ciudadana, Consejería de Medio Ambiente y Ordenación del Territorio, Junta de Andalucía. The authors would like to thank the EBD-CSIC monitoring team and Doñana ICTS-RBD, who provided logistic and technical support.

Institutional Review Board Statement: Not applicable.

Informed Consent Statement: Not applicable.

Data Availability Statement: Not applicable.

Acknowledgments: J.R.R. acknowledges J.E.C. Flux and D. Fraguglione for sending their stimulating papers on lagomorph biology, and to his wife, Angélica Catalán, for her continuous assistance while producing this research, and also for her patient work with some of the figures of the paper. J. Castroviejo gave us valuable information about Iberian hares' movements when the marshland was flooding. A reduced version of the results of the first period of study (1983–1985) was presented as a poster in the Symposium on Lagomorph Population Dynamics and Relevance to Management (IV International Theriological Congress; 1985, Edmonton, Canada). We are in debt to J.A. Litvaitis, who kindly revised earlier versions of the manuscripts, and made suggestions to improve its content and readability.

Conflicts of Interest: The authors declare no conflict of interest.

Appendix A

Tables A1 and A2. Relative abundances (individuals/10 km) of European rabbits and Iberian hares at two vegetation zones (A1: Ecotone scrubland- marshland, A2: Scrubland)of Doñana National Park, as obtained by roadside counts at sunset and night on the same transect as in Table 1, in two recent intervals of continuous population monitoring. For each month, the mean ± SE is given, n = sample size.

Table A1. Ecotone scrubland-marshland.

Year	Month	Rabbits		Hares	
		Sunset	Night	Sunset	Night
2005	April	0.00	0.70	0.35	0.35
	June	0.70	0.70	0.00	1.76
	September	0.00	0.00	0.00	0.00
Mean ± SE		**0.23**	**0.47 ± 0.02**	**0.12**	**0.70 ± 0.54**
2006	March	0.35	0.00	0.00	0.35
	June	0.00	1.60	0.00	1.40
	September	0.35	0.00	0.00	0.00
Mean ± SE		**0.23 ± 0.12**	**0.35**	**0.00**	**0.58 ± 0.42**
2007	March	0.35	1.06	0.00	6.33
	June	0.35	0.00	0.32	3.52
	September	0.00	1.06	0.00	2.81
Mean ± SE		**0.23 ± 0.12**	**0.71 ± 0.35**	**0.12**	**4.22 ± 1.10**
3-year mean ± SE		**0.23**	**0.51 ± 0.11**	**0.08 ± 0.04**	**1.84 ± 1.19**
2014	March	0.00	0.00	0.00	2.11
	September	0.00	0.35	0.00	0.35
Mean ± SE		**0.00**	**0.18**	**0.00**	**1.23 ± 0.88**
2015	March	0.00	0.00	0.00	5.27
	September	0.00	0.35	0.00	0.00
Mean ± SE		**0.00**	**0.18**	**0.00**	**2.64**
2016	March	0.00	0.00	0.00	3.87
	September	0.00	0.00	0.00	2.11
Mean ± SE		**0.00**	**0.00**	**0.00**	**2.99 ± 0.88**
3-year mean ± SE		**0.00**	**0.00**	**0.00**	**2.29 ± 0.54**

Table A2. Scrubland.

Year	Month	Rabbits		Hares	
		Sunset	Night	Sunset	Night
2005	April	0.00	0.57	0.00	0.00
	June	0.19	0.19	0.00	0.00
	September	0.00	0.00	0.00	0.00
Mean ± SE		**0.06**	**0.25 ± 0.17**	0.00	0.00
2006	March	0.00	0.00	0.00	0.00
	June	0.38	0.00	0.00	0.38
	September	0.38	0.00	0.00	0.00
Mean ± SE		**0.25 ± 0.13**	**0.00**	0.00	**0.13**
2007	March	0.00	0.19	0.00	0.19
	June	0.19	0.00	0.00	0.57
	September	0.00	0.38	0.00	0.76
Mean ± SE		**0.06**	**0.19 ± 0.11**	**0.00**	**0.51 ± 0.17**
3-year mean ± SE		**0.13 ± 0.06**	**0.15 ± 0.08**	**0.00**	**0.21 ± 0.15**
2014	March	0.00	0.00	0.00	0.00
	September	0.00	0.77	0.00	0.77
Mean ± SE		**0.00**	**0.39**	0.00	**0.39**
2015	March	0.00	0.19	0.00	0.19
	September	0.19	0.19	0.00	0.00
Mean ± SE		**0.10**	**0.19**	**0.00**	**0.10**
2016	March	0.00	0.00	0.00	0.19
	September	0.00	0.19	0.00	0.19
Mean ± SE		**0.00**	**0.10**	**0.00**	**0.19**
3-year mean ± SE		**0.03**	**0.22 ± 0.09**	**0.00**	**0.22 ± 0.09**

References

1. Amarasekare, P. Competitive coexistence in spatially structured environments: A synthesis. *Ecol. Lett.* **2003**, *6*, 1109–1122. [CrossRef]
2. Cheeseman, A.E.; Ryan, S.J.; Whipps, C.M.; Cohen, J.B. Competition alter seasonal resource selection and promotes use of invasive shrubs by an inperiled native cottontail. *Ecol. Evol.* **2018**, *8*, 11122–11133. [CrossRef]
3. Woodgate, Z.; Distiller, G.; O´Riain, M.J. Hare today gone tomorrow: The role of interspecific competition in shaping riverine rabbit occurrence. *Endanger. Species Res.* **2021**, *44*, 351–361. [CrossRef]
4. Lombardi, L.; Fernández, N.; Moreno, S.; Villafuerte, R. Habitat-related differences in rabbit (*Oryctolagus cuniculus*) abundance, distribution, and activity. *J. Mammal.* **2003**, *84*, 26–36. [CrossRef]
5. Terborg, J.; Estes, J.A. *Trophic Cascades: Predators, Prey, and the Changing Dynamics of Nature*; Island Press: Washington, DC, USA, 2010.
6. Viviano, A.; Mori, E.; Fattorini, N.; Mazza, G.; Lazzeri, L.; Panichi, A.; Strianese, L.; Mohamed, W. Spatiotemporal overlap between the European Brown Hare and its potential predators and competitors. *Animals* **2021**, *11*, 562. [CrossRef]
7. Moreno, S.; Beltrán, J.F.; Cotilla, I.; Kuffner, B.; Laffitte, R.; Jordán, G.; Ayala, J.; Quintero, M.C.; Jiménez, A.; Castro, F.; et al. Long-term decline of the European wild rabbit (*Oryctolagus cuniculus*) in south-western. *Wildl. Res.* **2007**, *34*, 652–658. [CrossRef]
8. Delibes-Mateos, M.; Delibes, M.; Ferreras, P.; Villafuerte, R. Key role of the European rabbit in the conservation of the Western Mediterranean basin hotspot. *Conserv. Biol.* **2008**, *22*, 1106–1117. [CrossRef] [PubMed]
9. Monterroso, P.; Garrote, G.; Serronha, A.; Santos, E.; Delibes-Mateos, M.; Abrantes, J.; Pérez de Ayala, R.; Silvestre, F.; Carvalho, J.; Vasco, I.; et al. Disease-mediated bottom-up regulation: An emergent virus affects a keystone species, and alters the dynamics of trophic webs. *Sci. Rep.* **2016**, *6*, 36072. [CrossRef]
10. Ferreras, P.; Travaini, A.; Zapata, S.C.; Delibes, M. Short responses of mammalian carnivores to a sudden collapse of rabbits in Mediterranean Spain. *Basic Appl. Ecol.* **2011**, *12*, 116–124. [CrossRef]
11. Calvete, C.; Estrada, R.; Villafuerte, R.; Osácar, J.J.; Lucientes, J. Epidemiology of viral haemorrhagic disease and myxomatosis in a free-living population of wild rabbits. *Vet. Rec.* **2002**, *150*, 776–782. [CrossRef] [PubMed]
12. Rothschild, M.A.; Marsh, H. Increase of hares (*Lepus europaeus* Pallas) at Ashton Wold with a note on the reduction in numbers of the brown rat (*Rattus norvegicus* Berkenhout). *Proc. Zool. Soc. Lond.* **1956**, *127*, 441–445. [CrossRef]
13. Rothschild, M. A further note in the increase of hares (*Lepus europaeus*) in France. *Proc. Zool. Soc. Lond.* **1958**, *131*, 328–329.
14. Rothschild, M.A. Increase of hares at Ashton Wold. *Proc. Zool. Soc. Lond.* **1961**, *137*, 634–635.
15. Flux, J.E.C. Hare reproduction in New Zealand. *N. Z. J. Agric.* **1964**, *109*, 483–486.

16. Broekhuizen, S. The position of the wild rabbit in the life system of the European hare. In *Proceedings of the XII Congress of the International Union of Game Biologists*; Direcção-Geral dos Recursos Florestais Serviço de Inspecção da Caça e Pesca: Lisboa, Portugal, 1975; pp. 75–79.
17. Broekhuizen, S.; Kemmers, R. The stomach worm, *Graphidium strigosum* (Dujardin) Railliet and Henry in the European hare, *Lepus europaeus* Pallas. In *Ecology and Management of European Hare Populations*; Pielowski, Z., Pucek, Z., Eds.; Polish Hunt. Ass.: Warsaw, Poland, 1976; pp. 157–171.
18. Gibb, J.A.; Flux, J.E.C. Why New Zealand should not use myxomatosis in rabbit control operations. *Search* **1983**, *14*, 41–43.
19. Fraguglione, D. Compétition interspécifique entre le lièvre commun (*Lepus europaeus* Pallas 1778) et le lapin de garenne (*Oryctolagus cuniculus* Linné 1758). *Diana* **1960**, *10*, 211–212.
20. Fraguglione, D. Les enemis du lièvre commun (*Lepus europaeus* P.). *Diana* **1960**, *3*, 54–57.
21. Flux, J.E.C. Field observations of behaviour in the genus Lepus. In Proceedings of the World Lagomorph Conference, Guelph, ON, Canada, 12–16 August 1979; pp. 377–394.
22. Keith, L.B. Population dynamics of hares. In Proceedings of the World Lagomorph Conference, Guelph, ON, Canada, 12–16 August 1979; pp. 395–440.
23. Flux, J.E.C. Prospects for hare farming in New Zealand. *N. Z. Agric. Sci.* **1981**, *15*, 24–29.
24. Katona, K.; Bíró, Z.; Hahn, I.; Kertész, M.; Altbäcker, V. Competition between European hare and European rabbit in a lowland area, Hungary: A long-term ecological study in the period of rabbit extinction. *Folia Zool.* **2004**, *53*, 255–268.
25. Le Gall-Reculé, G.; Zwingelstein, F.; Fages, M.P.; Bertagnoli, S.; Gelfi, J.; Aubineau, J.; Roobrouk, A.; Botti, G.; Lavazza, A.; Marchandeu, S. Characterisation of a non-pathogenic and non-protective infectious rabbit lagovirus related to RHDV. *Virology* **2011**, *410*, 395–402. [CrossRef]
26. Le Gall-Reculé, G.; Lavazza, A.; Marchandeu, S.; Bertagnoli, S.; Zwingelstein, F.; Cavadini, P.; Martinelli, N.; Lombardi, G.; Guérin, J.L.; Lemaitre, E.; et al. Emergence of a new lagovirus related to Rabbit Haemorrhagic Disease Virus. *Vet. Res.* **2013**, *44*, 81. [CrossRef] [PubMed]
27. Le Pendu, J.; Abrantes, J.; Bertagnoli, S.; Guitton, J.S.; Le Gall-Reculé, G.; Lopes, A.M.; Marchandeau, S.; Alma, F.; Almeida, A.M.; Alves, P.C.; et al. Proposal for a new unified classification system and nomenclature of lagoviruses. *J. Gen. Virol.* **2017**, *98*, 1658–1666. [CrossRef]
28. Delibes-Mateos, M.; Ferreira, C.; Carro, F.; Escudero, M.A.; Gortázar, C. Ecosystem Effects of Variant Rabbit Hemorrhagic Disease Virus, Iberian Peninsula. *Emerg. Infect. Dis.* **2014**, *20*, 2166–2168. [CrossRef] [PubMed]
29. Carro, F.; Soriguer, R. Long-term patterns in Iberian hare population dynamics in a protected area (Doñana National Park) in southwestern Iberian Peninsula: Effects of weather conditions and plant cover. *Integr. Zool.* **2016**, *12*, 49–60. [CrossRef] [PubMed]
30. García-Bocanegra, I.; Camacho-Sillero, L.; Risalde, M.A.; Dalton, K.P.; Caballero-Gómez, J.; Agüero, M.; Gómez-Guillamón, F. First outbreak of myxomatosis in Iberian hares (*Lepus granatensis*). *Transbound. Emerg. Dis.* **2019**, *66*, 2204–2208. [CrossRef]
31. Grant, P.R. Experimental studies of competitive interaction in a two-species system. III. *Microtus* and *Peromyscus* species in enclosures. *J. Anim. Ecol.* **1971**, *40*, 323–350. [CrossRef]
32. Grant, P.R. Competition between species of small mammals. In *Populations of Small Mammals under Natural Conditions*; Snyder, D.P., Ed.; Pymatuning Lab. of Ecol., Univ. of Pittsburg: Linesville, PA, USA, 1978; Volume 5, pp. 38–51.
33. Dickman, C.R.; Woodside, D.P. A test of competition model with reference to three species of small mammals in south-eastern Australia. *Oecologia* **1983**, *60*, 127–134. [CrossRef]
34. Pettorelli, N.; Coulson, T.; Durante, S.M.; Gaillard, J.M. Predation, individual variability and vertebrate population dynmics. *Oecologia* **2011**, *167*, 305–314. [CrossRef]
35. Diamond, J. Overview: Laboratory experiments, field experiments, and natural experiments. In *Community Ecology, Diamond J, TJ Case*; Harper & Row: New York, NY, USA, 1986; pp. 3–22.
36. Valverde, J.A. *An Ecological Sketch of the Coto Doñana*; British Birds: London, UK, 1958; Volume 51, pp. 1–23.
37. Aguilar-Amat, J.; Montes del Olmo, C.; Ramírez Díaz, L.; Torres Martínez, A. *Mapa Ecológico del Parque Nacional de Doñana*; ICONA: Sevilla, Spain, 1979.
38. Green, A.; Bustamante, J.; Janss, G.F.E.; Fernández-Zamudio, R.; Díaz-Paniagua, C. Doñana Wetlands (Spain). In *The WetlandBook*; Finlayson, C., Milton, G., Prentice, R., Davidson, N., Eds.; Springer: Dordrecht, The Netherlands, 2016; pp. 1126–1136.
39. Allier, C.; González, F.; Ramírez, L. Mapa ecológico de la Reserva Biológica de Doñana. In *CSIC, División de Ciencias*; Estación Biológica de Doñana: Sevilla, Spain, 1974.
40. Rogers, P.M.; Myers, K. Animal distributions, landscape classification and wildlife management, Coto Doñana. *J. Appl. Ecol.* **1980**, *17*, 545–565. [CrossRef]
41. Neff, J.D. The pellet-group count technique for big game trend, census, and distribution: A review. *J. Wildl. Manag.* **1968**, *32*, 597–614. [CrossRef]
42. Lawrence, M.J.; Brown, R.W. *Mammals of Britain, Their Tracks, Trails and Signs*; Blandford Press: London, UK, 1974.
43. Bang, P.; Dahlström, P. *Huellas y Señales de los Animales de Europa*; Ediciones Omega, S.A.: Barcelona, Spain, 1975.
44. Overton, W.S. Estimating the numbers of animals in wildlife populations. In *Wildlife Management Techniques*, 3rd ed.; Giles, R.H., Jr., Ed.; The Wildlife Society: Washington, DC, USA, 1971; pp. 403–456.
45. Salzmann-Wandeler, I.; Salzmann, H.C. *Erste Erfahrungen Bei Feldhaszen Zählungen Mit Scheinwerfern*; Naturhistorisches Museum Bern: Bern, Switzerland, 1973; Volume 5, pp. 1–16.

46. Zar, J.H. *Biostatistical Analysis*, 5th ed.; Prentice Hall, Inc.: Englewood Cliffs, NJ, USA, 2010.
47. Barnes, R.F.W.; Tapper, S.C.; Williams, J. Use of pastures by brown hares. *J. Appl. Ecol.* **1983**, *20*, 179–185. [CrossRef]
48. Homolka, M. Daily activity pattern of the European hare (*Lepus europaeus*). *Folia Zool.* **1986**, *35*, 33–42.
49. Barnes, R.F.W.; Tapper, S.C. A method for counting hares by spotlight. *J. Zool.* **1985**, *206*, 273–276. [CrossRef]
50. Frylestam, B. Estimating by spotlight the population density of the European hare. *Acta Theriol.* **1981**, *26*, 419–427. [CrossRef]
51. Frylestam, B. European hare. In *CRC Handbook of Census Methods for Terrestrial Vertebrates*; Davis, E.D., Ed.; CRC Press: Boca Raton, FL, USA, 1982; pp. 142–144.
52. Wywialowski, A.; Stottart, L.C. Estimation of jack rabbit density methodology makes a difference. *J. Wildl. Manag.* **1988**, *52*, 57–59. [CrossRef]
53. Rogers, P.M.; Myers, K. Ecology of the European wild rabbit, *Oryctolagus cuniculus* (L.) in Mediterranean habitats. I. Distribution in the landscape of the Coto Doñana, S. Spain. *J. Appl. Ecol.* **1979**, *16*, 691–703. [CrossRef]
54. Sale, P.F. Overlap in resource use, and interspecific competition. *Oecologia* **1974**, *17*, 245–256. [CrossRef]
55. Rosenweig, M.L. A theory of habitat selection. *Ecology* **1981**, *62*, 327–335. [CrossRef]
56. Vidus-Rosin, A.; Lizier, L.; Meriggi, A.; Serrano-Pérez, L. Habitat selection and segregation by two sympatric lagomorphs: The case of the European hare (*Lepus europaeus*) and Eastern cottontails (*Sylvilagus floridanus*) in northern Italy. *Acta Theriol.* **2012**, *57*, 295–304. [CrossRef]
57. Rowland, M.M.; White, G.C.; Karlen, E.M. Use of pellet group plots to measure trends in deer and elk populations. *Wildl. Soc. Bull.* **1984**, *12*, 147–155.
58. Alvarez, F.; Braza, F.; de Azcárate, T.; Aguilera, E.; Martín, R. Circadian activity rhythms in a vertebrate community of Doñana. In *Actas XV Congr. Nac. de Fauna Cinegética y Silvestre*; Trujillo: Cáceres, Spain, 1983; pp. 379–387.
59. Villafuerte, R.; Kufner, M.B.; Delibes, M.; Moreno, S. Environmental factors influencing the seasonal daily activity of the European rabbit (*Oryctolagus cuniculus*) in a Mediterranean area. *Mammalia* **1993**, *57*, 341–347. [CrossRef]
60. Carothers, J.H.; Jaksic, F.M. Time as niche difference: The role of interference competition. *Oikos* **1984**, *42*, 403–406. [CrossRef]
61. Soriguer, R.C. El conejo: Papel ecológico y estrategia de vida en los ecosistemas mediterráneos. In *XV Congreso Internacional de Fauna Cinegética y Silvestre*; Trujillo: Cáceres, Spain, 1981; pp. 517–542.
62. Soriguer, R.C. Consideraciones sobre el efecto de los conejos y los grandes herbívoros en los pastizales de la Vera de Doñana. *Doñana Acta Vert.* **1983**, *10*, 155–168.
63. Pimm, S.L.; Pimm, J.W. Resource use, competition and resource availability in Hawaian honeycreepers. *Ecology* **1982**, *63*, 1468–1480. [CrossRef]
64. Paine, R.T. Food web complexity and species diversity. *Am. Nat.* **1966**, *100*, 65–75. [CrossRef]
65. Delibes, M. Ecología alimenticia del águila imperial ibérica (*Aquila adalberti*) en el Coto Doñana durante la crianza de los pollos. Doñana. *Acta Vert.* **1978**, *5*, 35–60.
66. Delibes, M. Feeding ecology of the Spanish Lynx in the Coto Doñana. *Acta Theriol.* **1980**, *25*, 309–394. [CrossRef]
67. Carro, F.; Soriguer, R.C.; Beltrán, J.F.; Andreu, A.C. Heavy flooding effects on home range and habitat selection of free-ranging Iberian hares (*Lepus granatensis*) in Doñana National Park (SW Spain). *Acta Theriol.* **2011**, *53*, 375–382. [CrossRef]
68. Villafuerte, R. Riesgo de Predación y Estrategias Defensivas Del Conejo, *Oryctolagus cuniculus*, en el Parque. Nacional de Doñana. Ph.D. Thesis, Universidad de Córdoba, Córdoba, Spain, 1994.
69. Moreno, S.; Villafuerte, R.; Delibes, M. Cover is safe during the the day but dangerous at night: The use of vegetation by the European wild rabbit. *Can. J. Zool.* **1996**, *74*, 1656–1660. [CrossRef]
70. Carro, F. Historia natural de la liebre ibérica (*Lepus granatensis* Rosenhauer 1856) en el Parque Nacional de Doñana. Ph.D. Thesis, University of Santiago de Compostela, Santiago de Compostela, Spain, 2005.
71. Negro, J.J.; Ferrer, M. The near extinction of two large European predators: Super Specialists pay a price. *Conserv. Biol.* **2004**, *18*, 344–349.
72. Palomares, F.; Delibes, M.; Ferreras, P.; Gaona, P. Mesopredator release and prey abundance: A reply to Litvaitis and Villafuerte. *Cons. Biol.* **1996**, *10*, 678–679. [CrossRef]
73. Tilman, D. Resource competition and community structure. In *Monographs Population Biology, Volume 17*; Princeton University Press: New York, NY, USA, 1982.
74. Taylor, R.J. *Predation*; Chapman and Hall: London, UK, 1984.
75. Bakker, E.S.; Reiffers, R.C.; Olff, H.; Gleichman, J.M. Experimental Manipulation of Predation Risk and Food Quality: Effect on Grazing Behaviour in a Central-Place Foraging Herbivore. *Oecologia* **2005**, *146*, 157–167. [CrossRef]
76. Monterroso, P.; Alves, P.C.; Ferreras, P. Catch me if you can: Diel activity patterns of mammalian prey and predators. *Ethology* **2013**, *119*, 1044–1056. [CrossRef]
77. Beltrán, J.F.; Delibes, M. Environmental determinants of circadian activity of free ranging Iberian lynxes. *J. Mammal.* **1994**, *75*, 382–393. [CrossRef]

land

MDPI

Article

AbundanceR: A Novel Method for Estimating Wildlife Abundance Based on Distance Sampling and Species Distribution Models

Xinhai Li [1,2,*,†], Ning Li [3,†], Baidu Li [4], Yuehua Sun [1] and Erhu Gao [5]

1 Key Laboratory of Animal Ecology and Conservation Biology, Institute of Zoology, Chinese Academy of Sciences, Beijing 100101, China; sunyh@ioz.ac.cn
2 College of Life Sciences, University of Chinese Academy of Sciences, Beijing 100049, China
3 Institute of Applied Ecology, Nanjing Xiaozhuang University, Nanjing 211171, China; lining@njxzc.edu.cn
4 Lassonde School of Engineering, York University, 4700 Keele Street, Toronto, ON M3J 1P3, Canada; baidu@my.yorku.ca
5 Academy of Inventory and Planning, National Forestry and Grassland Administration, Beijing 100714, China; gaoerhu2022@126.com
* Correspondence: lixh@ioz.ac.cn; Tel.: +86-10-64807898
† These authors contributed equally to this work.

Abstract: Appropriate field survey methods and robust modeling approaches play an important role in wildlife protection and habitat management because reliable information on wildlife distribution and abundance is important for conservation planning and actions. However, accurately estimating animal abundance is challenging in most species, as usually only a small proportion of the population can be detected during surveys. Species distribution models can predict the habitat suitability index, which differs from species abundance. We designed a method to adjust the results from species distribution models to achieve better accuracy for abundance estimation. This method comprises four steps: (1) conducting distance sampling, recording species occurrences, and surveying routes; (2) performing species distribution modeling using occurrence records and predicting animal abundance in each quadrat in the study area; (3) comparing the difference between field survey results and predicted abundance in quadrats along survey routes, adjusting model prediction, and summing up to obtain total abundance in the study area; (4) calculating uncertainty from three sources, i.e., distance sampling (using detection rate), species distribution models (using R squared), and differences between the field survey and model prediction [using the standard deviation of the ratio (observation/prediction) at different zones]. We developed an R package called abundanceR to estimate wildlife abundance and provided data for the Tibetan wild ass (*Equus kiang*) based on field surveys at the Three-River-Source National Park, as well as 29 layers of environmental variables covering the terrestrial areas of the planet. Our method can provide accurate estimation of abundance for animals inhabiting open areas that can be easily observed during distance sampling, and whose spatial heterogeneity of animal density within the study area can be accurately predicted using species distribution models.

Keywords: abundance; distance sampling; population density; R package; species distribution models; wildlife survey; uncertainty

Citation: Li, X.; Li, N.; Li, B.; Sun, Y.; Gao, E. AbundanceR: A Novel Method for Estimating Wildlife Abundance Based on Distance Sampling and Species Distribution Models. *Land* **2022**, *11*, 660. https://doi.org/10.3390/land11050660

Academic Editors: Juan F. Beltrán, Pedro Abellán, John Litvaitis and Francesco Maria Angelici

Received: 10 February 2022
Accepted: 27 April 2022
Published: 29 April 2022

1. Introduction

Species abundance is fundamental information for ecological research, and various methods are used for its estimation [1–3]. Direct counting, including spotlight, track, and roadkill counts, can provide initial information on the abundance of several species, such as water birds and deer [4,5]. Catch-per-unit effort is often used for regularly captured species, such as commercial fish [6,7]. The mark-recapture technique has been widely used and it is suitable for populations in a small region when target species can be easily captured [8,9].

Camera-based models are playing an important role in abundance estimation in recent years [10,11]. The spatially explicit capture-recapture technique uses the camera trap data of species with identical characteristics and can provide good abundance estimation [12,13]. If individuals of the target species are not distinguishable, the random encounter model [14], more advanced time-to-event estimates and space-to-event estimates [10], and the movement-based method [15] are appropriate for abundance estimation. Mark-recapture and camera traps are appropriate for small regions.

At a large spatial scale, distance sampling [16] or regular line transects are suitable for species that can be directly observed, yet the results cannot be expanded to unsurveyed areas [16,17] because species are not evenly distributed [18]. The advantage of distance sampling is that it estimates the detection function, quantifying the relationship between the probability of detection and the animal-to-observer distance so that it provides a measure of survey uncertainty. The N-mixture model can quantify both the detection rate (using repeated surveys) and the contribution of environmental variables so that it can be extended to unsurveyed areas [19], yet it is not capable of handling a large number of environmental variables, which limits its ability for abundance estimation.

Species distribution models (SDMs) have been widely used to estimate species distribution in unsurveyed areas [20], as some SDMs can quantify the association of species occurrences and a large number of environmental variables [21,22] so as to predict animal density at unsurveyed areas. Boyce and McDonald [23] suggested that animal abundance could be estimated by summing the probabilities of presence calculated by SDMs, but such estimations are inaccurate unless the population is at carrying capacity or in an ideal free distribution [24]. In most cases, species distribution is constrained by environmental variables, which can be estimated by SDMs, yet animal populations are seldom at equilibrium nor have an ideal-free distribution [24]. Currently, there is no reliable method for estimating the abundance of animals occurring in a large region where only a small part is surveyed.

To estimate animal abundance in large areas, researchers usually applied specific methods that were only suitable for their target species. For example, Stauffer et al. used home range size and mean group size for the abundance estimation of wolves (*Canis lupus*), which was a scale-up process from occupancy to abundance [25]. Santos et al. used nesting beach monitoring data such as nest counts and clutch frequency to estimate the abundance of marine turtles [26]. Teton et al. used natural markings (no ear-tags or neck-bands) to identify individuals for mark-resight population estimation of invasive wild pigs (*Sus scrofa*) [27]. Shertzer et al. estimated abundance of an open population, gray triggerfish (*Balistes capriscus*), by pairing two models, a Markovian mode using data from telemetry tags for movement rate, and a Lincoln-Petersen abundance estimator model modified to account for mortality and movement [28]. Howard et al. used age-at-harvest data and auxiliary information such as estimated survival rates, harvest probabilities, and hunter effort to estimate mountain lion (*Puma concolor*) abundance in Arizona, USA [29].

In this study, we developed a more general method to estimate species abundance in a large region based on distance sampling and SDMs. We used SDMs to quantify the heterogeneity of spatial distribution to estimate animal abundance in unsurveyed areas and compare the SDM-predicted abundance on survey routes with field survey results with the aim of adjusting model prediction. The uncertainty of distance sampling, species distribution modeling, and spatial heterogeneity of the observation-prediction ratio was considered for the abundance estimation. All algorithms were coded in R [30] and can be installed from the GitHub repository.

2. Methods

The data used to estimate animal abundance were obtained from distance sampling. Both species occurrence data and survey routes were required. We provided survey data of the Tibetan wild ass (*Equus kiang*) in the Three-River-Source National Park as an example. Random forest was selected as the SDM for predicting the number of individuals in each quadrat in the entire study area. We compared the SDM prediction with field survey results.

By adjusting the prediction, we obtained a better estimation of animal abundance. To install the package and load it, the following code is used:

```
library(devtools) # or using library(remotes)
install_github("Xinhai-Li/abundanceR")
library(abundanceR)
```

2.1. Data Requirements

We used an example dataset to show the data needed to estimate animal abundance using the package abundanceR.

The example data used were from our surveys. We applied a distance sampling protocol [16] to conduct field surveys in the Three-River-Source National Park, which is the central part of the Qinghai-Tibet Plateau, comprising the headwaters of the Yangtze, Yellow, and Lancang rivers, with an area of 960 × 560 km^2 (92–102° E, 32–37° N) (Figure 1) and an average elevation of approximately 4500 m. We recorded 159 Tibetan wild ass occurrences on the survey routes. The survey results were recorded as GPX or KLM files, including information on species names, counts (number of individuals at one site), vertical distance to observers, latitude and longitude of observers, and date and time (Table 1). Most of the survey routes were on unpaved roads, vehicle speed was 15–30 km/h, and we kept recording while driving.

```
data(kiang) # load the data of distance sampling for the Tibetan wild ass (kiang).
head(kiang) # show the first six rows (Table 1).
```

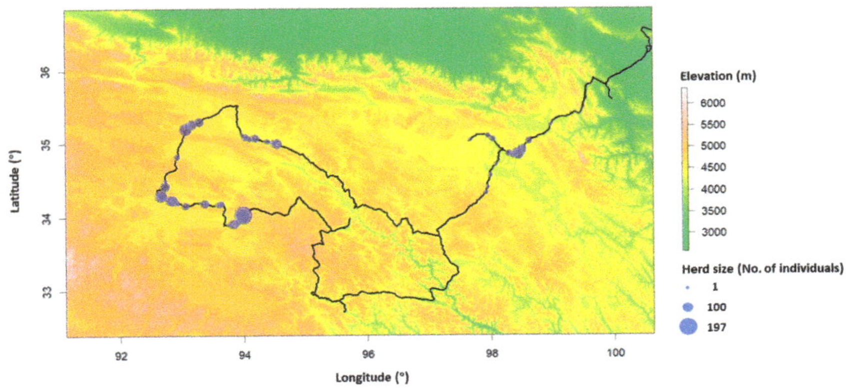

Figure 1. The study area in the Three-River-Source National Park. The 159 blue points indicate Tibetan wild ass occurrences of 1039 individuals, and point size indicates species group size ranging from 1 to 197. The black line is the survey route. The background is elevation.

Table 1. The survey data (the first six rows) for the Tibetan wild ass (*Equus kiang*) following the distance sampling protocol *.

Species	Size	Distance	Side	Lat	Lon	Elev	Date	Time
kiang	9	130	e	34.83078	98.37612	4217	17 July 2017	13:28:43
kiang	32	150	e	34.84620	98.44160	4223	17 July 2017	13:22:53
kiang	7	600	e	34.85080	98.29750	4225	17 July 2017	13:37:12
kiang	8	350	e	34.85908	98.45139	4232	17 July 2017	13:18:01
kiang	1	210	e	34.87584	98.47288	4236	17 July 2017	13:15:24
kiang	3	200	e	34.89577	98.49666	4244	17 July 2017	13:12:01

* In the table, "size" is the group size, i.e., number of individuals at the site (occurrence); "distance" is the distance between the observer and animals; "Side" indicates the direction of the animals from the road; "Elev" is the elevation of the observer (we assumed the animals stay at the same elevation). The names "species", "size", "distance", "Lat" and "Lon" cannot be changed due to case sensitivity. The variables "Side", "Elev", "Date" and "Time" are not needed in the abundance estimation.

2.2. Detection Functions in Distance Sampling

We used the R package Distance [31] to estimate the detection function, which quantifies the relationship between the probability of detection and the distance between animals and observers. In the package Distance, three key functions are provided: "hn" is a half-normal function (default), "hr" gives a hazard-rate function, and "unif" provides a uniform function. The package also provides three adjustment terms to tune the detection functions: "cos" gives a cosine term (default), "herm" gives a Hermite polynomial term, and "poly" provides a simple polynomial term. The combination of the three key functions and three adjustment terms is nine, and in abundanceR, we provided a function for all nine combinations and selected the best detection function with the lowest AIC value (Table S1). Based on the selected detection function, we designed to output the average detection rate across the entire range of distances between observers and animals. The average detection rate is an index of survey uncertainty, as the uncertainty decreases with an increase in the detection rate.

The maximum distance that the Tibetan wild ass can be detected is 1500 m, yet we set the truncation distance to 500 m, i.e., the animals outside of that range were ignored. As such, the detection range on each side was 500 m, and the width of the line transect was 1 km; thus, the field observations were comparable with the model prediction at the same spatial scale, i.e., 1-km^2 quadrats. The R code for selecting the best detection functions and adjustment terms and estimating the detection probability (Figure 2) is as follows:

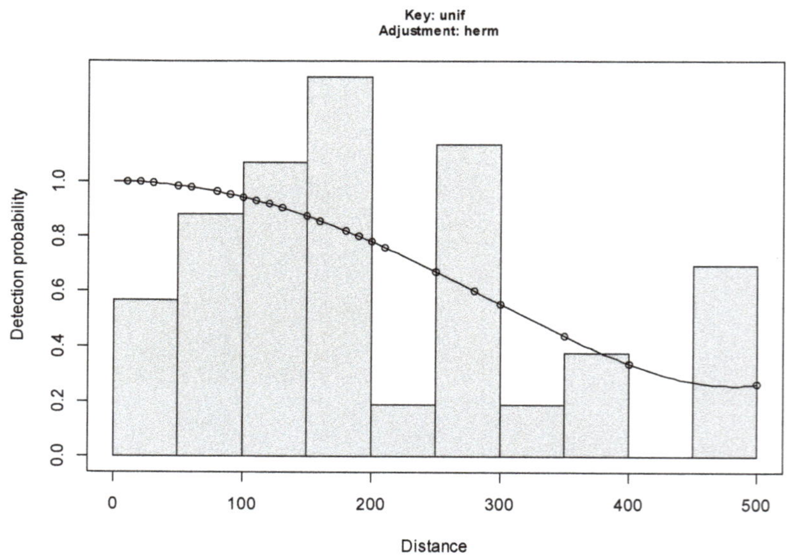

Figure 2. The detection function of distance sampling for the Tibetan wild ass in the Three-River Source National Park. The unit of distance is the meter.

```
library(abundanceR) # load the package
# calculate the AIC values for the 9 combinations of detection functions and adjustment terms
AICs = distanceSampling(kiang[kiang$distance <= 500,]) # truncation range is 500 m
AICs = AICs[!is.na(AICs$AIC),] # remove null values in case no AIC value is calculated
# The selected detection function
ds.kiang <- ds(kiang, key = AICs$Key [1], adjustment = AICs$Adjustment [1],
convert.units = 0.001, truncation = 500)
SM = summary(ds.kiang) # the results of distance sampling
Average.p = SM$ds$average.p # average detection rate across the distance range, 0.647
```

```
survey.uncertainty = 1 − Average.p # survey uncertainty
# Show the detection function (Figure 2)
plot(ds.kiang, main = paste("Key:", AICs$Key [1], "\n", "Adjustment:", AICs$Adjustment [1],
    sep = " "))
```

2.3. Environmental Variables for Species Distribution Models

We used 29 environmental variables and animal count data to develop species distribution models. The environmental variables included 19 climate variables [32], elevation [33], human footprint index [34], land cover [35], wetlands [36], as well as solar radiation, wind speed, and water vapor pressure for January and July [32] (Table S2).

All 29 variables were raster layers covering the study area (Figure 1) with a 1-km^2 resolution. We compiled data in R. grd format, which is a stack of 29 raster layers. The file size was 69 MB, covering an area of 537,600 km^2. The compressed file (BioClim.zip, 23 MB) is too large for a package, so we uploaded it to the author's GitHub repository: https://github.com/Xinhai-Li/abundanceR. To help users access environmental variables for their own study areas, we provided 29 raster layers for all terrestrial regions in the world (Figure S1). The file size was 105 GB, and the compressed size was 8 GB. The file (var29.zip) can be downloaded from the Baidu Cloud Disk™ for users in Mainland China at https://pan.baidu.com/s/1noU8A7WcsuYx0MSiQq6CeQ (access code: 1234); and can also be downloaded from the Google Drive at (https://drive.google.com/drive/folders/1bNh4SdikmjrOkgqE5VOVo86SozD2hvmT?usp=sharing). Users can use the cropLayers function to obtain environmental variables for any terrestrial area on Earth as follows:

```
library(raster)
BioClim <- brick('var29.grd') # load the 29 environmental variables
data(kiang); head(kiang) # load the species occurrences
# crop the global data to fit users' study area
BioClim = cropLayers(kiang, buffer = 0.2, Envlayers = BioClim)
# the argument, buffer = 0.2, defines the extents of environmental variables are larger than
# that of species occurrences by 0.2 degree at each side (north, south, east, and west) (Figure 3).
```

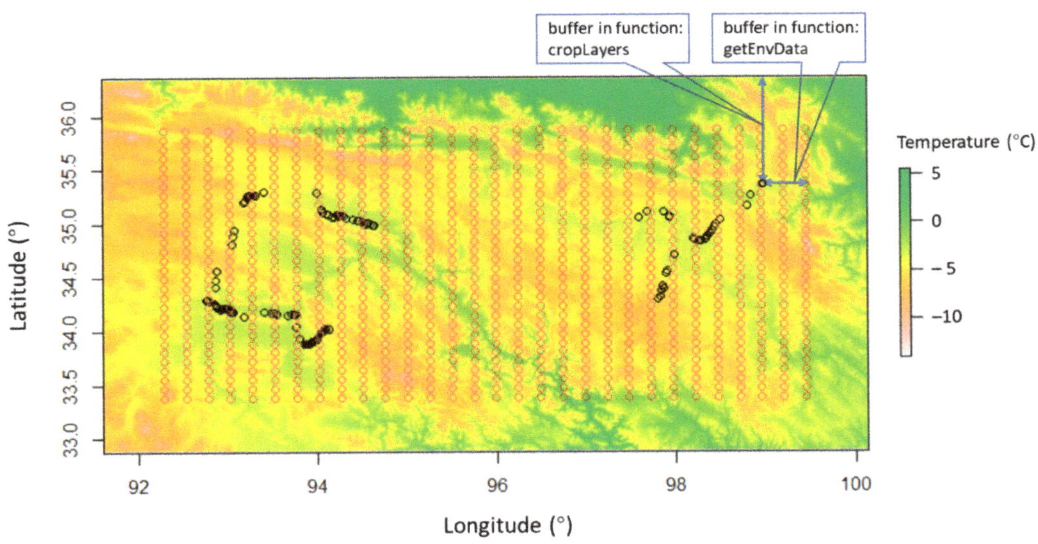

Figure 3. The study area (using annual mean temperature as the background), Tibetan wild ass presence data (black circles), and the pseudo-absence points (red circles). The argument buffer is demonstrated for the cropLayers and getEnvData functions.

2.4. Predicting Species Abundance Using Species Distribution Models

Among various algorithms, such as the generalized linear model, generalized additive model, support vector machines, random forest, Maxent, multiple adaptive regression spline, and artificial neural networks, we selected random forests because of their high performance [21,37–39].

Random forests require absence data for species distribution modeling. Therefore, we used the getEnvData function to generate pseudo-absence data in the range of occurrences (Figure 3). The dependence variable is the count of the Tibetan wild ass along the survey routes and evenly distributed pseudo-absence data (count 0) in the study area (Figure 3).

```
Data = getEnvData(kiang, buffer = 0.5, absence = 30, Envlayers = BioClim)
# the argument, buffer = 0.5, defines the extents of pseudo-absence points are larger than
# that of species occurrences by 0.5 degree at each side (Figure 3).
# the argument, absence = 30, defines the number of pseudo-absence points is 30*30 = 900
plot(BioClim[[1]], xlab = "Longitude", ylab = "Latitude") # annual mean temperature
points(Data$Lon[Data$Name == "absent"], Data$Lat[Data$Name == "absent"], col = 'red')
points(Data$Lon[Data$Name == "kiang"], Data$Lat[Data$Name == "kiang"])
```

We used the na.roughfix function provided by the randomForest package [40] to replace null values with mean values, because in certain places some variables such as human footprint index has null values while other variables have valid values. The code is: library(randomForest)

```
no.col = ncol(Data) # 33
Dat.fill <- na.roughfix(Data[,2:(no.col-4)]) # no.col-4: use 27 variables. no.col-2: using 29 variables #
    including landcover and wetland
#
# Build the species distribution model
RF <- randomForest(Dat.fill[, 2: ncol(Dat.fill)], Dat.fill[, 1], ntree = 1000,
importance = TRUE, na.action = na.roughfix)
RF # shows the proportion of variance of species count explained by environmental variables.
model.uncertainty = 1-max(RF$rsq) # model uncertainty #0.254
```

The species count can be predicted for every quadrat of the environmental variables, based on the association between species occurrence and environmental variables. Among the 29 environmental variables, land cover and wetland caused unexpected results; therefore, we removed the two variables from the model.

```
# using the model (RF) and environmental data (BioClim) for prediction
pred = popSize(BioClim[[1:27]], RF) # use the first 27 layers
# Show the predicted animal density
plot(pred)
# Change the color
plot(log(1 + log(1 + pred)), xlab = "Longitude", ylab = "Latitude", main = ",
col = colorRampPalette(c("grey90", "green", "yellow", "red"))(12))
# Add species occurrences
points(kiang$Lon, kiang$Lat, pch = 16, cex = log(kiang$size)/2, col = adjustcolor("red", 0.5))
```

The predicted abundance (Figure 4) was log-log transformed to obtain a better color effect. The number of individuals (kiang$Count) varied from 1 to 197; therefore, we used log transformation to compress the difference.

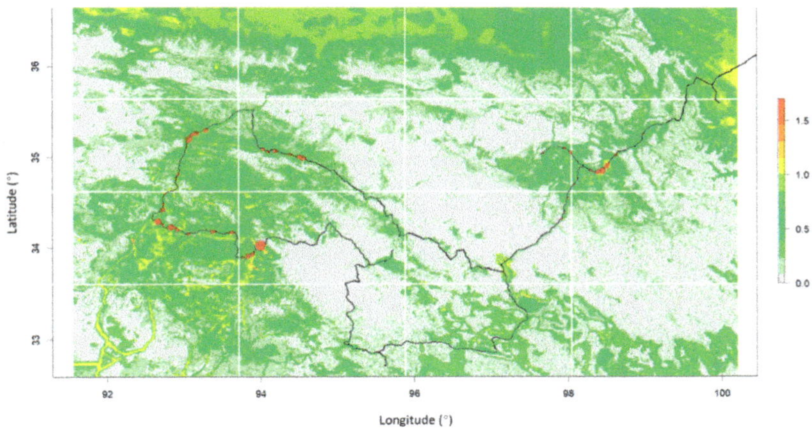

Figure 4. The predicted abundance (log-log transformed) of the Tibetan wild ass in the Three-River-Source National Park. The red circles indicate 159 Tibetan wild ass occurrences of 1039 individuals, and circle size indicates species group size ranging from 1 to 197. The black line is the survey route. The study area was divided into 16 zones (separated by white lines), and the observation-prediction ratios were compared among the 16 zones to estimate adjustment uncertainty.

2.5. Adjusting Model Prediction

The animal abundance predicted by species distribution models is usually biased [24]. Therefore, we adjusted for model bias using survey results. We loaded the survey route from a GIS shapefile and selected the track points from the route. The track points were 1 km apart (Figure 5). The predicted animal abundance in each quadrat was extracted using the trackPoints function. By comparing the predicted animal abundance along the survey route and the observed number of individuals during the survey, we obtained a ratio of prediction bias, which can adjust the model prediction.

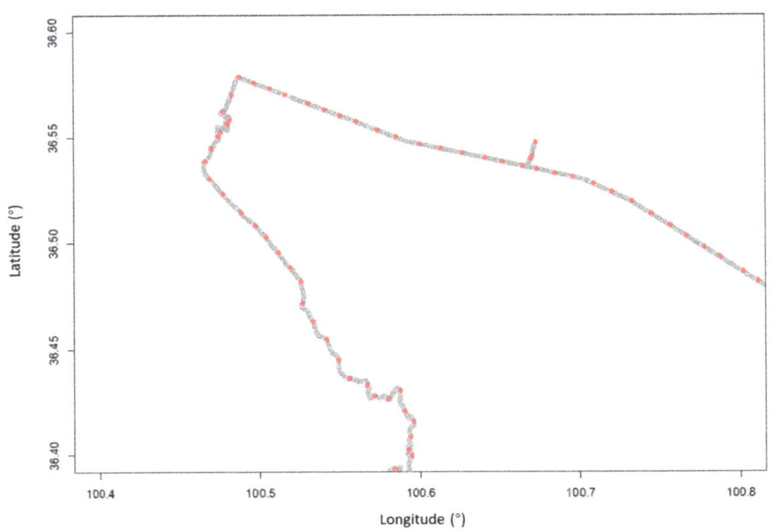

Figure 5. One segment of the survey route for the Tibetan wild ass (grey lines) and the selected track points (red points) for extracting predicted animal abundance using the trackPoints function.

The observed animals were also biased because we missed some individuals during the survey. Therefore, we used the average detection rate to adjust for this bias.

```
library(sp)
data(shape); plot(shape) # survey route
tracks = trackPoints(shape) # derive track points with 1 km interval from the survey route
```

The processes of obtaining the original animal abundance estimation, comparing the prediction and observation and making adjustments, are coded below:

```
# Total number of individuals for the raster
pop_ori = cellStats(pred, stat = 'sum', na.rm = TRUE) # 499,705.8
pre <- extract(pred, tracks) # predicted animal abundance on survey routes
pop_pre = sum(pre, na.rm = T) # The predicted number of individuals on the route, 4585
kiang = kiang[kiang$distance <= 500, ] # keep occ within 500 m to match quadrat of 1 km
pop_obs = sum(kiang$size) # observation, 449
pop_obs_adj = pop_obs/detection # distance sampling adjustment, 694
adjust = pop_pre/pop_obs_adj # SDM adjustment, 6.6
# Adjusted animal abundance
pop_est = cellStats(pred, stat = 'sum', na.rm = TRUE)/adjust # 75598
```

We designed a function estPopSize to perform all the above calculations as follows:

```
EST = estPopSize(pred, tracks, kiang, Average.p) # 75598
```

The argument 'pred' is a raster layer of the predicted number of individuals in each quadrat. The argument 'tracks' are the track points on the survey route at 1-km intervals. The argument 'kiang' is a data frame with occurrences of the species recorded during distance sampling. The argument 'Average.p' is the mean detection rate of the distance sampling. This function provides four values: the original model prediction for animal abundance, the predicted animal abundance on the survey route, the animal abundance along the survey route based on field surveys, and the adjusted animal abundance in the study area.

2.6. Estimation Uncertainty

To evaluate the uncertainty of the predicted animal abundance, we considered three sources: distance sampling, species distribution modeling, and adjustment based on the ratio of observation and prediction.

Some individuals were missed during distance sampling. Such a situation always occurs, which results in survey uncertainty. Therefore, we used the proportion of missed individuals, which is 1-average-detection-rate, as the index of survey uncertainty. The R code is:

```
survey.uncertainty = 1 − Average.p # survey uncertainty, based on distance sampling
```

Species distribution models depend on the relationship between animal occurrence and environmental variables, i.e., the extent to which environmental variables constrain animal distribution. Accordingly, environmental variables strongly influence habitat specialists and weakly influence habitat generalists. We used the unexplained variance of the dependent variable, the animal counts at their occurrences, as the model uncertainty, which is $1 - R^2$. The ratio of predictions to observations varied in different regions. Therefore, we used the standard deviation of the ratio among regions as the adjustment uncertainty. The R code is:

```
model.uncertainty = 1 − max(RF$rsq) # model uncertainty, based on random forest
```

To calculate the adjustment uncertainty, we divided the survey routes into a number of segments. In the following code, we used argument grid = 4, where the study area was divided into 4 × 4 grids (Figure 4). We calculated the observation-prediction ratio within each grid and obtained the standard deviation of the ratio. The R code is:

adjust.uncertainty = spatialMatch(kiang, tracks, pred, grid = 4)

We assumed that the three indices of uncertainty are independent. Therefore, we multiplied them together as the overall uncertainty:

CI = estimated_abundance * rnorm(1, 1, survey.uncertainty) * rnorm(1, 1, model.uncertainty) * rnorm(1, 1, adjust.uncertainty)

We generated three normally distributed random numbers 100,000 times and obtained the confidence intervals from the 100,000 values.

However, the movement of the animal population would cause errors in abundance estimation [41,42]. Therefore, we compared the Tibetan wild ass distribution in summer over different years and checked the stability of its spatial distribution.

3. Results

We recorded 159 occurrences of the Tibetan wild ass surveyed in 2017 as example data. The mean group size was 6.5 (the minimum value is 1, and the maximum value is 197), and its standard deviation was 17.8. The total number of observed individuals was 1039, including 103 observations of 449 individuals within the 500 m range (Figure 1). The average detection rate of the distance sampling within 500 m was 64.7% (Figure 2). The detection function can be best-fitted using a uniform function with herm adjustment (a Hermite polynomial term) (Table 1). The actual number of individuals along the survey route was 449/64.7% = 693.7.

Based on the 159 occurrences and 27 environmental variables, the random forest model predicted that Tibetan wild ass abundance was 499,705 individuals (Figure 4). Along the survey route, the predicted abundance was 4585, which was larger by 4585/693.7 = 6.61 times than the number of observed individuals. Therefore, after adjustment, we estimated Tibetan wild ass abundance in the study area to be 75,598. We used the observed animal abundance on the survey route to predict animal abundance in the entire study area. However, most animals were not around the survey route (Figure 6).

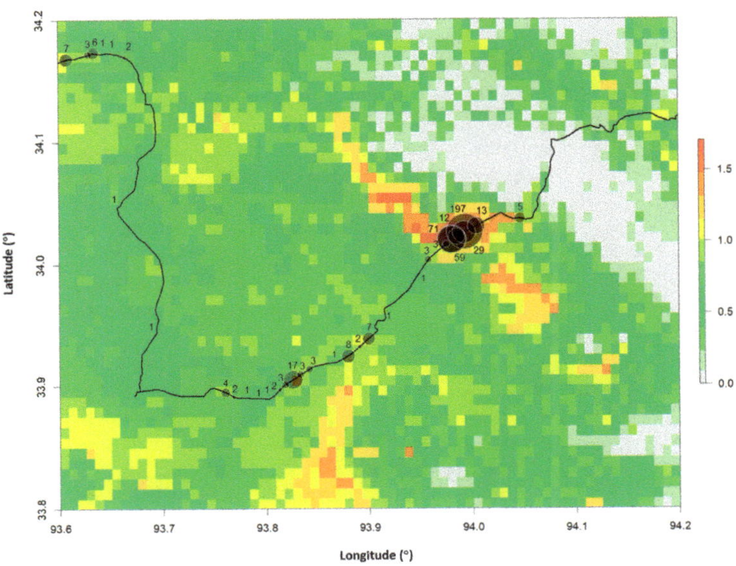

Figure 6. The predicted abundance (log-log transformed) of the Tibetan wild ass at Suojia Township, Zhiduo County in the Three-River-Source National Park. The resolution of the abundance prediction map is 1 km^2. The black circles represent Tibetan wild ass occurrence, and circle size indicates species group size ranging from 1 to 197. The black line is the survey route.

The average detection rate of distance sampling was 64.7%. The R^2 of the species distribution model (random forest) was 0.746. The standard deviation of the observation-prediction ratio among the different regions was 0.647. When the survey, model, and adjustment uncertainties were considered, the 95% confidence interval was 13,339–178,215, which is a very wide range. However, when we only consider the uncertainty of the field survey, the species distribution model, or the adjustment, the ranges of the confidence intervals were narrower (Table 2).

Table 2. The 95% confidence intervals (CIs) of the estimated animal abundance considering different error sources (field survey, species distribution model, and adjustment).

Source of Errors	Lower 95% CI	Upper 95% CI	R Code
survey, model, adjustment	13,339	178,215	CI(EST[4], survey.uncertainty, model.uncertainty, error.adjust)
model, adjustment	22,456	144,376	CI(EST[4], 0.001, model.uncertainty, adjust.uncertainty)
survey, adjustment	15,452	158,504	CI(EST[4], survey.uncertainty, 0.001, adjust.uncertainty)
survey, model	18,283	149,844	CI(EST[4], survey.uncertainty, model.uncertainty, 0.001)
adjustment	28,369	123,030	CI(EST[4], 0.001, 0.001, adjust.uncertainty)
model	37,986	113,024	CI(EST[4], 0.001, model.uncertainty, 0.001)
survey	23,622	127,671	CI(EST[4], survey.uncertainty, 0.001, 0.001)

We compared the distribution of the Tibetan wild ass in summers over different years based on our surveys, and found its distribution was very stable (Figure S2). Tibetan wild ass populations stay in their territory throughout the year, as they never migrate [43]. In contrast, the Tibetan gazelle (*Procapra picticaudata*) is more mobile than the Tibetan wild ass (Figure S2). The occurrences of the Tibetan fox significantly differed over the four days (Figure S2), and the reason is that it is difficult to detect the fox during distance sampling.

4. Discussion

The main features of our method are: (1). using distance sampling to quantify survey uncertainty; (2). applying a SDM (using the random forest algorithm) to predict the animal abundance in the whole study area including unsurveyed regions; (3). adjusting the predicted abundance based on survey results. As such, we can provide a better estimation than traditional methods. Although the N-mixture model can quantify survey uncertainty and the contribution of environmental variables, it is not designed for handling numerous variables to provide a reliable prediction map. Random forest can handle tens of variables (theoretically thousands of variable [38]) and it is convenient for spatial prediction using raster layers. To help researchers to use our method, we produced the abundanceR package to complete the estimation.

We also provided the data of 29 environmental variables covering global terrestrial areas to facilitate the analysis. The data are all publicly accessible, and we standardized their resolution and extension and created a stack file with 29 layers in R. We provided a function to crop the data based on the occurrence of the target animal, and we will upgrade the package to make the analysis more flexible. There are a number of methods using occurrence data to estimate animal abundance. If the home-range and mean group sizes are well established, animal occupancy can be directly scaled up to abundance [25], or a negative binomial distribution method can be used to obtain the global abundance of the target animal [44]. If the contribution of environmental variables is simple, with only linear, quadratic, and interaction effects, animal abundance can be estimated using linear models such as a binomial mixture model [45]. When the animal-environment association is complex with nonlinear, high-order, and interaction effects in a high-dimensional dataset, powerful machine learning algorithms such as random forest are preferable.

To estimate the confidence intervals of the estimated animal abundance, we multiplied the uncertainty of the field survey, the species distribution model, and the adjustment based on the observation-prediction ratio. We considered the uncertainty of field survey because the survey results from direct observation are usually biased [42]. We took into account the uncertainty of adjustment because the variance of detection among sites is often large [46], and the observation-prediction ratio varies in different regions. In the case of the Tibetan wild ass in the Three-River-Source National Park, the range of the confidence intervals is large because the species distribution model only explained half of the variance in animal count at the occurrences [43], and the observation-prediction ratio was different in different regions. If we have the distribution data of wolves and livestock, which strongly influence Tibetan wild ass distribution, the performance of the species distribution model would be substantially improved, and the observation-prediction ratio would be close to one across different regions.

5. Recommendations

Our model requires the target species to meet three conditions: (1) the animal lives in an open environment (e.g., grassland, wetland, etc.) where distance sampling is applicable so that the detection rate can be quantified using a detection function, and the uncertainty of the survey can be measured; (2) the species is large and visible (e.g., ungulates) and not elusive (e.g., carnivores); and (3) the target animal is a habitat specialist, so that its distribution is constrained by environmental variables and a properly fitted species distribution model can accurately predict animal abundance in unsurveyed areas. If users can adapt distance sampling for mountain areas [47], such as adjusting study areas from real mountain surface areas to vertical projected areas, they can use our model for mountain species such as the blue sheep (*Pseudois nayaur*) and argali (*Ovis ammon*).

To further minimize errors in our method for wildlife abundance estimation, users can: 1 carry out systematic field surveys covering a good gradient of the animal's habitat; 2 obtain key environmental variables, such as the distribution of predators, preys, and competing species, and various disturbing factors; 3 select animals with stable spatial distributions (avoiding the migration stage) so that survey data from different periods can be combined.

Supplementary Materials: The following supporting information can be downloaded at: https://www.mdpi.com/article/10.3390/land11050660/s1. Figure S1. The 29 environmental variables for species distribution modelling. The variables include the 19 climate variables [32], elevation [33], human footprint index [34], and solar radiation, wind speed, and water vapor pressure for January and July, respectively, as well as land cover [35] and wetland [36]. All the 29 layers have the spatial resolution of 1 km^2. Figure S2. The occurrences and group sizes of the Tibetan wild ass, Tibetan gazelle, and Tibetan fox surveyed on four different days on a 250-km road in the Three-River Source National Park. Table S1. The detection functions, adjustments and values of Akaike information criterion (AIC) in the distance sampling for the Tibetan wild ass at the Three-River Source National Park. Table S2. Variables used in the species distribution model for the Tibetan wild ass in the Three-River Source National Park. The variable names used in the models are in parentheses. The files (BioClim.grd and BioClim.gri) are the dataset of the 29 environmental variables covering the Three-River Source National Park.

Author Contributions: Conceptualization, X.L. and E.G.; methodology, X.L. and B.L.; validation, N.L.; formal analysis, X.L. and N.L.; investigation, X.L., B.L. and E.G.; data, X.L. and E.G.; writing—original draft preparation, X.L.; writing—review and editing, X.L., N.L., Y.S. and E.G.; visualization, X.L.; supervision, Y.S.; project administration, X.L.; funding acquisition, X.L. All authors have read and agreed to the published version of the manuscript.

Funding: This work was supported by the National Natural Science Foundation of China (No. 31970432 and 32171528), the Second Tibetan Plateau Scientific Expedition and Research Program (STEP, Grant No. 2019QZKK0501), the Third Xinjiang Scientific Expedition Project (Grant No. 2021XJKK1302), the Second National Wildlife Survey Project for Terrestrial Animals and the Key Subject of Ecology of Jiangsu Province (SUJIAOYANHAN(2022) No.2). The study is also supported by Alliance of

International Science Organization (ANSO) (Project ID: ANSO-CR- KP-2020-08) for ecosystem safety assessment and policy making at the China-Mongolia-Russia corridor at the one belt one road zone.

Institutional Review Board Statement: Not applicable.

Informed Consent Statement: Not applicable.

Data Availability Statement: All data used in this study are freely available to anyone. The example data and R code can be accessed at https://github.com/Xinhai-Li/abundanceR. The global environmental variables at the terrestrial areas (a 8 GB compressed file, var29.zip) can be downloaded at https://pan.baidu.com/s/1noU8A7WcsuYx0MSiQq6CeQ (access code is: 1234) for users in Mainland China; and can also be downloaded from the Google Drive at (https://drive.google.com/drive/folders/1bNh4SdikmjrOkgqE5VOVo86SozD2hvmT?usp=sharing).

Acknowledgments: We thank staff of Qinghai Forestry Department and the Three-River-Source National Park for their help during the field surveys. We are grateful to Yushan Wang, Ba Zhou, Qianqian Luo, Xiaojia Zhu, Boyi Wang for carrying out the field surveys.

Conflicts of Interest: The authors declare that they have no conflict of interest.

References

1. Iijima, H. A review of wildlife abundance estimation models: Comparison of models for correct application. *Mammal Study* **2020**, *45*, 177–188. [CrossRef]
2. Seber, G.A.F. A review of estimating animal abundance. *Int. Stat. Rev.* **1992**, *60*, 129–166. [CrossRef]
3. Schwarz, C.J.; Seber, G.A.F. Estimating animal abundance: Review III. *Stat. Sci.* **1999**, *14*, 427–456. [CrossRef]
4. Marchowski, D.; Jankowiak, Ł.; Ławicki, Ł.; Wysocki, D. Waterbird counts on large water bodies: Comparing ground and aerial methods during different ice conditions. *PeerJ* **2018**, *6*, e5195. [CrossRef]
5. Haus, J.M.; Eyler, T.B.; Bowman, J.L. A spatially and temporally concurrent comparison of popular abundance estimators for white-tailed deer. *Northeast. Nat.* **2019**, *26*, 305–324. [CrossRef]
6. Harley, S.J.; Myers, R.A.; Dunn, A. Is catch-per-unit-effort proportional to abundance? *Can. J. Fish. Aquat. Sci.* **2001**, *58*, 1760–1772. [CrossRef]
7. Delargy, A.J.; Lambert, G.I.; Kaiser, M.J.; Hiddink, J.G. Potential highly variable catch efficiency estimates complicate estimation of abundance. *Fish. Res.* **2022**, *245*, 106138. [CrossRef]
8. Pradel, R. Utilization of capture-mark-recapture for the study of recruitment and population growth rate. *Biometrics* **1996**, *52*, 703–709. [CrossRef]
9. Besbeas, P.; Freeman, S.N.; Morgan, B.J.T.; Catchpole, E.A. Integrating mark-recapture-recovery and census data to estimate animal abundance and demographic parameters. *Biometrics* **2002**, *58*, 540–547. [CrossRef]
10. Loonam, K.E.; Ausband, D.E.; Lukacs, P.M.; Mitchell, M.S.; Robinson, H.S. Estimating abundance of an unmarked, low-density species using cameras. *J. Wildl. Manag.* **2021**, *85*, 87–96. [CrossRef]
11. Moeller, A.K.; Lukacs, P.M.; Horne, J.S. Three novel methods to estimate abundance of unmarked animals using remote cameras. *Ecosphere* **2018**, *9*, 15. [CrossRef]
12. Borchers, D.L.; Efford, M.G. Spatially explicit maximum likelihood methods for capture–recapture studies. *Biometrics* **2008**, *64*, 377–385. [CrossRef] [PubMed]
13. Jolly, G.M. Explicit estimates from capture-recapture data with both death and immigration-stochastic model. *Biometrika* **1965**, *52*, 225–247. [CrossRef] [PubMed]
14. Rowcliffe, J.M.; Field, J.; Turvey, S.T.; Carbone, C. Estimating animal density using camera traps without the need for individual recognition. *J. Appl. Ecol.* **2008**, *45*, 1228–1236. [CrossRef]
15. Li, X.; Tian, H.; Piao, Z.; Wang, G.; Xiao, Z.; Sun, Y.; Gao, E.; Holyoak, M. CameratrapR: An R package for estimating animal density using camera trapping data. *Ecol. Inform.* **2022**, *69*, 101597. [CrossRef]
16. Buckland, S.T.; Anderson, D.R.; Burnham, K.P.; Laake, J.L. *Distance Sampling: Estimating Abundance of Biological Populations*; Chapman and Hall: London, UK, 1993; pp. i–xii, 1–446.
17. Krebs, C.J. *Ecological Methodology*; Benjamin/Cummings: Menlo Park, CA, USA, 1999; Volume 620.
18. Hothorn, T.; Mueller, J.; Schroeder, B.; Kneib, T.; Brandl, R. Decomposing environmental, spatial, and spatiotemporal components of species distributions. *Ecol. Monogr.* **2011**, *81*, 329–347. [CrossRef]
19. Royle, J.A. N^{-} mixture models for estimating population size from spatially replicated counts. *Biometrics* **2004**, *60*, 108–115. [CrossRef]
20. Guisan, A.; Zimmermann, N.E. Predictive habitat distribution models in ecology. *Ecol. Model.* **2000**, *135*, 147–186. [CrossRef]
21. Li, X.; Ma, L.; Hu, D.; Ma, D.; Li, R.; Sun, Y.; Gao, E. Potential range shift of snow leopard in future climate change scenarios. *Sustainability* **2022**, *14*, 1115. [CrossRef]
22. Mi, C.; Huettmann, F.; Guo, Y.; Han, X.; Wen, L. Why choose Random Forest to predict rare species distribution with few samples in large undersampled areas? Three Asian crane species models provide supporting evidence. *Peerj* **2017**, *5*, e2849. [CrossRef]

23. Boyce, M.S.; McDonald, L.L. Relating populations to habitats using resource selection functions. *Trends Ecol. Evol.* **1999**, *14*, 268–272. [CrossRef]
24. Boyce, M.S.; Johnson, C.J.; Merrill, E.H.; Nielsen, S.E.; Solberg, E.J.; van Moorter, B. Can habitat selection predict abundance? *J. Anim. Ecol.* **2016**, *85*, 11–20. [CrossRef] [PubMed]
25. Stauffer, G.E.; Roberts, N.M.; Macfarland, D.M.; Van Deelen, T.R. Scaling occupancy estimates up to abundance for wolves. *J. Wildl. Manag.* **2021**, *85*, 1410–1422. [CrossRef]
26. Santos, A.J.B.; Vieira, D.H.G.; Bellini, C.; Corso, G.; Ceriani, S.A.; Fuentes, M.M.P.B. Using data from nesting beach monitoring and satellite telemetry to improve estimates of marine turtle clutch frequency and population abundance. *Mar. Biol.* **2021**, *168*, 1–16. [CrossRef]
27. Teton, B.S.; Lewis, J.S.; Wright, C.T.; White, M.; Young, H. Using natural pelt patterns to estimate population abundance with mark-resight models. *Wildl. Soc. Bull.* **2020**, *44*, 695–704. [CrossRef]
28. Shertzer, K.W.; Bacheler, N.M.; Pine, W.E., III; Runde, B.J.; Buckel, J.A.; Rudershausen, P.J.; MacMahan, J.H. Estimating population abundance at a site in the open ocean: Combining information from conventional and telemetry tags with application to gray triggerfish (*Balistes capriscus*). *Can. J. Fish. Aquat. Sci.* **2020**, *77*, 34–43. [CrossRef]
29. Howard, A.L.; Clement, M.J.; Peck, F.R.; Rubin, E.S. Estimating mountain lion abundance in arizona using statistical population reconstruction. *J. Wildl. Manag.* **2020**, *84*, 85–95. [CrossRef]
30. R Core Team. *R: A Language and Environment for Statistical Computing*; R Foundation for Statistical Computing: Vienna, Austria, 2021.
31. Miller, D.L.; Rexstad, E.; Thomas, L.; Marshall, L.; Laake, J.L. Distance Sampling in R. *J. Stat. Softw.* **2019**, *89*, 1–28. [CrossRef]
32. Fick, S.E.; Hijmans, R.J. WorldClim 2: New 1km spatial resolution climate surfaces for global land areas. *Int. J. Climatol.* **2017**, *37*, 4302–4315. [CrossRef]
33. Rabus, B.; Eineder, M.; Roth, A.; Bamler, R. The shuttle radar topography mission - a new class of digital elevation models acquired by spaceborne radar. *Isprs J. Photogramm. Remote Sens.* **2003**, *57*, 241–262. [CrossRef]
34. Sanderson, E.W.; Jaiteh, M.; Levy, M.A.; Redford, K.H.; Wannebo, A.V.; Woolmer, G. The Human Footprint and the Last of the Wild. *Bioscience* **2002**, *52*, 891–904. [CrossRef]
35. Loveland, T.R.; Reed, B.C.; Brown, J.F.; Ohlen, D.O.; Zhu, J.; Yang, L.; Merchant, J.W. Development of a global land cover characteristics database and IGBP DISCover from 1-km AVHRR data. *Int. J. Remote Sens.* **2000**, *21*, 1303–1330. [CrossRef]
36. Lehner, B. *Global Lakes and Wetlands Database, Level 3*; World Wildlife Fund US: Grang, Switzerland, 2004.
37. Iverson, L.R.; Prasad, A.M.; Matthews, S.N.; Peters, M. Estimating potential habitat for 134 eastern US tree species under six climate scenarios. *For. Ecol. Manag.* **2008**, *254*, 390–406. [CrossRef]
38. Breiman, L. Random forests. *Mach. Learn.* **2001**, *45*, 5–32. [CrossRef]
39. Valavi, R.; Elith, J.; Lahoz-Monfort, J.J.; Guillera-Arroita, G. Modelling species presence-only data with random forests. *Ecography* **2021**, *44*, 1731–1742. [CrossRef]
40. Liaw, M.A. *Package 'Randomforest'*; University of California, Berkeley: Berkeley, CA, USA, 2018.
41. Ketz, A.C.; Johnson, T.L.; Monello, R.J.; Mack, J.A.; George, J.L.; Kraft, B.R.; Wild, M.A.; Hooten, M.B.; Hobbs, N.T. Estimating abundance of an open population with an N-mixture model using auxiliary data on animal movements. *Ecol. Appl.* **2018**, *28*, 816–825. [CrossRef] [PubMed]
42. Ganley, L.C.; Brault, S.; Mayo, C.A. What we see is not what there is: Estimating North Atlantic right whale Eubalaena glacialis local abundance. *Endanger. Species Res.* **2019**, *38*, 101–113. [CrossRef]
43. Li, X.; Gao, E.; Li, B.; Zhan, X. Estimating abundance of Tibetan wild ass, Tibetan gazelle and Tibetan antelope using species distribution models and distance sampling. *Sci. Sin. Vitae* **2019**, *49*, 151–162. [CrossRef]
44. Figueiredo, M.S.L.; Grelle, C.E.V. Predicting global abundance of a threatened species from its occurrence: Implications for conservation planning. *Divers. Distrib.* **2009**, *15*, 117–121. [CrossRef]
45. Kery, M. Estimating abundance from bird counts: Binomial mixture models uncover complex covariate relationships. *Auk* **2008**, *125*, 336–345. [CrossRef]
46. Dodd, C.K.; Dorazio, R.M. Using counts to simultaneously estimate abundance and detection probabilities in a salamander community. *Herpetologica* **2004**, *60*, 468–478. [CrossRef]
47. Pérez, J.M.; Serrano, E.; Alpízar-Jara, R.; Granados, J.; Soriguer, R. The potential of distance sampling methods to estimate abundance of mountain ungulates: Review of usefulness and limitations. *Pirineos* **2002**, *157*, 15–24. [CrossRef]

land

MDPI

Article

Estimating Road Mortality Hotspots While Accounting for Imperfect Detection: A Case Study with Amphibians and Reptiles

Noah Hallisey [1], Scott W. Buchanan [2], Brian D. Gerber [1], Liam S. Corcoran [1] and Nancy E. Karraker [1,*]

[1] Department of Natural Resources Science, University of Rhode Island, Kingston, RI 02881, USA; halliseyn@uri.edu (N.H.); bgerber@uri.edu (B.D.G.); liam_corcoran@uri.edu (L.S.C.)

[2] Rhode Island Department of Environmental Management, Division of Fish and Wildlife, 277 Great Neck Road, West Kingston, RI 02892, USA; scott.buchanan@dem.ri.gov

* Correspondence: nkarraker@uri.edu

Abstract: Wildlife road mortality tends to aggregate spatially at locations commonly referred to as road mortality hotspots. Predictive models can be used to identify locations appropriate for mitigation measures that reduce road mortality. However, the influence of imperfect detection (e.g., false absences) during road mortality surveys can lead to inaccurate or imprecise spatial patterns of road mortality hotspots and suboptimal implementation of mitigation measures. In this research, we used amphibians and reptiles as a case study to address imperfect detection issues when estimating the probability of road mortality hotspots using occupancy detection modeling. In addition, we determined the survey effort needed to achieve a high probability of detecting large roadkill events. We also assessed whether vehicle travel reductions associated with the COVID-19 pandemic travel restrictions led to reductions in road mortality. We conducted surveys at 48 sites throughout Rhode Island, USA, from 2019–2021. In total, we observed 657 carcasses representing 19 of Rhode Island's 37 native species. Of the 19 native species, eight species of frogs, four species of salamanders, four species of snakes, and three species of turtles were observed. We documented a reduction in roadkill density and the proportion of dead versus live amphibians and reptiles in pandemic years (2020 and 2021), but we were unable to link reductions in roadkill density to reductions in traffic volume. Our model results indicated that large roadkill events were more likely to occur on roads near wetlands and with low traffic volume and were more likely to be detected as daily precipitation increased. We determined that there was a low probability of detecting large roadkill events, suggesting that imperfect detection influences detection of large roadkill events, and many were likely missed during our surveys. Therefore, we recommend using occupancy modeling to account for the influence of imperfect detection when estimating road mortality hotspots. This approach will more effectively guide the implementation of mitigation measures.

Keywords: occupancy modeling; road ecology; mitigation measures

Citation: Hallisey, N.; Buchanan, S.W.; Gerber, B.D.; Corcoran, L.S.; Karraker, N.E. Estimating Road Mortality Hotspots While Accounting for Imperfect Detection: A Case Study with Amphibians and Reptiles. *Land* **2022**, *11*, 739. https://doi.org/10.3390/land11050739

Academic Editors: Juan F. Beltrán, Pedro Abellán, John Litvaitis and Francesco M. Angelici

Received: 4 March 2022
Accepted: 5 May 2022
Published: 14 May 2022

1. Introduction

As road infrastructure continues to expand in the United States, so do the negative impacts of roads and vehicle traffic on ecosystems and wildlife. Among the most notable impacts is the widespread direct mortality of wildlife from vehicle traffic [1]. For example, it has been estimated that between 89 and 340 million birds are struck and killed on roads annually in the United States [2]. Road mortality exerts significant impacts on amphibians and reptiles as well, with amphibians often accounting for 60–90% of all roadkill observations [3,4]. Herpetofauna play important ecological roles, acting as both predator and prey, supporting energy transfer between aquatic and terrestrial ecosystems, and serving as indicator species for the health of aquatic ecosystems [5,6]. With human-associated stressors, including road mortality, contributing to amphibian and reptile population declines worldwide, there is concern about these declines disrupting food webs and ecosystem functioning [7–9].

Herpetofauna possess ecological and behavioral characteristics that make them highly susceptible to road mortality. The close relationship between the periodicity of life history events and local weather conditions combines to create observable patterns in road mortality [10,11]. For example, amphibians make frequent road crossings during annual migrations between habitats to breed and forage, such as regular crossings made by the northern leopard frog (*Lithobates pipiens*) between upland and wetland habitats, increasing the risk of road mortality [12–14]. In Indiana (USA), increased levels of amphibian road mortality have been observed in the summer months from May through July, during periods when many species frequently cross roads to breed and forage [3]. Reptiles are also vulnerable because of seasonal behaviors. During movements to nesting sites, female turtles may cross roads to nest in the loose sandy areas adjacent to roads, thereby increasing their vulnerability to motor vehicle impacts [15]. Many snake species overwinter communally and emerge from hibernacula in large numbers in the spring. The resulting mass movements can result in large, localized mortality events [16]. Weather conditions also influence susceptibility to road mortality. For amphibians, increased road mortality has been observed in the evenings with warmer temperatures and precipitation during the breeding season [17,18]. For reptiles, a higher risk of road mortality occurs in the spring and early summer when temperatures increase, and many species are crossing roads to reach nesting sites [10].

Road surfaces constructed of asphalt and concrete absorb and retain heat, attracting snakes, frogs, and salamanders to roads to bask and thermoregulate [8,19]. For diurnal species, heat absorbed by roads on sunny days makes road surfaces ideal for basking. That heat, absorbed during the day, is retained and continues to radiate into the evening hours, serving as an important source of heat for nocturnal species on cool nights. For both diurnal and nocturnal species, the use of roads for thermoregulation can increase vulnerability to road traffic. Many amphibians are small, making them hard for drivers to see and avoid. They move slowly, and some species may remain immobile in response to traffic, which can increase the risk of being killed [20,21]. In fact, some snakes and other reptiles are targeted by drivers [22,23].

Road mortality can ultimately lead to population declines in amphibians and reptiles [12]. This can alter the demographic and genetic composition of populations and fragment breeding populations [8,24], which can further exacerbate population declines and lead to extirpation [25]. For example, an annual road mortality of >10% in adult spotted salamanders (*Ambystoma opacum*) may be enough to lead to population declines and potentially extirpation [8]. More concerning, annual road mortality >5% in terrestrial and large-bodied turtles in the northeastern United States may threaten local populations [26]. For turtles, life history traits, such as delayed sexual maturity and low fecundity, make them highly susceptible to the impacts of road mortality [27].

Road mortality does not occur randomly but is often spatially clustered at locations with specific landscape and road features, commonly referred to as road mortality hotspots [28]. For herpetofauna, the presence of habitats, the proximity of habitats to roads, and traffic volume are strong indicators of locations with high levels of road mortality [3,4,29]. In one study [30], sections of roads within 100 m of a wetland were determined to be the most important indicator of road mortality hotspots for amphibians and reptiles. For amphibians, road mortality hotspots occur on sections of roads with wetlands directly adjacent to or bisected by roads [31]. For turtles, road sections within close proximity to bodies of water were the best predictors of road mortality hotspots [32]. Road characteristics, including road width and traffic volume, may also serve as important predictors of road mortality hotspots. Wide roads generally have multiple lanes to support high levels of traffic, representing a longer distance and a longer time needed for herpetofauna to cross, increasing the chance of being killed [33]. Traffic volume, a measure of the number of vehicles on a road over a given period, usually daily, is another important road feature associated with road mortality hotspots. Although the risk of being killed increases with traffic volume [13], some studies have observed the highest levels of road mortality on roads with a lower traffic volume. For example, roads with low (350–470 vehicles/day) and moderate (1900–2900 vehicles/day) traffic volumes had the highest levels of amphibian mortality in Poland [29]. Roads with lower traffic volumes present less vehicle disturbance for animals, making them more likely locations for thermoregulation, thus increasing vulnerability to road mortality [12]. In addition, roads with the highest levels of traffic volume that have been in use for decades, such as major interstates and highways, may have initially had higher levels of road mortality when first constructed. However, high levels of road mortality lead to population declines near such roads, which results in decreased mortality over time [7,34]. Roads with a higher traffic volume tend to also occur in highly developed areas with less habitat for amphibians and reptiles. High traffic volumes may act as a behavioral deterrent that keeps some species from attempting to cross a road, and thus, more frequent road crossings may occur on low traffic volume roads [35].

Road surveys are used to identify road mortality hotspots by recording live and dead individuals on roads during high-risk periods [36]. Common survey methods include driving or walking along roadways. Driving surveys can be used along large road networks, but this method results in lower detection rates and missed carcasses on the road, thereby underestimating road mortality levels [36]. For example, it was estimated that the number of roadkill observed during driving surveys is 12–16 times lower than the actual mortality rate [37]. Walking surveys have higher detection rates and generate more precise estimates of road mortality but are time consuming and cover less roadways than driving surveys [12,36]. A recent study comparing methods observed that 75% of amphibian carcasses recorded during walking surveys were missed during driving surveys [38]. The frequency of surveys is also important and can impact hotspot identification, with weekly or longer intervals between surveys reducing the accuracy of identifying hotspots for amphibians and reptiles [39]. Regardless, techniques commonly used to identify hotspots rely on counts of roadkills recorded during surveys, which are often assumed to be underestimated due to imperfect detection during surveys [3,40].

Since large roadkill events tend to occur on only a few nights during the year and under certain weather conditions, the timing of surveys is important for capturing peaks in road mortality, especially since carcasses may remain on the road only for brief periods following a large event [41]. In other words, there may be a high risk of road mortality along a section of road; however, due to the timing of the survey, a low level of roadkill is observed; this leads to observers failing to detect road mortality hotspots. Several studies focusing on herpetofauna road mortality have recognized that the number of roadkill observed during surveys was likely underestimated due to imperfect detection during surveys, which likely influenced documented spatial patterns of road mortality [3,18]. Imperfect detection can lead to the false absence (or presence) of hotspots and spatial bias in road mortality patterns, potentially directing management efforts and mitigation measures that are ineffective in reducing road mortality [39,40,42].

Predictive models can be used to identify road mortality hotspots, as well as potential hotspots not previously surveyed, by investigating relationships between the numbers of animals killed on roads and particular landscape and road features [30,43]. Studies have used spatial clustering techniques, such as Getis-Ord-Gi [44–46], to identify the clustering of roadkill associated with road mortality hotspots. However, the techniques commonly used rely on roadkill count data and do not address the influence of imperfect detection during surveys on roadkill counts, which can lead to inaccurate or imprecise spatial patterns of road mortality hotspots. Occupancy models, a predictive modeling approach that incorporates detection probability to estimate occurrence—the probability of a species being present at a location but not detected during surveys—can be used to address imperfect detection during surveys to more precisely predict spatial patterns of road mortality [40,42,47]. Occupancy models incorporate site characteristics (e.g., the presence of a species' habitat) as well as the environmental conditions during surveys (e.g., precipitation) to estimate the probability of a species being present at a location and observed during a survey (i.e., detection or non-detection of a species). Applied to wildlife road mortality, occupancy is a proxy for determining the risk of road mortality, and detection probability is the likelihood of a carcass being detected during a survey, given that it occurs [42]. Estimating detection probability requires multiple visits to a site, and since multiple surveys may be required to identify a road mortality hotspot, the sampling design used for road surveys can be integrated to correct for imperfect detection during surveys [42]. As occupancy models account for detection probability, they can be used to correct for false absences by identifying locations with the greatest risk of road mortality rather than those with the highest observations of roadkill, guiding mitigation measures to locations where road mortality is most likely to occur [42]. Occupancy detection models are commonly used to estimate site occupancy and species distribution and to monitor wildlife populations [48,49]. However, few studies have used occupancy detection models to address imperfect detection in road mortality hotspot models with the goals of improving the identification of locations with the greatest risk of road mortality and enhancing recommendations for locations where mitigation measures may be most appropriate in reducing road mortality [39].

New England states (USA), including Rhode Island, are characterized by ongoing development, including road construction and high human population densities. Road mortality is likely an important threat for native amphibian and reptile populations on Rhode Island, but limited data are available. Using amphibians and reptiles as a case study, we employed an occupancy modeling framework to account for imperfect detection issues in road mortality hotspot probability to better identify and prioritize locations where mitigation measures would be most effective in reducing road mortality. In addition, we used our estimates of large roadkill event detectability to determine the survey effort needed to achieve a high probability of detecting large roadkill events.

2. Materials and Methods

2.1. Study Site

Rhode Island is the smallest state in the United States, with a total land area of 3100 km^2. However, it is ranked among the top states for both human population density and road density [50]. Rhode Island harbors 18 species of native amphibians (10 salamanders, 8 frogs) and 19 species of native reptiles (12 snakes, 7 non-marine turtles), many of which are suspected to be undergoing population declines because of anthropogenic stressors, including habitat loss and road mortality [51].

2.2. Study Design

We conducted road mortality surveys from late April to mid-July from 2019 to 2021. Surveys were undertaken on two-laned paved roads with traffic volumes ranging from 15–6000 vehicles per day. From roads with those characteristics, we randomly selected 48 road sections (each 200 m in length) throughout western Rhode Island (see Figure A1 in Appendix A) that were separated by at least 500 m to reduce spatial autocorrelation. Half of the sites were located within 100 m of a wetland (hotspots), and half of them were located farther than 100 m from a wetland (coldspots). On each visit, we conducted walking surveys at one group of six sites. Weather conditions varied among survey nights, which included nights with and without precipitation. We conducted surveys between 20:00–01:30 h on nights with air temperatures $\geq$5.5 °C.

2.3. Road Mortality Surveys

A survey consisted of two or more surveyors walking the length of each 200-m road section, one on each side of the road, scanning the surface and adjacent area for live or dead amphibians and reptiles. At the end of each road section, surveyors switched sides and walked back to the beginning of the road section, continuing to scan for live or dead animals. We used a mobile application, Herp Observer Road Edition, created by the Rhode

Island Department of Environmental Management, to document amphibians and reptiles found on or adjacent to the road along at each site. This application was developed within the data collection platform Survey123 (ESRI, Redlands, CA, USA) and was used to record the species, date, time, geographic coordinates, condition (alive or dead-on road), and any notes on the observation along with a photo of each carcass. Carcasses that were highly decomposed were recorded at the genus level. Once recorded, carcasses were removed to reduce double counting on the return walk during each survey. Any live amphibians and reptiles encountered on the road during surveys were recorded and moved off the road in the direction in which they were headed. We documented environmental conditions during surveys using the North American Amphibian Monitoring Program protocol [52]. Air temperature was recorded during surveys, and daily precipitation was obtained from the nearest Community Collaborative Rain, Hail, & Snow Network rain gauge [53].

2.4. Pandemic-Associated Road Mortality

Due to the COVID-19 pandemic, we were forced to curtail our survey effort and surveyed half the number of sites (12 sites) in 2020, as in 2019 and 2021 (24 sites each year). We saw this as an opportunity to determine if reductions in road mortality occurred during the COVID-19 pandemic, as fewer vehicles were likely traveling on roads due to travel-related restrictions and an economy-wide shut down. In late March 2020, the Rhode Island Governor implemented statewide mandates in an attempt to reduce the spread of COVID-19. These mandates included businesses ordering their employees to work remotely from home, schools switching to online learning, and many local businesses and restaurants closing or offering only road-side or take-out services, all of which reduced the number of vehicles traveling on roads. In addition, some towns enacted a "stay at home" order that directed residents to remain at home and travel only for necessities, such as groceries or to go to work. Rhode Island has a thriving tourism industry and receives an influx of out-of-state travelers during the spring and summer months. To curb interstate travel, the governor required that all out-of-state travelers enter the state for an extended period quarantine for 14 days, which likely deterred many tourists from crossing into Rhode Island. As many of the COVID-19 pandemic mandates to curb travel were established during a key time of year, when many of Rhode Island's amphibians and reptiles are highly active, we aimed to determine if potential traffic reductions during the COVID-19 pandemic led to a reduction in road mortality. Using the same survey technique, we conducted surveys at the six sites with the highest roadkill totals in 2019 and the six sites we planned to survey during our full survey season in 2020. Then, in 2021, we conducted additional surveys at the six sites surveyed in 2020 as part of our full survey season. We then compared two metrics, roadkill rates as carcasses observed per kilometer and the percentage of live versus dead amphibians and reptiles between years (pre-pandemic, pandemic season 2020, and pandemic season 2021), to determine if a reduction in road mortality had occurred.

2.5. Occupancy Model Development

In this study, we defined occupancy as the probability of a large roadkill event occurring at a site. Detection probability was defined as the probability of detecting a large roadkill event during a survey, given that one occurred at the site. We defined a large roadkill event as observing five or more amphibians and/or reptiles dead at a site during a survey. Five or more carcasses represented the top 10% of the sites with the highest roadkill observations per survey. We considered testing a second scenario for roadkill events of $\geq$10 carcasses being observed at a site during a survey; however, roadkill events of $\geq$10 individuals were only detected at 11.7% (n = 7) of all sites. Given the infrequency of these events ($\geq$10 carcasses on a given survey), reliable estimates using the occupancy modeling approach were not feasible. Due to the low number of roadkill observations by taxonomic group, all amphibian and reptile roadkill were aggregated regardless of species at each site. This decision was supported by previous research demonstrating that road mortality hotspots for amphibians and reptiles overlapped [30]. Live amphibians and reptiles on the road were documented but were not included in the analyses. We used the 'unmarked' package in R (Version 4.1.1) to estimate occupancy and detection probability for large roadkill events across all sites. The 'unmarked' package uses maximum likelihood estimates to predict occupancy and detection probability from observed data for detection or non-detection of a species.

Geospatial data for Land Use and Land Cover (2011) [54] and Rhode Island Department of Transportation Roads (2016) [55] were acquired from the Rhode Island Geographic Information System. The Land Use and Land Cover data were classified using the Anderson Level III classification system [56]. In addition, we acquired data from the National Wetlands Inventory [57]. We used ArcPro (ESRI, Version 2.8.2, Redlands, CA) to merge land use and land cover data with the National Wetlands Inventory data. The percentage of land cover surrounding a road was calculated by buffering each transect by 100 m and dividing the area of each land use class within the buffer by the total buffer area. A scale of 100 m was chosen, as it has been identified as an important scale associated with amphibian and reptile road mortality hotspots [30]. We also assessed percent wetland and forest within 500 m of a road; however, due to our limited sample size, we were forced to eliminate these models due to poor model fit and reliability of model estimates. Current comprehensive traffic volume data were not available within the state, so we used Functional Class Description Codes from the Federal Highway Administration [58] as a proxy for traffic volume levels, which assigns classes to roads based on traffic volume ranges. Repeated visits to a site are required when using occupancy modeling to determine detection history, which is a pattern of detection or non-detection of a species at a site, or in this case, detection of a large roadkill event. Therefore, in each year, we conducted five surveys at each site starting in mid-April and ending in mid-July. An additional survey was conducted at each transect in 2021; however, they were not included in our occupancy model as the surveys were conducted prior to the start of the surveying period in other years (2019 and 2020) and would have introduced sampling bias into model estimates for occupancy and detection probability. Using roadkill observation data from the five surveys conducted per sampling season at the 48 transects, we generated a detection history matrix for the surveyed transects. Sites surveyed over multiple years as part of our pandemic-associated road mortality study were included as spatial replicates, which included 12 extra sites in our detection history matrix. In total, our detection matrix consisted of 60 transects, with five surveys per transect.

Using survey detection history, we developed a model that assumed an unvarying influence of site and survey covariates to estimate constant occupancy and detection probability across all sites. We then developed a model set consisting of combinations of different covariates (Table 1) to estimate occupancy and detection probability for all sites. Covariates included percent wetland area and forest within a 100-m buffer around a transect, traffic volume classified using the Functional Class Description Code (see Table A1 in Appendix A), temperature during each survey, 24-h precipitation recorded from the nearest rain gauge, and day of year starting from 1 January of the survey year. For occupancy, we tested percentage wetland and forest within 100 m of a road, and road classification, as these characteristics have been found to be important indicators of road mortality hotspots for amphibians and reptiles [30]. For detection probability, we tested daily temperature, precipitation, and time of year, as amphibians and reptiles are more active under certain weather conditions and during specific times of the year [41]. We also tested detection probability against road classification, as carcasses' persistence can vary across road types, which can influence the detectability of roadkill [59].

We limited the number of covariates included in the models to those that we expected to influence road mortality based on published literature and with the goals of maximizing reliability of model estimates and reducing complexity of the models. Models were compared using Akaike's Information Criterion, a metric from information theory in which a set of candidate models is developed a priori based on prior knowledge and models are evaluated based on model fit and complexity [60]. Model averaging was then used to estimate occupancy and detection probability for all sites. Finally, using the simplest model that assumed an unvarying influence of covariates on occupancy and detection probability, we determined the survey effort needed to detect a large roadkill event at least once at occupied sites as:

$$p^* = 1 - (1 - p)^s$$

where (p^*) is the probability of detecting at least one large roadkill event at a transect assuming one occurs there, (p) is the estimated per survey detection probability, and (s) is the number of surveys. Using this, we were able to estimate the number of surveys necessary to reach a cumulative detection probability of 0.85 and 0.95 for detecting a large roadkill event. This allowed us to determine the survey effort (i.e., number of surveys) needed to reach a high probability of detecting a large roadkill event.

Table 1. Covariates used in model development to assess occupancy and detection probability of road mortality hotspots in Rhode Island, 2019–2021.

Covariate	Definition	Data Type
Perc_Wetland_100m	% wetland within 100-m of a road	Continuous
Perc_Forest_100m	% forest within 100-m of a road	Continuous
F_Class_Code	Road classification used as a proxy for traffic volume	Categorical
Temp	Temperature recorded during surveys	Continuous
Rain	Precipitation in previous 24-h recorded from nearest weather gauge	Continuous
Julian_Date	Day of year from 1 January	Continuous

3. Results

Between April 2019 and July 2021, we surveyed 48 sites that covered 9.6 km of roadway. With repeated visits to each site, we surveyed a cumulative linear distance of 64.8 km of roadway over 54 nights between 2019 and 2021. Mean air temperature during surveys was 15.1 °C (SD = 5.63 °C, range = 5.6–22.6 °C), and mean daily precipitation in the 24 h preceding surveys was 0.31 cm (SD = 0.60 cm, range = 0–3.3 cm). This equated to a roadkill rate of 10.1 carcasses per km surveyed. We recorded 657 roadkill observations, of which 19% were too damaged to be identified at the genus or species level. The largest roadkill event we observed while surveying was 33 carcasses. Of Rhode Island's 37 native species, roadkill observations represented 19 (51%), of which 82% were frogs, 7% were salamanders, 6% were turtles, and 5% were snakes. Of the 19 native species, eight species of frogs (42%), four species of salamanders (21%), four species of snakes (21%), and three species of turtles (16%) were represented. We observed a greater number of roadkill amphibians and reptiles at the expected hotspots (70%) than at the expected coldspots (30%).

Regarding the potential effects of the COVID-19 pandemic on road mortality, we did not observe a reduction in roadkill density (carcasses/km) during the 2020 surveys compared to 2019 (pre-pandemic). However, in 2021, we observed a reduction in road mortality rates compared to the other years (2019 and 2020). We also observed a reduction in the percentage of road-killed amphibians and reptiles in 2020 and 2021 as compared to 2019 (Table 2).

Table 2. Results of pre-pandemic (2019), pandemic season 2020, and pandemic season 2021 surveys of road mortality in Rhode Island.

Year	Distance Surveyed (km)	Roadkill Density (Carcasses/km)	Number Dead on Road (%)	Number Live on Road (%)	Total Number on Road
2019	24	10.4	249 (88)	33 (12)	282
2020	12	15.3	185 (71)	74 (29)	259
2021	28.8	7.74	223 (83)	46 (17)	269

We developed and evaluated 16 models (see Table A2 in Appendix A) for occupancy and detection probability of large roadkill events. We found that the percentage of wetland cover within 100 m of a road and traffic volume (as estimated by road classification) were the most supported covariates (Table 3). While these covariates were most supported, model results indicated that there was no statistically significant relationship between percentage wetland cover within 100-m of a road and traffic volume and occurrence of large roadkill events. Model average results indicated that percent wetland positively influenced estimated site occupancy, while percent forest negatively influenced site occupancy (Figure 1). Estimated occupancy was highest at sites on roads with low traffic volume (15–400 vehicles/day). We found that precipitation in the preceding 24 h and road classification most strongly influenced detection probability (Table 3). We found that detection probability increased ($p < 0.001$) as daily precipitation increased, and detection probability decreased ($p < 0.001$) as traffic volume decreased (Figure 2).

Table 3. Results of the most supported models for occupancy and detection at the 100-m scale for Rhode Island, USA, 2019–2021. Intercept is the mean value for occupancy, and detection, Perc_Forest_100m is the percent forest within 100 m of a road, F_Class_Code is the Functional Class Description code for traffic volume, Rain is the cumulative precipitation 24 h prior to a survey, Estimate is the model parameter estimates, SE is the standard error, Z is the latent value, and $p(>|z|)$ is p-value.

Covariate	Estimate	SE	Z	$p(>\|z\|)$
Occupancy models				
Intercept	2.51	1.565	1.60	0.109
Perc_Forest_100m	1.28	1.152	1.11	0.210
F_Class_Code	1.07	1.07	1.25	0.210
Detection models				
Intercept	−2.157	0.261	−8.25	<0.001
Rain	0.778	0.162	4.79	<0.001
F_Class_Code	−0.785	0.228	−3.44	<0.001

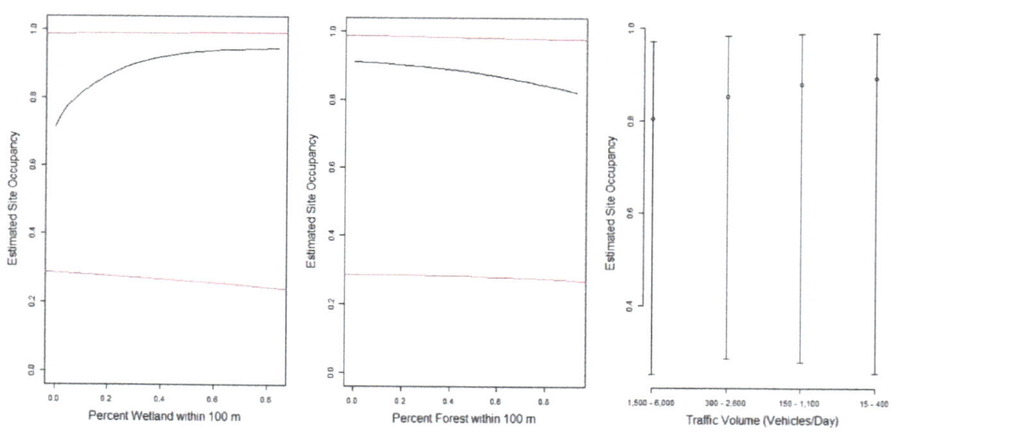

Figure 1. Marginal effect of covariates on estimated site occupancy from the model-averaged results. Black lines represent model estimates for occupancy, and red lines represent upper and lower confidence intervals. Plots hold other covariates at their mean value.

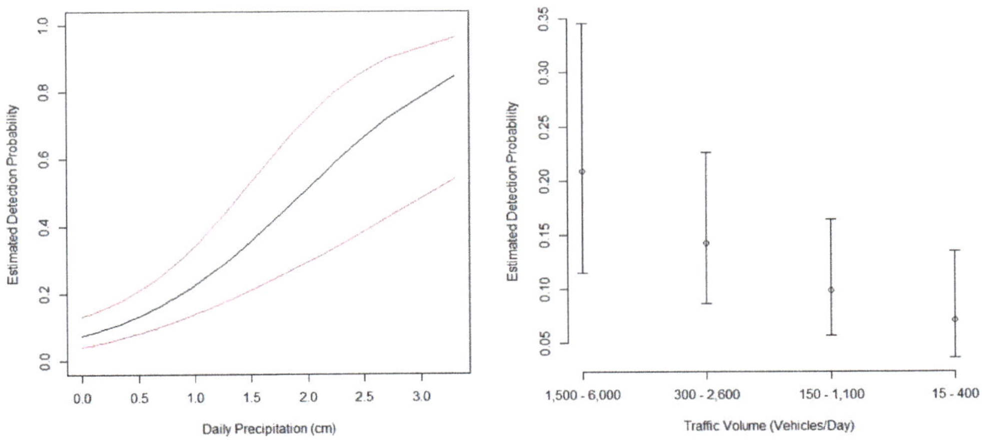

Figure 2. Marginal effect of covariates on estimated detection probability from the model-averaged results. Black lines represent model estimates for detection probability, and red lines represent upper and lower confidence intervals. Plots hold other covariates at their mean value.

At 40% of sites, we observed at least one large roadkill event, which represented the naïve occupancy or proportion of sites with at least one large roadkill event during the surveys. Using estimates of occupancy and detection probability from the model-averaged results, we estimated a site occupancy of 0.87, meaning that a large roadkill event occurred at 87% of surveyed sites. We estimated a detection probability of 0.22, or a 22% chance of detecting a large roadkill at a site given that a large roadkill event occurs there. Using these estimates, the effort required to reach a high probability of detecting a roadkill event occurring with 85% certainty is 7 surveys and 95% certainty is 12 surveys (Figure 3). If a large roadkill event was not detected by the 12th survey, then a large roadkill event likely did not occur at the site.

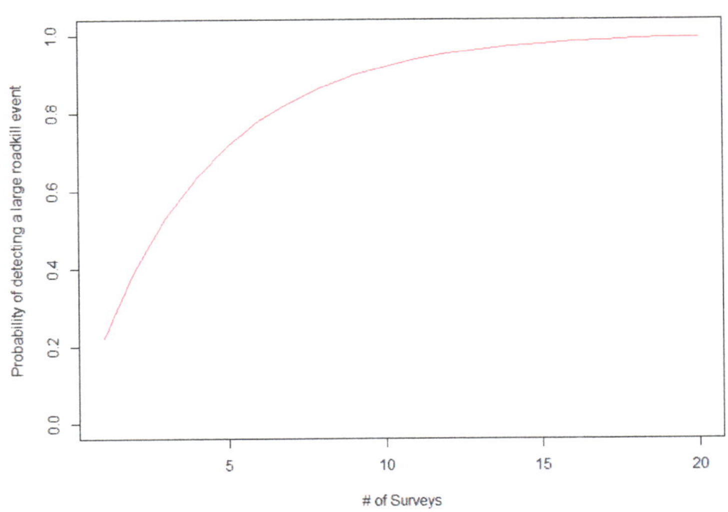

Figure 3. Number of surveys versus the probability of detecting at least one large roadkill event (≥5 amphibian or reptile carcasses) at occupied sites in Rhode Island, USA, 2019–2021.

4. Discussion

Our study represents a model approach, using amphibians and reptiles as a case study, to address imperfect detection in road mortality studies. This study was the first to assess amphibian and reptile road mortality in Rhode Island, a U.S. state undergoing rapid change that includes increasing human populations and associated land and road development. Overall, the roadkill rates we observed were similar to those of other studies documenting amphibian and reptile road mortality, which reported 2–8 carcasses/km/day [3,36]. Corroborating the results of others [3,7], amphibians made up the majority (89%) of all roadkill observations. This is likely due to their high abundance near roadways, in addition to having several life history characteristics that predispose them to road mortality, such as frequent road crossings and not actively avoiding vehicles once in the road [14,61]. Higher numbers of frog carcasses compared to salamander carcasses may reflect differences in abundance between the two groups or challenges with detection based on the small sizes of some salamanders. In addition, frogs tend to travel longer distances and make more frequent road crossings. Reptiles represented a smaller proportion of all roadkill observations. Turtles and many snakes are diurnal, and it is likely that their carcasses did not persist on the road long enough due to predation to be observed during our night surveys. Reptiles are also more active during the day, and it is possible that there were more incidences of road mortality; however, due to the timing of surveys, these carcasses were not observed. However, all reptiles observed in this study were found dead on the road, suggesting that there is a higher risk of road mortality for reptiles, as was found by other studies that determined a high risk of road mortality for reptiles crossing roads, especially for turtles [26,62]. This is of particular concern, as the low fecundity and delayed sexual maturity of many turtle species can make populations highly vulnerable to the impacts of road mortality, leading to local population declines and extirpation [27]. For example, at two sites that both intersected with large wetland complexes, we observed multiple spotted turtle (*Clemmys guttata*) carcasses. The spotted turtle is a Species of Greatest Conservation Need in Rhode Island due to population declines [51] and is a candidate for listing under the federal Endangered Species Act. The higher proportion of roadkill observations at hotspots versus coldspots supports previous evidence that amphibian and reptile road mortality happens more frequently on sections of roads near wetlands.

4.1. Pandemic-Associated Road Mortality

We documented conflicting results in our assessment of changes in road mortality between 2019 and the two subsequent years. We did not document a lower carcass density on roads in 2020, the first year of the pandemic, but carcass density was lower on roads in 2021. Differences in carcass densities in these years could have been due to variation in the timing of surveys and environmental conditions (i.e., precipitation) during surveys influencing the level of roadkill we observed [63]. However, we recorded a lower proportion of dead amphibians and reptiles relative to live animals during surveys in 2020 (pandemic season) and 2021 (post-pandemic) compared with those in 2019 (pre-pandemic). Combining these two metrics, our findings may have resulted from a reduced risk of road mortality, potentially because of fewer vehicles traveling on roads. Our results are similar to findings in Maine (U.S.A), where reduced traffic volumes were associated with a reduced risk of road mortality for frogs in 2020, as compared with 2019 [63]. It is likely that fewer vehicles were traveling on roads due to statewide mandates regarding travel, as many restaurants and businesses were closed, and many residents were following recommendations to travel only if necessary. However, traffic volume data were not available for 2019–2021 in Rhode Island, so we cannot confidently conclude that decreased percentages of dead amphibians and reptiles at survey sites during the pandemic (2020, 2021) resulted from reduced traffic volume. Overall, the results of our surveys in 2020 and 2021 (post-pandemic) suggest that there was likely a reduction in roadkill risk, but we cannot clearly attribute this reduction to changes in traffic volumes during the pandemic.

4.2. Occupancy Modeling Development

In our study area, larger amounts of wetland areas surrounding a road increased the probability of the occurrence of a large roadkill event. This supports findings from other studies [3,30] that documented increases in road mortality at road sections with wetlands adjacent to or bisected by a road. We also found that roads with lower traffic volume (15–400 vehicles/day) were associated with the highest occurrence of large roadkill events. This is likely due to roads with a lower traffic volume occurring in less developed areas that contained more habitat for amphibians and reptiles, thereby supporting larger populations that make more frequent road crossings. These results are similar to findings from other studies that reported higher levels of amphibian mortality on roads with lower traffic volume [7,34] and may be due to several reasons. First, roads with a higher traffic volume may result in carcasses being more quickly destroyed before they are observed during a survey. Second, many roads with high traffic volume on Rhode Island occur in areas with high urban development that lack habitat for amphibians and reptiles. In addition, lower abundances of amphibians have been observed near high traffic volume roads, which could contribute to lower levels of mortality [17]. Amphibians and reptiles may also avoid crossing roads with high traffic volumes due to the increased disturbance caused by vehicles. For detection probability, our results indicated that at sites where large roadkill events occurred, the probability of detecting the event during a survey increased on roads with a higher traffic volume. Despite this, we observed that large roadkill events were less likely to occur on high traffic volume roads. In other words, on roads with high traffic volume, there is a low probability that a large roadkill event will occur. However, should a large roadkill event occur, there is a high probability that it will be detected during a survey. This is likely due to the higher traffic levels increasing the risk of road mortality. Regardless of scale, large roadkill events were more likely to be detected during surveys with higher precipitation in the preceding 24 h. During periods of increased precipitation, amphibians and reptiles more frequently cross roads to breed and forage, thereby increasing the risk of mortality [17]. As indicated by our results, occupancy and detection probability varied at sites, depending on the surrounding habitat and timing of surveys.

Although several studies have used occupancy modeling to examine roadkill risk [42], this study is among the first to use occupancy modeling to identify locations where large amphibian and reptile roadkill events are most likely to occur. Importantly, our study has addressed the influence of imperfect detection during surveys on spatial patterns of road mortality, a challenge noted in several studies attempting to identify road mortality hotspots using roadkill counts [38,40,41,59]. The results of this study are specific to our study area. However, the developed modeling framework could be applied to other regions by those interested in better targeting mitigation measures for herpetofaunal road mortality. Using occupancy modeling, we were able to address imperfect detection during surveys to generate more reliable estimates of the occurrence of large roadkill events across the sites we surveyed. As indicated by the results of our occupancy analyses, large roadkill events occurred at a greater number of sites (52%) than were observed during road surveys (31%). This is likely due to the low probability (16.9%) of detecting a large roadkill event during a single survey, suggesting that imperfect detection influences our ability to detect large roadkill events. Such events are likely to be missed, and several studies have indicated that this may be a factor limiting the identification of road mortality hotspots [36,40]. Factors such as timing of surveys (e.g., surveying on a dry evening vs. an evening when it rains) or missing carcasses due to their size or being destroyed by cars or scavenged potentially contributed to imperfect detection of roadkill during our surveys [37,38,59]. By addressing imperfect detection during surveys, we were able to identify and prioritize the locations most appropriate for mitigation measures that reduce road mortality.

Mitigation measures, including infrastructure that keeps amphibians and reptiles off roads, can be costly and are most effective when implemented at locations with the greatest risk of road mortality [42,64]. When implemented appropriately, mitigation measures can be highly effective in reducing road mortality [64–66]. Using the results of our models, locations can be identified and targeted for implementing mitigation measures. As we have demonstrated, the influence of imperfect detection on spatial patterns of road mortality (e.g., non-detection leading to false absence) has the potential to misguide the implementation of mitigation measures, reducing their effectiveness in preventing road mortality [39]. To implement mitigation measures most effectively, conservation biologists and land managers should consider the following approach when addressing herpetofaunal road mortality. We recommend conducting surveys on sections of roads with low traffic near habitat for amphibians and reptiles, where large roadkill events are most likely to happen. In addition to survey location, the timing of surveys is also important, as surveying under ideal weather conditions more precisely captures spatial patterns of road mortality [59]. As our results indicate, precipitation has a strong influence on the probability of detecting a large roadkill event. Therefore, we recommend conducting surveys either during or immediately after a rain event, as large roadkill events are more likely to be detected.

As we have demonstrated, there is a low probability of detecting a large roadkill event (17.5%) from a single survey. Therefore, it is likely that locations where large roadkill events occurred were missed by our survey efforts. Given the low detectability of large roadkill events, multiple surveys may need to be conducted before considering a location for mitigation measures that reduce road mortality. However, while increased survey effort would increase the detectability of large roadkill events, conducting more surveys may have marginal gains in identifying high-occurrence roadkill locations and are time-and resource-intensive [67]. Instead of conducting more surveys to identify locations where mitigation measures are most appropriate, a balance of survey design (i.e., number of surveys) and the modeling technique we have applied can be used to correct for imperfect detection. Using our approach, fewer surveys can be conducted, and occupancy modeling can be used to address imperfect detection during surveys to identify and prioritize locations where mitigation measures would be most effective in reducing road mortality.

5. Conclusions

Using amphibians and reptiles as a case study, we developed an approach for assessing the influence of imperfect detection on spatial patterns of road mortality from hotspot models. As our results indicated, there is a low probability of detecting a large roadkill event for amphibians and reptiles, and both the location and timing of surveys should be considered when addressing amphibian and reptile road mortality. The approach we have developed can be used to address imperfect detection, allowing for more cost-effective survey design by guiding survey effort (i.e., the number of surveys conducted) and when surveys should be conducted to better capture the spatial patterns of road mortality hotspots. Importantly, our modeling approach can be used to correct for imperfect detection issues during surveys to allow for the prioritization of locations based on those with the highest probability of a large roadkill event occurring. Using this information, mitigation measures can be implemented at locations where they are more effective in preventing road mortality, thereby reducing the impacts of roads and traffic on amphibians and reptiles.

Supplementary Materials: The following are available online at https://www.mdpi.com/article/10.3390/land11050739/s1, Table S1: Distribution of mortality by species for roadkill surveys conducted in Rhode Island, USA, 2019–2021.; Table S2: Detection history matrix used to code for large roadkill events for amphibians and reptiles along surveyed sites in Rhode Island, USA, 2019–2021. Large roadkill events ($\geq$5 carcasses during a site survey (<5 carcasses during a site survey) are coded as 0.; Table S3: Site specific covariate values for precent wetland and percent forest within 100-m of a road and functional class description code used to estimate occupancy and detection probability of large roadkill events for amphibians and reptiles in Rhode Island, USA, 2019–2021.; Table S4: Survey specific covariate values for temperature during surveys, 24-hr precipitation, and time of year used to estimate detection probability of large roadkill events for amphibians and reptiles in Rhode Island, USA, 2019–2021.

Author Contributions: Conceptualization, N.H., N.E.K. and S.W.B.; methodology, N.E.K., S.W.B. and N.H.; formal analysis, N.H. and B.D.G.; resources, N.E.K. and S.W.B.; data curation, N.H.; writing—original draft preparation, N.H. and N.E.K.; writing—review and editing, N.H., N.E.K., S.W.B., B.D.G. and L.S.C.; visualization, N.H.; supervision, N.E.K.; project administration, N.E.K. All authors have read and agreed to the published version of the manuscript.

Funding: This research received no external funding.

Institutional Review Board Statement: The animal study protocol was approved by the Institutional Animal Care and Use Committee of the University of Rhode Island (protocol number AN1920-008, approval date 25 March 2019).

Informed Consent Statement: Not applicable.

Data Availability Statement: Supplementary spatial data for this article can be found as Supplementary Materials.

Acknowledgments: We are grateful for all the undergraduate and graduate student volunteers from the University of Rhode Island who assisted with road mortality surveys. We thank the Rhode Island Department of Environmental Management for providing support and assistance with data collection.

Conflicts of Interest: The authors declare no conflict of interest.

Appendix A

Table A1. Federal Highway Administration functional class description codes used to estimate traffic volume for roads in Rhode Island.

Code	Functional Class Description	Traffic Volume (Vehicles/Day)
1	Interstate	12,000–34,000
2	Other Freeways & Expressways	4000–18,500
3	Other Principal Arterial	2000–8500
4	Minor Arterial	1500–6000
5	Major Arterial	300–2600
6	Minor Collector	15–1100
7	Local	15–40

Table A2. Akaike's Information Criterion (AIC) table for all occupancy models ranked by AIC score, which shows the relationship between large roadkill events and measured covariates in Rhode Island, USA, 2019–2021; psi is occupancy parameter, p is detection parameter, Delta is the Delta AIC, AICwt is the Akaike weight for each model, cumltvWt is the cumulative AIC weight, and K is the number of model parameters.

Model	K	AIC	Delta	AICwt	cumltvWt
psi (Perc_Wetland_100m, F_Class_Code), p (Rain, F_Class_Code)	6	187.71	0.00	0.35	0.35
psi (Perc_Wetland_100m), p (Rain, F_Class_Code)	5	188.81	1.10	0.23	0.55
psi (Perc_Forest_100m), p (Rain, F_Class_Code)	5	189.71	1.93	0.13	0.69
psi (Perc_Wetland_100m), p (Temperature, Rain)	5	190.15	2.43	0.01	0.79
psi (Perc_Forest_100m), p (Temperature, Rain)	5	190.40	2.68	0.092	0.88
psi (Perc_Wetland_100m), p (Rain, Julian_Date)	5	191.24	3.53	0.060	0.95
psi (Perc_Forest_100m), p (Rain, Julian_Date)	5	191.77	4.06	0.046	0.99
psi (Perc_Forest_100m, F_Class_Code), p(Rain)	5	195.92	8.21	0.0058	1.00
psi (Perc_Wetland_100m, F_Class_Code), p(Rain)	5	197.39	9.67	0.0028	1.00
psi (Perc_Forest_100m, F_Class_Code), p (Temp., F_Class_Code)	6	205.98	18.27	0.000038	1.00
psi (Perc_Wetland_100m), p (Temperature, F_Class_Code)	5	206.43	18.72	0.000030	1.00
psi (Perc_Forest_100m), p (Temperature, F_Class_Code)	5	207.07	19.36	0.000022	1.00
psi (Perc_Wetland_100m, F_Class_Code), p (Temp., F_Class_Code)	6	207.29	19.58	0.000020	1.00
psi (Perc_Wetland_100m, F_Class_Code), p(Temperature)	5	212.19	24.48	0.0000017	1.00
psi (Perc_Forest_100m, F_Class_Code), p(Temperature)	5	212.84	25.13	0.0000060	1.00
psi (.), p (.)	2	213.26	25.55	0.0000010	1.00

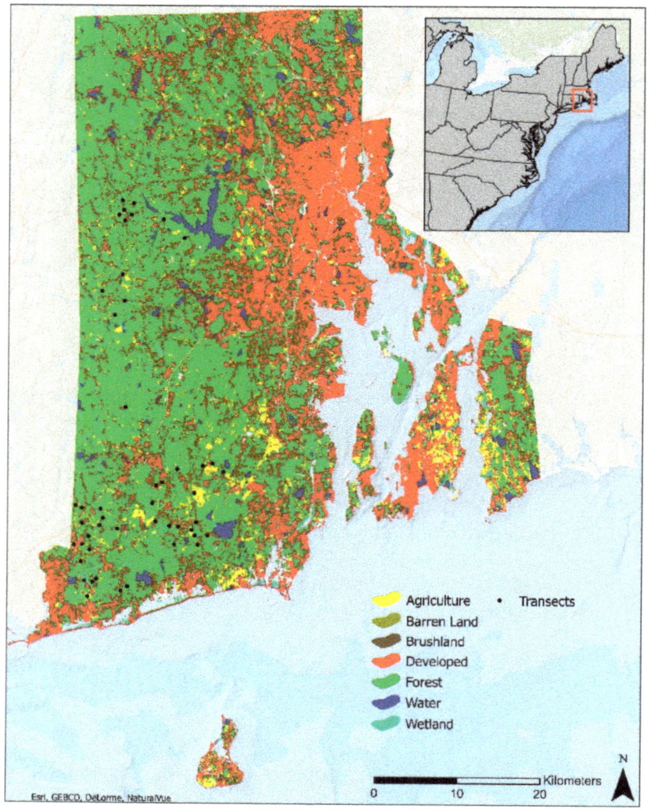

Figure A1. Land use and land cover classification within the study area and the locations of study sites in Rhode Island (USA) from 2019–2021.

References

1. Hill, J.E.; DeVault, T.L.; Belant, J.L. Cause-specific mortality of the world's terrestrial vertebrates. *Glob. Ecol. Biogeogr.* **2019**, *28*, 680–689. [CrossRef]
2. Loss, S.R.; Will, T.; Marra, P.P. Estimation of bird-vehicle collision mortality on US roads. *J. Wildl. Manag.* **2014**, *78*, 763–771. [CrossRef]
3. Glista, D.J.; De Vault, T.L.; De Woody, J.A. Vertebrate road mortality predominantly impacts amphibians. *Herpetol. Conserv. Biol.* **2008**, *3*, 77–87.
4. Matos, C.; Sillero, N.; Argaña, E. Spatial analysis of amphibian road mortality levels in northern Portugal country roads. *Amphib. Reptil.* **2012**, *33*, 469–483. [CrossRef]
5. Guyer, C.; Bailey, M.A. Amphibians and Reptiles of Longleaf Pine Communities. In Proceedings of the Tall Timbers Fire Ecology Conference, Tall Timbers Research Station, Tallahassee, FL, USA, 1993. Available online: https://www.talltimbers.org/wp-content/uploads/2018/09/139-GuyerandBailey1993_op.pdf (accessed on 3 March 2022).
6. Welsh Jr, H.H.; Ollivier, L.M. Stream amphibians as indicators of ecosystem stress: A case study from California's redwoods. *Ecol. Appl.* **1998**, *8*, 1118–1132. [CrossRef]
7. Fahrig, L.; Pedlar, J.H.; Pope, S.E.; Taylor, P.D.; Wegner, J.F. Effect of road traffic on amphibian density. *Biol. Conserv.* **1995**, *73*, 177–182. [CrossRef]
8. Gibbs, J.P.; Shriver, W.G. Can road mortality limit populations of pool-breeding amphibians? *Wetl. Ecol. Manag.* **2005**, *13*, 281–289. [CrossRef]
9. Beaudry, F.; Demaynadier, P.G.; Hunter, M.L. Non-marine turtle plays important functional roles in Indonesian ecosystems. *Ecol. Evol.* **2020**, *10*, 9613–9623.
10. Beaudry, F.; Demaynadier, P.G.; Hunter, M.L., Jr. Identifying hot moments in road-mortality risk for freshwater turtles. *J. Wildl. Manag.* **2010**, *74*, 152–159. [CrossRef]

11. Cureton, J.C.; Deaton, R. Hot moments and hot spots: Identifying factors explaining temporal and spatial variation in turtle road mortality. *J. Wildl. Manag.* **2012**, *76*, 1047–1052. [CrossRef]
12. Colino-Rabanal, V.J.; Lizana, M. Herpetofauna and roads: A review. *Basic Appl. Herpetol.* **2012**, *26*, 5–31. [CrossRef]
13. Hels, T.; Buchwald, E. The effect of road kills on amphibian populations. *Biol. Conserv.* **2001**, *99*, 331–340. [CrossRef]
14. Bouchard, J.; Ford, A.T.; Eigenbrod, F.; Fahrig, L. Behavioral responses of northern leopard frogs (*Rana pipiens*) to roads and traffic: Implications for population persistence. *Ecol. Soc.* **2009**, *14*, 23. [CrossRef]
15. Steen, D.A.; Aresco, M.J.; Beilke, S.G.; Compton, B.W.; Condon, E.P.; Dodd, C.K.; Forrester, H.; Gibbons, J.W.; Greene, J.L.; Johnson, G.; et al. Relative vulnerability of female turtles to road mortality. *Anim. Conserv.* **2006**, *9*, 269–273. [CrossRef]
16. Jochimsen, D.M.; Peterson, C.R.; Harmon, L. Influence of ecology and landscape on snake road mortality in a sagebrush-steppe ecosystem. *Anim. Conserv.* **2014**, *17*, 583–592. [CrossRef]
17. Gravel, M.; Mazerolle, M.J.; Villard, M.-A. Interactive effects of roads and weather on juvenile amphibian movements. *Amphib. Reptil.* **2012**, *33*, 113–127. [CrossRef]
18. Zhang, W.; Shu, G.; Li, Y.; Xiong, S.; Liang, C.; Li, C. Daytime driving decreases amphibian roadkill. *PeerJ* **2018**, *6*, e5385. [CrossRef]
19. Andrews, K.M.; Gibbons, J.W. How do highways influence snake movement? Behavioral responses to roads and vehicles. *Copeia* **2005**, *2005*, 772–782. [CrossRef]
20. Brehme, C.S.; Hathaway, S.A.; Fisher, R.N. An objective road risk assessment method for multiple species: Ranking 166 reptiles and amphibians in California. *Landsc. Ecol.* **2018**, *33*, 911–935. [CrossRef]
21. Mazerolle, M.; Huot, M.; Gravel, M. Behavior of amphibians on the road in response to car traffic. *Herpetologica* **2005**, *61*, 380–388. [CrossRef]
22. Crawford, B.A.; Andrews, K.M. Drivers' attitudes toward wildlife-vehicle collisions with reptiles and other taxa. *Anim. Conserv.* **2016**, *19*, 444–450. [CrossRef]
23. Ashley, E.P.; Kosloski, A.; Petrie, S.A. Incidence of intentional vehicle–reptile collisions. *Hum. Dimens. Wildl.* **2007**, *12*, 137–143. [CrossRef]
24. Carr, L.W.; Fahrig, L. Effect of road traffic on two amphibian species of differing vagility. *Conserv. Biol.* **2001**, *15*, 1071–1078. [CrossRef]
25. Steen, D.A.; Gibbs, J.P. Effects of roads on the structure of freshwater turtle populations. *Conserv. Biol.* **2004**, *18*, 1143–1148. [CrossRef]
26. Gibbs, J.P.; Shriver, W.G. Estimating the effects of road mortality on turtle populations. *Conserv. Biol.* **2002**, *16*, 1647–1652. [CrossRef]
27. Howell, H.J.; Seigel, R.A. the effects of road mortality on small, isolated turtle populations. *J. Herpetol.* **2019**, *53*, 39–46. [CrossRef]
28. Clevenger, A.P.; Chruszcz, B.; Gunson, K.E. Spatial patterns and factors influencing small vertebrate fauna road-kill aggregations. *Biol. Conserv.* **2003**, *109*, 15–26. [CrossRef]
29. Orłowski, G.; Ciesiołkiewicz, J.; Kaczor, M.; Radwańska, J.; Żywicka, A. Species composition and habitat correlates of amphibian roadkills in different landscapes of south-western Poland. *Pol. J. Ecol.* **2008**, *56*, 659–671.
30. Langen, T.A.; Ogden, K.M.; Schwarting, L.L. Predicting hot spots of herpetofauna road mortality along highway networks. *J. Wildl. Manag.* **2009**, *73*, 104–114. [CrossRef]
31. Patrick, D.A.; Gibbs, J.P.; Popescu, V.D.; Nelson, D.A. Multi-scale habitat-resistance models for predicting road mortality "hotspots" for turtles and amphibians. *Herpetol. Conserv. Biol.* **2012**, *7*, 407–426.
32. Langen, T.A.; Gunson, K.E.; Scheiner, C.A.; Boulerice, J.T. Road mortality in freshwater turtles: Identifying causes of spatial patterns to optimize road planning and mitigation. *Biodivers. Conserv.* **2012**, *21*, 3017–3034. [CrossRef]
33. Gu, H.; Dai, Q.; Wang, Q.; Wang, Y. Factors contributing to amphibian road mortality in a wetland. *Curr. Zool.* **2011**, *57*, 768–774. [CrossRef]
34. Sutherland, R.W.; Dunning, P.R.; Baker, W.M. Amphibian encounter rates on roads with different amounts of traffic and urbanization. *Conserv. Biol.* **2010**, *24*, 1626–1635. [CrossRef] [PubMed]
35. Jacobson, S.L.; Bliss-Ketchum, L.L.; de Rivera, C.E.; Smith, W.P. A behavior-based framework for assessing barrier effects to wildlife from vehicle traffic volume. *Ecosphere* **2016**, *7*, e01345. [CrossRef]
36. Langen, T.A.; Machniak, A.; Crowe, E.K.; Mangan, C.; Marker, D.F.; Liddle, N.; Roden, B. Methodologies for surveying herpetofauna mortality on rural highways. *J. Wildl. Manag.* **2007**, *71*, 1361–1368. [CrossRef]
37. Slater, F.M. An assessment of wildlife road casalties–the potential discrepancy between numbers counted and numbers killed. *Web Ecol.* **2002**, *3*, 33–42. [CrossRef]
38. Ogletree, K.A.; Mead, A.J. What roadkills did we miss in a driving survey? A comparison of driving and walking surveys in Baldwin County, Georgia. *Ga. J. Sci.* **2020**, *78*, 8.
39. Santos, S.M.; Marques, J.T.; Lourenço, A.; Medinas, D.; Barbosa, A.M.; Beja, P.; Mira, A. Sampling effects on the identification of roadkill hotspots: Implications for survey design. *J. Environ. Manag.* **2015**, *162*, 87–95. [CrossRef]
40. Teixeira, F.Z.; Coelho, A.V.P.; Esperandio, I.B.; Kindel, A. Vertebrate road mortality estimates: Effects of sampling methods and carcass removal. *Biol. Conserv.* **2013**, *157*, 317–323. [CrossRef]
41. Degregorio, B.A.; Hancock, T.E.; Kurz, D.J.; Yue, S. How quickly are road-killed snakes scavenged? Implications for underestimates of road mortality. *J. North Carol. Acad. Sci.* **2011**, *127*, 184–188. [CrossRef]

42. Santos, R.A.L.; Mota-Ferreira, M.; Aguiar, L.M.S.; Ascensão, F. Predicting wildlife road-crossing probability from roadkill data using occupancy-detection models. *Sci. Total Environ.* **2018**, *642*, 629–637. [CrossRef] [PubMed]
43. Malo, J.E.; Suárez, F.; Díez, A. Can we mitigate animal–vehicle accidents using predictive models? *J. Appl. Ecol.* **2004**, *41*, 701–710. [CrossRef]
44. Garrah, E.; Danby, R.K.; Eberhardt, E.; Cunnington, G.M.; Mitchell, S. Hot spots and hot times: Wildlife road mortality in a regional conservation corridor. *Environ. Manag.* **2015**, *56*, 874–889. [CrossRef] [PubMed]
45. Healey, R.M.; Atutubo, J.R.; Kusrini, M.D.; Howard, L.; Page, F.; Hallisey, N.; Karraker, N.E. Road mortality threatens endemic species in a national park in Sulawesi, Indonesia. *Glob. Ecol. Conserv.* **2020**, *24*, e01281. [CrossRef]
46. Shilling, F.M.; Waetjen, D.P. Wildlife-vehicle collision hotspots at US highway extents: Scale and data source effects. *Nat. Conserv.* **2015**, *11*, 41. [CrossRef]
47. MacKenzie, D.I.; Nichols, J.D.; Royle, J.A.; Pollock, K.H.; Bailey, L.L.; Hines, J.E. *Occupancy Estimation and Modeling: Inferring Patterns and Dynamics of Species Occurrence*; Elsevier: Amsterdam, The Netherlands, 2017.
48. Pavlacky, D.C., Jr.; Blakesley, J.A.; White, G.C.; Hanni, D.J.; Lukacs, P.M. Hierarchical multi-scale occupancy estimation for monitoring wildlife populations. *J. Wildl. Manag.* **2012**, *76*, 154–162. [CrossRef]
49. Pellet, J.; Schmidt, B.R. Monitoring distributions using call surveys: Estimating site occupancy, detection probabilities and inferring absence. *Biol. Conserv.* **2005**, *123*, 27–35. [CrossRef]
50. U.S. Census Bureau. Guide to State and Local Census Geography—Rhode Island. 2010. Available online: https://www.census.gov/geographies/reference-files/2010/geo/state-local-geo-guides-2010/rhode-island.html (accessed on 20 October 2021).
51. Rhode Island Department of Environmental Management (RIDEM). *Rhode Island Wildlife Action Plan*; Rhode Island Department of Environmental Management (RIDEM): Providence, RI, USA, 2015.
52. Weir, L.; Royle, J.; Nanjappa, P.; Jung, R.E. Modeling anuran detection and site occupancy on North American Amphibian Monitoring Program (NAAMP) routes in Maryland. *J. Herpetol.* **2005**, *39*, 627–639.
53. Colorado Climate Center. Community Collaborative Rain, Hail, and Snow Network. 2021. Available online: https://www.cocorahs.org/ (accessed on 18 October 2021).
54. Rhode Island Geographic Information System (RIGIS). Land Use and Land Cover. 2011. Available online: http://www.rigis.org (accessed on 16 March 2021).
55. Rhode Island Geographic Information System (RIGIS). Rhode Island Department of Transportation Roads (2016). 10 March 2016. Available online: http://www.rigis.org (accessed on 16 March 2021).
56. Anderson, J.R. *A Land Use and Land Cover Classification System for Use with Remote Sensor Data*; US Government Printing Office: Washington, DC, USA, 1976; Volume 964.
57. United States Fish and Wildlife Service (USFWS). National Wetlands Inventory. 2021. Available online: https://www.fws.gov/wetlands/ (accessed on 2 March 2019).
58. Federal Highway Administration (FHWA). Highway Functional Classification Concepts, Criteria and Procedures. 2013. Available online: https://dot.sd.gov/media/documents/HwyFunctionalClassification.pdf (accessed on 3 July 2020).
59. Santos, S.M.; Carvalho, F.; Mira, A. How long do the dead survive on the road? Carcass persistence probability and implications for road-kill monitoring surveys. *PLoS ONE* **2011**, *6*, e25383. [CrossRef]
60. Anderson, D.; Burnham, K. *Model Selection and Multi-Model Inference: A Practical Information-Theoretic Approach*; Springer: Berlin/Heidelberg, Germany, 2004; pp. 35–37.
61. Mazerolle, M.J. Amphibian road mortality in response to nightly variations in traffic intensity. *Herpetologica* **2004**, *60*, 45–53. [CrossRef]
62. Aresco, M.J. Mitigation measures to reduce highway mortality of turtles and other herpetofauna at a north Florida lake. *J. Wildl. Manag.* **2005**, *69*, 549–560. [CrossRef]
63. LeClair, G.; Chatfield, M.W.H.; Wood, Z.; Parmelee, J.; Frederick, C.A. Influence of the COVID-19 pandemic on amphibian road mortality. *Conserv. Sci. Pract.* **2021**, *3*, e535. [CrossRef] [PubMed]
64. Beebee, T.J. Effects of road mortality and mitigation measures on amphibian populations. *Conserv. Biol.* **2013**, *27*, 657–668. [CrossRef] [PubMed]
65. Gonçalves, L.O.; Alvaares, D.J.; Teixeira, F.Z.; Schuck, G.; Coelho, I.P.; Esperandio, I.B.; Anza, J.; Beduschi, J.; Bastazini, V.A.G.; Kindel, A. Reptile road-kills in Southern Brazil: Composition, hot moments and hotspots. *Sci. Total Environ.* **2018**, *615*, 1438–1445. [CrossRef]
66. Glista, D.J.; DeVault, T.; DeWoody, J. A review of mitigation measures for reducing wildlife mortality on roadways. *Landsc. Urban Plan.* **2009**, *91*, 1–7. [CrossRef]
67. Shannon, G.; Lewis, J.S.; Gerber, B.D. Recommended survey designs for occupancy modelling using motion-activated cameras: Insights from empirical wildlife data. *PeerJ* **2014**, *2*, e532. [CrossRef]

MDPI
St. Alban-Anlage 66
4052 Basel
Switzerland
Tel. +41 61 683 77 34
Fax +41 61 302 89 18
www.mdpi.com

Land Editorial Office
E-mail: land@mdpi.com
www.mdpi.com/journal/land

www.ingramcontent.com/pod-product-compliance
Lightning Source LLC
Chambersburg PA
CBHW061619220926
51850CB00005B/551

* 9 7 8 3 0 3 6 5 4 8 7 1 5 *